Texts and Monographs in Physics

Springer
Berlin
Heidelberg
New York
Barcelona
Hong Kong
London
Milan
Paris
Singapore
Tokyo

Physics and Astronomy   ONLINE LIBRARY

http://www.springer.de/phys/

Texts and Monographs in Physics

Series Editors: R. Balian   W. Beiglböck   H. Grosse   E. H. Lieb
N. Reshetikhin   H. Spohn   W. Thirring

A selection of titles:

Sergei V. Ketov

Quantum Non-linear Sigma-Models

From Quantum Field Theory
to Supersymmetry,
Conformal Field Theory,
Black Holes and Strings

With 51 Figures

 Springer

2000

Professor Sergei V. Ketov

Institut für Theoretische Physik
Universität Hannover
Appelstrasse 2
30167 Hannover, Germany

Editors

Roger Balian

CEA
Service de Physique Théorique de Saclay
91191 Gif-sur-Yvette, France

Wolf Beiglböck

Institut für Angewandte Mathematik
Universität Heidelberg, INF 294
69120 Heidelberg, Germany

Harald Grosse

Institut für Theoretische Physik
Universität Wien
Boltzmanngasse 5
1090 Wien, Austria

Elliott H. Lieb

Jadwin Hall
Princeton University, P.O. Box 708
Princeton, NJ 08544-0708, USA

Nicolai Reshetikhin

Department of Mathematics
University of California
Berkeley, CA 94720-3840, USA

Herbert Spohn

Zentrum Mathematik
Technische Universität München
80290 München, Germany

Walter Thirring

Institut für Theoretische Physik
Universität Wien
Boltzmanngasse 5
1090 Wien, Austria

Library of Congress Cataloging-in-Publication Data applied for.
Die Deutsche Bibliothek - CIP-Einheitsaufnahme.
Ketov, Sergej V.: Quantum non-linear sigma-models : from quantum field theory to supersymmetry, conformal field theory, black holes and strings / S. V. Ketov. - Berlin ; Heidelberg ; New York ; Barcelona ; Hong Kong ; London ; Milan ; Paris ; Singapore ; Tokyo : Springer, 2000 (Texts and monographs in physics) (Physics and astronomy online library)
ISBN 3-540-67461-6

ISSN 0172-5998
ISBN 3-540-67461-6 Springer-Verlag Berlin Heidelberg New York

Springer-Verlag Berlin Heidelberg New York
a member of BertelsmannSpringer Science+Business Media GmbH

© Springer-Verlag Berlin Heidelberg 2000
Printed in Germany

Typesetting:
Cover design: *design & production* GmbH, Heidelberg
Printed on acid-free paper SPIN: 10720911   55/3141/tr - 5 4 3 2 1 0

To Tatiana, Michael and Denise

Preface

The idea for this book came to me after visiting DESY and CERN in 1996 and 1997. For a long time, two-dimensional Non-Linear Sigma-Models (NLSM) served as a useful laboratory for the study of perturbative and non-perturbative properties of four-dimensional non-Abelian gauge theories, since they share many remarkable features like renormalizability, asymptotic freedom, solitons, confinement, etc. [1]. For instance, the low-energy effective physics of pions in four-dimensional Quantum Chromodynamics (QCD) receives the most natural description in terms of the principal NLSM whose solitonic solutions (skyrmions) can be interpreted as baryons [2]. In fact, NLSM are also important for spontaneous symmetry breaking, extended supersymmetry and supergravity, conformal field theory, gravity and string theory. This book is entirely devoted to recent applications of NLSM in various dimensions.

In the late 1980s and while in Russia, I wrote the book [3] entitled 'Nonlinear Sigma-Models in Quantum Field Theory and Strings', which was eventually published in Russian by the Nauka Publishers in 1992. This book is *not* a translation of my earlier book into English, though it shares about one third of its content with the Russian edition. The main additions include the two-dimensional Wess-Zumino-Novikov-Witten (WZNW) models in conformal field theory and strings, gauging NLSM isometries, four-dimensional NLSM with $N = 2$ extended supersymmetry in the context of Seiberg-Witten theory and M-theory, $N = 2$ strings and D-brane dynamics.

This book is *not* a collection of all known facts about NLSM. For instance, any extended discussion of NLSM solitonic solutions and their low-energy scattering, as well as many standard applications of NLSM in condensed matter and low-energy physics of hadrons, were intentionally excluded, since there is already extensive monographic literature on these, see e.g., [4, 5, 6, 7, 8] and references therein. The book is not merely aimed at providing the formal mathematical background to the field theory of NLSM and their quantization. In fact, I have adopted a more 'applied' approach for my presentation that is oriented towards practitioners in quantum field theory, supersymmetry, gravity and modern string theory. The book content is heavily based on my original papers, so that the selected material considerably reflects my own research interests in the past. Nevertheless, this book is not just a collection of my papers, and it does not duplicate any existing review. Although extensive, the list of literature at the end of the book cannot be considered an exhaustive bibliography on NLSM by any means. I would like to apologize to those authors whose contributions escaped my attention or were not mentioned.

This book cannot replace an introduction to quantum field theory, conformal field theory or string theory, though I have done my best to make it readable for those who are merely familiar with the foundations of quantum field theory and classical general relativity, and who are interested in various applications of the NLSM techniques. Therefore, this book should certainly be accessible to Ph.D. students wishing to do research in either quantum field theory, strings, supersymmetry, conformal field theory or related areas of mathematical physics, as well as to those readers interested in phenomenologically oriented applications of the formalism presented here.

Acknowledgements. I am grateful to Luis Alvarez-Gaumé, Jan Ambjorn, Jonathan Bagger, Ioannis Bakas, Joseph Buchbinder, Francois Delduc, Norbert Dragon, Efim Fradkin, Jim Gates Jr., Marc Grisaru, Murat Günaydin, Chris Hull, Evgeny Ivanov, Olaf Lechtenfeld, Ulf Lindström, Dieter Lüst, Alexei Morozov, Werner Nahm, Hermann Nicolai, Burt Ovrut, Werner Rühl, Norisuke Sakai, John Schwarz, Kellog Stelle, Arkady Tseytlin, Igor Tyutin, Cumrun Vafa and Bernard de Wit for helpful discussions. A significant part of the book is based on my lecture course on non-linear sigma-models, which was given at the ITP Hannover in 1999. I am grateful to all participants of the 'Graduiertenkolleg' in Hannover for numerous discussions and suggestions.

The camera-ready manuscript for the layout of this book was prepared with the help of the LATEX macro package *cl2emono* provided by Springer-Verlag Heidelberg. I am also grateful to Wolf Beiglböck, Brigitte Reichel-Mayer and Petra Treiber of Springer-Verlag in Heidelberg for pleasant collaboration.

This book was made possible by the generous financial support of the Deutsche Forschungsgemeinschaft, the Volkswagen Stiftung and NATO. Last but not least, the book might never have appeared without crucial moral support from my family and friends.

Hannover, July 2000   Sergei V. Ketov

Contents

1. Introduction

The Non-Linear Sigma-Model (NLSM) is a field theory whose field takes its values in a Riemannian manifold $\mathcal{M}$. The NLSM field values can therefore be considered as a set of coordinates in the internal manifold $\mathcal{M}$ whose metric is field-dependent. The NLSM format is thus the very general field-theoretical concept whose apparent geometrical meaning is the main reason for many successful applications of NLSM in field theory, string theory and statistical mechanics.

The relevance of NLSM in Quantum Field Theory (QFT) originates from the paramount importance of symmetry principles in fundamental physics. The roots of our understanding of elementary particle interactions and gravity go back to the well-known fundamental theoretical discoveries of Maxwell, Einstein and Weyl in classical field theory. The symmetry principles as the key tools for a construction of phenomenologically relevant field theories were further developed by Yang and Mills [9], Higgs [10], t'Hooft [11, 12], Coleman, Wess and Zumino [13], and many others. It is certainly true that the fundamental physics at distances well below the Planck scale is described by the so-called Standard Model (SM), that is the local QFT based on the gauge symmetry $SU(3) \times SU(2) \times U(1)$. The SM unifies the theory of strong interactions — the so-called Quantum Chromodynamics (QCD) — with the Weinberg-Salam-Glashow gauge theory of electroweak interactions, and it is consistent with all known experimental data. However, despite the remarkable success of the perturbative approach to the asymptotically-free QCD [14, 15] in explaining, e.g., deep inelastic scattering experiments and jet production in high-energy reactions, a satisfactory QCD-based theoretical explanation of the most obvious experimental evidence, namely, the *confinement* of quarks and gluons inside hadrons, is still absent. The reason is, of course, the non-perturbative nature of the confinement whose formal (analytic) proof in a non-Abelian gauge field theory requires non-perturbative theoretical tools for a derivation of the quantum low-energy effective action. Until recently, the lattice approach was the only one used to study (at least, numerically) large-distance phenomena from first principles, since the evaluation of the QCD effective action by the field theory methods was a very formidable task. It was only recently realized by Seiberg and Witten [16, 17] that supersymmetry and duality may help to solve this problem and, perhaps, explain confinement in QCD as well. If, as many theoretical studies suggest, super-symmetry is relevant to particle theory, the building blocks of ultimate QFT should be four-dimensional supersymmetric gauge field theories. It is often much easier to do quantum calculations in a supersymmetric field theory than in its bosonic sector, which is yet another advantage of supersymmetry due to cancellations between Feynman graphs involving bosons and fermions. For instance, in $N = 2$ extended supersymmetry, the vector gauge particles are in one irreducible supermultiplet with Higgs particles. Hence, the non-linearity of the effective gauge field equations of motion implies by supersymmetry the non-linear structure of the effective equations of motion for the scalar Higgs fields, i.e. a (special) NLSM. Similarly, the low-energy effective action of $N = 2$ scalar matter (given by a NLSM) is severely constrained also by $N = 2$ supersymmetry.

A unification of gravity, supersymmetry and gauge theories leads to *su-pergravity* theories with the number of supersymmetries ranging from $N = 1$ to $N = 8$. When $N \geq 4$, the minimal supergravity multiplet contains scalars, so that the non-linearity of the gravity action implies a non-linearity of the scalar kinetic terms, i.e. a NLSM as the part of the extended supergravity Lagrangian. Unfortunately, quantum Einstein gravity is non-renormalizable [18, 19], which is also the case for all supergravity theories, simply because there is no mechanism that could be responsible for otherwise miraculous cancellations of severe ultraviolet divergences expected in their multi-loop Feynman graphs [20, 21, 22]. Quantum theory of gravity should therefore be a non-local quantum field theory, while the only known consistent ap-proach (at least, perturbatively) is provided by the theory of *superstrings*. Ultimate (yet unknown) non-perturbative theory already called *M (or U) theory* may be given by a unified theory of superstrings and their solitons (called *branes*). Modern superstring theory naturally accommodates many (if not all) concepts of field theory, including gauge symmetry and super-symmetry, supergravity and higher dimensions, effective actions and NLSM. For instance, the string action in a non-trivial gravitational background takes the form of a NLSM or its generalization. In particular, the fundamental role of conformal invariance in perturbative string theory results in deep connec-tions between Conformal Field Theory (CFT), strings and NLSM, while they go over to the supersymmetric case too. Extended supersymmetry adds new non-trivial relations between NLSM, complex geometry and topology. The remarkable correspondence between the two-dimensional NLSM renormal-ization group β-functions and the perturbative low-energy effective action of strings and superstrings is one of the topics to be discussed in this book.

The geometrical nature of the interaction in NLSM results in the geomet-rical nature of their counterterms, the existence of topologically non-trivial field configuratons (solitons) and asymptotic freedom, dynamical generation of vector bosons, i.e. all the familiar features of four-dimensional quantum gauge theories [1, 23]. The NLSM also provides a useful field-theoretical laboratory for studying some two-dimensional, exactly solvable systems on a lattice, such as the Ising model or the Heisenberg antiferromagnet, in statistical mechanics. Some particular $O(n)$ two-dimensional NLSM are frequently used in condensed matter physics, e.g. in connection with antiferromagnetic spin chains [24] and the quantum Hall effect [25]. The effective Lagrangian for superfluid He 3 is also described by a NLSM [26].

Yet another strong argument for studying NLSM comes from *spontaneous* symmetry breaking which is crucial for phenomenological applications of QFT. The spontaneously broken symmetries are not realized as the symmetry transformations of physical states, and, in particular, they do not leave the vacuum state invariant. In field theory, a spontaneously broken symmetry is always associated with the degeneracy of vacuum states. In this book, I also consider the situation when some global (rigid) continuous symmetries, or supersymmetry, are spontaneously broken. Then, according to the well-known Goldstone theorem [27], the spectrum of physical states always has one massless (Nambu-Goldstone) particle for each broken symmetry generator. The spontaneously broken symmetries also unambiguously determine the (highly non-linear) effective action of the Nambu-Goldstone fields. If these fields are scalars, their (Goldstone) low-energy effective action often appears to be a NLSM. General non-linear Goldstone actions may thus be considered as the natural generalizations of the NLSM actions, even if the former have no scalars at all. The best example is provided by the (supersymmetric) Born-Infeld action describing the non-linear generalization of Maxwell (super)electrodynamics.

The main purpose of this book is to explore formal properties of NLSM in two and four dimensions, and outline their applications in gauge theories, strings and CFT. Chapter 2 is devoted to the most general bosonic two-dimensional NLSM. Their classical and quantum properties are discussed and their three-loop renormalization group β-functions calculated. We discuss a few specific NLSM too. In Chap. 3 the most general supersymmetric NLSM in two dimensions are considered, and their four-loop β-functions are calculated. The maximally extended (linearly realized) supersymmetry in NLSM is discussed at length in Chap. 4 by using superspace. In Chap. 5 we consider the so-called Wess-Zumino-Novikov-Witten (WZNW) models (or NLSM on group manifolds), their gauging and supersymmetrization, in the context of two-dimensional CFT, as well as topological quantization of the WZNW coefficient. In Chap. 6 we briefly discuss the relation between NLSM and strings. In Chap. 7 the chiral (or *heterotic*) NLSM are defined, and their perturbative anomalies are calculated. Since all NLSM in four dimensions are non-renormalizable, their interpretation in QFT is different from that of their two-dimensional counterparts. The four-dimensional NLSM naturally appear as *effective* actions, not as fundamental ones. This is illustrated in Chap. 8 in the context of $N = 2$ supersymmetric gauge field theories, which is based on the availability of their exact solutions [16, 17]. Finally, in Chap. 9, we introduce some generalizations of the NLSM concept. First, we demonstrate the dynamical generation of particles in the non-compact four-dimensional NLSM with the Asymptotically Locally Euclidean (ALE) target spaces. Second, we propose the NLSM interpretation of the supersymmetric Born-Infeld-Nambu-Goto actions that describe, in particular, the gauge-fixed Dirichlet-branes. Third, we introduce the renormalizable (fourth-order) NLSM in four dimensions at the fundamental level, as well as their supersymmetric generalizations. These fourth-order NLSM are quantized, while some of them are shown to be asymptotically free.

Implicit summation over repeated indices is assumed throughout the book, as well as the fundamental units $c = \hbar = 1$. The two-digit numbering of equations is carried out section-wise in each individual chapter, while a third-digit prefix, if any, always refers to another chapter. Similarly, sections are always denoted by two-digit numbers, whereas three-digit numbers are used for subsections.

With this book I would like to convince the reader, by means of a variety of explicit examples, that the NLSM concept is very useful for many applications in quantum field theory and beyond, so that he can use this knowledge in his own research and teaching.

2. Classical Structure and Renormalization

In this chapter we define the general bosonic Non-Linear Sigma-Model (NLSM) action (without a scalar potential) and discuss its symmetries and quantization. The two-dimensional (2d) NLSM are distinguished by their renormalizability and the existence of a Wess-Zumino-Novikov-Witten (WZNW) term of second order in the spacetime derivatives. The covariant background field method and on-shell renormalization are developed, and the three-loop Renormalization Group (RG) β-functions of the general two-dimensional NLSM are derived. The 2d WZNW model and the 2d Freedman-Townsend NLSM are defined, and their perturbative RG β-functions are calculated. We also briefly consider classical duality transformations in NLSM, gauging NLSM isometries, and the Skyrme model of pions and nucleons in four dimensions.

2.1 Bosonic NLSM: Definition and Examples

Consider a set of coupled scalar fields $\phi^a(x^\mu)$, $a = 1, 2, \ldots, D$, in a d-dimensional flat spacetime Σ, $\mu = 0, 1, 2, \ldots, d-1$, with the action

$$S[\phi] = \frac{1}{2\lambda^2} \int_\Sigma d^d x \, g_{ab}(\phi) \partial^\mu \phi^a \partial_\mu \phi^b \, , \qquad (1.1a)$$

where $\partial^\mu = \eta^{\mu\nu}\partial_\nu$, $\partial_\nu = \partial/\partial x^\nu$ and $\eta^{\mu\nu} = \mathrm{diag}(+, -, \ldots, -)$ is the Minkowski metric. We also define the Euclidean version of (1.1a),

$$S_{\mathrm{E}}[\phi] = \frac{1}{2\lambda^2} \int_{\Sigma_{\mathrm{E}}} d^d x \, g_{ab}(\phi) \partial^\mu \phi^a \partial_\mu \phi^b \, , \qquad (1.1b)$$

where $\mu = 1, 2, \ldots, d$ now. The coupling constant λ^2 was explicitly written down in (1.1) since it is going to be the loop-counting parameter (instead of $\hbar$) in our quantum calculations.

The field theory (1.1) is called the NLSM with the metric $g_{ab}(\phi)$. It is usually assumed that $g_{ab}(\phi)$ is a *positive-definite* field-dependent matrix, in order to ensure the absence of negative norm states. [1] The known mechanisms

---

[1] Non-compact NLSM, however, appear in 4d, $N \geq 4$ extended supergravities (Sect. 9.4), while they are also relevant for a dynamical generation of Higgs and vector bosons (Sect. 9.1).

for the construction of compact NLSM (without ghosts) are: (i) imposing constraints on the NLSM scalars so that they take their values in a compact (usually symmetric) space, and (ii) gauging some internal symmetries of NLSM, which implies spontaneous symmetry breaking. Both approaches are discussed throughout the book from various viewpoints, while in this section we confine ourselves to the simplest examples only.

From the *field* theory point of view, the NLSM metric $g_{ab}(\phi)$ is just a set of given functions of ϕ. After being expanded in powers of ϕ,

$$g_{ab}(\phi) = g_{ab}(0) + \partial_c g_{ab}(0)\phi^c + \tfrac{1}{2}\partial_c\partial_d g_{ab}(0)\phi^c\phi^d + \dots, \qquad (1.2)$$

the action (1.1) thus represents a field theory with a generically *infinite* number of coupling constants $\{g_{ab}(0), \partial_c g_{ab}(0), \partial_c\partial_d g_{ab}(0), \dots\}$. From the *string* theory point of view (Chap. 6), the two-dimensional space Σ represents a string world-sheet, whereas the NLSM metric is supposed to be identified with the 'truly' spacetime metric representing the gravitational background where the string propagates.

The NLSM action (1.1) can also be geometrically interpreted, when considering the fields ϕ^a themselves as the coordinates of the internal Riemannian manifold $\mathcal{M}$ with metric $g_{ab}(\phi)$. Indeed, the action (1.1) is formally invariant under the (infinitesimal) field reparametrizations

$$\phi^{a\prime} = \phi^a + \xi^a(\phi) \qquad (1.3)$$

that can be interpreted as the coordinate diffeomorphisms of $\mathcal{M}$ provided that the metric g_{ab} transforms as a second-rank tensor with respect to its indices, i.e.

$$g_{ab}{}'(\phi') = \frac{\partial\phi^c}{\partial\phi^{a\prime}}\frac{\partial\phi^d}{\partial\phi^{b\prime}}g_{cd}(\phi) \ . \qquad (1.4)$$

The transformation (1.4) is a transformation of the fields *and* the coupling constants so that, generally speaking, there is no Noether current that could be associated with this symmetry. Two NLSM are physically equivalent as field theories if they are related via the field redefinition (1.3) alone, which implies the scalar transformation law for the metric,

$$g_{ab}{}'(\phi') = g_{ab}(\phi) \ . \qquad (1.5)$$

The NLSM field theory is thus defined on the equivalence classes of the NLSM metrics, with the equivalence relation being defined by (1.5). Nevertheless, due to the possible geometrical interpretation of the NLSM action (1.1) with the metric subject to the tensor transformation law (1.4), it is sometimes desirable to keep this formal diffeomorphism invariance manifest in quantum theory. For instance, this is quite natural in string theory [28, 29, 30] (see also Sect. 6.1).

It is easy to check that both prescriptions (1.4) and (1.5) are only compatible when

$$L_\xi g_{ab} = 0 \; , \tag{1.6}$$

where L_ξ is the Lie derivative, i.e. when ξ^a is a Killing vector. In this case the NLSM action (1.1) has an isometry or an internal symmetry that leads to the conserved Noether current and the corresponding conserved charge.

A NLSM can have enough isometries to reduce the infinite number of coupling constants in (1.2) to a single one. For example, the $O(n)$ NLSM is defined by the Euclidean action

$$S[n] = \frac{1}{2\lambda^2} \int d^d x \, \partial^\mu n \cdot \partial_\mu n \; , \tag{1.7}$$

where the real scalar fields $n(x^\mu)$ are subject to the constraint

$$n \cdot n = 1 \; , \qquad n = \{n^a\} \; . \tag{1.8}$$

After solving the constraint (1.8) in terms of some independent field variables $\{\phi\}$, and substituting the result $n = n(\phi)$ back into the action (1.7) one gets a NLSM action of the form (1.1). In particular, the $O(3)$ NLSM model can be interpreted as the continuum limit of an isotropic ferromagnet in statistical mechanics. The two-dimensional $O(3)$ model is known to be integrable as a classical field theory [31]. Its exact factorizable S-matrix is also known [32].

Another useful mechanism of generating NLSM with isometries is so-called *gauging*, [2] the simplest example being provided by the CP^n models [33] with the action

$$S[n] = \frac{1}{2\lambda^2} \int d^d x \, (D^\mu n)^* \cdot D_\mu n \; , \tag{1.9}$$

in terms of $(n+1)$ complex scalar fields $n = \{n^a\}$ subject to the constraint

$$n^* \cdot n = 1 \; , \tag{1.10}$$

and the covariant derivative

$$D_\mu n = (\partial_\mu + iA_\mu)n \; . \tag{1.11}$$

The action (1.9) is obviously invariant under Abelian gauge transformations

$$n^a(x) \rightarrow \exp[i\Lambda(x)]n^a(x) \; , \qquad A_\mu \rightarrow A_\mu - \partial_\mu \Lambda \; , \tag{1.12}$$

with the gauge parameter $\Lambda(x)$. Since there are no kinetic terms for the gauge field A_μ in the action (1.9), this gauge field can be easily eliminated from the action via its algebraic equation of motion in terms of the NLSM fields n, as $A_\mu = in^* \cdot \partial_\mu n$. The resulting action also has the form (1.1) simply because the action (1.1) with an arbitrary NLSM metric is the most general (2d Lorentz-, parity- and $\mathcal{M}$ diffeomorphism-invariant) NLSM action.

---

[2] See Sect. 2.8 for the general gauging procedure.

The *two-dimensional* (2d) NLSM is special since its fields ϕ, its metric g_{ab} and, hence, its coupling constants are all dimensionless. In quantum theory, this implies that the 2d NLSM is renormalizable by index of divergence, i.e. its ultraviolet counterterms are of the same (mass) dimension two as the NLSM Lagrangian itself. Since we already know that (1.1) is the most general (parity conserving) form of the 2d kinetic terms, all the UV counterterms can therefore be absorbed into the metric as its 'quantum' deformations. This is known as the (generalized) on-shell renormalizability [34, 35]. Being non-linear, it should be distinguished from the usual (multiplicative) renormalizability that is only the case for a *symmetric* NLSM space $\mathcal{M}$ [36, 37, 38].

In this book we only consider an *on-shell* renormalization of NLSM, i.e. modulo its highly non-linear field renormalization, [3] and calculate its perturbative RG β-functions. A formal proof of NLSM renormalizability can be found e.g., in [39, 40]. A discussion of the general structure of NLSM renormalization is available in [41, 42, 43].

Yet another special feature of two dimensions is the possibility of adding the so-called *generalized* WZNW term to the action (1.1). It reads

$$S_{\mathrm{gWZNW}}[\phi] = -\frac{1}{2\lambda^2} \int \mathrm{d}^2x \, h_{ab}(\phi)\varepsilon^{\mu\nu}\partial_\mu\phi^a\partial_\nu\phi^b \, , \tag{1.13}$$

where $h_{ab}(\phi)$ represents an arbitrary two-form on $\mathcal{M}$,

$$h = h_{ab}(\phi)\mathrm{d}\phi^a \wedge \mathrm{d}\phi^b \, . \tag{1.14}$$

The WZNW term (1.13) is not invariant under some reflection symmetries of the action (1.1), both in spacetime and, perhaps, in the NLSM target manifold, so that (1.13) cannot appear as a counterterm in the 2d quantum field theory to be defined by the action (1.1) alone. The action (1.13) is only changed by a surface term under the gauge transformations

$$\delta h = \mathrm{d}\beta \, , \quad \text{or} \quad \delta h_{ab} = \partial_{[a}\beta_{b]} \, , \tag{1.15}$$

where the one-form $\beta(\phi)$ is a gauge parameter.

Let us now compactify our Euclidean spacetime to a sphere, or just consider only those NLSM maps $\phi: \Sigma_E \to \mathcal{M}$ that have the property $\phi \to$ const. at $|x| \to \infty$. Let $\hat{\phi}$ be the image of our compactified spacetime in $\mathcal{M}$ under the map ϕ, which should be topologically equivalent to the sphere. According to Stokes theorem, we can rewrite (1.13) to the form

$$S_{\mathrm{gWZNW}}[\phi] = \frac{1}{\lambda^2} \int_{\hat{\phi}} h = \frac{1}{\lambda^2} \int_B H \, , \tag{1.16}$$

where the three-form

---

[3] In fact, there are extra dimensional parameters in an off-shell quantized NLSM, which lead to extra counterterms whose form is different from (1.1) — see Subsect. 2.4.2.

$$H = dh \tag{1.17}$$

and the three-dimensional ball $B \in \mathcal{M}$, whose boundary is $\partial B = \hat{\phi}$, have been introduced. The NLSM fields are supposed to be extended from the boundary S^2 to its interior B, while the NLSM action should not depend upon it.[4] We are now in a position to slightly change the rules of the game, and *define* the WZNW term by (1.16) instead of (1.13), assuming that the corresponding 2d action exists globally [44, 45]. The three-form H is supposed to be closed but it may not be exact. Then (1.17) is still valid locally (the latter is always the case due to the Poincaré lemma), but it may not be valid globally if H belongs to a non-trivial third cohomology class. Therefore, the action (1.13) may not be globally well defined or, equivalently, it may be dependent upon the choice of B. Insisting on the globally defined action (1.13) in quantum theory leads to topological quantization of the coupling constant in front of the WZNW term (Subsect. 5.1.3). The most important example is provided by the standard WZNW model [46, 47, 48] which is simply the NLSM on a Lie group manifold $\mathcal{M} = G$. Let us introduce the G-group valued field $U(x)$ as

$$U(x) = \exp[i\phi(x)] , \quad \phi(x) = \phi^i(x)t^i , \tag{1.18}$$

where t^i are the Lie group generators satisfying the relations

$$[t^i, t^j] = 2\mathrm{i} f^{ijk} t^k , \quad \mathrm{tr}(t^i t^j) = 2\delta^{ij} , \tag{1.19}$$

and f^{ijk} are the corresponding Lie algebra structure constants. The 2d WZNW action [46, 47, 48] is defined by

$$S_{\mathrm{WZNW}}[U] = \frac{1}{4\lambda^2} \int d^2x \, \mathrm{tr}(\partial^\mu U \partial_\mu U^{-1}) + \Gamma\,[U] , \tag{1.20}$$

where $\Gamma[U]$ is the so-called WZ functional having the form (1.16). In terms of the Lie algebra valued fields ϕ, the first kinetic term of (1.20) takes the NLSM form (1.1) with a calculable metric $g_{ab}(\phi)$. The calculation of the WZNW metric $g_{ab}(\phi)$, and the associated *vielbein* $V_a^i(\phi)$ defined by the relations

$$g_{ab} = \delta^{ij} V_a^i V_b^j , \quad g^{ab} V_a^i V_b^j = \delta^{ij} , \tag{1.21}$$

can be found in Sect. 2.5, see also [49, 50]. Their explicit forms are not going to be used in this section.

The three-form H defining the WZNW functional Γ in (1.16) can be most conveniently written down in terms of the vielbein (1.21),

$$H_{abc} = \eta f^{ijk} V_a^i V_b^j V_c^k , \quad \text{where} \quad \eta \equiv \frac{n\lambda^2}{3\pi} . \tag{1.22}$$

It is the coefficient n introduced in (1.22) that should be an integer, in order the WZNW action (1.20) be globally well defined in quantum theory (Subsect. 5.3.1). Similarly, the Riemannian curvature associated with the WZNW group metric takes the simple form in terms of the vielbein, namely,

---

[4] We assume, of course, that the extended fields are smooth and differentiable.

$$R_{abcd} = f^{ijm} f^{klm} V_a^i V_b^j V_c^k V_d^l \; . \tag{1.23}$$

The quantum WZNW model at the RG fixed point $\eta^2 = 1$ is fully conformally invariant [51, 52], and it is known to be equivalent (as a CFT) to a free fermionic theory. This phenomenon is called non-Abelian bosonization in two dimensions [53, 54] (see, e.g., [45] for more).

2.2 Covariant Background-Field Method

A naive quantization of the scalar field theory (1.1) may be done along the standard lines of QFT [27, 55, 56]. One introduces a source $J^a(x)$ for each NLSM component field $\phi^a(x)$, and then one defines the quantum generating functional of Green's functions in the form of a path integral

$$Z[J] = \exp\{iW[J]\} = \int [\mathrm{d}\phi] \exp\{i(S + \phi \cdot J)\} \; , \tag{2.1}$$

where we have used the condensed notation by considering the coordinates x^μ on an equal footing with the NLSM field index. It is assumed that the path integral (2.1) is actually defined by the standard perturbative expansion, while the associated momentum integrals are supposed to be evaluated in Euclidean space after a Wick rotation [27, 55, 56].

The quantum generating functional Γ of the 1-Particle Irreducible (1PI) Green functions is related to the generating functional W of the connected Green functions via the Legendre transform [27, 55, 56]

$$\Gamma[\bar{\phi}] = W[J(\bar{\phi})] - \bar{\phi} \cdot J[\bar{\phi}] \; , \tag{2.2}$$

where $\bar{\phi}$ is the so-called *mean* field,

$$\bar{\phi} = \frac{\delta W}{\delta J} \; . \tag{2.3}$$

It follows from (2.2) and (2.3) that

$$\exp(i\Gamma[\bar{\phi}]) = \int [\mathrm{d}\phi] \exp\left\{i\left(S[\phi] - (\phi - \bar{\phi}) \cdot \frac{\delta\Gamma}{\delta\bar{\phi}}\right)\right\} \; . \tag{2.4}$$

Let us now define the quantum fluctuating field π simply as a difference between the full (classical + quantum) field ϕ and the background (classical) mean field $\bar{\phi}$, i.e.

$$\phi^a(x) = \bar{\phi}^a(x) + \pi^a(x) \; . \tag{2.5}$$

Equation (2.4) then takes the form

$$\exp(i\Gamma[\bar{\phi}]) = \int [\mathrm{d}\pi] \exp\left\{i\left(S[\bar{\phi} + \pi] - \pi \cdot \frac{\delta\Gamma}{\delta\bar{\phi}}\right)\right\} \; . \tag{2.6}$$

The subtraction in the exponential factor removes the vertices linear in the quantum field π, which results in the 1PI Feynman graphs without external π-lines in the perturbative expansion of (2.6).

One can also define a new generating functional $\overline{\Gamma}[\bar{\phi}, \bar{\pi}]$ depending upon yet another mean field $\bar{\pi}$, when one starts with the initial NLSM action whose field argument takes the form (2.5). One introduces the generating functional $\overline{W}[\bar{\phi}, J]$ with respect to the quantum π-fields as usual,

$$\exp\left(i\overline{W}[\bar{\phi}, J]\right) = \int [\mathrm{d}\pi]\, \exp\left\{i\left(S[\bar{\phi} + \pi] + \pi \cdot J\right)\right\} , \qquad (2.7)$$

and then one performs a Legendre transform with respect to the argument J,

$$\overline{\Gamma}[\bar{\phi}, \bar{\pi}] = \overline{W}[\bar{\phi}, J] - \bar{\pi} \cdot J . \qquad (2.8)$$

It is not difficult to check that [57]

$$\overline{\Gamma}[\bar{\phi}, 0] = \Gamma[\bar{\phi}] . \qquad (2.9)$$

Hence, the standard 1PI functional $\Gamma[\bar{\phi}]$ is obtained from the background-dependent generating functional $\overline{\Gamma}[\bar{\phi}, \bar{\pi}]$ simply by omitting the Feynman diagrams with external quantum π-lines.

When being applied to the NLSM (1.1), the naive quantization above has, however, two drawbacks. First, it is not covariant with respect to the NLSM reparametrizations, which leads to non-tensor vertices in the expansion of the background-quantum splitted NLSM action $S[\bar{\phi} + \pi]$ with respect to the quantum field π. As a result, the quantum effective action turns out to be non-covariant too. Second, our considerations above do not take into account renormalization (Sect. 2.3).

The non-covariance can be avoided when using a *non-linear* background-quantum splitting known as the covariant background-field method for NLSM [58, 59, 60, 61, 62, 63]. It is based on another (covariant) definition of the quantum effective action, which is different from the standard one off-shell, being equivalent on-shell. The covariance is achieved by using geodesics in the internal manifold $\mathcal{M}$, which are defined by the equation

$$\frac{\mathrm{d}^2}{\mathrm{d}s^2}\rho^a(x, s) + \Gamma^a_{bc}[\rho]\frac{\mathrm{d}}{\mathrm{d}s}\rho^b\frac{\mathrm{d}}{\mathrm{d}s}\rho^c = 0 , \qquad (2.10)$$

where s is the parameter of the geodesic connecting ϕ and $\phi + \pi$: [5]

$$\rho(x, s = 0) = \phi(x) , \qquad \rho(x, s = 1) = \phi(x) + \pi(x) , \qquad (2.11)$$

and Γ^a_{bc} are the Christoffel symbols in terms of the metric g_{ab}. Let us now define the tangent vector along the geodesic,

---

[5] Whenever it does not lead to confusion, we omit the bar over the classical background field ϕ here and in what follows, in order to simplify our notation.

$$\xi_s^a = \frac{\mathrm{d}}{\mathrm{d}s}\rho^a , \qquad \xi_s^a\big|_{s=0} = \xi^a , \tag{2.12}$$

and consider ξ^a as our fundamental quantum fields, instead of π^a. An expansion of the action $S[\phi + \pi(\xi)]$ in powers of ξ fields will be covariant since, unlike π, the field ξ is a vector with respect to diffeomorphisms on $\mathcal{M}$. As regards the two-dimensional NLSM with a generalized WZNW term, one has

$$S[\phi + \pi(\xi)] = \frac{1}{2\lambda^2} \int \mathrm{d}^2x \ \{g_{ab}[\phi + \pi(\xi)]\eta^{\mu\nu} + h_{ab}[\phi + \pi(\xi)]\varepsilon^{\mu\nu}\}$$

$$\times \partial_\mu[\phi + \pi(\xi)]^a \partial_\nu[\phi + \pi(\xi)]^b = S[\phi] + S_1 + S_2 + \dots , \tag{2.13}$$

where

$$S_n = \frac{1}{n!} \frac{\mathrm{d}^n}{\mathrm{d}s^n} S[\rho(s)]\bigg|_{s=0} . \tag{2.14}$$

Since the dependence of π upon ξ is non-linear, the 1PI diagrams with respect to π are, in general, not 1PI with respect to ξ, and vice versa. For an actual calculation of S_n the following obvious relations are useful:

$$\frac{\mathrm{d}}{\mathrm{d}s}\partial_\mu\rho^a = \partial_\mu\xi_s^a , \qquad \frac{\mathrm{d}}{\mathrm{d}s}g_{ab} = \xi_s^c\partial_c g_{ab} , \qquad \partial_\mu g_{ab} = \partial_\mu\rho^c\partial_c g_{ab} . \tag{2.15}$$

For instance, one easily finds that

$$S_1 = \frac{1}{2\lambda^2} \int \mathrm{d}^2x \ \{-2\xi^a g_{ab}[\phi]\left(D_\mu\partial^\mu\phi^b - H_{cd}^b\varepsilon_{\mu\nu}\partial^\nu\phi^d\partial^\mu\phi^c\right)\} , \tag{2.16}$$

where we have used integration by parts, omitted a surface term, and introduced the covariant derivative D_μ as

$$(D_\mu\xi_s)^a = (\delta^{ab}\partial_\mu + \Gamma_{bc}^a\partial_\mu\rho^c)\xi_s^b . \tag{2.17}$$

It should be noted that we can use, in fact, *any* connection for a covariant expansion of the action, not just the one of (2.17). We sometimes find it convenient to use the following connection with the torsion H:

$$(\widehat{D}_\mu\xi_s)^a = (D_\mu\xi_s)^a - H_{bc}^a[\rho]\varepsilon_{\mu\nu}\partial^\nu\rho^c\xi_s^b . \tag{2.18}$$

For example, the first term $S_1[\rho]$ then takes the compact form

$$S_1(s) = \frac{1}{2\lambda^2} \int \mathrm{d}^2x \ 2g_{ab}(\widehat{D}_\mu\xi_s)^a\partial^\mu\phi^b . \tag{2.19}$$

It is clear from (2.14) and (2.19) that all the other terms S_n in the expansion (2.13) should also depend upon the torsion potential h only via its field strength $H = \mathrm{d}h$ that can be called the NLSM *torsion* because of the way it enters (2.18) and (2.19).

To explicitly calculate S_n in (2.13), it is quite useful to use the differential operator $D(s)$ with a covariant completion instead of the naive differentiation with respect to s. One has

$$D(s)V_a = \frac{\mathrm{d}}{\mathrm{d}s}V_a - \Gamma^c_{ab}[\rho(s)]\xi^b_s V_c ,$$

$$D(s)V^a = \frac{\mathrm{d}}{\mathrm{d}s}V^a + \Gamma^a_{bc}[\rho(s)]\xi^b_s V^c ,$$

(2.20)

for any co- and contravariant vectors V_a and V^a, respectively. In particular, when acting with $D(s)$ on a function of $\rho(s)$ only, one has $D(s) = \xi^a_s D_a$. The following identities are quite useful for explicit calculations:

$$D(s)\xi^a_s = 0 ,$$

$$D(s)\partial_\mu \rho^a = (D_\mu \xi_s)^a ,$$

(2.21)

$$D^2(s)\partial_\mu \rho^a = R^a{}_{bcd}\xi^b_s \xi^c_s \partial_\mu \rho^d .$$

It is now easy to calculate the second term S_2 from (2.19). It reads

$$S_2 = \frac{1}{2\lambda^2}\int \mathrm{d}^2 x \left\{(\widehat{D}_\mu \xi)^2 + \widehat{R}_{abcd}\xi^b \xi^c (\eta^{\mu\nu} - \varepsilon^{\mu\nu})\partial_\mu \phi^a \partial_\nu \phi^d\right\} ,$$

(2.22)

where we have introduced the generalized curvature tensor $\widehat{R}_{abcd}$ with the torsion H as follows:

$$\widehat{R}_{abcd} = R_{abcd}(\widehat{\Gamma})$$

$$\equiv R_{abcd}(\Gamma - H)$$

(2.23)

$$= R_{abcd} - D_c H_{abd} + D_d H_{abc} + H_{acf}H^f{}_{bd} - H_{adf}H^f{}_{bc} .$$

Here R_{abcd} is the standard Riemann-Christoffel curvature tensor, $R_{abcd} = g_{ae}(\partial_c \Gamma^e_{db} - \partial_d \Gamma^e_{cb} + \Gamma^e_{df}\Gamma^f_{bc} - \Gamma^e_{cf}\Gamma^f_{bd})$, and

$$\Gamma^a_{bc} = g^{ad}\Gamma_{bc,d} , \quad \Gamma_{ab,c} = \frac{1}{2}\left(\frac{\partial g_{ac}}{\partial \phi^b} + \frac{\partial g_{bc}}{\partial \phi^a} - \frac{\partial g_{ab}}{\partial \phi^c}\right).$$

(2.24)

Equations (2.19) and (2.22) naturally lead to the conjecture that all the terms S_n might be expressed in terms of the covariant derivative $\widehat{D}_\mu$ and the generalized curvature $\widehat{R}_{abcd}$ alone, i.e. without a non-minimal (explicit) dependence upon the torsion H_{abc}. In fact, this is not the case. We find

$$S_3 = \frac{1}{2\lambda^2}\int \mathrm{d}^2 x \left\{\left(\tfrac{1}{3}D_g R_{abcd} + \tfrac{4}{3}H^f_{ga}D_b H_{cdf}\right)\partial_\mu \phi^a \partial^\mu \phi^d \xi^b \xi^c \xi^g\right.$$

$$+ \left(\tfrac{1}{3}D_a D_b H_{cdf} + \tfrac{2}{3}H_{agh}H^g_{bd}H^h_{fc} + \tfrac{2}{3}R_{gabc}H^g_{fd}\right)\partial_\mu \phi^c \varepsilon^{\mu\nu}\partial_\nu \phi^d \xi^f \xi^a \xi^b$$

$$+ \tfrac{4}{3}(R_{abcd} - H^f_{ab}H_{fcd})\partial_\mu \phi^a \widehat{D}^\mu \xi^d \xi^b \xi^c$$

$$\left. + \tfrac{4}{3}D_a H_{dbc}\partial_\mu \phi^b \varepsilon^{\mu\nu}\widehat{D}_\nu \xi^c \xi^a \xi^d + \tfrac{2}{3}H_{abc}\widehat{D}_\mu \xi^a \varepsilon^{\mu\nu}\widehat{D}_\nu \xi^b \xi^c\right\} .$$

(2.25a)

It can be easily verified that (2.25a) cannot be put into the minimal form for an arbitrary Riemannian NLSM manifold $\mathcal{M}$ with a generic 'torsion' H_{abc}.

Similarly, we find

$$
\begin{aligned}
S_4 =\frac{1}{2\lambda^2} \int \mathrm{d}^2x \Big\{ &\big(\tfrac{1}{12} D_a D_b R_{cdfg} + \tfrac{1}{3} R^k{}_{dfc} R_{kbag} - \tfrac{1}{2} D_a D_b H_{cfh} H^h_{dg} \\
&-\tfrac{1}{2} R^k{}_{abc} H_{kfh} H^h_{dg} + \tfrac{1}{6} R^k{}_{abh} H_{kcf} H^h_{dg}\big) \partial_\mu\phi^c \partial^\mu\phi^g \xi^a \xi^b \xi^d \xi^f \\
&+ \big(\tfrac{1}{2} D_a R_{bcdf} - H^g_{bc} D_a H_{gdf}\big) \partial_\mu\phi^b \widehat{D}^\mu \xi^f \xi^a \xi^c \xi^d + \tfrac{1}{3} R_{abcd} \widehat{D}_\mu \xi^a \\
&\times \widehat{D}^\mu \xi^d \xi^b \xi^c + \big(\tfrac{1}{12} D_a D_b D_c H_{gdf} + \tfrac{1}{2} D_a H^k_{gd} R_{kbcf} + \tfrac{1}{2} D_a H_{hbk} \\
&\times H^h_{cd} H^k_{gf} - \tfrac{1}{3} H^k_{gd} D_a R_{kbcf}\big) \partial_\mu\phi^d \varepsilon^{\mu\nu} \partial_\nu\phi^f \xi^g \xi^a \xi^b \xi^c \\
&+ \big(\tfrac{1}{2} D_a D_b H_{fcd} + \tfrac{1}{2} H_{kdf} R^k{}_{abc} + \tfrac{1}{2} R^k{}_{abd} H_{kcf}\big) \partial_\mu\phi^c \varepsilon^{\mu\nu} \widehat{D}_\nu \xi^d \xi^a \xi^b \xi^f \\
&+ \tfrac{1}{2} D_a H_{dbc} \widehat{D}_\mu \xi^b \varepsilon^{\mu\nu} \widehat{D}_\nu \xi^c \xi^a \xi^d \Big\} \ .
\end{aligned}
$$

$$(2.25b)$$

The next term S_5 in the covariant background-field NLSM expansion (2.13) reads as follows:

$$
\begin{aligned}
S_5 =\frac{1}{2\lambda^2} \int \mathrm{d}^2x \Big\{ &\big[\tfrac{1}{60} D_a D_b D_c R_{gdfh} - \tfrac{7}{30} R_{bgck} D_a R^k{}_{dfh} \\
&- \tfrac{2}{15} H^k_{fh} D_a D_b D_c H_{gdk} - \tfrac{2}{15} D_a R_{kbcg} H^k_{dl} H^l_{fh} \\
&- \tfrac{2}{5} R_{kbcg} D_a H^k_{dl} H^l_{fh} - \tfrac{1}{30} D_a R_{kbcl} H^k_{gd} H^l_{fh} \\
&+ \tfrac{2}{15} R_{kbcl} H^k_{gd} D_a H^l_{fh}\big] \xi^a \xi^b \xi^c \xi^d \xi^f \partial^\mu\phi^g \partial_\mu\phi^h \\
&+ \big[\tfrac{2}{15} D_a D_b R_{fcdg} + \tfrac{4}{15} R_{gabk} R^k{}_{cdf} + \tfrac{2}{5} D_a D_b H_{cfk} H^k_{dg} \\
&+ \tfrac{4}{15} R_{hab[f} H^h{}_{k]d} H^k{}_{cg}\big] \xi^a \xi^b \xi^c \xi^d \widehat{D}_\mu \xi^f \partial^\mu\phi^g \\
&+ \tfrac{1}{6} D_a R_{dcbf} \xi^a \xi^b \xi^c \widehat{D}_\mu \xi^d \widehat{D}^\mu \xi^f
\end{aligned}
$$

$$
\begin{aligned}
+ \big[&\tfrac{1}{60} D_a D_b D_c D_d H_{fgh} + \tfrac{1}{10} D_a D_b R_{kcdg} H^k_{fh} + \tfrac{1}{10} R^k{}_{abg} H_{kmc} R^m{}_{dfh} \\
+ &\tfrac{7}{30} R_{mabk} R^k{}_{cdg} H^m_{fh} - \tfrac{1}{5} D_a D_b H_{ckm} H^k_{dg} H^m_{fh} \\
- &\tfrac{2}{15} R^k{}_{abm} H_{kcl} H^l_{dg} H^m_{fh} - \tfrac{2}{15} D_a R_{kbdg} D_c H^k_{fh} \\
+ &\tfrac{1}{5} D_a D_b H^k_{fg} R_{kcdh}\big] \xi^a \xi^b \xi^c \xi^d \xi^f \partial_\mu\phi^g \varepsilon^{\mu\nu} \partial_\nu\phi^h + \big[\tfrac{2}{15} D_a D_b D_c H_{dfh} \\
+ &\tfrac{2}{15} D_a R_{mbch} H^m_{df} + \tfrac{4}{15} D_a R_{mbcf} H^m_{dh} - \tfrac{2}{15} D_a H^k_{bh} R_{fcdk} \\
+ &\tfrac{2}{5} D_a H^k_{bf} R_{hcdk}\big] \xi^a \xi^b \xi^c \xi^d \widehat{D}^\mu \xi^f \varepsilon_{\mu\nu} \partial^\nu\phi^h + \big[\tfrac{1}{5} D_a D_b H_{cdf} \\
- &\tfrac{2}{15} R_{dabk} H^k_{cf}\big] \xi^a \xi^b \xi^c \widehat{D}_\mu \xi^d \varepsilon^{\mu\nu} \widehat{D}_\nu \xi^f \Big\} \ .
\end{aligned}
$$

$$(2.25c)$$

The explicit form of S_6 can be found in [63]. It is straightforward to calculate the next terms in the expansion (2.13), either by using the identities given above or, equivalently, in normal coordinates. [6] The general structure of the covariant background-field expansion is given by

$$
\begin{aligned}
S_n = \frac{1}{2\lambda^2} \int \mathrm{d}^2 x \, \Big\{ & \Pi^{(n,2)}_{(a_1\cdots a_n)(b_1 b_2)} \xi^{a_1} \cdots \xi^{a_n} \partial^\mu \phi^{b_1} \partial_\mu \phi^{b_2} \\
& + \Pi^{(n,1)}_{(a_1\cdots a_{n-1})b_1 b_2} \xi^{a_1} \cdots \xi^{a_{n-1}} \widehat{D}_\mu \xi^{b_1} \partial^\mu \phi^{b_2} \\
& + \Pi^{(n,0)}_{(a_1\cdots a_{n-2})(b_1 b_2)} \xi^{a_1} \cdots \xi^{a_{n-2}} \widehat{D}_\mu \xi^{b_1} \widehat{D}^\mu \xi^{b_2} \\
& + E^{(n,2)}_{(a_1\cdots a_n)[b_1 b_2]} \xi^{a_1} \cdots \xi^{a_n} \partial_\mu \phi^{b_1} \varepsilon^{\mu\nu} \partial_\nu \phi^{b_2} \\
& + E^{(n,1)}_{(a_1\cdots a_{n-1})b_1 b_2} \xi^{a_1} \cdots \xi^{a_{n-1}} \widehat{D}_\mu \xi^{b_1} \varepsilon^{\mu\nu} \partial_\nu \phi^{b_2} \\
& + E^{(n,0)}_{(a_1\cdots a_{n-2})[b_1 b_2]} \xi^{a_1} \cdots \xi^{a_{n-2}} \widehat{D}_\mu \xi^{b_1} \varepsilon^{\mu\nu} \widehat{D}_\nu \xi^{b_2} \Big\} ,
\end{aligned}
\tag{2.26}
$$

where the Π's and E's stand for certain tensor combinations of the curvature, torsion and their covariant derivatives, in a self-explanatory notation. In general, those tensors cannot be written down in minimal form, i.e. in terms of $\widehat{R}_{abcd}$ and $\widehat{D}_a$ alone.

In quantum theory, we employ a Dimensional Regularization (DR) [65], in order to renormalize 2d NLSM. In the presence of a WZNW term, the standard DR prescription [65] has to be extended since the WZNW term only exists in two dimensions. It is also desirable to keep the Abelian gauge invariance (1.15) in the process of renormalization of NLSM. At this point, it is enough to notice that we first define the covariant background-quantum splitting and, hence, all the covariant NLSM vertices in two dimensions, and then continue to $2 - 2\epsilon$ dimensions (Sect. 2.3).

We are now in a position to define the covariant NLSM quantum generating functional. It is just the naive source terms in (2.1) and (2.7) that are non-covariant. The quantum effective action defined by (2.6) is non-covariant too, even after changing the variables $\pi = \pi(\xi)$, simply because there is the non-vector field π^a in front of $\delta\Gamma/\delta\bar{\phi}$. This observation also suggests a way to modify (2.6), namely, by replacing the non-covariant field π by the covariant (vector) field ξ^a (see also [66]),

$$
\exp\left(\mathrm{i}\Gamma_\mathrm{V}[\bar{\phi}]\right) = \int [\mathrm{d}\xi] \, \exp\left\{ \mathrm{i} \left(S[\bar{\phi} + \pi(\xi)] - \xi \cdot \frac{\delta\Gamma_\mathrm{V}}{\delta\bar{\phi}} \right) \right\} .
\tag{2.27}
$$

However, it has yet to be proved that (2.27) results in the 1PI-diagrams to all orders of quantum perturbation theory. A consistent approach is possible by introducing a non-standard source term [67],

---

[6] Closed formulae for the tensor coefficients in any finite order n can be found by using the identities of [64].

$$K[\phi] = \int \mathrm{d}^2 x \, K(x; \phi(x)) \; . \tag{2.28}$$

The quantum generating functional

$$\exp\left(\mathrm{i}W[K]\right) = \int [\mathrm{d}\phi] \, \exp\left\{\mathrm{i}\left(S[\phi] + K[\phi]\right)\right\} \tag{2.29}$$

is obviously reparametrization-invariant under the condition that K transforms as a scalar. In fact, (2.28) implies an infinite number of sources related to all powers of the NLSM field. In practical calculations, it is convenient to use only one source, namely, the one associated with the covariant quantum field ξ itself. This amounts to a definition of the NLSM quantum generating functional W in the form

$$\exp\left(\mathrm{i}W[\bar\phi, J]\right) = \int [\mathrm{d}\xi] \, \exp\left\{\mathrm{i}\left(S[\bar\phi + \pi(\xi)] + \xi \cdot J\right)\right\} \; . \tag{2.30}$$

The Feynman rules defined by (2.30) are manifestly covariant, when using the covariant background-field method. The associated generating functional of 1PI Green functions is obtained from that of connected Green functions in (2.30) along the standard lines of QFT. Let us define the mean field

$$\bar\xi(x) = \frac{\delta W}{\delta J(x)} \; , \tag{2.31}$$

and perform a Legendre transform

$$\overline{\Gamma}[\bar\phi, \bar\xi] = W[\bar\phi, J(\bar\xi)] - \bar\xi \cdot J(\bar\xi) \; . \tag{2.32}$$

It follows that

$$\exp\left(\mathrm{i}\overline{\Gamma}[\bar\phi, \bar\xi]\right) = \int [\mathrm{d}\xi] \, \exp\left\{\mathrm{i}\left(S[\bar\phi + \pi(\xi)] - (\xi - \bar\xi)\frac{\delta\overline{\Gamma}}{\delta\bar\xi}\right)\right\} \; . \tag{2.33}$$

Equation (2.33) can be rewritten in the form of a 'double' expansion with respect to the classical and quantum NLSM fields [28],

$$\exp\left(\mathrm{i}\overline{\Gamma}[\bar\phi, \bar\xi]\right) = \int [\mathrm{d}\xi'] \, \exp\left\{\mathrm{i}\left(S[\bar\phi + \pi(\bar\xi + \xi')] - \xi'\frac{\delta\overline{\Gamma}}{\delta\bar\xi}\right)\right\} \; , \tag{2.34}$$

where we have introduced $\xi' = \xi - \bar\xi$.

When using (2.34) in quantum calculations, for example, in a calculation of the NLSM renormalization group β-functions, it is important to know whether it is enough to calculate 1PI Feynman diagrams *without* external quantum lines, in accordance with the main idea of the background field method. In the case of a *linear* background-quantum splitting (2.5), the answer is in the affirmative – see (2.9). However, in the case of a *non-linear* background-quantum splitting, like that in the NLSM covariant background-field method above, the situation is much more complicated. As was shown, e.g., in [68] without using a particular parametrization of a 2d NLSM, the NLSM renormalization is not multiplicative, being accompanied by a highly non-linear field renormalization. This should have been expected since the 2d NLSM field has a vanishing canonical dimension. Moreover, it follows from a study of the NLSM Ward identities [67] that $\overline{\Gamma}[\bar{\phi}, \bar{\xi}]$ is *not* the same as the expansion of $\Gamma[\bar{\phi}] = \overline{\Gamma}[\bar{\phi}, \bar{\xi} = 0]$ in normal coordinates. Hence, if $\Delta L[\bar{\phi}]$ are all the counterterms needed to remove the divergences in $\Gamma[\bar{\phi}]$, the expansion of $\Delta L[\bar{\phi}]$ in normal coordinates may not remove the divergences in $\overline{\Gamma}[\bar{\phi}, \bar{\xi}]$. Therefore, generally speaking, the non-linear renormalization of the quantum NLSM fields $\bar{\xi}$ should not be ignored in quantum calculations. This is indeed essential for a correct account of both Ultraviolet (UV) and Infrared (IR) divergences, as well as for a cancellation of the mixed divergent terms proportional to $\epsilon^{-p}[\ln(m^2/\mu^2)]^q$, where μ is the renormalization scale and m is the IR-regulating mass, and (p, q) are positive integers. As was shown in [67], the naive BPHZ procedure (or R-operation) in application to $\Gamma[\bar{\phi}]$, simply by normally expanding its counterterms in perturbation theory, does lead to incorrect results beyond the three-loop level. When using our prescription (2.28)–(2.30), a calculation of the 1PI Feynman diagrams without external quantum lines is, nevertheless, enough for computing the on-shell NLSM quantities like the RG β-functions [67].

In the DR method, the UV divergences of the 1PI diagrams in the l-loop approximation of quantum perturbation theory comprise the $1/\epsilon^p$-divergences with the poles of order p ranging from 1 to l. The simple-pole divergences are the most important ones (in fact, they just determine the NLSM β-functions, see Sect. 2.3), while the higher-order divergences are essentially determined by them via the renormalization group pole equations [34, 69, 70]. Since we are only interested in calculating the covariant NLSM β-functions on-shell (i.e. modulo the equations of motion), we simply ignore the complications associated with the NLSM field renormalization, by restricting ourselves to a calculation of the simple pole $(1/\epsilon)$ divergences only, similarly to the approach adopted in [71, 72, 73].

We still have to fix the measure in the functional integral (2.30) or (2.34). Our choice is rather obvious, namely,

$$[\mathrm{d}\xi] = \prod_x \sqrt{g(\phi)} \prod_{a=1}^{D} \mathrm{d}\xi^a(x) \, , \tag{2.35}$$

where $g = \det g_{ab}$. When using the vielbein V_a^i defined by (1.21) that relates the base manifold $\mathcal{M}$ to its target (flat) space $T(\mathcal{M})$, we have

$$\xi^a = V^{ai}\xi^i \, , \qquad \xi^i = V_a^i\xi^a \, . \tag{2.36}$$

This allows us to rewrite the measure (2.35) in the 'canonical' form

$$[\mathrm{d}\xi] = \prod_x \prod_{i=1}^{D} \mathrm{d}\xi^i(x) \ . \tag{2.37}$$

The covariant derivatives (2.17) and (2.19) then take the form

$$\begin{aligned} V_a^i(D_\mu \xi)^a &= (\delta^{ij}\partial_\mu + A_\mu^{ij})\xi^j \equiv (\nabla_\mu \xi)^i \ , \\ V_a^i(\hat{D}_\mu \xi)^a &= (\delta^{ij}\partial_\mu + A_\mu^{ij} + \varepsilon_{\mu\nu}B^{\nu ij})\xi^j \equiv (\hat{\nabla}_\mu \xi)^i \ , \end{aligned} \tag{2.38}$$

in terms of the corresponding spin-connection A_μ^{ij} and torsion B_μ^{ij}. The leading (background-independent) kinetic terms of the quantum ξ^i fields in the NLSM action (2.13) then have the standard (free) form.

2.3 Regularization and Quantum Ambiguities

Being a massless scalar field theory, the quantum NLSM (1.1) suffers from both UV- and IR-divergences in two dimensions. The simplest way to regularize the IR-divergences is to add a mass term of the form [7]

$$S_m = -\frac{m^2}{2\lambda^2} \int \mathrm{d}^2x \, g_{ab}(\phi)\phi^a \phi^b \ , \tag{3.1}$$

as the IR-regulator. It is convenient to use dimensional regularization (DR) against UV divergences. In particular, a free regularized propagator of the quantum ξ-fields then takes the form

$$\mathrm{i}\,\langle \xi^i(x)\xi^j(y)\rangle \equiv \delta^{ij}G(x-y) = \mathrm{i}\lambda^2\delta^{ij} \int \frac{\mathrm{d}^d p}{(2\pi)^d} \frac{\exp[-\mathrm{i}p \cdot (x-y)]}{p^2 - m^2 + \mathrm{i}0} \ , \tag{3.2}$$

where $d = 2 - 2\epsilon$. The total effect of (3.1) on the NLSM Feynman diagrams thus amounts to replacing $p^2 \to p^2 - m^2$ in the denominators of momentum integrals. In practice, it is more convenient to insert m^2 factors only into the propagators that, otherwise, would lead to IR-singularities, for example [71, 72, 73]:

$$\int \mathrm{d}^d k \mathrm{d}^d q \frac{k \cdot q}{k^2 q^2 (k-q)^2} \to \int \mathrm{d}^d k \mathrm{d}^d q \frac{k \cdot q}{k^2 q^2 [(k-q)^2 - m^2]} \ . \tag{3.3}$$

It is easy to check that the different ways of inserting the m^2-regulator merely change the finite part of a given Feynman diagram. This happens because the IR-divergenes in a 2d NLSM are logarithmic at most. In other words, the UV-divergences are not affected. As an example, let us just compare the two integrals in the two-dimensional limit:

---

[7] See e.g., [74, 75] for an introduction to renormalization theory.

$$\lim_{\epsilon \to 0} \left[\int d^d k d^d q d^d p \, \delta^{(d)}(k+q+p)(k \cdot p) \right.$$

$$\left. \times \left\{ \frac{1}{k^2 p^2 (q^2 - m^2)} - \frac{1}{(k^2 - m^2)(p^2 - m^2)(q^2 - m^2)} \right\} \right] = 0 . \tag{3.4}$$

The situation becomes more involved after taking into account the need to subtract subdivergences in multi-loop diagrams, but even in this case the conclusion turns out to be the same. See [76, 77] for more about IR-divergences in NLSM.

Since a WZNW term contains the antisymmetric Levi-Civita symbol $\varepsilon^{\mu\nu}$ that is not defined away from two dimensions, the DR prescription in d dimensions has to be properly extended. The actual appearance of $\varepsilon^{\mu\nu}$ in perturbative calculations implies the need for a further definition of only two relevant contractions in d dimensions, namely, $\varepsilon^{\lambda\nu}\eta_{\nu\mu}\varepsilon^{\mu\rho}$ and $\varepsilon^{\mu\nu}\varepsilon^{\rho\sigma}$. For symmetry reasons, they should be of the form

$$\begin{aligned} \varepsilon^{\lambda\nu}\eta_{\nu\mu}\varepsilon^{\mu\rho} &= \psi(\epsilon)\eta^{\lambda\rho} , \\ \varepsilon^{\mu\nu}\varepsilon^{\rho\sigma} &= \omega(\epsilon)[\eta^{\mu\sigma}\eta^{\nu\rho} - \eta^{\mu\rho}\eta^{\nu\sigma}] , \end{aligned} \tag{3.5}$$

where $\eta^{\mu\nu}$ is a formal Minkowski tensor in d dimensions, whereas $\psi(\epsilon)$ and $\omega(\epsilon)$ are arbitrary analytic functions that have to satisfy the normalization condition $\psi(0) = \omega(0) = 1$. We should, therefore, expect that the perturbative RG β-functions are dependent upon those functions, i.e. upon the renormalization prescription chosen for their calculation. This happens to be the case indeed, beyond the one-loop level (Sect. 2.4). In perturbation theory we only have to deal with power series

$$\omega(\epsilon) = 1 + 2\omega_1 \epsilon + 4\omega_2 \epsilon^2 + \ldots, \tag{3.6}$$

since (3.5) implies

$$\psi(\epsilon) = (d - 1)\omega(\epsilon) = (1 - 2\epsilon)\omega(\epsilon) . \tag{3.7}$$

The NLSM l-loop RG β-function is thus going to be dependent upon $(l-1)$ arbitrary parameters $(\omega_1, \ldots, \omega_{l-1})$. This dependence upon the renormalization prescription can be removed by imposing some extra conditions on the NLSM renormalization, the most popular one being related with the 2d conformal invariance at a RG fixed point in two dimensions [69, 78, 79, 80]. The symbol $\varepsilon^{\mu\nu}$ is then considered as an essentially two-dimensional object satisfying the identity

$$\varepsilon^{\mu\nu}\varepsilon^{\rho\sigma} = \bar{\eta}^{\mu\sigma}\bar{\eta}^{\nu\rho} - \bar{\eta}^{\mu\rho}\bar{\eta}^{\nu\sigma} , \tag{3.8}$$

where $\bar{\eta}^{\mu\nu}$ is the Minkowski metric in two dimensions. Let us introduce $\hat{\eta}^{\mu\nu}$ as a formal difference [79]

$$\hat{\eta}^{\mu\nu} = \bar{\eta}^{\mu\nu} - \eta^{\mu\nu} . \tag{3.9}$$

It follows that

$$\widehat{\eta}^{\mu}{}_{\mu} = 2 - d = 2\epsilon , \quad \text{and} \quad \widehat{\eta}^{\mu\nu} = \left(\frac{2-d}{d}\right)\eta^{\mu\nu} . \tag{3.10}$$

Hence, we have

$$\varepsilon^{\mu\nu}\varepsilon^{\rho\sigma} = \frac{4}{d^2}[\eta^{\mu\sigma}\eta^{\nu\rho} - \eta^{\mu\rho}\eta^{\nu\sigma}] . \tag{3.11}$$

Equations (3.5) and (3.11) now imply that

$$\omega^{HVB}(\epsilon) = \frac{1}{(1-\epsilon)^2} ; \quad \omega_1^{HVB} = 1 , \quad \omega_2^{HVB} = \frac{3}{4} . \tag{3.12}$$

Equation (3.12) will be referred to in what follows as the Hooft-Veltman-Bos (HVB) renormalization prescription [69, 79].

A difference between any two prescriptions for the function $\omega(\epsilon)$ generically amounts to a finite transformation of a bare NLSM metric g_{ab}^B and a bare NLSM torsion potential h_{ab}^B. The RG β-functions to be calculated by using different renormalization prescriptions are therefore related to each other by a finite renormalization, in essentially the same way as they would be related when using different regularizations (cf. [81]).

In the DR method the NLSM fields ϕ^a remain massless in d dimensions. Hence, the bare metric g_{ab}^B and torsion $H_{abc}^B = \partial_{[a}h_{bc]}^B$ can be expressed in terms of the renormalized quantities as (cf. [35])

$$g_{ab}^B = \mu^{2\epsilon}\left\{g_{ab}^R + \sum_{n=1}^{\infty}\sum_{l=n}^{\infty}\frac{1}{(2\epsilon)^n}g_{ab}^{(n,l)}(g^R, H^R)\right\} ,$$

$$h_{ab}^B = \mu^{2\epsilon}\left\{h_{ab}^R + \sum_{n=1}^{\infty}\sum_{l=n}^{\infty}\frac{1}{(2\epsilon)^n}H_{ab}^{(n,l)}(g^R, H^R)\right\} , \tag{3.13}$$

where μ is the RG scale parameter of dimension of mass. When using the covariant background field method (Sect. 2.2), the quantum corrections $g_{(ab)}^{(n,l)}$ and $H_{[ab]}^{(n,l)}$ are all tensors in $\mathcal{M}$, which are polynomial in the curvature, torsion and their covariant derivatives. In particular, all the l-loop corrections have the factor $(\lambda^2)^{l-1}$. [8] The scale (μ) dependence of the renormalized metric and torsion can be determined by standard renormalization methods, simply from the condition of the scale independence of the corresponding bare quantities.

The (generalized) RG β-functions are defined by the relations [9]

$$\mu\frac{d}{d\mu}g_{ab} = \beta_{(ab)}^g(g, H) , \quad \mu\frac{d}{d\mu}h_{ab} = \beta_{[ab]}^H(g, H) , \tag{3.14}$$

and

---

[8] These factors are normally suppressed in this section.

[9] We omit the index R for renormalized quantities.

$$\beta_{ab} \equiv \beta^g_{(ab)} + \beta^H_{[ab]} \ . \tag{3.15}$$

The scaling property of the l-loop counterterms under rigid conformal transformations,

$$g^{(n,l)}_{ab} \rightarrow \Lambda^{(l-1)} g^{(n,l)}_{ab} \ , \quad H^{(n,l)}_{ab} \rightarrow \Lambda^{(l-1)} H^{(n,l)}_{ab} \ , \tag{3.16}$$

implies for the residue of the simple ϵ-pole in (3.13) that

$$\beta^g_{(ab)} = -\sum_{l=1}^{\infty} l g^{(1,l)}_{(ab)} \ , \quad \text{and} \quad \beta^H_{[ab]} = -\sum_{l=1}^{\infty} l H^{(1,l)}_{[ab]} \ . \tag{3.17}$$

Similarly, when equating the residues at the higher-order ϵ-poles in (3.13), one gets the generalized RG pole equations [69, 70]. The latter can be used to determine the higher-order (in ϵ) UV-divergences from the known RG β-function. This fact can be useful either as an alternative to direct perturbative calculations or as a check of them.

On dimensional reasons, the covariant l-loop counterterm should have the form

$$\Delta S^{(l)} = \frac{1}{2} \int d^2 x \, T^{(l)}_{ab} (\eta^{\mu\nu} - \varepsilon^{\mu\nu}) \partial_\mu \phi^a \partial_\nu \phi^b \ , \tag{3.18}$$

where

$$T^{(l)}_{ab} = \sum_{n=1}^{l} \frac{1}{(2\epsilon)^n} T^{(n,l)}_{ab} (g, H) \ . \tag{3.19}$$

For instance, in the three-loop approximation, one gets

$$\beta_{ab} = T^{(1,1)}_{ab} + 2 T^{(1,2)}_{ab} + 3 T^{(1,3)}_{ab} \ . \tag{3.20}$$

Yet another source of quantum ambiguities in the perturbative RG β-functions is their dependence upon the way of subtracting subdivergences in multi-loop Feynman graphs. For example, there exist several calculations of the two-loop NLSM β-function with a WZNW term [35, 82, 83, 84, 85, 86], whose renormalization prescriptions as well as the results are different. It is, therefore, important to understand all the renormalization ambiguities in the NLSM β-functions and find some independent consistency conditions.

A natural consistency condition on the 2d NLSM RG β-functions is given by the Zamolodchikov c-theorem [87] (see Subsect. 6.1.2). As is well known in string theory (Chap. 6), the propagation of a (test) closed bosonic string in curved spacetime can be described by the quantum conformally invariant 2d NLSM of the type (1.1). [10] The effective string equations of motion are given by the vanishing NLSM RG β-functions to be calculated by 2d quantum field theory methods. The perturbative NLSM β-functions, playing the role of the

---

[10] The inclusion of an antisymmetric (Kalb-Ramond) tensor field in string theory leads to an NLSM with a generalized WZNW term.

effective string equations of motions, should therefore be derivable from the low-energy string effective action to be calculated from the bosonic string scattering amplitudes (string S-matrix). This implies the existence of an action whose variational equations are just the NLSM RG β-functions. It is the existence of such action that is the statement of the c-theorem (Sect. 6.1). Unfortunately, the original proof of the c-theorem [87] does not tell us much about the form of the action. Since the low-energy string effective action is only constrained by the perturbative string S-matrix, it is only defined on-shell, i.e. modulo reparametrizations of the NLSM metric and torsion potential. The on-shell nature of the relation between the string S-matrix and the effective field theory action should always be kept in mind when comparing QFT calculations in d dimensions with string theory calculations in D dimensions (see Chap. 6 for more).

The existence of the Zamolodchikov action is, however, not enough to remove all quantum ambiguities in the NLSM RG β-functions. Yet another (related) condition is naturally provided by two-dimensional CFT [45]. Being considered at the RG fixed point, the quantum WZNW model is known to be conformally invariant as a quantum 2d field theory [52, 45]. This fact can be used for a non-perturbative calculation of the first derivative (with respect to the coupling constant) of the WZNW β-function in the fixed point by CFT methods [52]. [11] Being expanded in powers of the coupling constant, the CFT result [52] can be compared with independent calculations in quantum perturbation theory, i.e. it can serve as yet another consistency condition on the perturbative RG β-functions of the general NLSM.

In Sect. 2.4 we demonstrate that both consistency conditions mentioned above can be efficiently applied to fix the renormalization ambiguities in 2d NLSM perturbative RG β-functions.

2.4 NLSM Renormalization Group β-Functions

The standard arguments, based on dimensional reasons and covariance, allow us to ignore the minimal connection of the NLSM covariant derivatives in calculating UV divergences. Indeed, the potential UV counterterms can only depend upon the connection via its field strength of dimension two (in units of mass), whose invariant square is already of dimension four. This simple argument is supported by explicit calculations [28].

In Subsect. 2.4.1 we give the one-loop results for the general 2d NLSM with a WZNW term [88]. The one-loop RG β-function in the purely metric case (1.1) has been known since 1971 at least [58]. The two-loop RG β-function is calculated in Subsect. 2.4.2. The results of our three-loop calculations with a non-vanishing torsion are presented in Subsect. 2.4.3.

---

[11] Of course, the RG β-function vanishes at the RG fixed point by definition.

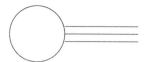

Fig. 2.1. Feynman graph contributing to the NLSM one-loop RG β-function. The simple lines stand for the ξ-propagators (3.2), whereas the triple lines denote the background-dependent vertices

2.4.1 One-Loop Results

There is only one Feynman graph at the one-loop level (Fig. 2.1). The corresponding counterterm is given by

$$\Delta S_1 = -\tfrac{1}{2} I_1 \int d^2 x \, \widehat{R}_{ab} (\eta^{\mu\nu} - \varepsilon^{\mu\nu}) \partial_\mu \phi^a \partial_\nu \phi^b \,, \tag{4.1}$$

where I_1 is the tadpole integral

$$
\begin{aligned}
I_1 &\equiv \frac{G(0)}{\lambda^2} = i \int \frac{d^d p}{(2\pi)^d} \frac{1}{p^2 - m^2} = \frac{\Gamma(\epsilon)}{(4\pi)^{d/2}(m^2)^\epsilon} \\
&= \frac{1}{4\pi\epsilon} - \frac{1}{\pi} \ln m + \text{finite terms} \,.
\end{aligned}
\tag{4.2}
$$

The subtraction procedure in application to the tadpole integral is trivial,

$$I_1 + \text{subtractions} = \frac{1}{4\pi\epsilon} \,. \tag{4.3}$$

Hence, the one-loop RG β-function is given by

$$\beta_{ab}^{(1)} = \frac{1}{2\pi} \widehat{R}_{ab} = \frac{1}{2\pi} \left(R_{ab} - H_a^{gh} H_{bgh} + D^c H_{abc} \right) \,. \tag{4.4}$$

Note that adding a mass term to the NLSM action (1.1) also leads to a non-covariant one-loop counterterm

$$\Delta S_1^{(m)} = \frac{m^2}{2} I_1 \int d^2 x \left\{ \tfrac{1}{3} R_{ab} \phi^a \phi^b - \phi^a \Gamma_{acd} g^{cd} \right\} \,. \tag{4.5}$$

We consider the mass parameter m as the IR-regulator, and we take the limit $m \to 0$ after renormalization.

Equation (4.4) is the key formula for analysing the one-loop renormalization behaviour of effective charges. As an example, consider the multiplicatively renormalizable NLSM on a symmetric space $\mathcal{M}$ [44, 89]. Then the three-form H is co-closed, while the WZNW coupling constant is not renormalized at all. If $\mathcal{M}$ has p simple factors, and $\{\lambda_i\}$ are the corresponding coupling constants for each factor, the one-loop renormalization group equations with the β-function (4.4) take the form

$$\mu \frac{d}{d\mu} \lambda_i^{-2} = \frac{a_i}{2\pi} - \frac{c_i}{(2\pi)^3} \frac{\lambda_1^{2\alpha_1} \lambda_2^{2\alpha_2} \cdots \lambda_p^{2\alpha_p}}{\lambda_i^2} \,, \tag{4.6}$$

where (a_i, c_i, α_i) are positive model-dependent numbers. In particular, in the simplest case of $p = 1$, one gets the only possible form of RG equation,

$$\mu \frac{d}{d\mu} \lambda^{-2} = \frac{a}{2\pi} - \frac{c}{(2\pi)^3} \lambda^4 , \qquad (4.7)$$

whereas, if $p = 2$, there are two possibilities, either

$$\mu \frac{d}{d\mu} \lambda_1^{-2} = \frac{a_1}{2\pi} - \frac{c_1}{(2\pi)^3} \lambda_2^4 , \quad \mu \frac{d}{d\mu} \lambda_2^{-2} = \frac{a_2}{2\pi} - \frac{c_2}{(2\pi)^3} \lambda_1^4 , \qquad (4.8a)$$

or

$$\mu \frac{d}{d\mu} \lambda_1^{-2} = \frac{a_1}{2\pi} - \frac{c_1}{(2\pi)^3} \lambda_2^4 , \quad \mu \frac{d}{d\mu} \lambda_2^{-2} = \frac{a_2}{2\pi} - \frac{c_2}{(2\pi)^3} \lambda_1^2 \lambda_2^2 . \qquad (4.8b)$$

Here are the values of the coefficients for some low-dimensional symmetric spaces $\mathcal{M}$ [44]:

$$S^3 = O(4)/O(3) : \quad a = 2, \quad c = 2n^2 ,$$
$$S^1 \times S^2 : \quad a_1 = 0, \quad c_1 = n^2/8 , \quad a_2 = 1 , \quad c_2 = c_1 , \qquad (4.9)$$
$$S^1 \times S^1 \times S^1 : \quad a_i = 0, \quad c_i = n^2/8\pi^2 .$$

One easily finds that in the simplest case of $p = 1$ there is a trivial fixed point at $\lambda = 0$ and an IR-stable fixed point at $\lambda^2 = 2\pi \sqrt{a/c}$. The solution to the RG equation (4.8a) is given by [44]

$$S - S_0 = z^{-1} + \tfrac{1}{2} \ln \left(\frac{z-1}{z+1} \right) , \qquad (4.10)$$

where the new variables,

$$S \equiv a \sqrt{a/c} \ln \mu , \quad z \equiv \frac{\lambda^2}{2\pi} \sqrt{c/a} , \qquad (4.11)$$

have been introduced. Similarly, in the case of $p = 2$, we find [89]

$$\frac{a_2}{\lambda_1^2} - \frac{a_1}{\lambda_2^2} = \frac{1}{(2\pi)^2} (c_1 \lambda_2^2 - c_2 \lambda_1^2) + \text{const.} , \qquad (4.12)$$

whose fixed point is

$$\lambda_1^2 = 2\pi \sqrt{a_2/c_2} , \quad \lambda_2^2 = 2\pi \sqrt{a_1/c_1} . \qquad (4.13)$$

2.4.2 Two-Loop Results

In the two-loop approximation, we should use the covariant background-quantum field expansion of the NLSM action (Sect. 2.2) that is supposed to be free from the one-loop subdivergences in quantum perturbation theory. In other words, we should start from the one-loop renormalized action by adding the one-loop NLSM counterterm (4.1) to be expanded up to the second order in the quantum ξ-fields. The initial background-quantum splitted NLSM action has now to be expanded up to the fourth order in the quantum fields.

The covariant background expansion of the one-loop NLSM counterterm reads [82]

$$
\begin{aligned}
\Delta S_{1|2}^{(c)} &= -\tfrac{1}{2} I_1 \int d^2x \left\{ \widehat{R}_{ab}(\eta^{\mu\nu} - \varepsilon^{\mu\nu})\widehat{D}_\mu \xi^a \widehat{D}_\nu \xi^b + (D_f \widehat{R}_{ab} + \widehat{R}_{ac} H_{bf}^c) \right. \\
&\quad \times (\eta^{\mu\nu} - \varepsilon^{\mu\nu})\xi^f \widehat{D}_\mu \xi^a \partial_\nu \phi^b + (D_f \widehat{R}_{ab} - H_{af}^c \widehat{R}_{ac})(\eta^{\mu\nu} - \varepsilon^{\mu\nu})\partial_\mu \phi^a \\
&\quad \times \widehat{D}_\nu \xi^b \xi^f + \xi^f \xi^g \partial_\mu \phi^a \partial_\nu \phi^b (\eta^{\mu\nu} - \varepsilon^{\mu\nu}) \left(\tfrac{1}{2} D_g D_f \widehat{R}_{ab} + \tfrac{1}{2} \widehat{R}_{pb} R^p{}_{gfa} \right. \\
&\quad \left. \left. + \tfrac{1}{2} \widehat{R}_{ap} R^p{}_{gfb} + H_{af}^c \widehat{R}_{cd} H_{gb}^d + H_{bg}^c D_f \widehat{R}_{ac} - H_{ag}^c D_f \widehat{R}_{cb} \right) \right\} .
\end{aligned}
$$
(4.14)

Similarly, the one-loop mass counterterm (4.5) has to be expanded too,

$$
\Delta S_{1|2}^{(m)} = \tfrac{m^2}{2} I_1 \int d^2x \, \tfrac{1}{3} R_{ab} \xi^a \xi^b .
$$
(4.15)

It can be shown that the non-covariant UV-divergent contributions from the one-loop mass counterterm, and the NLSM mass term (3.1) to be expanded up to the fourth order in the quantum fields as well, do not lead to invariant UV counterterms of dimension two, so that they can be ignored on-shell altogether [82]. The full list of relevant two-loop Feynman graphs is given in Fig. 2.2. It is straightforward to calculate their UV divergences, so that we restrict ourselves to the results with a few comments.

The α-counterterm is given by

$$
\begin{aligned}
\alpha &= -\tfrac{1}{2}\lambda^2 I_1^2 \int d^2x \left\{ \left(\Delta R_{df} - \tfrac{1}{12} D_d D_f R + \tfrac{1}{2} R_{da} R_f^a - \tfrac{1}{6} R_d{}^{ab}{}_f R_{ab} \right. \right. \\
&\quad + \tfrac{1}{2} R_d{}^{abc} R_{fabc} + \tfrac{1}{2} H_d^{ab} \Delta H_{fab} + \tfrac{1}{2} R_d^a H_{abc} H_f^{bc} - \tfrac{1}{6} H_{dab} H_{fc}^b R^{ac} \\
&\quad \left. + H_{da}^c D^{(a} D^{b)} H_{fbc} + \tfrac{3}{4} R_d{}^{abc} H_{bch} H_{af}^h - \tfrac{1}{4} R^{ahbc} H_{dbc} H_{haf} \right) \partial_\mu \phi^d \partial^\mu \phi^f \\
&\quad + \left(\tfrac{1}{4} D^b \Delta H_{dfb} - \tfrac{3}{4} R^{bca}{}_d D_a H_{bcf} - \tfrac{3}{4} H_{abd} D^c R^{ab}{}_{cf} + \tfrac{1}{12} R^{ac} D_a H_{cdf} \right. \\
&\quad \left. \left. + \tfrac{1}{24} H_{df}^a D_a R - H_{bd}^c H_{ahf} D^b H_c^{ah} + \tfrac{1}{2} D_a H_{bc}^a H_d^{ch} H_{hf}^b \right) \partial_\mu \phi^d \varepsilon^{\mu\nu} \partial_\nu \phi^f \right\} ,
\end{aligned}
$$
(4.16)

where Δ is the covariant d'Alembertian, $\Delta = D^a D_a$.

The momentum integral arising from the first β-diagram in Fig. 2.2 takes the form

$$
\int \frac{d^d p\, d^d q}{(2\pi)^{2d}} \frac{p_\mu p_\nu}{(p^2 - m^2)([k+q]^2 - m^2)([p+q]^2 - m^2)} ,
$$
(4.17)

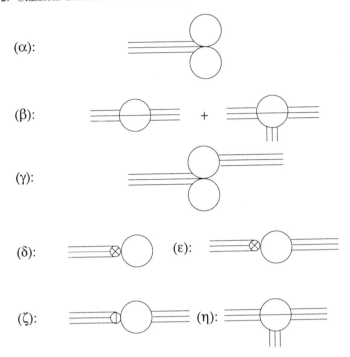

(α):

(β): $+$

(γ):

(δ): (ε):

(ζ): (η):

Fig. 2.2. Feynman graphs contributing to the NLSM two-loop RG β-function. The small loops denote the one-loop divergences

and it is only divergent at the vanishing external momentum k_μ. A simple calculation yields

$$\beta_1 = -\frac{2\lambda^2}{3d}I_1^2\int \mathrm{d}^2x \left\{-(R_{a(bc)d} - H^f_{a(b}H_{c)df})(R_m{}^{bcd} - H^{bh}_m H^{cd}_h)\right.$$
$$\times \partial^\mu\phi^a\partial_\mu\phi^m + D_{(b}H_{d)ac}D^b H^{cd}_m\partial_\lambda\phi^a(\varepsilon^{\lambda\nu}\eta_{\nu\mu}\varepsilon^{\mu\rho})\partial_\rho\phi^m$$
$$\left.+2(R_{a(bc)d} - H^f_{a(b}H_{c)df})D^b H^{dc}_m\varepsilon^{\mu\rho}\partial_\mu\phi^a\partial_\rho\phi^m\right\}\ .$$

(4.18)

A tedious calculation of the two remaining β-type diagrams in Fig. 2.2 yields [82]

$$\beta_2 + \beta_3 = -\lambda^2\omega(\epsilon)\left[\frac{10-7d}{18d}\right]I_1^2\int \mathrm{d}^2x \left\{\widehat{D}_\mu H_{abc}\widehat{D}^\mu H^{abc}\right.$$

$$\left.+6H_{dag}H^g_{bc}\widehat{D}_f H^{abc}\partial_\mu\phi^d(\eta^{\mu\nu} - \varepsilon^{\mu\nu})\partial_\nu\phi^f\right\}\ .$$

(4.19)

In the calculation of the γ-counterterm,

$$\gamma = \tfrac{1}{6}\lambda^2(I_1^2 + 2mI_1J)\int \mathrm{d}^2x\,\widehat{R}_{abcd}R^{bc}(\eta^{\mu\nu} - \varepsilon^{\mu\nu})\partial_\mu\phi^a\partial_\nu\phi^d\ ,$$

(4.20)

we used the identity

$$\lim_{x \to y} i \langle \partial_\rho \xi(x) \partial^\rho \xi(y) \rangle = i \int \frac{d^d p}{(2\pi)^d} \frac{p^2}{p^2 - m^2} = m^2 I_1 . \tag{4.21}$$

The γ- counterterm is only IR-divergent. The notation J in (4.20) stands for the UV-convergent integral

$$J \equiv i \int \frac{d^2 p}{(2\pi)^2} \frac{1}{(p^2 - m^2)^2} = -\frac{1}{(2\pi)^2} \int d^2 k_{\mathrm{E}} \frac{1}{(k_{\mathrm{E}}^2 + m^2)^2} . \tag{4.22}$$

Similarly, we obtain [82]

$$\delta = \tfrac{1}{4}\lambda^2 I_1^2 \int d^2 x \, (\eta^{\mu\nu} - \varepsilon^{\mu\nu}) \partial_\mu \phi^d \partial_\nu \phi^f \left[\Delta \widehat{R}_{df} + R_d^a \widehat{R}_{af} + R_f^a \widehat{R}_{da} \right.$$

$$\left. -2 H_{dba} H_{fc}^a \widehat{R}^{bc} + 2 H_f^{ab} D_a \widehat{R}_{db} - 2 H_d^{ab} D_a \widehat{R}_{bf} \right] , \tag{4.23a}$$

$$\varepsilon = -\tfrac{1}{2}\lambda^2 (I_1^2 + m^2 I_1 J) \int d^2 x \, \widehat{R}_{a(bc)d} \widehat{R}^{bc} (\eta^{\mu\nu} - \varepsilon^{\mu\nu}) \partial_\mu \phi^a \partial_\nu \phi^d , \tag{4.23b}$$

$$\zeta = -\tfrac{1}{6}\lambda^2 I_1 J \int d^2 x \, \widehat{R}_{abcd} R^{bc} (\eta^{\mu\nu} - \varepsilon^{\mu\nu}) \partial_\mu \phi^a \partial_\nu \phi^d , \tag{4.23c}$$

$$\eta = \tfrac{1}{4}\lambda^2 \omega(\epsilon)(I_1^2 + m^2 I_1 J) \int d^2 x \, \widehat{R}_{abcd} H_{fg}^b H^{cfg} (\eta^{\mu\nu} - \varepsilon^{\mu\nu}) \tag{4.23d}$$

$$\times \partial_\mu \phi^a \partial_\nu \phi^d .$$

One still has to extract the simple-pole divergences by rewriting all the two-loop counterterms in 'canonical' form, see (3.18) and (3.19). The two-loop contributions to the RG β-function are determined by (3.20). We find [82]

$$\beta_1 : \quad -\frac{2}{3(2\pi)^2} \widehat{R}_{d(ab)c} \widehat{R}^{c(ab)}{}_f - \frac{(\omega_1 - 1)}{(2\pi)^2} \left\{ \tfrac{4}{3} \widehat{R}_{[c(\underline{ab})d]} \widehat{R}^{[c(\underline{ab})}{}_{f]} \right.$$

$$\left. + \widehat{R}_{[abcd]} \widehat{R}^{[abc}{}_{f]} \right\} ,$$

$$\beta_2 + \beta_3 : \quad -\frac{(5 - 4\omega_1)}{9(2\pi)^2} \left\{ \widehat{D}_a H_{abc} \widehat{D}_f H^{abc} + 6 H_{dag} H_{bc}^g \widehat{D}_f H^{abc} \right\} \tag{4.24}$$

$$-\frac{(\omega_1 - 1)}{9(2\pi)^2} \widehat{R}_{[abcd]} \widehat{R}^{[abc}{}_{f]} ,$$

$$\eta : \quad \frac{\omega_1}{(2\pi)^2} \widehat{R}_{dabf} H_{gh}^a H^{bgh} ,$$

after extensive use of the identities

$$\widehat{R}^{[c}{}_{(ab)}{}^{d]} = -D_{(a} H_{b)}^{cd} , \quad \widehat{R}_{[abcd]} = 2 H_{[ab}^h H_{cd]h} , \tag{4.25a}$$

$$\widehat{D}_d H_{abc} \widehat{D}_f H^{abc} = \tfrac{9}{16} \left(\widehat{R}_{[abc]d} + \widehat{R}_{d[abc]} \right) \left(\widehat{R}^{[abc]}{}_f + \widehat{R}_f{}^{[abc]} \right)$$

$$- \tfrac{9}{4} \widehat{R}_{[abcd]} \widehat{R}^{[abc}{}_{f]} - \tfrac{9}{4} \left(\widehat{R}^{abc}{}_{[d} + \widehat{R}_{[d}{}^{abc} \right) \widehat{R}_{[f]abc]} \ . \tag{4.25b}$$

As is clear from (4.24) and (4.25), the generic NLSM two-loop RG β-function, $\beta = \beta_1 + \beta_2 + \beta_3 + \eta$, cannot be written down without an explicit appearance of the torsion H_{abc}, i.e. in terms of the generalized curvature $\widehat{R}_{abcd}$ and the generalized covariant derivative $\widehat{D}_a$ alone [82].

It is also easy to verify that all the two-loop divergences proportional to $m^2 I_1 J$ actually cancel. This gives an important internal consistency check of our calculations. The remaining two-loop double-pole divergences ($\sim I_1^2$), after the use of Bianchi identities for the curvature and torsion, can be put into the form [82]

$$\Sigma_{\text{double}} = - \tfrac{1}{2} \lambda^2 I_1^2 \int \mathrm{d}^2 x \left\{ - \left(-\tfrac{1}{4} \Delta_{\mathrm{L}} \widehat{R}_{df} + H^a_{db} \left[H_{fca} \widehat{R}^{(bc)} \right. \right. \right.$$

$$\left. \left. + \tfrac{3}{2} D^b \widehat{R}_{[af]} \right] \right) \partial^\mu \phi^d \partial_\mu \phi^f + \left(\tfrac{3}{4} D^b D_{[b} \widehat{R}_{df]} \right. \tag{4.26}$$

$$\left. \left. + \tfrac{1}{2} \widehat{R}^{(ab)} D_a H_{bdf} - H^{ab}_d D_a \widehat{R}_{(bf)} \right) \partial^\mu \phi^d \varepsilon^{\mu\nu} \partial_\nu \phi^f \right\} \ ,$$

where we have introduced the Lichnerovitch Laplacian,

$$\Delta_{\mathrm{L}} \widehat{R}_{(df)} = \Delta \widehat{R}_{(df)} + 2 R^a_d \widehat{R}_{(af)} - 2 R_d{}^{ab}{}_f \widehat{R}_{(ab)} \ . \tag{4.27}$$

In deriving (4.26) we systematically ignored all terms proportional to

$$\int \mathrm{d}^2 x \left[D_{(d} D_{f)} R + H^a_{df} D_a R \right] \partial^\mu \phi^d (\eta^{\mu\nu} - \varepsilon^{\mu\nu}) \partial_\nu \phi^f \ , \tag{4.28}$$

since they represent the diffeomorphisms of $\mathcal{M}$.

Yet another consistency check is provided by the generalized RG pole equations [34, 70] that relate the residue at the two-loop double-pole counterterm to the one-loop RG β-function. It is enough to notice that (4.26) can be rewritten in the form

$$\sim I_1^2 \int \mathrm{d}^2 x \left\{ \frac{\delta \widehat{R}_{df}}{\delta g_{ab}} + \frac{\delta \widehat{R}_{df}}{\delta h_{ab}} \right\} \widehat{R}_{ab} (\eta^{\mu\nu} - \varepsilon^{\mu\nu}) \partial_\mu \phi^d \partial_\nu \phi^f \ . \tag{4.29}$$

If we were only interested in calculating the two-loop RG β-function above, we could only take into account the simple-pole divergences from the very beginning, by using the relation

$$I_1^p + \text{subtractions} = \frac{1}{(4\pi\epsilon)^p} \tag{4.30}$$

that is valid for any positive integer p. There are only two topologically different types of two-loop graphs (Fig. 2.3) in the background-field method.

(a)   **(b)**

Fig. 2.3. Topological structures of the two-loop Feynman graphs in the background-field method: **(a)** 'eight' and **(b)** 'fish'

Any 'eight'-type diagram UV-divergence is proportional to I_1^2 without a d-dependent pre-factor. All the 'eight'-type diagrams are therefore irrelevant for our purposes, since they do not contribute to the $1/\epsilon$-divergences and, hence, to the RG β-functions as well, after subtracting their subdivergences according to (4.30).

The UV divergences of the 'fish'-type graphs are proportional to $f(\epsilon)I_1^2$ with some ϵ-dependent pre-factor $f(\epsilon)$. Subtracting the subdivergences results in

$$f(\epsilon)I_1^2 \longrightarrow \left[f(0) + \epsilon \left. \frac{\partial f}{\partial \epsilon} \right|_{\epsilon=0} \right] \frac{1}{(4\pi\epsilon)^2} \; . \tag{4.31}$$

The 'fish'-type diagrams thus normally contribute to the RG β-functions if $\partial f / \partial \epsilon|_0 \neq 0$. One ultimately arrives at the same contributions to NLSM two-loop RG β-functions. In Subsect. 2.4.3 we describe the results of the simplified procedure for the three-loop RG β-functions of the bosonic NLSM with a generalized WZNW term.

Yet another useful lesson of our two-loop calculations comes from the last β-type diagram of Fig. 2.2 that does contribute to the RG β-functions. Therefore, it would be incorrect to ignore the non-minimal torsion in the generalized covariant derivatives $\widehat{D}_\mu$ by referring to the naive covariance argument with respect to them. The reason why this is not correct with respect to the non-minimal connection with torsion, though it is correct with respect to the minimal connection without torsion, is due to the fact that the minimal connection does not explicitly appear in the covariant vertices, unlike the torsion itself.

Let us now calculate a Zamolodchikov action by considering the NLSM perturbative two-loop RG β-functions given by the sum of (4.4) and (4.24) as the variational equations of the action in question, with respect to the metric and torsion potential. The very existence of the invariant action, i.e. the integrability condition on the NLSM β-functions, is highly non-trivial in this context.

Let us assume the desired invariant action to be of the form

$$I_Z[g, H] = \int d^D \phi \sqrt{g} \, \mathcal{L}(g, H) \; , \tag{4.32}$$

where the ϕ^a now play the role of the D-dimensional (Euclidean) spacetime coordinates, while the Lagrangian $\mathcal{L}$ is supposed to be defined modulo a total derivative. Varying the action (4.32) yields

$$\frac{\delta I_Z}{\delta g^{ab}} = W_{ab} + \tfrac{1}{2} g_{ab} \mathcal{L} \; , \qquad \frac{\delta I_Z}{\delta h^{ab}} = 0 \; . \tag{4.33}$$

The invariance of the action (4.32) with respect to the $\mathcal{M}$ diffeomorphisms implies the relation

$$D^b W_{ab} - H_a^{bc} \frac{\delta I_Z}{\delta h^{bc}} = -\tfrac{1}{2} D_a \mathcal{L} . \tag{4.34}$$

Hence, the integrability condition on the RG β-functions can be written down in the form [86, 90]

$$D^b \beta_{(ab)} - H_a^{bc} \beta_{[bc]} = -\tfrac{1}{2} D_a \mathcal{L} . \tag{4.35}$$

Given the NLSM β-functions β, this implies the existence of a scalar $\mathcal{L}$ satisfying (4.35). For instance, as regards the one-loop β-functions (4.4), one easily finds [91]

$$D^b \beta^{(1)}_{(db)} - H_d^{bc} \beta^{(1)}_{[bc]} = \frac{1}{2\pi} D_d \left(\tfrac{1}{2} R - \tfrac{1}{6} H^2 \right) , \tag{4.36}$$

where R is the scalar curvature and

$$H^2 \equiv H_{abc} H^{abc} . \tag{4.37}$$

Similarly, as regards the two-loop contribution (4.24) to the NLSM β-functions in the HVB-prescription (3.12), we find

$$D^b \beta^{(2)}_{(db)} - H_d^{bc} \beta^{(2)}_{[bc]} = \frac{1}{2(2\pi)^2} D_d \left[-\tfrac{1}{8} R_{abcd} R^{abcd} + \tfrac{1}{4} R_{abcd} \right.$$

$$\left. \times H^{abh} H_h^{cd} + \tfrac{1}{4} (H^2)^{ab} (H^2)_{ab} - \tfrac{1}{12} H_{abc} H^{agf} H^b_{fk} H^{ck}_g \right] , \tag{4.38}$$

where we have introduced the notation

$$(H^2)^{ab} = H^{agh} H^b_{gh} . \tag{4.39}$$

Therefore, the Zamolodchikov action in the two-loop approximation reads ($\lambda^2 = 1$) [86, 3]

$$I_Z = \tfrac{1}{2\pi} \int d^D \phi \sqrt{g} \left[-R + \tfrac{1}{3} H^2 - \tfrac{1}{8\pi} \left(\tfrac{1}{2} R_{abcd} R^{abcd} - R_{abcd} H^{abh} H_h^{cd} \right. \right.$$

$$\left. \left. - (H^2)^{ab} (H^2)_{ab} + \tfrac{1}{3} H_{abc} H^{adf} H^b_{fk} H^{ck}_d \right) \right] . \tag{4.40}$$

It is easy to calculate the two-loop RG β-function of the WZNW model from the general result (4.24) by using (1.22) and (1.23). We find [82]

$$\beta_{ij} = \frac{Q}{2\pi} (1 - \eta^2) \delta_{ij} + \frac{\lambda^2 Q^2}{2(2\pi)^2} (1 - \eta^2)(1 - \eta^2 - 2\omega_1 \eta^2) \delta_{ij} , \tag{4.41}$$

where we have introduced the second Casimir operator eigenvalue in the adjoint representation of the Lie group G,

$$f_{ijk}f_{mjk} = Q\delta_{im} \ . \tag{4.42}$$

The NLSM β-function (4.41) is obviously symmetric in (ij), which means the absence of a renormalization of the WZNW term. This should have been expected because of the topological nature of this term (Subsect. 5.1.3).

Note that the generalized curvature of the WZNW model is given by

$$\widehat{R}_{abcd} = (1 - \eta^2)R_{abcd} \ . \tag{4.43}$$

This is in agreement with the fact that the WZNW model has a fixed point at $\eta^2 = 1$. Let us now compare the first derivative of the two-loop RG β-function at the fixed point with the exact non-perturbative result of CFT [52],

$$\left.\frac{\partial\beta_\lambda}{\partial\lambda^2}\right|_{\eta^2=1} = \frac{2Q}{Q+n} = \frac{2Q}{n}\left[\frac{1}{1+Q/n}\right] \ , \tag{4.44}$$

where the β-function β_λ has been defined by the RG equation

$$\beta_\lambda \equiv \mu\frac{\mathrm{d}}{\mathrm{d}\mu}\lambda^2(\mu) \ . \tag{4.45}$$

The β-functions β_{ij} and β_λ are obviously related,

$$\beta_{ij} = \mu\frac{\mathrm{d}}{\mathrm{d}\mu}\left[\frac{1}{\lambda^2(\mu)}\delta_{ij}\right] = -\frac{\delta_{ij}}{\lambda^4}\beta_\lambda \ . \tag{4.46}$$

We thus have [82]

$$\beta_\lambda = -\frac{\lambda^4 Q}{2\pi}(1-\eta^2) - \frac{\lambda^6 Q^2}{2(2\pi)^2}(1-\eta^2)(1-\eta^2-2\omega_1\eta^2) \ . \tag{4.47}$$

Comparing the two-loop (perturbative) result (4.47) with the perturbative expansion

$$\left.\frac{\partial\beta_\lambda}{\partial\lambda^2}\right|_{\eta^2=1} = \frac{2Q}{n} - \frac{2Q^2}{n^2} + \frac{2Q^3}{n^3} + O(Q^4/n^4) \tag{4.48}$$

of the non-perturbative expression (4.44) gives rise to exact agreement only in the case of the HVB renormalization prescription ($\omega_1 = 1$). Hence, it is the HVB renormalization prescription that is compatible with the 2d conformal invariance. The next correction of order $O(Q^3/n^3)$ in (4.48) is going to be used for verifying the three-loop RG β-function of the WZNW model in Subsect. 2.4.3.

2.4.3 Three-Loop Results

In the three-loop approximation, the UV divergences of all the three-loop graphs (Fig. 2.4) can be reduced to the 'flower'-type one ($\sim I_1^3$), corresponding to the α-graphs in Fig. 2.4, by the use of partial integration that generically yields some d-dependent pre-factors. Subtraction of subdivergences is

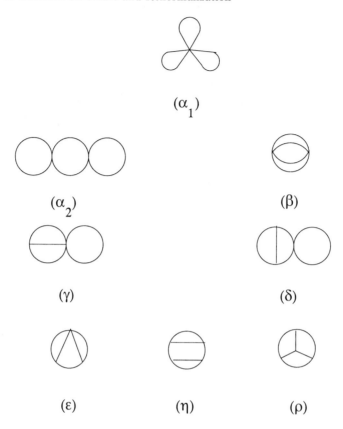

(α_1)

(α_2) (β)

(γ) (δ)

(ε) (η) (ρ)

Fig. 2.4. Topology of the three-loop Feynman graphs in the background-field method

then reduced to the trivial procedure (4.30). Though being non-minimal in subtracting the subdivergences, this approach allows us to treat all the three-loop UV divergences on an equal footing by reducing them to a few basic divergent integrals [92].

For instance, any three-loop graph of the type α_1 or α_2 in Fig. 2.4 has the UV divergence proportional to I_1^3 *without* a d-dependent pre-factor. Hence, after subtracting its subdivergences, it only gives rise to a $1/\epsilon^3$ pole divergence that can be simply ignored in calculating the RG β-functions (Sect. 2.3).

The UV divergences of the β-type diagrams in Fig. 2.4 without an external $\partial\phi$-dependence can be reduced to the integral (Fig. 2.5)

$$I_W^{(1)} = i^3 \int \frac{d^d p \, d^d q \, d^d s}{(2\pi)^{3d}} \frac{s \cdot (k - p - q - s) p \cdot q}{s^2 (k - p - q - s)^2 p^2 q^2} \quad . \tag{4.49}$$

This integral is IR-convergent. After expanding the integrand in powers of the external momenta k_μ, we find

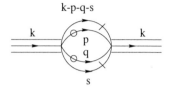

Fig. 2.5. The three-loop Feynman graph corresponding to the integral $I_W^{(1)}$. Dashes and circles denote contracted momenta on the internal lines

$$I_W^{(1)} \longrightarrow k^2 \frac{\epsilon^2}{d} I_E^{(1)} , \tag{4.50}$$

where we have introduced the basic three-loop integral

$$
\begin{aligned}
I_E^{(1)} &= i^3 \int \frac{d^d p\, d^d q\, d^d s}{(2\pi)^{3d}} \; \frac{p \cdot (p+s)\, q \cdot (q+s)}{p^2 (p+s)^2 q^2 (q+s)^2 (s^2 - m^2)} \\
&= \frac{4}{(4\pi)^{3d/2} (m^2)^{2\epsilon}} \; \frac{\Gamma(3\epsilon)\Gamma^2(\epsilon)\Gamma^3(1-\epsilon)\Gamma(1-3\epsilon)}{\Gamma^2(1-2\epsilon)} .
\end{aligned}
\tag{4.51}
$$

The β-type diagrams in Fig. 2.4 with the manifest $(\partial\phi)^2$-dependence lead to another UV-divergent integral (Fig. 2.6),

$$
\begin{aligned}
I_W^{(2)} &= i^3 \int \frac{d^d p\, d^d q\, d^d s}{(2\pi)^{3d}} \; \frac{(-)s \cdot (p+q+s)\, q \cdot (p+q)}{(p+q+s)^2 s^2 p^2 (q^2 - m^2)} \\
&= \frac{-2\epsilon}{(4\pi)^{3d/2} (m^2)^{3\epsilon}} \; \frac{\Gamma^2(\epsilon)\Gamma(2\epsilon)\Gamma(1-\epsilon)}{d} .
\end{aligned}
\tag{4.52}
$$

It can be shown that the UV divergences of all the β-type graphs can be reduced to a linear combination of I_1^3, $I_W^{(1)}$ and $I_W^{(2)}$, by the use of integration by parts only [92]. As regards the γ- and δ-type graphs (Fig. 2.4), all their UV-divergences can be similarly reduced to I_1^3 multiplied by generically d-dependent factors, which is simply a consequence of the two-loop analysis of the previous subsection. All the ε-type graph UV-divergences can be reduced to the two basic ones pictured in Fig. 2.7 [92].

The UV-divergence of the graph (1) in Fig. 2.7 is just given by (4.51), whereas the graph (2) in Fig. 2.7 yields

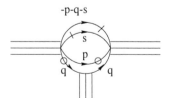

Fig. 2.6. The three-loop Feynman graph whose UV-divergence is given by $I_W^{(2)}$

(1)   **(2)**

Fig. 2.7. The three-loop graphs whose UV-divergences are given by $I_E^{(1)}$ and $I_E^{(2)}$, respectively

$$
\begin{aligned}
I_E^{(2)} &= \mathrm{i}^3 \int \frac{\mathrm{d}^d p\, \mathrm{d}^d q\, \mathrm{d}^d s}{(2\pi)^{3d}} \frac{q \cdot (p+s) p \cdot (q+s)}{p^2 q^2 (s^2 - m^2)(p+s)^2 (q+s)^2} \\
&= \frac{d}{(4\pi)^{3d/2}(m^2)^{3\epsilon}} \frac{\Gamma(3\epsilon)\Gamma^2(\epsilon)\Gamma^3(1-\epsilon)\Gamma(1-3\epsilon)}{(d-1)\Gamma^2(1-2\epsilon)} \qquad (4.53) \\
&= \frac{d}{4(d-1)} I_E^{(1)} \; .
\end{aligned}
$$

All the η- and ρ-type graphs in Fig. 2.4 do not lead to new types of UV divergences. Integrating by parts in them always results in either a tadpole or a cancellation of one of the propagators, thus effectively reducing the calculation of their UV-divergences to the other types in Fig. 2.4. Their actual treatment is simplified by using the equations [92]

$$
\begin{aligned}
I_{\mu\nu} &= \mathrm{i} \int \frac{\mathrm{d}^d p}{(2\pi)^d} \frac{p_\mu (s+p)_\nu}{p^2 (s+p)^2} = \left[\left(\eta_{\mu\nu} - \frac{s_\mu s_\nu}{s^2} \right) \frac{1}{2(d-1)} + \frac{s_\mu s_\nu}{2s^2} \right] \mathrm{Int}(s) \\
&= \left[\frac{\eta_{\mu\nu}}{2(d-1)} - \frac{\epsilon s_\mu s_\nu}{(d-1)s^2} \right] \mathrm{Int}(s) \; ,
\end{aligned}
$$
$$(4.54a)$$
$$
\mathrm{Int}(s) \equiv I_\mu{}^\mu(s) = \mathrm{i} \int \frac{\mathrm{d}^d q}{(2\pi)^d} \frac{q \cdot (q+s)}{q^2 (s+q)^2} = \frac{2}{(4\pi)^{d/2}(s^2)^\epsilon} \frac{\Gamma(\epsilon)\Gamma^2(1-\epsilon)}{\Gamma(1-2\epsilon)} \; ,
$$
$$(4.54b)$$

and

$$
\mathrm{i} \int \frac{\mathrm{d}^d q}{(2\pi)^d} \frac{1}{(q^2)^{(n-1)\epsilon}(q^2 - m^2)} = \frac{\Gamma(n\epsilon)\Gamma(1-n\epsilon)}{(4\pi)^{d/2}(m^2)^{n\epsilon}\Gamma(1-\epsilon)} \; , \qquad (4.55)
$$

where $n \in \mathbf{Z}$. Note also the useful identities

$$
\ln \Gamma(1+\epsilon) = -\epsilon C + \sum_{k=2}^{\infty} \frac{(-1)^k}{k} \zeta(k)\epsilon^k \; , \quad \Gamma(\epsilon)\Gamma(-\epsilon) = -\frac{\pi}{\epsilon \sin \pi\epsilon} \; , \qquad (4.56)
$$

where $\zeta(x)$ in the Riemann function and $C = 0.5772157\ldots$ is the Euler constant, as well as some of their elementary consequences

$$
\epsilon \Gamma(\epsilon) = \exp(-C\epsilon) + \tfrac{1}{2}\zeta(2)\epsilon^2 - \tfrac{1}{3}\zeta(3)\epsilon^3 + O(\epsilon^4) \; ,
$$

$$
[\epsilon \Gamma(\epsilon)][-\epsilon \Gamma(-\epsilon)] = \frac{\pi\epsilon}{\sin \pi\epsilon} = 1 + \zeta(2)\epsilon^2 + O(\epsilon^4) \; , \qquad (4.57)
$$

where, in particular, $\zeta(2) = \pi^2/6$.

Equations (4.56) and (4.57) are enough to perform the subtraction of subdivergences in the basic three-loop integrals I_W and $I_E^{(1,2)}$. We find

$$I_W^{(1)} + \text{subtractions} = \frac{2}{3(4\pi)^3 \epsilon} \ ,$$

$$I_W^{(2)} + \text{subtractions} = \frac{1}{(4\pi)^3}\left[-\frac{1}{2\epsilon^2} + \frac{1}{4\epsilon} \right] \ ,$$

$$I_E^{(1)} + \text{subtractions} = \frac{1}{3(4\pi)^3 \epsilon^3} \ ,$$

$$I_E^{(2)} + \text{subtractions} = \frac{1}{6(4\pi)^3}\left[\frac{1}{\epsilon^3} + \frac{1}{\epsilon^2} + \frac{2}{\epsilon} \right] \ .$$

(4.58)

In particular, any dependence upon the IR-regulator m^2 disappears after the subtractions, as it should. It also follows from the third line of (4.58) that we can replace $I_E^{(1)}$ by $\frac{1}{3}I_1^3$ in our calculations of the three-loop RG β-functions. This implies that the integral $I_E^{(1)}$ is not independent. In fact, its UV-divergence can be reduced to the 'flower'-type ($\sim I_1^3$) by the use of integration by parts [92], which simultaneously provides a consistency check of our subtraction approach.

All the UV-divergent three-loop graphs can be divided into four groups, according to the structure of their vertices, with the latter being defined by the covariant background-field method (see Sect. 2.2, and (2.26), in particular):

• one vertex $\Pi^{(l,2)}$ or $E^{(l,2)}$ and all the other of type $\Pi^{(p,0)}$ and/or $E^{(q,0)}$. All the integrations by parts can be done on the internal lines only, i.e. inside the diagram and with the vanishing momenta on the external lines;

• two vertices of type $\Pi^{(n,1)}$ and/or $E^{(n,1)}$ and, hence, all the other of type $\Pi^{(k,0)}$ and/or $E^{(m,0)}$. One can still integrate by parts without taking into account the external momenta;

• one vertex $\Pi^{(n,1)}$ or $E^{(n,1)}$ and all the other of type $\Pi^{(k,0)}$ and/or $E^{(m,0)}$. A linear dependence upon external momenta should come out in the process of performing integrations by parts;

• all the vertices of type $\Pi^{(k,0)}$ and/or $E^{(m,0)}$. The quadratic dependence upon the external momenta should be extracted, i.e. two derivatives should ultimately act on the external lines after integrating by parts.

The basic equation for our three-loop calculation of the RG β-functions of NLSM reads

$$\int d^2x \, \beta_{ab}^{(3-\text{loop})}(\eta^{\mu\nu} - \varepsilon^{\mu\nu})\partial_\mu \phi^a \partial_\nu \phi^b = 2 \cdot 3 \cdot 2\epsilon \Delta S^{(1,3)}$$

$$= 12\epsilon \sum_{(3-\text{loop})} (\text{all the } 1/\epsilon - \text{counterterms}) \ . \tag{4.59}$$

We use the covariant background-field method, the corresponding Feynman rules and the Wick theorem to define the quantum perturbative expansion. We then calculate the UV divergences of each graph with the help of the equations given above. Though being straightforward, the three-loop calculations with a non-vanishing torsion are extremely tedious, so that we refer the interested reader to our original paper [92] for details. At the end of this subsection, we give a summary of the three-loop results [92].

Let us first consider the purely metric NLSM, i.e. with the vanishing torsion $H_{abc} = 0$. Its three-loop RG β-function was calculated by several authors [93, 94, 95]. We find ($\lambda^2 = 1$)

$$
\beta_{df}^{(3)} = \frac{2}{(4\pi)^3} \left\{ -\frac{9}{4} D_{(a} R_{\underline{dbc})g} D^{(a} R_f{}^{bc)g} - \frac{5}{6} D_d R_{a(bc)g} D_f R^{a(bc)g} \right.
$$

$$
- \frac{2}{3} R_{d(ab)f} R^a{}_{(gh)k} R^{b(gh)k} + 24 R_d{}^{(ab)c} g^{mn} \left[D_{(m} D_n R_{\underline{c}ab)f} \right.
$$

$$
\left. + 2 R_{f(ab\underline{k}} R^k{}_{mn)c} \right] + 2 R_{d(ab)c} R^c_k R_f{}^{(ab)k} + \frac{1}{4} R^{abck} \left[-5 R_{abdn} R^n{}_{fck} \right.
$$

$$
\left. + 24 R_{d(ak)n} R^n{}_{(bc)f} - 32 R_{dbcn} R^n{}_{kaf} \right] + 18 R^{abck} R_{abdn} R^n{}_{fck}
$$

$$
\left. + (d \leftrightarrow f) \right\} \ .
$$

(4.60)

When using the identities

$$
D_{(a} R_{\underline{dbc})g} D^{(a} R_f{}^{bc)g} = \frac{1}{3} \left[D_a R_{dbcg} + 2 D_c R_{dabg} \right] D^a R_f{}^{bcg} \ ,
$$

$$
D_d R_{a(bc)g} D_f R^{a(bc)g} = \frac{3}{2} \left[D_a R_{dbcg} - 2 D_c R_{dabg} \right] D^a R_f{}^{bcg} \ ,
$$

$$
R_{d(ab)c} R_f{}^{(ab)c} = \frac{3}{4} R_{dabc} R_f{}^{abc} \ ,
$$

$$
R_d{}^{(ab)c} g^{mn} \left[D_{(m} D_n R_{\underline{c}ab)f} + 2 R_{f(ab\underline{k}} R^k{}_{mn)c} \right] = \frac{1}{8} R_d{}^{abc} \Delta R_{fabc}
$$

$$
+ \frac{1}{12} R_d{}^{abc} \left[12 D_a D_b R_{cf} - 2 D_b D_f R_{ac} - \frac{5}{2} R^k_a R_{fkbc} - \frac{1}{2} R^k_f R_{kabc} \right.
$$

$$
\left. + 4 R^k_b R_{f(ak)c} \right] + \frac{1}{6} R_d{}^{abc} \left[2 R^k{}_{fcm} R_{a(km)b} + \frac{9}{2} R^{km}{}_{af} R_{kmbc} \right] \ , \quad (4.61)
$$

we can rewrite (4.60) in the form

$$
\beta_{df}^{(3)} = \frac{1}{2(4\pi^2)^3} \left\{ D^a R_d{}^{bcg} \left[D_a R_{fbcg} - 2 D_c R_{fgab} \right] + 4 R_d{}^{abc} R_f{}^{kmc} R_{kabm} \right.
$$

$$
- \frac{4}{3} R_{dabf} R^a{}_{(mn)k} R^{b(mn)k} + 2 R_{cbad} \left[24 D^a D^b R^c_f - 2 D^b D_f R^{ac} \right.
$$

$$
\left. - 5 R^{ak} R_{fkbc} - R^k_f R_{kabc} + 8 R^{bk} R_{f(ak)c} \right] + 4 \Delta \left[R_d{}^{(ab)c} R_{f(ab)c} \right]
$$

$$
\left. + (d \leftrightarrow f) \right\} \ .
$$

(4.62)

The first line of this equation can be rewritten in the form

$$\beta_{df}^{(3),G} = \frac{1}{(2\pi)^3} \left\{ \tfrac{1}{8} D^c R_{dabg} D_c R_f{}^{abg} - \tfrac{1}{16} D_d R_{abcg} D_f R^{abcg} \right.$$
$$\left. + \tfrac{1}{2} R_d{}^{abc} R_f{}^{kmc} R_{kabm} - \tfrac{3}{8} R_{dabf} R^a{}_{mnk} R^{bmnk} \right\} \tag{4.63}$$

which is just the result of [93]. The difference between (4.62) and (4.63) is proportional to R_{ab} or $R_{amnp} R_b{}^{mnp}$ which are the lower-order contributions to the NLSM β-function at the one- and two-loop level, respectively. Therefore, the reason for the difference is the use of different renormalization schemes in the NLSM quantum perturbation theory. Generally speaking, any term in the l-loop contribution to the RG β-function of NLSM, which vanishes together with the lower loop-order contributions, represents a perturbative ambiguity related to the choice of renormalization or the subdivergence subtraction procedure. In other words, all those terms are dependent upon an interpretation of the perturbative result. In the string perturbation theory (Chap. 6), they are totally arbitrary since they amount to the (spacetime) D-dimensional NLSM metric reparametrizations.

The curvature-cubed terms in the low-energy string effective action, which is simultaneously the Zamolodchikov action for the three-loop contribution to the purely metric NLSM β-function, were calculated in [96] from the 3- and 4-point on-shell amplitudes describing the tree-level graviton scattering in closed bosonic string theory. The corresponding string graviton (Euclidean) effective action reads [96]

$$I^{MT}[g] = -\frac{1}{2\pi} \int d^D \phi \sqrt{g} \left\{ R + \frac{1}{8\pi} G_2 + \frac{1}{4\pi^2} \left[\tfrac{1}{48} T_3 + \tfrac{1}{24} G_3 \right] + \dots \right\}, \tag{4.64}$$

where the dots stand for higher-order terms, and

$$T_3 \equiv R_{cd}{}^{ab} R_{mn}{}^{cd} R_{ab}{}^{mn}, \tag{4.65}$$

while G_2 and G_3 are proportional to the Gauss-Bonnet combinations up to the Ricci tensor dependent terms,

$$G_2 = R_{abcd}^2 - 4R_{ab}^2 + R^2, \quad G_3 = T_3 - 2R_{cd}{}^{ab} R_{bn}{}^{dm} R^n{}_{am}{}^c. \tag{4.66}$$

Calculation of the variational equations from the action (4.64), modulo the Ricci-tensor dependent terms and the terms proportional to the metric (they can only contribute to the dilaton β-function [97], see also Subsect. 7.1.3), yields the following string theory prediction for the three-loop terms in the NLSM RG β-function:

$$\beta_{df}^{(3),\text{MT}} = -\frac{3}{(2\pi)^3}\left[\frac{1}{48}+\frac{1}{24}\right]\{-D_c R_{dabg}D^a R_f{}^{cbg} - R_d{}^{abc}R_{fckm}R^{km}{}_{ab}$$

$$-2R_{dabc}R_{fkm}{}^c R^{kabm} + (d \leftrightarrow f)\} + \frac{3}{(2\pi)^3}\left[\frac{1}{24}\right]\{D_a R_{dbgc}$$

$$\times D^c R_f{}^{gba} - R^{kabm}D_k D_m R_{dabf} + 3R_d{}^{abc}R^m{}_{cfk}R^k{}_{abm}$$

$$-R_d{}^{abc}R^m{}_{cfk}R^k{}_{bam} + (d \leftrightarrow f)\} \ .$$

$$(4.67)$$

After substituting the identity

$$R^{kabm}D_k D_m R_{dabf} = D_f\left(\tfrac{1}{2}R^{kabm}D_k R_{dabm}\right) + D^a R_f{}^{bcg}\left(D_c R_{dabg}\right.$$

$$-\tfrac{1}{2}D_a R_{dbcg}\left.\right) + \tfrac{1}{4}R_d{}^{cab}R_{fckm}R^{km}{}_{ab}$$

$$+ \tfrac{1}{2}R_{dabf}R^a{}_{mnk}R^{bmnk} - R_{dabc}R_{fkm}{}^c R^{akmb}$$

$$(4.68)$$

into (4.67) and omitting the diffeomorphism terms, we find exact agreement with (4.63). This means that our three-loop RG β-function in the purely metric NLSM case satisfies the (Zamolodchikov) integrability condition. Simultaneously, it provides yet another consistency check of the perturbative QFT calculations in the 2d NLSM.

The general results [92] for the NLSM three-loop RG β-functions with torsion also give the three-loop RG β-function of the WZNW model (see also [98, 99]). There are, in fact, further simplifications in the case of the WZNW model, because many covariant vertices in (2.26) vanish,

$$E^{(3,1)} = \Pi^{(3,2)} = E^{(4,2)} = \Pi^{(4,1)} = E^{(4,1)} = E^{(4,0)} = E^{(5,1)} = 0 , \quad (4.69)$$

and there are extra identities,

$$D_a H_{bcd} = D_f R_{abcd} = \widehat{D}_\mu H_{abc} = \widehat{D}_\mu R_{abcd} = \widehat{D}_\mu \widehat{R}_{abcd} = 0 , \qquad (4.70a)$$

and

$$H^f_{[ab}H_{c]df} = \widehat{R}_{[abc]d} = 0 , \qquad (4.70b)$$

where we have used (1.22) and (1.23), and the Jacobi identity for the structure constants of the Lie group G,

$$f^m_{[ij}f_{k]lm} = 0 . \qquad (4.71)$$

In order to simplify further our notation in the WZNW case, we use the indices of the (flat) tangent space to the NLSM (group) manifold G. The only non-vanishing WZNW covariant vertices in three-loop perturbation theory are given by

$$\Pi^{(5,1)}_{(fh\underline{d}bc)a} = \tfrac{4}{15}(1-\eta^2)f^{pa}_{(f}f^{kp}_{h}f^{nk}_{b}f^{dn}_{c)} \ ,$$

$$\Pi^{(4,2)}_{(abdf)(cg)} = \tfrac{1}{3}(1-\eta^2)f^{(a}_{m(c}f^{b}_{g)p}f^{d}_{mn}f^{f)}_{np} \ ,$$

$$\Pi^{(3,1)}_{abcd} = \tfrac{4}{3}(1-\eta^2)f^{h}_{ab}f_{cdh} \ , \qquad (4.72)$$

$$\Pi^{(4,0)}_{abcd} = \tfrac{1}{3}f^{h}_{ab}f_{cdh} \ , \qquad E^{(3,0)}_{abc} = \tfrac{2}{3}\eta f_{abc} \ ,$$

$$E^{(3,2)}_{(abf)[cd]} = \tfrac{2}{3}\eta(\eta^2-1)f^{(a}_{gh}f^{b}_{g[c}f^{f)}_{d]h} \ ,$$

where the underlined indices are always excluded from (anti)symmetrization. In the WZNW case we have, in addition, the identities

$$f_{afm}f^{m}_{gk}f^{ka}_{c} = \tfrac{1}{2}Qf_{fgc} \ , \qquad (4.72a)$$

$$(f^{hk}_{b}f^{rl}_{b})(f^{c}_{ah}f^{c}_{jr})(f^{t}_{jk}f^{t}_{il}) = 0 \ , \qquad (4.72b)$$

$$\delta^{fh}f^{ap}_{(b}f^{pk}_{c}f^{kn}_{f}f^{nd}_{h)} = -\tfrac{3}{4}Qf^{n}_{a(b}f^{n}_{c)d} - \tfrac{1}{6}f_{kap}f^{hp}_{(b}f^{nk}_{c)}f^{n}_{hd} \ , \qquad (4.72c)$$

which are quite useful in simplifying the general results [92]. We find

$$\beta^{(3-\text{loop})}_{ij} = \delta_{ij}\frac{\lambda^4 Q^3}{(4\pi)^3}(1-\eta^2)\left[q_0 + (1-\eta^2)q_2 + (1-\eta^4)q_4\right] \ , \qquad (4.73)$$

where the numerical coefficients are given by [92, 98]

$$q_0 = \left(24 - \frac{4}{3}\right)\omega_2 - 4\omega_1(\omega_1 + 1) - 1 \ ,$$

$$q_2 = \frac{1}{2^3 \cdot 3^4}\left(-7857 + 4038\omega_1 + 2212\omega_2\right) \ , \qquad (4.74)$$

$$q_4 = \frac{1}{2^3 \cdot 3^4}\left(-1053 - 1446\omega_1 - 1564\omega_2 + 2592\omega_1^2\right) \ .$$

The β-function (4.73) is obviously consistent with the non-renormalization of the WZNW term, as well as the existence of the exact fixed point at $\eta^2 = 1$.

Equation (4.74) is considerably simplified when one uses the HVB-prescription (3.12) which implies

$$\omega_1^{\text{HVB}} = 1 \ , \quad \omega_2^{\text{HVB}} = 3/4 \ , \quad 4\omega_2^{\text{HVB}} - 2\omega_1^{\text{HVB}} - 1 = 0 \ . \qquad (4.75)$$

Substituting (4.75) into (4.74) yields

$$q_0^{\text{HVB}} = 8 \ , \qquad q_2^{\text{HVB}} = -10/3 \ , \qquad q_4^{\text{HVB}} = -5/3 \ . \qquad (4.76)$$

The perturbative value of the coefficient q_0 in (4.76) is exactly the same as that from the non-perturbative result (4.44) after the use of (4.48).

The three-loop RG β-functions of the general NLSM with non-vanishing torsion are very complicated, while they depend upon the renormalization prescription used in actual calculations [92]. The generic structure of the three-loop contribution is

$$
\begin{aligned}
\beta_{df}^{(3)} \sim\ & \widehat{R}R^2 + (\widehat{\nabla}R)^2 + R^2\widehat{\nabla}H + R(\widehat{\nabla}R)H + (\widehat{\nabla}^2 R)(\widehat{\nabla}H) + \widehat{R}RH^2 \\
& + R(\widehat{\nabla}H)^2 + (\widehat{\nabla}R)(\widehat{\nabla}H)H + RH(\widehat{\nabla}^2 H) + (\widehat{\nabla}^2 R)H^2 \\
& + RH^2(\widehat{\nabla}H) + (\widehat{\nabla}R)H^3 + (\widehat{\nabla}H)^3 + (\widehat{\nabla}^2 H)(\widehat{\nabla}H)H \\
& + (\widehat{\nabla}^3 H)H^2 + \widehat{R}H^4 + (\widehat{\nabla}H)^2 H^2 + H^4(\widehat{\nabla}H) \ .
\end{aligned}
$$

$$(4.77)$$

Since writing down the explicit general results would take several pages, we refer the interested reader to our paper [92] for more details.

To calculate the correction to the Zamolodchikov action, corresponding to the general three-loop contribution to the NLSM β-functions with non-vanishing torsion, we have to fulfil the following program:

- write down the most general diffeomorphism-invariant (scalar) *Ansatz* for the Zamolodchikov action at the given order, with undetermined numerical coefficients, in terms of the curvature, torsion and their covariant derivatives;
- take into account all the ambiguities related to arbitrary field reparametrizations of the NLSM metric and torsion potential;
- compare the variational equations following from the *Ansatz* with the NLSM three-loop β-functions calculated from the 2d perturbative QFT modulo the ambiguous terms;
- fix the coefficients of the *Ansatz* for the Zamolodchikov action from the comparison of its equations of motion with the calculated NLSM β-functions.

In addition, one should also take into account the fact that the equations of motion following from the Zamolodchikov action are 'weakly' or 'on-shell' equivalent to the NLSM β-functions. Their off-shell relation, if any, may only exist via the so-called *K-matrix* [100, 101] which is, in fact, a non-trivial (field-dependent) differential operator in higher orders of the NLSM perturbation theory. To avoid this complication, we use the lower order equations of motion, $\beta^{(1)} = \beta^{(2)} = 0$, to simplify our three-loop calculations.

There exists another (string-related) approach to the calculation of the Zamolodchikov action [42, 102, 103], which is based on a perturbative calculation of the *dilaton* RG β-function of the NLSM in curved 2d space (string world-sheet) Σ (see Sects. 6.1 and 6.3 for the corresponding string theory actions). In this approach Zamolodchikov's action can be understood as the conformal anomaly or, equivalently, as the central extension in the Virasoro algebra associated with the string propagating in a curved spacetime $\mathcal{M}$ (see also Subsect. 7.1.3 for more examples).

The most general *Ansatz* for the three-loop effective action can be found in [101, 104], and it has more than 60 independent scalar terms. After taking into account the field redefinition ambiguities, it is possible to reduce the number of invariant terms down to 21. One finds [101, 104]

$$
\begin{aligned}
I_Z = \ & -\frac{1}{32(2\pi)^2}\int \mathrm{d}^D\phi\sqrt{g}\,\bigl\{ a_1 R_{abcd}R^{cd}{}_{pq}R^{pqab} + a_2 R_{abcd}R^{apc}{}_t R^{btd}{}_p \\
& + a_3 R_{abcd}H^{apq}H^b_{pt}H^c_{qs}H^{dts} + a_4 R_{abcd}H^{abp}H^{cdk}H_{ptq}H^{tq}_k \\
& + a_5 R_{abcd}H^{atk}H^{bsm}H^{cd}_s H_{tkm} + a_6 R_{abcd}H^{atk}H^{bs}_t H^{cdp}H_{ksp} \\
& + a_7 R_{abcd}H^{atk}H^b_{sm}H^c_{tk}H_{tkm} + a_8 R_{abcd}R^{abpq}H^{cdt}H_{pqt} \\
& + a_9 R_{abcd}R^{abpq}H^c_{pt}H^{dt}_q + a_{10}R_{abcd}R^{apcq}H^b_{pk}H^{dk}_q \\
& + a_{11}R_{abcd}R^{abct}H^{dpq}H_{tpq} + a_{12}R_{abcd}R^{apqt}H^{cd}_p H^b_{qt} \\
& + a_{13}H_{abc}H^{apq}H^{bct}H_{pqs}H_{tkm}H^{skm} \\
& + a_{14}H_{abc}H^{apq}H^{bct}H_{pts}H_{qkm}H^{skm} \\
& + a_{15}H_{abc}H^{apq}H^{bct}H_{psm}H^{sk}_q H^m_{tk} + a_{16}H_{abc}H^{apq}H^{bmt}H^{cks}H_{pmk}H_{qts} \\
& + a_{17}H_{abs}H^{apq}H^b_{pt}H^{cms}H_{qmk}H^{tk}_s + a_{18}H_{abc}H^{ap}_k D^b H^{cms}D^k H_{msp} \\
& + a_{19}H^2(DH)^2 + a_{20}H^2(DH)^2 + a_{21}R_{abcd}D^a H^b_{kt}D^c H^{dkt} \bigr\} \ ,
\end{aligned}
$$

$$(4.78)$$

where, as regards the 19th and 20th terms, one may take any two linearly independent combinations of

$$
\begin{array}{ll}
H_{abc}H^{apq}D_p H^b_{ms}D_q H^{cms} \ , & H_{abc}H^{ams}D_k H^b_{mt}D^k H^{ct}_s \ , \\[4pt]
H_{abc}H^{pqt}D^a H^{bck}D_t H_{pqk} \ , & H_{abc}H^{pqt}D^a H_{pqk}D_t H^{bck} \ .
\end{array}
$$

$$(4.79)$$

The action (4.78) agrees with the three-loop RG β-function [92] in the HVB-prescription, if the coefficients $\{a_i\}$ are given by [104, 105, 106]

$$a_1 = -2\,, \quad a_2 = -8/3\,, \quad a_3 = -2/3\,, \quad a_4 = 3/2\,, \quad a_5 = 9/2\,, \quad a_6 = 5/3\,,$$

$$a_7 = 9/8\,, \quad a_8 = 7/4\,, \quad a_9 = -4/3\,, \quad a_{10} = 2/9\,, \quad a_{11} = -3\,, \quad a_{12} = 2\,,$$

$$a_{13} = 880/81\,, \quad a_{14} = -281/27\,, \quad a_{15} = 1045/81\,, \quad a_{16} = -2626/27\,,$$

$$a_{17} = -2383/27\,, \quad a_{18} = a_{21} = 3/2\,, \quad a_{19} = a_{20} = 0\,. \tag{4.80}$$

The coefficients $(a_1, a_2, a_3, a_8, a_9, a_{10}, a_{11}, a_{12}, a_{16}, a_{17}, a_{18}, a_{21})$ are invariant under arbitrary reparametrizations of the metric and the torsion potential.

Partial results about the four-loop RG β-functions in the purely metric NLSM are obtained in [107].

2.5 Buscher Duality and the Freedman-Townsend Model

Two different field theories of second order in derivatives are *dual* to each other if they can both be obtained from the same first-order action. This definition implies the classical equivalence of dual field theories. The dual field theories are known to have the same S-matrix [108]. In the case of NLSM, a duality transformation generically changes the geometry of the NLSM field manifold $\mathcal{M}$.

Let us consider a 2d NLSM coupled to 2d gravity. An action of this NLSM can be interpreted as the action of a closed bosonic string propagating in the spacetime background of its massless excitations (Chap. 6). Let us assume that the $(n+1)$-dimensional NLSM target space (or spacetime) $\mathcal{M}$ parametrized by (x^i, x^0) has an isometry. The string NLSM action then reads (cf. Sect. 6.1)

$$
S = \frac{1}{4\pi\alpha'} \int_\Sigma \mathrm{d}^2\xi \left\{ \sqrt{\gamma}\gamma^{\mu\nu} g_{ab}(x)\partial_\mu x^a \partial_\nu x^b \right.
$$
$$
\left. + \varepsilon^{\mu\nu} h_{ab}(x)\partial_\mu x^a \partial_\nu x^b + \alpha' \sqrt{\gamma} R^{(2)} \Phi(x) \right\} , \tag{5.1}
$$

where α' is the string coupling constant, $R^{(2)}$ is the two-dimensional scalar curvature constructed out of the two-dimensional metric $\gamma_{\mu\nu}(\xi)$ on Σ, $\gamma \equiv \det(\gamma_{\mu\nu})$, while $h_{[ab]}$ is the NLSM torsion potential known as the Kalb-Ramond field, and $\Phi(x)$ is a scalar known as the *dilaton* field in string theory. We also assume that the metric g_{ab}, the torsion potential h_{ab} and the dilaton Φ are all independent of one coordinate x^0. [12]

The corresponding first-order action can be easily written down by using Lagrange multipliers [109],

$$
S_1 = \frac{1}{4\pi\alpha'} \int_\Sigma \mathrm{d}^2\xi \left\{ \sum_{i,j=1}^n \left[\sqrt{\gamma}\gamma^{\mu\nu} \left(g_{00} V_\mu V_\nu + 2g_{0i} V_\mu \partial_\nu x^i + g_{ij} \partial_\mu x^i \partial_\nu x^j \right) \right. \right.
$$
$$
\left. \left. + \varepsilon^{\mu\nu} \left(h_{0i} V_\mu \partial_\nu x^i + h_{ij} \partial_\mu x^i \partial_\nu x^j \right) \right] + \varepsilon^{\mu\nu} \widehat{x}^0 \partial_\mu V_\nu + \alpha' \sqrt{\gamma} R^{(2)} \Phi(x) \right\} . \tag{5.2}
$$

It is easy to verify that the action (5.2) reduces to (5.1) after varying it with respect to the Lagrange multiplier $\widehat{x}^0$. This also implies that $V_\mu = \partial_\mu x^0$.

The dual action $\widetilde{S}$ is obtained from (5.2) after excluding the vector vields V_μ according to their algebraic equations of motion,

---

[12] Given a manifold with an isometry, there always exists a coordinate frame where the isometry amounts to a (one) coordinate independence.

$$\widetilde{S} = \frac{1}{4\pi\alpha'} \int_{\Sigma} \mathrm{d}^2\xi \left\{ \sqrt{\gamma}\gamma^{\mu\nu}\widetilde{g}_{ab}(x)\partial_\mu\widetilde{x}^a\partial_\nu\widetilde{x}^b \right.$$
$$\left. +\varepsilon^{\mu\nu}\widetilde{h}_{ab}(x)\partial_\mu\widetilde{x}^a\partial_\nu\widetilde{x}^b + \alpha'\sqrt{\gamma}R^{(2)}\varPhi(x) \right\} , \tag{5.3}$$

where the dual metric and the dual torsion potential are given by [109]

$$\widetilde{g}_{00} = 1/g_{00} , \quad \widetilde{g}_{0i} = h_{0i}/g_{00} , \quad \widetilde{g}_{ij} = g_{ij} - (g_{0i}g_{0j} - h_{0i}h_{0j})/g_{00} , \tag{5.4a}$$

and

$$\widetilde{h}_{0i} = -\widetilde{h}_{i0} = g_{0i}/g_{00} , \quad \widetilde{h}_{ij} = h_{ij} + (g_{0i}h_{0j} - h_{0i}g_{0j})/g_{00} . \tag{5.4b}$$

Though the duality transformation does not change classical dynamics, it nevertheless affects quantum conformal properties of the NLSM at its RG fixed point. For instance, the one-loop conformal invariance conditions [91] (see also Sect. 6.1),

$$\frac{1}{\alpha'}\frac{n-25}{48\pi^2} + \frac{1}{16\pi^2}\left\{4(\nabla\varPhi)^2 - 4\nabla^2\varPhi - R - \tfrac{1}{3}H^2\right\} = 0 ,$$

$$\widehat{R}_{(ab)} + 2\nabla_{(a}\nabla_{b)}\varPhi = 0, \quad \widehat{R}_{[ab]} + 2H^c_{ab}\nabla_c\varPhi = 0 , \tag{5.5}$$

are going to be invariant under the duality transformation only if the dilaton field gets shifted as [109]

$$\varPhi \to \widetilde{\varPhi} = \varPhi - \tfrac{1}{2}\log g_{00} . \tag{5.6}$$

A duality transformation may completely change the form of an action in field theory. For example, the gauge field theory of an antisymmetric tensor in 4d is known to be equivalent to the so-called *principal* (chiral) 4d NLSM [110] (see below). Given the interacting, Lie algebra-valued, antisymmetric tensor gauge fields, $B = t^k B^k_{\mu\nu}(x)\mathrm{d}x^\mu \wedge \mathrm{d}x^\nu$, $k = 1, 2, \ldots, \dim\mathcal{G}$, whose field strength is $G = \mathrm{d}B$, their gauge-invariant Lagrangian is given by [110]

$$\mathcal{L} = -\tfrac{1}{8}G^{i\mu}\widetilde{K}_{\mu\nu}{}^{ij}G^{j\nu} , \tag{5.7}$$

where we have introduced the notation

$$K^{ij\mu\nu} \equiv \eta^{\mu\nu}\delta^{ij} - gf^{ijk}\varepsilon^{\mu\nu\rho\sigma}B^k_{\rho\sigma} , \quad \widetilde{K}_{\mu\rho}{}^{ik}K^{\rho\nu kj} = \delta_\mu{}^\nu\delta^{ij} , \tag{5.8}$$

f^{ijk} are the structure constants of the Lie algebra $\mathcal{G}$, $\varepsilon^{\mu\nu\rho\sigma}$ is the 4d Levi-Civita symbol, and g is a coupling constant.

Equation (5.7) can be rewritten in the form

$$\mathcal{L} = -\tfrac{1}{8}A^i_\mu K^{\mu\nu ij}A^j_\nu , \tag{5.9}$$

in terms of the vector potential

$$A^i_\mu = \widetilde{K}_{\mu\nu}{}^{ij} G^{\nu j} . \tag{5.10}$$

Varying (5.9) with respect to $B^i_{\mu\nu}$ yields the equations of motion

$$\frac{\delta \mathcal{L}}{\delta B^i_{\mu\nu}} = \tfrac{1}{2} \varepsilon^{\mu\nu\rho\sigma} F^i_{\rho\sigma} \equiv \widetilde{F}^{i\mu\nu} = 0 , \tag{5.11}$$

where we have introduced the standard non-Abelian vector field strength

$$F^i_{\rho\sigma} = \partial_\rho A^i_\sigma - \partial_\sigma A^i_\rho + g f^{ijk} A^j_\rho A^k_\sigma . \tag{5.12}$$

The first-order action reads [110]

$$\mathcal{L}_1 = -\tfrac{1}{8} B^i_{\mu\nu} \varepsilon^{\mu\nu\rho\sigma} F^i_{\rho\sigma} + \tfrac{1}{8} A^i_\mu A^{\mu i} , \tag{5.13}$$

where both fields A^i_μ and $B^i_{\mu\nu}$ are considered to be independent. Now, on the one hand, varying the first-order action with respect to A^i_μ yields the field equation whose solution is given by (5.10). Substituting the solution into the first-order action gives us back the original theory (5.7). On the other hand, varying the first-order action with respect to the Lagrange multiplier $B^i_{\mu\nu}$ yields the constraint

$$F^i_{\rho\sigma} = 0 . \tag{5.14}$$

Equation (5.14) means that the non-Abelian gauge field A^i_μ is a pure gauge,

$$A^i_\mu = 2M^i_n(\theta) \partial_\mu \theta^n , \tag{5.15}$$

where we have introduced the local independent field coordinates $\theta^n(x)$ on the group manifold, and the vielbein $M^i_n(\theta)$ satisfying the Maurer-Cartan equation

$$\frac{\partial M^i_n(\theta)}{\partial \theta^m} - \frac{\partial M^i_m(\theta)}{\partial \theta^n} + 2g f^{ijk} M^j_m(\theta) M^k_n(\theta) = 0 . \tag{5.16}$$

Substituting the general solution (5.15) of the constraint (5.14) back into the first-order action (5.13) results in the dual NLSM Lagrangian

$$\widetilde{\mathcal{L}} = \tfrac{1}{2} g_{mn}(\theta) \partial^\mu \theta^m \partial_\mu \theta^n , \tag{5.17}$$

with the NLSM metric

$$g_{mn}(\theta) = 4 M^i_m(\theta) M^i_n(\theta) . \tag{5.18}$$

The dual field theory (5.17) is called the principal (chiral) NLSM. Its classical equivalence to the Freedman-Townsend model (5.7) can be extended to the quantum equivalence in the sense that the Euclidean partition functions of both theories coincide. In particular, the vacuum expectation values of their quantum energy-momentum tensors are the same [111].

Let us now consider the two-dimensional (2d) analogue of the Freedman-Townsend model whose first-order Lagrangian is [112]

$$\mathcal{L}_1 = -\tfrac{1}{8} B^a \varepsilon^{\mu\nu} G^a_{\mu\nu} - \tfrac{1}{8}(A^a_\mu)^2 \; , \tag{5.19}$$

where we have chosen $g = 1$ for simplicity, and

$$G^a_{\mu\nu} = \partial_\mu A^a_\nu - \partial_\nu A^a_\mu + f^{abc} A^b_\mu A^c_\nu \; . \tag{5.20}$$

The dual second-order actions can be generated from (5.19) by varying it with respect to either A^a_μ or B^a, respectively, and substituting the solution of the variational equation back into the first-order action. For simplicity, we only consider the case $G = SU(2)$ with the structure constants $f^{abc} = \varepsilon^{abc}$, $a, b, \ldots = 1, 2, 3$.

On the one hand, varying (5.19) with respect to A^a_μ yields the algebraic equations of motion for the vector fields,

$$2\partial_\mu B^a \varepsilon^{\mu\lambda} + (\eta^{\lambda\mu}\delta^{ab} + \varepsilon^{\lambda\mu} B^c \varepsilon^{abc}) A^b_\mu = 0 \; . \tag{5.21}$$

It is not difficult to check that the field-dependent matrix

$$(L^{-1})_{\alpha\lambda}{}^{cd} = \frac{1}{1 + B^2} \left(\eta_{\alpha\lambda}\delta^{cd} + \eta_{\alpha\lambda} B^c B^d - \varepsilon_{\alpha\lambda} B^e \varepsilon^{cde} \right) \tag{5.22}$$

is the inverse of the field-dependent matrix

$$L^{\lambda\mu,ab} = \eta^{\lambda\mu}\delta^{ab} + \varepsilon^{\lambda\mu} B^c \varepsilon^{abc} \tag{5.23}$$

that appears in (5.21), in front of A^b_μ. Hence, we find from (5.21) that

$$A^b_\alpha = \frac{1}{1 + B^2} \left(\varepsilon_\alpha{}^\mu \partial_\mu B^b + \varepsilon_\alpha{}^\mu B^b B^a \partial_\mu B^a + \partial_\alpha B^a B^c \varepsilon^{abc} \right) \; , \tag{5.24}$$

where we have introduced the notation

$$\varepsilon_\alpha{}^\mu = \eta_{\alpha\nu}\varepsilon^{\nu\mu} \; , \quad B^2 = B_1^2 + B_2^2 + B_3^2 \; . \tag{5.25}$$

Substituting (5.24) back into the first-order action (5.19) results in the second-order Lagrangian

$$\mathcal{L}[B] = g_{ab}\partial^\mu B^a \partial_\mu B^b + h_{ab}\varepsilon^{\mu\nu}\partial_\mu B^a \partial_\nu B^b \; , \tag{5.26}$$

of the NLSM with torsion, whose metric and the torsion potential are given by [13]

$$g_{ab}(B) = \frac{\delta_{ab} + B_a B_b}{1 + B^2} \; , \quad h_{ab}(B) = \frac{-\varepsilon_{abc} B_c}{1 + B^2} \; . \tag{5.27}$$

On the other hand, after varying (5.19) with respect to B_a first, we arrive at the constraint

---

[13] The (up and down) positions of the flat $SU(2)$ indices of the B-field are irrelevant, while it matters for the 'world' indices of the NLSM metric, torsion and curvature tensors.

$$G^a_{\mu\nu} = 0 \ , \tag{5.28}$$

analogous to (5.14). Substituting its solution in the form (5.15) back into (5.19) now yields the dual NLSM with vanishing torsion. In order to calculate the dual metric (5.18), we apparently have to solve the Maurer-Cartan equations (5.16) for the corresponding vielbein. Their solution in the $SU(2)$ case reads

$$M^a_n(\theta) = \tfrac{i}{2}\mathrm{Tr}(t^a U^{-1}\partial_n U) = -\tfrac{1}{2}\int_0^1 du\,\mathrm{Tr}(t^a U^{-u} t_n U^u) \ , \tag{5.29}$$

where the trace goes over the $SU(2)$ group matrices. Here we use the notation

$$U(\theta) = \exp(i\theta \cdot t) \ , \quad \theta \cdot t = \sum_{a=1}^{3} \theta_a t_a \ . \tag{5.30}$$

Using the elementary identities

$$(\theta \cdot t)^{2k+1} = (\theta^2)^k \theta \cdot t \ , \quad (\theta \cdot t)^{2k+2} = (\theta^2)^k (\theta \cdot t)^2 \ ,$$

$$\mathrm{Tr}(t_a t_b) = 2\delta_{ab}, \quad \mathrm{Tr}(t_a t_b t_c) = i\varepsilon_{abc} \ , \quad \mathrm{Tr}(t_a t_b t_c t_d) = \delta_{ab}\delta_{cd} + \delta_{ad}\delta_{bc} \ ,$$

$$\mathrm{Tr}\left[t_a(\theta \cdot t)^2 t_b(\theta \cdot t)\right] = 0 \ , \quad \mathrm{Tr}\left[t_a(\theta \cdot t)^2 t_b(\theta \cdot t)^2\right] = 2\theta^2 \theta_a \theta_b \ , \tag{5.31}$$

where $\theta^2 = \theta_1^2 + \theta_2^2 + \theta_3^2$, we find

$$
\begin{aligned}
U^{-u} &= 1 - \frac{i\sin Ru}{R}(\theta \cdot t) + \frac{\cos Ru - 1}{R^2}(\theta \cdot t)^2 \ , \\
U^{+u} &= 1 + \frac{i\sin Ru}{R}(\theta \cdot t) + \frac{\cos Ru - 1}{R^2}(\theta \cdot t)^2 \ ,
\end{aligned}
\tag{5.32}
$$

where $R \equiv \sqrt{(\theta^2)}$. Substituting (5.32) into (5.29) and integrating over u yield

$$M_{ab}(\theta) = -\frac{\sin R}{R}\delta_{ab} - \frac{\cos R - 1}{R^2}\varepsilon_{abc}\theta_c - \frac{R - \sin R}{R^3}\theta_a\theta_b \ . \tag{5.33}$$

It follows from (5.18) and (5.33) that the dual NLSM metric is

$$g_{ab}(\theta) = \frac{4(2\cos R - \cos 2R - 1)}{R^4}(R^2\delta_{ab} - \theta_a\theta_b) + \frac{4\theta_a\theta_b}{R^2} \ . \tag{5.34}$$

At the end of this section we calculate the two-loop RG β-functions for the dual 2d Freedman-Townsend NLSM actions, in the case of the $SU(2)$ gauge group. Given the NLSM action,

$$S = \tfrac{1}{2}\int d^2x\,\left[g_{ij}(\phi)\partial^\mu\phi^i\partial_\mu\phi^j + \varepsilon^{\mu\nu}b_{ij}(\phi)\partial_\mu\phi^i\partial_\nu\phi^j\right] \ , \tag{5.35}$$

with the torsion

$$H_{ijk} = \tfrac{1}{2}\left(\partial_i b_{jk} + \partial_k b_{ij} + \partial_j b_{ki}\right) \ . \tag{5.36}$$

the one-loop RG β-functions are given by (Subsect. 2.4.1)

$$(2\pi)\beta^{(1)}_{(ij)} = R_{ij} + H_i^{kn}H_{jkn} , \quad R_{ij} = R^k{}_{ikj} ,$$

$$(2\pi)\beta^{(1)}_{[ij]} = \nabla^k H_{kij} .$$

(5.37)

Here $R^i{}_{jkl}$ and ∇_i are the standard Riemannian curvature tensor and the covariant derivative, respectively, defined in terms of the NLSM metric g_{ab} without torsion.

In our calculations of (5.37) we used the REDUCE program for analytic calculations on computer [113]. As regards the 2d Freedman-Townsend model (5.26) with the metric and the torsion potential of (5.27), we find [112]

$$(2\pi)\beta^{(1)}_{(ij)} = \frac{1}{(1+B^2)^2} \left\{ g_{ij} \left[\tfrac{1}{2}(3+B^2) + B^2 - 3 \right] - \delta_{ij}(B^2+3) \right\} ,$$

$$(2\pi)\beta^{(1)}_{[ij]} = -\frac{2b_{ij}}{(1+B^2)^2} .$$

(5.38)

As regards the dual, purely metric NLSM (5.34), let us first rewrite it in the form

$$g_{ab}(\theta) = 4 \left[\frac{k}{R^4}(\theta_a\theta_b - R^2\delta_{ab}) + \frac{\theta_a\theta_b}{R^2} \right] ,$$

(5.39)

where we have introduced the notation

$$k \equiv \cos(2R) - 2\cos R + 1 , \quad R \equiv \sqrt{\theta_1^2 + \theta_2^2 + \theta_3^2} , \quad a,b,\ldots = 1,2,3 . \quad (5.40)$$

The first line of (5.37) yields in this case [112]

$$(2\pi)\widetilde{\beta}^{(1)}_{ab} = \frac{1}{2R^2} \left[\frac{k''+2}{R^2}(\theta_a\theta_b - R^2\delta_{ab}) + \frac{2kk'' - (k')^2}{k^2}\theta_a\theta_b \right] ,$$

(5.41)

where

$$k' \equiv \frac{dk}{dR} = -2\sin 2R + 2\sin R ,$$

$$k'' \equiv \frac{d^2k}{dR^2} = -4\cos 2R + 2\cos R .$$

(5.42)

The two-loop contributions to the RG β-functions of the general 2d NLSM having the form (5.35) are given by (Subsect. 2.4.2)

$$(4\pi)^2\beta^{(2)}_{(ij)} = -2R_{ilkm}R_j{}^{lkm} - 4H^{sk}{}_{(i}R_{j)klm}H_s^{lm} - 4R_{kijl}H^{kst}H^l{}_{st}$$

$$+ 2\nabla_l H_{imk}\nabla^l H_j^{mk} - \tfrac{2}{3}\nabla_i H_{klm}\nabla_j H^{klm}$$

$$- 4H_{jml}H_i^{mk}H_{kst}H^{lst} - 4H_{sil}H^{smk}H_{rjk}H^r{}_m{}^l ,$$

$$(4\pi)^2\beta^{(2)}_{[ij]} = 4\nabla^l H_{st[i}R_{j]l}{}^{st} - 4H_{lm[i}\nabla_{j]}H^{lsk}H^m_{sk}$$

$$+ 4H_{lm[i}\nabla^l H_{j]sk}H^{msk} - 4\nabla_l H_{ijk}H^{kst}H^l_{st} .$$

(5.43)

In the case of the NLSM described by (5.26) and (5.27) we find

$$
\begin{aligned}
(4\pi)^2 \beta^{(2)}_{(ij)} &= \tfrac{1}{2}(1+B^2)^{-5}\left\{ B_i B_j\left[-3(1+B^2)^4 - 24(1+B^2)^3\right.\right. \\
&\qquad \left. -48(1+B^2)^2 + 48(1+B^2) - 32\right] + \delta_{ij}\left[-3(1+B^2)^4\right. \\
&\qquad \left.\left. +80(1+B^2) - 32\right]\right\} , \\
(4\pi)^2 \beta^{(2)}_{[ij]} &= (1+B^2)^{-4} b_{ij}\left\{(1+B^2)^2 + 4(1+B^2) - 2\right\} .
\end{aligned}
$$

(5.44)

The two-loop contribution to the β-functions of the dual NLSM with the metric (5.34) or (5.39) is given by [112]

$$
\begin{aligned}
(4\pi)^2 \widetilde{\beta}^{(2)}_{(ab)} &= -\frac{1}{8R^2 k^3}\left\{\frac{1}{2R^2}\left[(2k''k - k'^2)^2 + (k'^2 + 4k)^2\right]\right. \\
&\qquad \left. \times(\theta_a\theta_b - R^2\delta_{ab}) + \frac{1}{k}(2k''k - k'^2)^2\theta_a\theta_b\right\} .
\end{aligned}
$$

(5.45)

2.6 Gauging NLSM Isometries

Let us go back to the general NLSM of the form (1.1). At each point ϕ^a, $a = 1, 2, \ldots, D$, of the NLSM target space $\mathcal{M}$, the *isometry* group G has an *isotropy* subgroup H_ϕ consisting of those transformations of G that leave the point ϕ^a fixed. The isotropy group H_ϕ is clearly a subgroup of $SO(D)$. The remaining symmetries of G move the point ϕ^a, being *non-linearly* realized. We follow [114] here. The infinitesimal action of G reads

$$
\delta\phi^a = \lambda^A k^a_A(\phi) , \tag{6.1}
$$

where $k^a_A(\phi)$, $A = 1, 2, \ldots, \dim G$, are Killing vectors on the manifold $\mathcal{M}$, and λ^A are constant parameters. The Killing vectors satisfy the Killing equation (1.6). Equation (6.1) can be rewritten in the form [114]

$$
\delta\phi^a = [\lambda^A k^b_A \partial/\partial\phi^b, \phi^a] \equiv L_{\lambda\cdot\kappa}\phi^a , \tag{6.2}
$$

where $L_{\lambda\cdot\kappa}$ is the Lie derivative along $\lambda^A k^a_A$. The group G is generated by the Killing vectors,

$$
[k_A, k_B]^a \equiv k^b_A k^a_{B,b} - k^b_B k^a_{A,b} = f_{AB}{}^C k^a_C , \tag{6.3}
$$

where the comma in a subscript denotes partial differentiation, and $f_{AB}{}^C$ are the structure constants of G. Accordingly, a finite G-transformation reads

$$
\phi^{a'} = \exp\left\{L_{\lambda\cdot\kappa}\right\}\phi^a , \tag{6.4}
$$

provided that G is a compact group. Otherwise, a general group element is the product of the exponentials (6.4) over all connected components of a non-compact group G.

The isotropy subgroup H_ϕ generically depends upon the point ϕ^a chosen, while it may also be trivial. Special (or adapted) local coordinates can always be introduced, with a given point ϕ^a being at the origin, $\phi^a = 0$, where H_ϕ acts *linearly*, i.e.

$$\delta\phi^a = i\lambda^X (T_X)^a{}_b \phi^b \ . \tag{6.5}$$

Here $(T_X)^a{}_b$ are the Hermitian generators of H_ϕ, and λ^X are constant parameters. Comparing (6.1) and (6.5) for the linearly realized subgroup H_ϕ in adapted coordinates yields

$$k_X^a(\phi) = i(T_X)^a{}_b \phi^b \ , \quad X = 1, 2, \ldots, \dim H_\phi \ . \tag{6.6}$$

The isotropy subgroup H_ϕ at a point ϕ^a in adapted coordinates can be gauged by the use of the standard (minimal) coupling procedure (Sect. 2.1). To gauge non-linear realizations leaving no point fixed, one needs a geometrical formulation valid in any coordinate system. Let us now consider gauging an arbitrary subgroup $\mathcal{G}$ of the isometry group G of $\mathcal{M}$, by promoting the parameters λ of $\mathcal{G}$ to arbitrary functions $\lambda(x)$ [114]. Accordingly, let $\mathcal{H}_\phi$ be the restriction of H_ϕ to $\mathcal{G}$, i.e. $\mathcal{H}_\phi = H_\phi \cap \mathcal{G}$.

The kinetic term (1.1) of the bosonic NLSM can be made gauge-invariant by promoting the partial derivative ∂_μ to the gauge-covariant derivative,

$$\partial_\mu \phi^a \to \nabla_\mu \phi^a = \partial_\mu \phi^a - A_\mu^B k_B^a = \partial_\mu \phi^a - [A_\mu^B k_B^b, \partial/\partial\phi^b, \phi^a] \ , \tag{6.7}$$

where the gauge connection A_μ^B for the group $\mathcal{G}$ has been introduced. Its gauge transformation law is given by

$$\delta A_\mu^B = \partial_\mu \lambda^B + f_{CD}{}^B A_\mu^C \lambda^D(x) \ . \tag{6.8}$$

In adapted coordinates for a linearly realized subgroup $\mathcal{H}_\phi$, (6.7) reduces to the standard minimal coupling,

$$\partial_\mu \phi^a \to \nabla_\mu \phi^a = \partial_\mu \phi^a - i A_\mu^X (T_X)^a{}_b \phi^b \ . \tag{6.9}$$

By construction, the covariant derivative ∇_μ transforms covariantly under a finite local transformation $\phi' = g\phi$, where $g \in \mathcal{G}$, i.e.

$$\nabla'_\mu = g \nabla_\mu g^{-1} \ . \tag{6.10}$$

The action of $\mathcal{G}$ divides $\mathcal{M}$ into gauge equivalence classes (*orbits*) consisting of points that can be transformed into each other or related by a gauge transformation. If $\mathcal{H}_\phi$ is the isotropy subgroup of the gauge group $\mathcal{G}$ at a point ϕ^a, the vacuum expectation value $\phi_0^a = \langle \phi^a \rangle$ spontaneously breaks the gauge group $\mathcal{G}$ to $\mathcal{H}_0 \equiv \mathcal{H}_{\phi_0}$. Accordingly, one NLSM scalar degree of freedom is absorbed by the gauge fields for each broken generator. All the NLSM scalars can, therefore, be divided into *unphysical* scalars that are the coordinates of the orbit passing through ϕ_0, and *physical* scalars that are the coordinates of the space of gauge equivalence classes (i.e. the orbits) [114].

A few comments are in order.

The symmetry given by some one-parameter subgroup of $\mathcal{G}$ can only have an unbroken phase if the corresponding Killing vector has a fixed point on $\mathcal{M}$. Since the fixed point structure of $\mathcal{M}$ is closely related to its topology, there may be topological obstructions to certain spontaneous symmetry breaking schemes $\mathcal{G}/\mathcal{H}$.

A scalar potential $V(\phi)$ may be added to the NLSM Lagrangian whose isometry symmetry $\mathcal{G}$ is supposed to be gauged, provided that the potential is $\mathcal{G}$-invariant, i.e.

$$\delta V = \lambda^A k_A^a \frac{\partial V}{\partial \phi^a} = 0 \ . \tag{6.11}$$

By a proper choice of the potential one can treat ϕ_0 as the critical point of the potential. The gauge kinetic terms may also be added to the gauged NLSM, though we are not going to consider them here (see, e.g., [114]).

2.7 Skyrme NLSM of Pions and Nucleons

According to the Goldstone theorem in QFT [27], the spectrum of physical states necessarily has massless (Nambu-Goldstone) particles associated with generators of any spontaneously broken (continuous) *global* symmetry. In field theory, a spontaneously broken symmetry is always related to the degeneracy of vacua. The Goldstone fields transform inhomogeneously under the action of the spontaneously broken symmetries. It has a nice geometrical description in terms of the coset G/H, with G being the spontaneously broken symmetry group and H the unbroken (vacuum stability) group [13, 115]. The highly non-linear (Goldstone) actions of the Nambu-Goldstone fields are unambiguously fixed by the broken symmetries too. In this section, we briefly discuss the Goldstone action given by the principal 4d NLSM that is associated with the spontaneously broken chiral symmetry in four-dimensional QCD, and its higher-derivative (Skyrme) generalization.

Let us begin with the fundamental QCD Lagrangian,

$$\mathcal{L}_{\mathrm{QCD}} = -\tfrac{1}{4}(F_{\mu\nu}^a)^2 + i\bar{q}_f^{\mathrm{L}}\gamma^\mu D_\mu q_f^{\mathrm{L}} + i\bar{q}_f^{\mathrm{R}}\gamma^\mu D_\mu q_f^{\mathrm{R}} \ , \tag{7.1}$$

where $F_{\mu\nu}^a$ is the non-Abelian gauge field strength of gluons, and D_μ is the gauge-covariant derivative. The local (colour) gauge symmetry group is $SU_c(3)$. We only take into account light quarks (u,d,s), so that the number of flavours (N_f) to be considered is supposed to be either two or three. The quark masses are ignored. The (classical) internal symmetry of the Lagrangian (7.1) is given by

$$U(3)_L \times U(3)_R \cong SU(3)_L \times SU(3)_R \times U(1)_B \times U(1)_A \ , \qquad (7.2)$$

since the chiral (or Weyl) quark spinor fields q_L and q_R are obviously allowed to transform independently. The non-chiral (baryon) symmetry $U(1)_B$ is still present in quantum theory, whereas the chiral symmetry,

$$U(1)_A : \qquad q_L \to e^{i\alpha} q_L \ , \quad q_R \to e^{-i\alpha} q_R \ , \qquad (7.3)$$

is known to be broken, due to a chiral anomaly [27]. Hence, the symmetry (7.2) is to be reduced to

$$G_{\text{global}} = SU(3)_L \times SU(3)_R \times U(1)_B \ . \qquad (7.4)$$

It is, however, still larger than the standard classification symmetry of hadrons,

$$G_{\text{phenomenological}} = SU(3) \times U(1)_B \ , \qquad (7.5)$$

so that there should exist a mechanism of spontaneous breaking of the chiral symmetry (7.4) down to (7.5) in QCD. The standard proposal is to assume the formation of the quark-antiquark pair condensate via a non-perturbative generation of the scalar colour-singlet vacuum expectation

$$\langle \bar{q}_f^L q_{f'}^R \rangle \sim U_{ff'} \in SU(3) \ . \qquad (7.6)$$

Though we do not understand in detail how the vacuum expectation value (7.6) may be derived from QCD, this (widely accepted) assumption is just enough for the chiral symmetry breaking in question, while the $SU(3)$ of (7.5) is then supposed to be identified with the diagonal subgroup $SU(3)_{\text{diagonal}} \equiv H$ of $SU(3)_L \times SU(3)_R \equiv G$. The Goldstone theorem now predicts the existence of eight Nambu-Goldstone scalar bosons in the adjoint of $SU(3)$. They can be identified with the exceptionally light pseudo-scalar mesons $(\pi, K, \bar{K}, \eta)$. Indeed, the experimental estimates are $m_\pi^2/m_N^2 \approx 0.022 \ll 1$, where m_N^2 is the nucleon mass,

It is remarkable that the low-energy effective (Goldstone) action (LEEA) of the pseudo-scalar mesons (= pions) can be determined from symmetry considerations alone, without knowing any details associated with the quark condensation (7.6) in the underlying QCD or using a large-N_c expansion. The most straightforward derivation of the LEEA would amount to a highly complicated integration over high-frequency modes in the QCD (non-Gaussian) path integral that is not even well defined yet. A rough approximation to the LEEA for pions can, however, be obtained on symmetry grounds, and it takes the form of the four-dimensional principal NLSM over the $SU(N_f)$ group manifold. Indeed, the effective (or collective) degrees of freedom should be associated with point-to-point deformations in the vacuum orientation, which suggests promoting the VEV (7.6) to the $SU(N_f)$-valued field $U(x)$, while its effective action $LEEA[U]$ should be invariant under the full global symmetry G. The quantum fluctuations of the field $U(x)$ are the low-energy mesons (pions) made up of quark-antiquark pairs.

Because of the constraint $UU^\dagger = 1$, there can be no scalar potential. The leading term in the (low-energy) momentum expansion of the effective action is of second order in spacetime derivatives, while the only invariant candidate is given by

$$\mathcal{L}_{\text{LEEA}}[U] = \frac{F_\pi^2}{16}\text{tr}\left(\partial_\mu U^\dagger \partial^\mu U\right) = -\frac{F_\pi^2}{16}\text{tr}\left(A_\mu A^\mu\right) , \qquad (7.7)$$

where we have introduced the Cartan one-form

$$A_\mu = U^\dagger \partial_\mu U \qquad (7.8)$$

that homogeneously transforms under the symmetry G and, therefore, is quite useful to define invariants. It is worth mentioning that the non-Abelian (Yang-Mills) field strength $F(A)$ to be constructed in terms of the vector field A_μ of (7.8) identically vanishes, $F_{\mu\nu}(A) = 0$, since (7.8) represents a 'pure gauge', while $U^{-1} = U^\dagger$. The dimensional constant F_π introduced in front of (7.7) is called the pion decay constant. Its experimental value is 186 MeV.

By construction, $\mathcal{L}_{\text{LEEA}}[U]$ at $N_f = 3$ is invariant under the non-linearly realized $SU(3)_\text{L} \times SU(3)_\text{R}$ rigid transformations

$$U(x) \to G_\text{L} U(x) G_\text{R} , \quad \text{where} \quad G_\text{L}, G_\text{R} \in SU(3) . \qquad (7.9)$$

The constrained field $U(x)$ can be represented in terms of real, unconstrained, Lie algebra-valued fields $\xi^a(x)$, by using the exponential map between the $SU(3)$ group and its Lie algebra,

$$U(x) = \exp\left\{2\text{i} \sum_a \xi^a(x) t_a\right\} , \qquad (7.10)$$

where we have introduced the Gell-Mann matrices t_a, $a = 1, 2, \ldots, 8$ as the $SU(3)$ generators, [14]

$$t_1 = \begin{pmatrix} 0 & 1 & 0 \\ 1 & 0 & 0 \\ 0 & 0 & 0 \end{pmatrix} , \quad t_2 = \begin{pmatrix} 0 & -\text{i} & 0 \\ \text{i} & 0 & 0 \\ 0 & 0 & 0 \end{pmatrix} , \quad t_3 = \begin{pmatrix} 1 & 0 & 0 \\ 0 & -1 & 0 \\ 0 & 0 & 0 \end{pmatrix} ,$$

$$t_4 = \begin{pmatrix} 0 & 0 & 1 \\ 0 & 0 & 0 \\ 1 & 0 & 0 \end{pmatrix} , \quad t_5 = \begin{pmatrix} 0 & 0 & -\text{i} \\ 0 & 0 & 0 \\ \text{i} & 0 & 0 \end{pmatrix} , \quad t_6 = \begin{pmatrix} 0 & 0 & 0 \\ 0 & 0 & 1 \\ 0 & 1 & 0 \end{pmatrix} ,$$

---

[14] In the case of two flavours, one has $a = 1, 2, 3$, while the generators t_a are Pauli matrices τ_a.

$$t_7 = \begin{pmatrix} 0 & 0 & 0 \\ 0 & 0 & -i \\ 0 & i & 0 \end{pmatrix} , \quad t_8 = \frac{1}{\sqrt{3}} \begin{pmatrix} 1 & 0 & 0 \\ 0 & 1 & 0 \\ 0 & 0 & -2 \end{pmatrix} . \tag{7.11}$$

The Gell-Mann (as well as Pauli) matrices are all Hermitian and traceless. In addition, they satisfy the relation

$$\mathrm{tr}(t_a t_b) = 2\delta_{ab} . \tag{7.12}$$

The standard parametrization leading to the canonically normalized kinetic terms for the pion fields is given by [27]

$$\sum_a \xi^a t_a = \frac{\sqrt{2}}{F_\pi} \begin{pmatrix} \frac{1}{\sqrt{2}}\pi^0 + \frac{1}{\sqrt{6}}\eta^0 & \pi^+ & K^+ \\ \pi^- & -\frac{1}{\sqrt{2}}\pi^0 + \frac{1}{\sqrt{6}}\eta^0 & K^0 \\ K^- & \bar{K}^0 & -\sqrt{2/3}\,\eta^0 \end{pmatrix} . \tag{7.13}$$

Substituting (7.10) into (7.7) results in the standard NLSM form (1.1) of the Goldstone action (7.7) in terms of the ξ-fields with the $SU(3)$ invariant NLSM metric $g_{ab}(\xi)$. This action is obviously invariant under the linearly realized $SU(3)_{\mathrm{diag}}$ transformations acting on the ξ-fields.

The symmetry (7.4) is, of course, only approximate in the real world where pions are not massless. Their masses may, nevertheless, be taken into account by introducing a scalar potential into the NLSM (7.7), for example, in the form

$$L_{\mathrm{mass}} = -\frac{F_\pi^2}{2} \mathrm{tr} \left[\hat{M}^2 (U^\dagger + U) \right] , \tag{7.14}$$

where $\hat{M}^2$ stands for a constant (mass squared) matrix. It is worth mentioning, however, that this naive approach may not work, especially, when strangeness ($N_{\mathrm{f}} > 2$) is included [116].

A revolutionary idea for incorporating baryons within the framework of the NLSM for mesons was proposed by Skyrme [2] who noticed the existence of topologically non-trivial classical NLSM solutions (='quasi-solitons'), having a finite energy but, unfortunately, being unstable against collapse to zero size. [15] In order to make the NLSM solitonic field configuratons stable, Skyrme [2] added a new term of fourth-order in derivatives into the pion NLSM Lagrangian. This term is now called the *Skyrme term*, while the (Skyrme) modified NLSM Lagrangian reads [2]

$$\mathcal{L}_{\mathrm{Skyrme}} = -\frac{F_\pi^2}{16} \mathrm{tr}\left(A_\mu A^\mu \right) + \frac{1}{32e^2} \mathrm{tr}\left([A_\mu, A_\nu]^2 \right) , \tag{7.15}$$

where e is yet another dimensionless coupling constant. The stability of the solitonic solutions to the field theory (7.15) is ensured by their (conserved) topological winding number that is naturally identified with the baryon number (see below).

---

[15] This follows, e.g., from a simple application of scaling arguments [117].

The extra (Skyrme) term in (7.15) is the *unique* invariant with four space-time derivatives and of second order in the time derivative, which leads to a *positive* Hamiltonian. The Skyrme term is, therefore, a viable candidate for the next-to-leading-order correction to the pion NLSM in the momentum expansion of the full pion effective action to be dictated by QCD. In fact, a four-dimensional WZNW term, which has four spacetime derivatives and is *linear* in the time derivative, also has to be added to the effective pion action [48]. The WZNW term is necessary for the Skyrme solitons to have the right quantum numbers of QCD baryons [48]. Indeed, (7.15) is separately invariant under $x \to -x$ and $U \to U^{\dagger}$, whereas QCD is merely invariant under their combination. The required modification of (7.15) is just provided by the (parity breaking) WZNW term that is usually written down as a five-dimensional topological density (cf. Subsect. 5.1.3),

$$\Gamma_{\text{WZNW}} = -\frac{in}{240\pi^2} \int_{B^5} \mathrm{d}^5 x \, \varepsilon^{\mu\nu\alpha\beta\gamma} \mathrm{tr}\left(A_\mu A_\nu A_\alpha A_\beta A_\gamma\right) \,, \qquad (7.16)$$

where the (compactified) 4d spacetime is supposed to be the boundary of B_5. The coefficient n must be an integer, in order to avoid the multi-valuedness of the action (7.16) in quantum theory (Subsect. 5.1.3).

To identify the value of n that is compatible with QCD, it is useful to couple the pion theory to electromagnetism [48]. Being applied to the pion decay, $\pi^0 \to \gamma\gamma$, the WZNW implies the effective interaction term

$$\mathcal{L}_{\text{int}} = \frac{ne^2}{48\pi^2 F_\pi} \pi^0 \varepsilon^{\mu\nu\lambda\rho} F_{\mu\nu} F_{\lambda\rho} \,. \qquad (7.17)$$

Equation (7.17) agrees with the standard amplitude from the triangle graph if and only if n coincides with the number of colors, $n = N_{\text{c}}$ [48].

Though the Skyrme model is merely a rough approximation to the real world, its mathematical elegance and surprising quantitative agreement with the experimental data about the low-lying baryons resulted in a quite remarkable variety of applications in nuclear and particle physics [8], as well as in mathematics [118]. The low-energy parameters $(F_\pi, e, \hat{M})$ of the Skyrme NLSM are, in principle, derivable from QCD, though the relevant calculations are unfeasable. As regards phenomenological studies, they imply, in particular, that $e \approx 5.45$ [119].

We now briefly discuss the simplest *static* solitonic solution to the Skyrme model with two flavours, $N_{\text{f}} = 2$, and baryon charge $B = 1$ [119, 120]. The solitonic solution in the $SU(3)$ Skyrme model is essentially the same since it is the $SU(2)$ subgroup of $SU(3)$ that has the same topology (S^3) as our (compactified) space $\mathbf{R}^3 \cup \infty$. In so-called natural units ($F_\pi/8e \approx 6$ MeV for the energy unit and $4/eF_\pi \approx 0.6$ fm for the unit of length), the static energy (i.e. the soliton mass) is given by [120]

$$M = \int \mathrm{d}^3 x \left\{ -\tfrac{1}{2}\mathrm{tr}\left(U^{\dagger}\partial_i U U^{\dagger}\partial_i U\right) - \tfrac{1}{16}\mathrm{tr}\left([U^{\dagger}\partial_i U, U^{\dagger}\partial_j U]^2\right) \right\} \,. \qquad (7.18)$$

The topological equivalence classes of Skyrme solitons are the homotopy classes of mappings $S^3 \to S^3$, with the third homotopy group being $\Pi_3(S^3) = \mathbf{Z}$. The corresponding topological charge (winding number) is given by

$$
\begin{aligned}
B &= \frac{1}{24\pi^2} \int d^3x \, \varepsilon^{ijk} \mathrm{tr}\left(U^\dagger \partial_i U U^\dagger \partial_j U U^\dagger \partial_k U\right) \\
&= \frac{1}{24\pi^2} \int d^3x \, \varepsilon^{ijk} \mathrm{tr}\left(A_i A_j A_k\right) \ .
\end{aligned}
\tag{7.19}
$$

Its physical interpretation is just given by the baryon number [2].

The Skyrme NLSM soliton of $B = 1$ (and of the lowest energy) is called a *skyrmion*. It can be parametrized by using the Skyrme *Ansatz*,

$$
U(x) = \exp\left\{iR(r)\hat{r} \cdot \boldsymbol{\tau}\right\} \ ,
\tag{7.20}
$$

in terms of a single radial function $R(r)$, where we have introduced the notaton $r = |x|$, $\hat{r} = x/r$, and $\boldsymbol{\tau}$ are Pauli matrices. The field configuration (7.20) is sometimes referred to as the '*hedgehog*'.

Substitution of the 'hedgehog' *Ansatz* (7.20) into the Skyrme NLSM equations of motion or, equivalently, plugging (7.20) into the static energy (7.17) and then considering the variational equation for the energy functional in terms of the radial function $R(r)$, result in an ordinary second-order differential equation [119],

$$
\left(\frac{\rho^2}{4} + 2\sin^2 F\right) F'' + \tfrac{1}{2}\rho F' + \sin 2F F'^2 - \tfrac{1}{4}\sin 2F - \frac{\sin^2 F \sin 2F}{\rho^2} = 0 \ , \tag{7.21}
$$

where we have introduced the dimensionless variable $\rho = eF_\pi r$ and defined $R(r) = F(\rho)$, as in [119, 120].

The static energy (in natural units) reads [120]

$$
\begin{aligned}
M = 4\pi \int_0^\infty d\rho \left[\rho^2 \left(\frac{dF}{d\rho} + \frac{\sin^2 F}{\rho^2}\right)^2 + 2\sin^2 F \left(\frac{dF}{d\rho} + 1\right)^2\right] \\
- 24\pi \int_0^\infty d\rho \sin^2 F \frac{dF}{d\rho} \ .
\end{aligned}
\tag{7.22}
$$

The first integral in (7.22) is positive definite. The second integral is topological, while it only depends upon $F(0)$ and $F(\infty)$. If the energy is to be finite, we have to assume that $F(\rho) \to 0$ as $\rho \to \infty$, while $F(0)$ then equals $B\pi$ because of (7.19). Hence, we arrive at the topological *BPS bound* for the soliton mass (or the static energy) [120]

$$
M \geq 12\pi^2 |B| \ .
\tag{7.23}
$$

It is easy to verify that this BPS pound *cannot* be saturated [120] since the first integral in (7.17) vanishes if and only if

$$\frac{\mathrm{d}F}{\mathrm{d}\rho} + 1 = 0 \tag{7.24}$$

and

$$\frac{\mathrm{d}F}{\mathrm{d}\rho} + \frac{\sin^2 F}{\rho^2} = 0 \tag{7.25}$$

are simultaneously valid, which is clearly impossible.

Though the solitonic solutions to (7.21) are not available in an analytic form, they can be investigated numerically [119]. For example, in the case of a skyrmion ($B = 1$) with the boundary conditions $F(0) = \pi$ and $F(\infty) = 0$, one easily finds the following asymptotic behaviour of the solution: $F(\rho) \approx \pi - \alpha\rho$ near the origin $\rho = 0$, and $F(\rho) \approx \kappa/\rho^2$ for large values of ρ, with some constants α and κ. Most of the skyrmion energy is, therefore, concentrated around $F = \frac{1}{2}\pi$, in agreement with the non-relativistic 'bag approximation' used in nuclear physics. Numerical integration roughly estimates the Skyrmion mass at $1.23 \times 12\pi^2$ in natural units or, equivalently, 850 MeV, which is close enough to the nucleon mass. Many applications of the Skyrme model in particle and nuclear physics can be found, e.g., in [8, 118] and references therein.

3. Supersymmetric NLSM

In this chapter we extend some of the general results of Chap. 2 to the supersymmetric NLSM. Though our presentation is self-contained, we would like to mention some basic references about supersymmetry [121, 122, 123, 124, 125, 126, 127, 128], supergravity [129, 130, 131], and superspace [132, 133, 134, 135, 136, 137], as well as some reviews [138, 139, 140, 141, 142, 143, 144, 145, 146, 147, 148]. A supersymmetry algebra in d dimensions is the Z_2 graded extension of the Poincaré algebra. In addition to the *even* Poincaré generators, it has *odd* supersymmetry generators that transform in a spinor representation of the Lorentz group. A simple supersymmetry has only one irreducible spinor representation of minimal dimension. If there are N such spinors, one has the N-extended supersymmetry. In two dimensions, a generic (p, q) supersymmetry algebra can have p chiral and q anti-chiral real spinor generators. The minimal 2d non-chiral supersymmetry is $N = (1, 1)$, while the real chiral $(1, 0)$ supersymmetry in 2d is often called *heterotic*.

It is straightforward to supersymmetrize the bosonic 2d NLSM action (2.5.35) in components, by the use of the Noether procedure of 'trial and error' [149]. One finds that the N-extended supersymmetry implies the existence of $(N - 1)$ covariantly constant complex structures on the NLSM target manifold $\mathcal{M}$, while the NLSM metric should be Hermitian with respect to each complex structure. The covariant constancy of the complex structures implies that they all commute with the holonomy group of $\mathcal{M}$, whose generators are built out of the curvature tensor components. Assuming non-vanishing curvature, this leads to certain constraints on the holonomy group in extended supersymmetry. If the holonomy group is irreducible, the number of supersymmetries is limited to $(p, q) \leq (4, 4)$ [149]. We are only going to consider $(1, 0)$, $(1, 1)$, $(2, 2)$ and $(4, 4)$ supersymmetry, if counted from the 2d point of view. [1] There are no restrictions on the holonomy of $\mathcal{M}$ in the cases of $(1, 0)$ and $(1, 1)$ supersymmetry, so that any 2d bosonic NLSM can be extended to the heterotic NLSM or the $(1, 1)$ supersymmetric NLSM.

In this chapter we use *superspace* where the most general non-chiral $N = 1$ supersymmetric NLSM in 2d can be easily defined. We investigate the $N = 1$ NLSM renormalization in quantum perturbation theory, along the lines of the bosonic case considered in Chap. 2. Superspace is generically

---

[1] See e.g., [150, 151] for the 2d NLSM actions with $(2, 1)$ supersymmetry.

defined by the quotient of the super-Poincaré group over its Lorentz subgroup. Supersymmetry representations are naturally carried by *superfields* (i.e. fields in superspace). The N-extended superfields define reducible representations of N-extended supersymmetry, so that they have to be properly constrained in extended superspace (Chap. 4).

The 2d heterotic $(1,0)$ NLSM will be considered separately (Chap. 7), in the context of heterotic strings. The $(2,2)$ supersymmetric NLSM gets the most natural (off-shell and manifestly supersymmetric) description in $(2,2)$ extended superspace, since its target space geometry always possesses a potential. For instance, in the purely metric case, the $(2,2)$ NLSM potential is just a Kähler potential [152, 153]. The purely metric $(4,4)$ supersymmetric 2d NLSM must have a hyper-Kähler metric, while its most natural, off-shell and manifestly supersymmetric description is provided by harmonic superspace where a hyper-Kähler potential exists (Chap. 4).

An extended supersymmetry in a given number of spacetime dimensions can be interpreted as a simple or a 'less extended' supersymmetry in higher dimensions, which is dictated by the dimensions of the irreducible spinor representations in various spacetime dimensions. The corresponding *dimensional reduction* (see, e.g., the original papers [154, 155], or [156, 157] and references therein) from higher to lower spacetime dimensions is a very powerful method of constructing field theories with extended supersymmetry. In Sect. 3.4 we exploit this relationship by considering the most general 4d (Kähler) NLSM having $N = 1$ supersymmetry, which gives rise to the $(2,2)$ supersymmetric NLSM after a dimensional reduction to 2d. In Chap. 4 we establish a similar correspondence between $N = 4$ supersymmetry in 2d and $N = 2$ supersymmetry in 4d, in order to explicitly construct some maximally supersymmetric 2d NLSM with remarkable geometry and highly non-trivial renormalization properties.

The 2d NLSM on homogeneous spaces and their (one-loop) renormalization are briefly discussed in Sect. 3.3. In Sect. 3.4 we consider gauging isometries in a 4d, $N = 1$ supersymmetric NLSM.

3.1 $N = 1$ NLSM Action and Background-Field Method

Let us define 2d Dirac gamma matrices as

$$\gamma^0 = \begin{pmatrix} 0 & -\mathrm{i} \\ \mathrm{i} & 0 \end{pmatrix} , \quad \gamma^1 = \begin{pmatrix} 0 & \mathrm{i} \\ \mathrm{i} & 0 \end{pmatrix} , \quad \gamma_3 = \gamma^0\gamma^1 = \begin{pmatrix} 1 & 0 \\ 0 & -1 \end{pmatrix} . \tag{1.1}$$

They have the properties

$$\{\gamma^\mu,\gamma^\nu\} = 2\eta^{\mu\nu} = 2\mathrm{diag}(+,-) , \quad \mathrm{tr}(\gamma^\mu\gamma^\nu) = 2\eta^{\mu\nu} , \quad \gamma_3^2 = 1 . \tag{1.2}$$

We raise and lower spinor indices with a 2d charge conjugation matrix C having the form

$$C_{\alpha\beta} = \begin{pmatrix} 0 & -i \\ i & 0 \end{pmatrix} , \quad C^{\alpha\beta} = \begin{pmatrix} 0 & i \\ -i & 0 \end{pmatrix} , \tag{1.3}$$

and the properties

$$C\gamma_\mu^{\mathrm{T}} C^{-1} = -\gamma_\mu , \quad C^{\mathrm{T}} = -C , \quad C^2 = -1 . \tag{1.4}$$

The Dirac conjugation is defined by

$$\bar{\theta} = \theta^\dagger \gamma^0 , \tag{1.5}$$

so that a 2d Majorana spinor θ satisfies the relation

$$\bar{\theta}^\alpha = C^{\alpha\beta} \theta_\beta . \tag{1.6}$$

Here are some useful identities for any two 2d Majorana spinors ψ and χ:

$$\bar{\psi}\chi = \bar{\chi}\psi , \quad \bar{\psi}\gamma_\mu\chi = -\bar{\chi}\gamma_\mu\psi , \quad \bar{\psi}\gamma_3\chi = -\bar{\chi}\gamma_3\psi . \tag{1.7}$$

The so-called *Fierz identity* for any three 2d Majorana spinors ψ_i, $i = 1, 2, 3$, reads

$$\psi_1(\bar{\psi}_2\psi_3) = -\tfrac{1}{2}\psi_3(\bar{\psi}_2\psi_1) - \tfrac{1}{2}\gamma^\mu\psi_3(\bar{\psi}_2\gamma_\mu\psi_1) - \tfrac{1}{2}\gamma_3\psi_3(\bar{\psi}_2\gamma_3\psi_1) . \tag{1.8}$$

Another useful identity for the 2d gamma matrices is

$$\mathrm{tr}(\gamma^\mu\gamma^\nu\gamma^\rho\gamma^\lambda) = 2\left(\eta^{\mu\nu}\eta^{\rho\lambda} - \eta^{\mu\rho}\eta^{\nu\lambda} + \eta^{\mu\lambda}\eta^{\nu\rho}\right) . \tag{1.9}$$

The supercovariant derivatives in a flat $N = (1,1)$ superspace $Z^M = (x^\mu, \theta^\alpha)$ are defined by

$$D_\alpha = \frac{\partial}{\partial\bar{\theta}^\alpha} - i(\slashed{\partial}\theta)_\alpha , \tag{1.10}$$

and they satisfy the algebra

$$\{D_\alpha, D_\beta\} = 2i\partial_{\alpha\beta} , \quad \{\bar{D}^\alpha, \bar{D}^\beta\} = 2i\partial^{\alpha\beta} . \tag{1.11}$$

Our notation for the ordinary 2d spacetime derivatives is

$$\partial_{\alpha\beta} = (\gamma^\mu)_\alpha{}^\delta C_{\delta\beta}\partial_\mu , \quad \partial^{\alpha\beta} = C^{\alpha\delta}(\gamma^\mu)_\delta{}^\beta \partial_\mu ,$$
$$\partial_{\alpha\beta}\partial^{\beta\gamma} = \delta_\alpha{}^\gamma \Box , \quad \Box = \partial^\mu\partial_\mu . \tag{1.12}$$

As regards the superspace covariant derivatives, we also define

$$D^2 = \tfrac{1}{2}\bar{D}^\alpha D_\alpha = \tfrac{1}{2}C^{\alpha\beta} D_\beta D_\alpha . \tag{1.13}$$

It is straightforward to verify the identities

$$D_\alpha D_\beta = i\partial_{\alpha\beta} + C_{\beta\alpha}D^2 \ , \qquad \bar{D}^\alpha \bar{D}^\beta = i\partial^{\alpha\beta} + C^{\beta\alpha}D^2 \ ,$$

$$(D^2)^2 = -\Box \ , \qquad \bar{D}\gamma_3 D = 0 \ ,$$

$$D_\alpha D^2 = -D^2 D_\alpha = -i\partial_{\alpha\beta}\bar{D}^\beta \ , \qquad \bar{D}^\alpha D_\beta D_\alpha = 0 \ ,$$

$$\bar{D}^\alpha D^2 = -D^2 \bar{D}^\alpha = i\partial^{\alpha\beta}D_\beta \ .$$

(1.14)

The Grassmann δ-function in superspace is defined by the relations

$$\int d^2\theta \, \delta^2(\theta) = 1 \ , \quad \delta^2(\theta) = \theta_2\theta_1 = \frac{1}{2i}\bar{\theta}^\alpha\bar{\theta}_\alpha \ , \tag{1.15}$$

where we have used the standard (Berezin) integration rules over the anti-commuting (Grassmann) superspace coordinates [158, 159, 160].

The Fourier-transformed superspace covariant derivatives $D_\alpha(p)$ satisfy the relations

$$D_\alpha^{(1)} \bar{D}_\beta^{(2)} \delta^2(\theta_1 - \theta_2) = \left[i\delta_{\alpha\beta} - \not{p}_{\alpha\beta}\delta^2(\theta_1 - \theta_2) \right] \exp(\bar{\theta}_1 \not{p}\theta_2) \ ,$$

$$\bar{D}D(p)\delta^2(\theta_1 - \theta_2) = 2i\exp(\bar{\theta}_1 \not{p}\theta_2) \ ,$$

$$D_\alpha^{(1)} \bar{D}_\beta^{(2)} \tfrac{1}{2}\bar{D}D(p)\delta^2(\theta_1 - \theta_2) = \left[i\not{p}_{\alpha\beta} - p^2\delta_{\alpha\beta}\delta^2(\theta_1 - \theta_2) \right] \exp(\bar{\theta}_1 \not{p}\theta_2) \ ,$$

$$\tfrac{1}{2}(\bar{D}D)(\bar{D}D)\delta^2(\theta_1 - \theta_2) = 2p^2\delta^2(\theta_1 - \theta_2) \ .$$

(1.16)

The notation and equations above are quite useful for quantum perturbative calculations in superspace (Sect. 3.2).

The $N = 1$ supersymmetric version of the general 2d NLSM with a WZNW term can be most easily constructed in superspace, simply by replacing the NLSM real scalar fields $\phi^a(x)$ by unconstrained real scalar superfields $\phi^a(x,\theta)$ having the form

$$\phi^a(x,\theta) = \phi^a(x) + \bar{\theta}\psi^a(x) + \tfrac{1}{2}\bar{\theta}\theta F^a(x) \ , \tag{1.17}$$

where the 2d physical (Majorana) spinor fields ψ^a and the auxiliary real scalar fields $F^a(x)$ have been introduced. Together with ϕ^a, they are usually referred to as the *components* of the superfield (1.17).

The superspace action of the 2d, $N = 1$ NLSM with torsion reads [88]

$$S_{\text{susy}} = \frac{1}{2\lambda^2} \int d^2x \, \frac{1}{2i} \int d^2\theta \, \{g_{ab}(\phi) - h_{ab}(\phi)\} \, \bar{D}\phi^a(1 + \gamma_3)D\phi^b \ . \tag{1.18a}$$

In components, i.e. after integrating over the anticommuting superspace coordinates, omitting a surface term and using the equations of motion for the auxiliary fields, one finds from (1.18a) that

$$S_{\text{susy}} = \frac{1}{2\lambda^2} \int d^2x \, \Big\{ g_{ab}\partial^\mu\phi^a\partial_\mu\phi^b + h_{ab}\varepsilon^{\mu\nu}\partial_\mu\phi^a\partial_\nu\phi^b + ig_{ab}\bar{\psi}^a\gamma^\mu\hat{D}_\mu\psi^b$$

$$+ \tfrac{1}{8}\hat{R}_{abcd}\bar{\psi}^a(1 + \gamma_3)\psi^c\bar{\psi}^b(1 + \gamma_3)\psi^d \Big\} \ ,$$

(1.18b)

where we have introduced the notation

$$\hat{D}_\mu \psi^a = \left[\delta^{ab} \partial_\mu + \Gamma^a_{bc} \partial_\mu \phi^c - H^a_{bc} \varepsilon_{\mu\nu} \partial^\nu \phi^c \right] \psi^b . \tag{1.19}$$

The covariant background-quantum field expansion of the superspace NLSM action (1.18a) goes along the lines of the bosonic case (Sect. 2.2). All one needs is to make the formal substitutions

$$\phi(x) \to \phi(x, \theta) , \qquad \int d^2 x \to \int d^2 x \frac{1}{2i} \int d^2\theta \equiv \frac{1}{2i} \int d^4 Z ,$$

$$\eta^{\mu\nu} \partial_\mu \cdots \partial_\nu \cdots \to \bar{D} \cdots D \cdots , \qquad \varepsilon^{\mu\nu} \partial_\mu \cdots \partial_\nu \cdots \to - \bar{D} \cdots \gamma_3 D \cdots . \tag{1.20}$$

In particular, the superspace analogue of (2.2.26) is given by

$$\begin{aligned} S_S^{(n)} = \frac{1}{4i\lambda^2} \int d^4 Z \Big\{ &\Pi^{(n,2)} (\xi)^n (\bar{D}\phi D\phi) + \Pi^{(n,1)} (\xi)^{n-1} (\bar{D}\xi D\phi) \\ &+ \Pi^{(n,0)} (\xi)^{n-2} (\bar{D}\xi \mathcal{D}\xi) + E^{(n,2)} (\xi)^n (\bar{D}\phi\gamma_3 D\phi) \\ &+ E^{(n,1)} (\xi)^{n-1} (\bar{D}\xi\gamma_3 D\phi) + E^{(n,0)} (\xi)^{n-2} (\bar{D}\xi\gamma_3 \mathcal{D}\xi) \Big\} , \end{aligned} \tag{1.21}$$

where the Π- and E-tensors are the same as in (2.2.26), whereas $(\mathcal{D}_\alpha \xi)^a$ is the NLSM covariant derivative in superspace,

$$(\mathcal{D}_\alpha \xi)^a \equiv D_\alpha \xi^a + \Gamma^a_{bc} D_\alpha \phi^c \xi^b - H^a_{bc} (\gamma_3 D)_\alpha \phi^c \xi^b . \tag{1.22}$$

Since we are only interested in calculating the RG β-functions of the quantized supersymmetric NLSM (1.18), we are allowed to omit in our background-field expansion all the terms proportional to the equations of motion, i.e. the $S^{(1)}$ contribution. The next terms in the covariant background-quantum expansion of (1.18a) are given by [161]

$$S_{\text{susy}}^{(2)} = \frac{1}{4i\lambda^2} \int d^4 Z \left\{ g_{ab} \bar{D}\xi^a \mathcal{D}\xi^b + \hat{R}_{abcd} \xi^b \xi^c \bar{D}\phi^a (1 + \gamma_3) D\phi^b \right\} ,$$

$$\begin{aligned} S_{\text{susy}}^{(3)} = \frac{1}{4i\lambda^2} \int d^4 Z \Big\{ &\frac{1}{6} \left(\hat{D}_a \hat{R}_{bcde} - 2\hat{R}_{abcf} H^f_{de} \right) \bar{D}\phi^b (1 - \gamma_3) D\phi^e \xi^a \xi^d \xi^c \\ &+ \frac{1}{3} (\hat{R}_{abcd} + \hat{R}_{dbca}) \bar{D}\phi^a \mathcal{D}\xi^d \xi^b \xi^c - \frac{1}{3} (\hat{R}_{abcd} - \hat{R}_{dbca}) \\ &\times (\bar{D}\phi^a \gamma_3 \mathcal{D}\xi^d) \xi^b \xi^c + \frac{2}{3} H_{abc} \xi^a (\mathcal{D}\xi^b \gamma_3 \mathcal{D}\xi^c) \Big\} , \end{aligned}$$

$$S_{\text{susy}}^{(4)} = \frac{1}{4i\lambda^2} \int d^4 Z \left\{ \frac{1}{24} \left(\widehat{D}_a \widehat{D}_b \widehat{R}_{cdef} + 3\widehat{R}_{cabg} \widehat{R}^g{}_{def} + \widehat{R}_{cabg} \widehat{R}_{fde}{}^g \right. \right.$$

$$\left. -4\widehat{D}_a \widehat{R}_{bcdg} H^g_{ef} + 4\widehat{R}_{cabg} H^g_{dh} H^h_{ef} \right) \bar{D}\phi^c (1 - \gamma_3) D\phi^f \xi^a \xi^b \xi^d \xi^e$$

$$+ \frac{1}{8} \left(\widehat{D}_a \widehat{R}_{bcde} + \widehat{D}_a \widehat{R}_{ecdb} + +2\widehat{R}_{eacf} H^f_{db} - 2H^f_{ea} \widehat{R}_{fcdb} \right) \bar{D}\phi^b \mathcal{D}\xi^e \xi^a$$

$$\times \xi^c \xi^d - \frac{1}{8} \left(\widehat{D}_a \widehat{R}_{bcde} - \widehat{D}_a \widehat{R}_{ecdb} - 2\widehat{R}_{eacf} H^f_{db} + 2H^f_{ea} \widehat{R}_{fcdb} \right) \xi^a \xi^c$$

$$\times \xi^d \bar{D}\phi^b \gamma_3 \mathcal{D}\xi^e + \frac{1}{6} R_{abcd} \xi^b \xi^c (\mathcal{D}\xi^a \mathcal{D}\xi^d) - \frac{1}{4} \widehat{R}_{abcd} \xi^b \xi^c (\bar{D}\xi^a \gamma_3 \mathcal{D}\xi^d) \right\}$$

$$\tag{1.23}$$

It is straightforward to calculate the other terms in the background field expansion of the supersymmetric NLSM action to any finite order in the quantum ξ-superfields, similarly to the bosonic case.

The action $S^{(2)}$ determines a superpropagator of the massless quantum ξ^a-superfields in the form

$$G^{ab}(Z_1, Z_2) = g^{ab} D^2 K(Z_1, Z_2) \equiv (-i\lambda^2) g^{ab} D^2 \Box^{-1} \delta^2(x_1 - x_2)\delta^2(\theta_1 - \theta_2) . \tag{1.24}$$

Adding a supersymmetric mass term

$$S_m = \frac{m}{2i\lambda^2} \int d^4 Z \, g_{ab}(\phi)\phi^a \phi^b \tag{1.25}$$

as the IR-regulator, and transforming the NLSM target space indices by the use of a vielbein (Sect. 2.2), result in the superpropagator

$$i \langle \xi^i(Z_1)\xi^j(Z_2) \rangle = \delta^{ij} G(Z_1, Z_2) , \tag{1.26}$$

where (after a formal continuation to d dimensions)

$$G(Z_1, Z_2) = \lambda^2 \int \frac{d^d p}{(2\pi)^d} \frac{\exp[-ip \cdot (x_1 - x_2)]}{p^2 - m^2 + i0} \left[\tfrac{1}{2}\bar{D}D(p) + m \right] \delta^2(\theta_1 - \theta_2) . \tag{1.27a}$$

Equation (1.27a) can be rewritten in the equivalent form [161]

$$G(Z_1, Z_2) = \lambda^2 \int \frac{d^d p}{(2\pi)^d} \frac{i \exp(\bar{\theta}_1 \not{p}\theta_2) + m\delta^2(\theta_1 - \theta_2)}{p^2 - m^2 + i0} \exp[-ip \cdot (x_1 - x_2)] . \tag{1.27b}$$

Of course, a renormalization of the supersymmetric NLSM (1.17) can also be considered in the component approach [44, 82]. In the component background-field method for NLSM, the spinor fields ψ^a are treated as additional quantum fluctuating fields transforming as vectors in the NLSM target space $\mathcal{M}$. The parallel transport of spinors along geodesics in $\mathcal{M}$ is governed by the equation

$$D(s)\psi^a(x;s) = \frac{\mathrm{d}}{\mathrm{d}s}\psi^a(x;s) + \Gamma^a_{bc}[\rho]\xi^b_s\psi^c(x;s) = 0 \ . \tag{1.28}$$

For instance, as regards the two-loop component calculations, it is enough to decompose the spinor kinetic term in (1.18b) up to second-order. By using the relations

$$D(s)\left[D_\mu\psi(s)\right]^a \ = \ R^a{}_{bcd}[\rho]\psi^b(s)\xi^c_s\partial_\mu\rho^d \ ,$$

$$D(s)\left[\widehat{D}_\mu\psi(s)\right]^a \ = \ R^a{}_{bcd}[\rho]\psi^b(s)\xi^c_s\partial_\mu\rho^d - H^a_{bc}[\rho]\psi^b(s)\varepsilon_{\mu\nu}(D^\nu\xi_s)^c \tag{1.29}$$

$$- \ D_d H^a_{bc}[\rho]\psi^b(s)\varepsilon_{\mu\nu}\partial^\nu\rho^c\xi^d_s \ ,$$

we find [82]

$$S^{(0)}_\psi \ = \ \frac{1}{2\lambda^2}\int \mathrm{d}^2x \ \left\{\mathrm{i}g_{ab}\bar\psi^a\gamma^\mu D_\mu\psi^b - \mathrm{i}H_{abc}\bar\psi^a\gamma^\mu\varepsilon_{\mu\nu}\partial^\nu\phi^c\psi^b\right\} \ ,$$

$$S^{(1)}_\psi \ = \ \frac{1}{2\lambda^2}\int \mathrm{d}^2x \ \left\{\mathrm{i}g_{ab}\bar\psi^a\gamma^\mu \left[R^b{}_{cdf}\psi^c\xi^d\partial_\mu\phi^f - H^b_{cd}\psi^c\varepsilon_{\mu\nu}(D^\nu\xi^d)\right.\right.$$

$$\left.\left.-D_f H^b_{cd}\psi^c\varepsilon_{\mu\nu}\partial^\nu\phi^d\psi^f\right]\right\} \ ,$$

$$S^{(2)}_\psi \ = \ \frac{1}{2\lambda^2}\int \mathrm{d}^2x \ \left\{ \tfrac{1}{2}\mathrm{i}g_{ab}\bar\psi^a\gamma^\mu \left[D_f R^b{}_{cdg}\xi^f\psi^c\xi^d\partial_\mu\phi^g + R^b{}_{cdf}\psi^c\xi^d(D_\mu\xi)^f\right.\right.$$

$$- \ 2D_f H^b_{cd}\xi^f\psi^c\varepsilon_{\mu\nu}(D^\nu\xi)^d - H^b_{cd}\psi^c\varepsilon_{\mu\nu}R^d{}_{mnp}\xi^m\xi^n\partial^\nu\phi^p$$

$$\left.\left.-D_g D_f H^b_{cd}\psi^c\varepsilon_{\mu\nu}\partial^\nu\phi^d\xi^f\xi^g\right] + \frac{1}{8}\widehat{R}_{abcd}\bar\psi^a(1+\gamma_3)\psi^c\bar\psi^b(1+\gamma_3)\psi^d\right\} . \tag{1.30}$$

The UV- and IR-regularized spinor propagator reads

$$\langle\psi^j(x)\bar\psi^k(y)\rangle = \mathrm{i}\lambda^2\delta^{jk}\int\frac{\mathrm{d}^dp}{(2\pi)^d}\frac{m-\gamma^\mu p_\mu}{p^2-m^2+\mathrm{i}0'}\exp\left[-\mathrm{i}p\cdot(x-y)\right] \ , \tag{1.31}$$

where the middle latin indices refer to the tangent space of $\mathcal{M}$.

3.2 The Supersymmetric NLSM β-Functions

In this section we present the results of our perturbative calculations in superspace and in components, as regards the RG β-functions of the 2d, $N = 1$ supersymmetric NLSM with a generalized WZNW term and the action (1.18), up to four loops [161].

3.2.1 One-Loop Results

The one-loop supersymmetric NLSM β-function is the same as the bosonic one (Subsect. 2.4.1), i.e. it is proportional to the generalized Ricci tensor (with torsion) [88]. This fact follows immediately when using the superspace approach with the NLSM superfield propagator (1.27). It is easy to verify in components that a fermionic loop with propagator (1.31) does not have any UV-divergences in 2d, which also implies that the one-loop supersymmetric NLSM β-functon coincides with the bosonic one.

3.2.2 Two-Loop Results

Like in the one-loop approximation, the two-loop calculations of the NLSM β-function can be done either in components or in superspace, while the actual calculations are much simpler in superspace, as may have been expected [82, 162]. We consider both approaches since it simultaneously provides a consistency check of our calculations.

In the component approach, in addition to the bosonic two-loop graphs depicted in Fig. 2.2, there are three additional two-loop diagrams with a fermionic loop, which are depicted in Fig. 3.1.

The analytic integral corresponding to the first ω-type graph in Fig. 3.1 reads

$$\int \frac{\mathrm{d}^d p \, \mathrm{d}^d q}{(2\pi)^{2d}} \frac{\mathrm{tr}\left\{ \gamma^\mu (\slashed{q} + m) \gamma^\nu (\slashed{p} + \slashed{q} + m) \right\}}{(p^2 - m^2)(q^2 - m^2)[(p+q)^2 - m^2]} \ . \tag{2.1}$$

We use the so-called Supersymmetric Dimensional Regularization (SDR) via dimensional reduction [163, 164, 165, 166] in our calculations. The SDR prescribes the use of the two-dimensional γ-algebra and its properties (1.2) for the supersymmetric (dimensionally regularized) NLSM Feynman graphs.

The UV counterterm for the first ω-graph in Fig. 3.1 is just the two-loop UV-divergence of the integral (2.1) with the opposite sign. We find

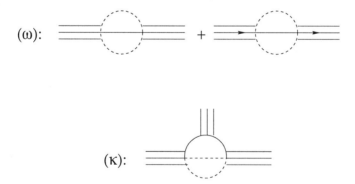

Fig. 3.1. Two-loop fermionic contributions to the supersymmetric RG β-function. The solid lines denote the NLSM scalar propagators (2.3.2), whereas the dashed lines denote the fermionic propagators (3.1.31)

$$\omega_1 = \lambda^2 \left(\frac{2-d}{4d}\right) I_1^2 \int \mathrm{d}^2x \left\{ (R_{abcd} - H^h_{ab} H_{hcd})(R^{abc}{}_f - H^{ab}_k H^{kc}_f) \right.$$

$$\times \partial^\mu \phi^d \partial_\mu \phi^f - D_a H_{bcd} D^a H^{bc}_f \partial^\mu \phi^d \partial_\mu \phi^f - 2(R_{abcd} - H^h_{ab} H_{hcd})$$

$$\left. \times D^c H^{ab}_f \partial_\mu \phi^d \varepsilon^{\mu\lambda} \partial_\lambda \phi^f \right\} .$$

$$(2.2)$$

Similarly, the UV-counterterm for the second ω-type graph in Fig. 3.1 reads

$$\omega_2 = \lambda^2 \left(\frac{2-d}{4d}\right) I_1^2 \int \mathrm{d}^2x \, \widehat{D}_\mu H_{abc} \widehat{D}^\mu H^{abc} . \qquad (2.3)$$

The κ-graph in Fig. 3.1 turns out to be UV-finite.

It is useful to rewrite the bosonic β_1-counterterm (2.4.18) in the form

$$\beta_1 = -\frac{\lambda^2}{2d} I_1^2 \int \mathrm{d}^2x \left\{ (R_{abcd} - H^h_{ab} H_{hcd})(R^{abc}{}_f - H^{ab}_k H^{kc}_f) \right.$$

$$\times \partial^\mu \phi^d \partial_\mu \phi^f - D_a H_{bcd} D^a H^{bc}_f \partial^\mu \phi^d \partial_\mu \phi^f - 2(R_{abcd} - H^h_{ab} H_{hcd})$$

$$\left. \times D^c H^{ab}_f \partial^\mu \phi^d \varepsilon_{\mu\nu} \partial^\nu \phi^f \right\} - \frac{\lambda^2}{18d} I_1^2 \int \mathrm{d}^2x \, \widehat{D}_\mu H_{abc} \widehat{D}^\mu H^{abc} .$$

$$(2.4)$$

The full bosonic NLSM counterterm is given by (2.4.24). The SDR prescription is compatible with the HVB prescription (2.3.12).

It is now easy to see that there are only two different background structures in (2.2) and (2.4) that contribute to the two-loop RG β-function, while their d-dependent coefficients in front of I_1^2 add up to

$$-\frac{1}{2d} + \frac{2-d}{4d} = -\frac{1}{4} ,$$
$$-\frac{1}{18d} + \frac{10 - 7d}{18d} + \frac{d-2}{4d} = -\frac{5}{36} ,$$

$$(2.5)$$

i.e. they have no d-dependence at all! This leads to the absence of any simple-pole $1/\epsilon$-divergences after subdivergence subtraction according to (2.4.30) in the two-loop component graphs and, hence, the vanishing two-loop contribution to the supersymmetric RG β-function [82, 162].

Let us now calculate the same two-loop contribution in superspace [82]. The relevant two-loop supergraphs are still given by Fig. 2.2, but in the superfield interpretation. In particular, the divergent supersymmetric contributions from the supergraphs $(\alpha_S, \gamma_S, \delta_S, \varepsilon_S, \zeta_S)$ can be immediately obtained from the bosonic results of Subsect. 2.4.2 after the substitution (1.20). The UV-divergence of the η_S-type supergraph also follows the same pattern in the HVB prescription. This means that the η_S-divergence merely results in the double-pole in ϵ after subtractions, so that it does not contribute to the supersymmetric RG β-function either. In addition, this also implies that the residue in the supersymmetric double-pole two-loop UV divergence is the same as that in the bosonic case. This should have been expected since the double-pole residue is controlled by the one-loop RG β-function via the generalized RG pole equations [34], while the one-loop RG β-function is the same in the bosonic and supersymmetric cases.

Compared to the bosonic case, the differences appear in the calculation of the β-type supergraphs. The superspace integral corresponding to the β_{S1} supergraph in Fig. 2.2 is given by

$$\int \frac{\mathrm{d}^d p \mathrm{d}^d q}{(2\pi)^{2d}} \mathrm{d}^2 \theta \, \frac{-\mathrm{i}(\not{p}_{\alpha\beta} + m\delta_{\alpha\beta}) + \delta^2(\theta) \left[p^2 \delta_{\alpha\beta} - m\not{p}_{\alpha\beta} - 2m^2 \delta_{\alpha\beta} \right]}{(p^2 - m^2)(q^2 - m^2)([p+q]^2 - m^2)} \, , \quad (2.6)$$

where all the exponential factors of the superpropagators (1.27b) disappeared due to momentum conservation at the vertices after using (1.16) and (1.27b). The UV-divergence of the integral (2.6) is given by

$$\beta_{S1} = -\frac{\lambda^2 I_1^2}{6\mathrm{i}} \int \mathrm{d}^4 Z \, \widehat{R}_{a(bc)d} \widehat{R}^{dbc}{}_m \bar{D} \phi^a (1 + \gamma_3) D \phi^m \, . \quad (2.7)$$

Similarly, we find

$$b_{S2} = -\frac{\lambda^2 I_1^2}{18\mathrm{i}} \int \mathrm{d}^4 Z \widehat{\bar{D}} H_{abc} \widehat{D} H^{abc} \, . \quad (2.8)$$

All the divergences proportional to mIJ cancel altogether, as in the bosonic case. The remaining two-loop divergences are thus proportional to the tadpole integral squared $(\sim I_1^2)$ *without* any d-independent pre-factors. Therefore, after the subdivergence subtractions, they do not contribute to the RG β-function, in agreement with the component results.

The vanishing two-loop contribution to the supersymmetric RG β-function of the general NLSM with torsion obviously implies the vanishing two-loop contribution to the RG β-function of the supersymmetric WZNW model [167] too (see also Sect. 5.2).

3.2.3 Three-Loop Results

All the topologically inequivalent three-loop supergraphs are classified as in the bosonic case, see Fig. 2.4. The covariant background-field dependence originates from the vertices and the possible (one or two) two-point insertions in any of the propagators. We take into account only UV-divergent supergraphs by using the generalized renormalizability of the supersymmetric NLSM. The classification of the UV-divergent supergraphs according to the structure of their vertices follows the bosonic case (Subsect. 2.4.3) too — see (1.21).

The γ_3-dependence inside supergraphs can always be eliminated by using the properties $\gamma_3^2 = 1$ and $\bar{D}\gamma_3 D = 0$. A divergent supergraph should have a D^2-factor for each loop, otherwise it vanishes. The rest of the superspace covariant derivatives appearing on the internal lines of a supergraph can be eliminated by using their *D-algebra* (1.14), which results in the usual momentum integrals.

Since only the simple-pole divergences in $1/\epsilon$ are relevant for computing the RG β-function, we systematically ignore all the higher-order divergences, like in the bosonic case. This means, in particular, that each supergraph with topological structure $(\alpha, \beta, \gamma, \delta)$ in Fig. 2.4 cannot contribute to the RG β-function in the supersymmetric case. Moreover, whenever such a supergraph appears in the process of doing the D-algebra, it can be dropped too. This is valid because no d-dependent pre-factors appear when using the SDR prescription, so that after the subdivergence subtraction we are left with the higher-order UV divergences only. The most striking difference from the bosonic case thus appears to be in the role of the β-type graphs in Fig. 2.4, which are crucial for the bosonic NLSM three-loop β-function whereas they don't contribute to the supersymmetric NLSM β-function. In fact, it is true for all the β-type graphs in any loop order that they do not contribute to the RG β-function in the supersymmetric case.

The two-point insertions do not require an independent treatment since the D^2-factors on their internal lines always cancel one of the propagators after integrating by parts and using the identity $(D^2)^2 = -\Box$.

Hence, we have to consider in detail all the three-loop supergraphs of the types $(\varepsilon, \eta, \rho)$ only, perform the D-algebra, and check the actual appearance, if any, of the simple-pole divergences after subdivergence subtraction in the resulting momentum integrals. In fact, there are only two 'dangerous' momentum structures at the three loop-level, which can contribute to the supersymmetric β-function (Fig. 3.2), so that the real question is whether they do appear in actual calculations or not.

It is straightforward to derive the following conclusions from a detailed graph-by-graph investigation [161]:

- the D-algebra and integration by parts in any ρ-type supergraph either create a tadpole or cancel one of the superpropagators, thus reducing the η-type supergraph to one of type ε (Fig. 2.4);
- the D-algebra and integration by parts in any η-type supergraph also create a tadpole or cancel one of the superpropagators between its one-loop subdiagrams, so that it is always reducible to ε-type diagrams;
- all the ε-type supergraphs can be similarly reduced to β-type graphs and the tadpole contributions.

(i)

(ii)

Fig. 3.2. The potentially 'dangerous' three-loop supergraphs (after performing the D-algebra) that have simple-pole UV divergences

Of course, all these statements apply to the divergent parts of the Feynman diagrams (we do not consider here their finite parts at all). As a result, we find that the UV divergences of all the three-loop supergraphs can actually be reduced to tadpole contributions only. After the trivial subdivergence subtraction in the three-loop tadpole according to (2.1.30) with $p = 3$, we thus get the $1/\epsilon^3$-divergences only, and, hence, the vanishing three-loop contribution to the supersymmetri NLSM β-functions. This result is also consistent with the already established fact about the vanishing two-loop contribution to the supersymmetric RG β-functions and the generalized RG pole equations. Therefore, the three-loop supersymmetric β-functions are still the same as the one-loop supersymmetric ones [161, 168], even in the presence of a generalized WZNW term. The absence of three-loop renormalization in the purely metric supersymmetric NLSM was established in [169]. The vanishing three-loop result in the general NLSM with torsion essentially appears due to the presence of three-point vertices in the tadpole-irreducible three-loop graphs of the type $(\varepsilon, \eta, \rho)$, and it cannot be naively generalized to higher orders of quantum perturbation theory in superspace.

As an immediate corollary of the general result above, it follows that the general $N = (1, 1)$ supersymmetric NLSM with a generalized WZNW term on the generalized Ricci-flat manifold $\mathcal{M}$ with $\widehat{R}_{ab} = 0$ has vanishing three-loop RG β-functions. In particular, this applies to the supersymmetric WZNW model (Sect. 5.2) too.

Yet another consistency check comes from superstring theory [170, 171, 172], because of the anticipated equivalence between the supersymmetric NLSM RG β-functions and the equations of motion of the massless superstring modes. Both are supposed to be obtained as the variational equations from some (Zamolodchikov) effective action (Chap. 6). The superstring effective action can be independently calculated from the superstring tree scattering amplitudes. In particular, the type-II and heterotic superstring tree scattering amplitudes do not have terms of sixth-order in momenta, which implies the absence of curvature-cubed terms in the corresponding supersting low-energy effective action [96]. This is clearly consistent with the vanishing three-loop RG β-functions in 2d supersymmetric NLSM.

3.2.4 Four-Loop Results

All the topological structures of four-loop supergraphs are schematically depicted in Figs. 3.3 and 3.4. It is only a small part of them that actually appears in the purely metric NLSM without torsion, namely, $(\alpha, \beta, \gamma, \delta, \mu, \nu)$ in Fig. 3.3 and (α, β, γ) in Fig. 3.4. The four-loop RG β-function of the purely metric 2d supersymmetric NLSM was calculated in [71, 72]. The results of [161] for the 2d, $N = 1$ NLSM with a non-vanishing WZNW term are given below.

For dimensional reasons, we can exclude contributions with more than two spinor derivatives on the external lines of a supergraph, as well as any

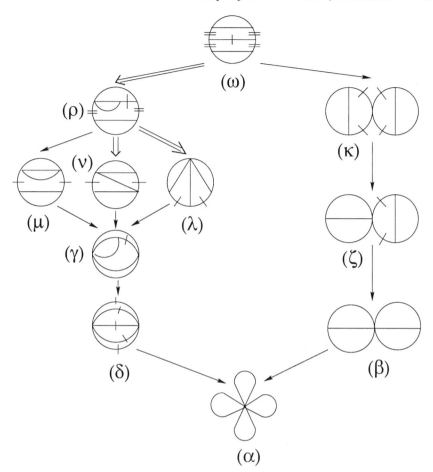

Fig. 3.3. The reduction of the four-loop UV divergences in the NLSM supergraphs that do not contribute to the RG β-functions

such terms that appear as the result of integration by parts. As in the previous subsections, we can also ignore all supergraphs with tadpoles since they cannot contribute to the RG β-function, as well as the supergraphs whose UV divergences can be reduced to tadpoles (Fig. 3.3).

Let us now classify all the four-loop supergraphs (see (1.21) for our notation) according to the degree of their superficial UV-divergence or, equivalently, according to the power of the $(D\phi)$-factor on their external lines *before* doing D-algebra and integration by parts, namely,

- with one vertex of type $\Pi^{(n,2)}$ or $E^{(n,2)}$ and all the other vertices of type $\Pi^{(k,0)}$ and/or $E^{(m,0)}$, or with two vertices of type $\Pi^{(n,1)}$ and/or $E^{(n,1)}$

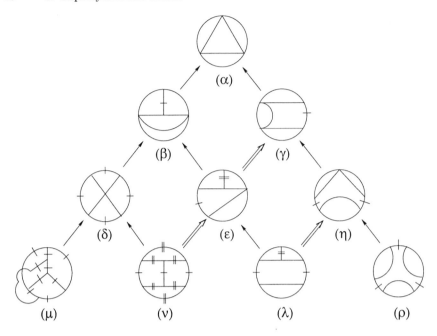

Fig. 3.4. The reduction of the four-loop UV divergences in the NLSM supergraphs that do contribute to the RG β-functions

and all the other vertices of type $\Pi^{(k,0)}$ and/or $E^{(m,0)}$, as the $(D\phi)^2$-type (logarithmically divergent) graphs;

• with one vertex of type $\Pi^{(n,1)}$ or $E^{(n,1)}$ and all the other vertices of type $\Pi^{(k,0)}$ and/or $E^{(m,0)}$, as the $(D\phi)^1$-type (linearly divergent) graphs;
• with all vertices of type $\Pi^{(k,0)}$ and/or $E^{(m,0)}$, as the $(D\phi)^0$-type (quadratically divergent) graphs.

Of course, this classification only applies before integration by parts, and it has nothing to do with the standard Feynman diagram classification according to the index of divergence (the latter is the same for all NLSM graphs).

As regards the purely metric supersymmetric NLSM, it was shown in [71, 72] that the UV divergences of all the four-loop graphs contributing to the RG β-functions, can be reduced to a single divergent integral,

$$\int \frac{\mathrm{d}^d k \, \mathrm{d}^d q \, \mathrm{d}^d r \, \mathrm{d}^d t}{(2\pi)^{4d}} \frac{k \cdot (t-k) q \cdot (t-q)}{k^2 (t-k)^2 q^2 (t-q)^2 (r^2+m^2)[(t-r)^2+m^2]} \ , \qquad (2.9)$$

whose momentum dependence on the internal lines is schematically pictured in Fig. 3.5.

After subdivergence subtraction the integral (2.9) takes the form

$$A_4 = \frac{4}{(4\pi)^4} \left[\frac{\zeta(3)}{\epsilon} - \frac{1}{6\epsilon^4} \right] \ . \qquad (2.10)$$

Fig. 3.5. The only irreducible four-loop supergraph topology that contributes to the β-functions of the supersymmetric NLSM with torsion

This result can be generalized to the NLSM with non-vanishing torsion (Fig. 3.4). On the one hand, after the D-algebra and integration by parts, the UV divergences of all the supergraphs depicted in Fig. 3.3 can be reduced to that of a multiple tadpole (α-type in Fig. 3.3). Therefore, all the supergraphs of Fig. 3.3. do not contribute to the RG β-functions. On the other hand, the UV divergences of all the supergraphs depicted in Fig. 3.4 can be reduced to the α-type in Fig. 3.4. The latter is just given by (2.9), or (2.10) after subdivergence subtraction.

The proof goes as follows [161]. Each graph in Figs. 3.3 and 3.4 represents, in fact, the whole bunch of supergraphs after specifying their vertices and the covariant derivatives on their external lines. The full set of four-loop supergraphs is generated by the covariant background-field expansion of the NLSM action and a quantum perturbation theory along standard lines (Wick theorem), which result in a few hundred independent UV-divergent diagrams. Then the D-algebra and integration by parts are applied to each four-loop supergraph separately, in order to reduce its UV-divergence to a simpler type (denoted by arrows in Figs. 3.3 and 3.4). The actual reduction procedure turns out to be rather individual for each supergraph, so that the very existence of the complete reduction to only two independent divergent structures is rather non-trivial (cf. the analysis of [173] in an arbitrary number of loops). One always needs a D^2-factor for each loop, in order to get a non-vanishing result after integration over the anticommuting superspace coordinates. The result always takes the form of a single full integration over the anticommuting superspace coordinates (this statement is known as the *non-renormalization theorem* in supersymmetry). To reduce the number of spinor derivatives on the internal lines of a supergraph to two at most, one uses the identities

$$D_\alpha^{Z_1} G(Z_1, Z_2) = -i\partial_{\alpha\beta}^{x_1} \bar{D}_{Z_1}^{\beta} K(Z_1, Z_2) ,$$

$$D_\alpha^{Z_1} D_\beta^{Z_2} G(Z_1, Z_2) = \left[-C_{\alpha\beta}\Box + +i\partial_{\alpha\beta}^{x_1} D^2 \right] K(Z_1, Z_2) ,$$

$$D_\alpha^{Z_1} D_\beta^{Z_2} K(Z_1, Z_2) = \left[i\partial_{\alpha\beta}^{x_1} + C_{\beta\alpha} D^2 \right] K(Z_1, Z_2) ,$$

$$D^2 G(Z_1, Z_2) \equiv \Box K(Z_1, Z_2) = \frac{1}{i\lambda^2} \delta^4(Z_1, Z_2) ,$$

(2.11)

where

$$G = D^2 K \equiv \frac{1}{i\lambda^2} \Box^{-1} \delta^4(Z_1, Z_2) .$$

(2.12)

In the course of the D-algebra, the total number of spinor derivatives in a given supergraph can always be reduced to eight. After integrating over the θ-coordinates, we get the $(D\phi)^2$-dependence on the external lines, as it should. Next, we integrate the remaining spacetime derivatives on the internal lines by parts, in order to form a $\Box$-operator which should cancel one of the propagators $\Box^{-1}$, thus reducing the corresponding internal line to a point (in x-space). This results in either a simpler supergraph type or an existing tadpole. Figures 3.3 and 3.4 give our results for the topologically different types of the four-loop supergraphs that actually represent hundreds of supergraphs in condensed form. In particular, the single arrows in both figures show what happens with a given graph after cancelling the propagator denoted by a single dash, the double arrows mean the same for the propagators denoted by two dashes, etc.

The ultimate four-loop UV-divergence is thus given by the A_4 integral of (2.10) in any case. Of course, each supergraph has its own background dependence that should be carefully calculated (see below). According to Fig. 3.3, all the supergraphs having vertices of more than fourth order in the quantum fields are irrelevant for computing the RG β-functions. Hence, the background-field expansion terms calculated in Sect. 3.1 are enough for an actual four-loop calculation. More details can be found in [161].

As an obvious consequence of our analysis, we find that the whole four-loop contribution to the RG β-functions of the general supersymmetric NLSM with non-vanishing torsion is proportional to the irrational number $\zeta(3)$, similarly to that in the purely metric case [71, 72]. This is consistent with the superstring low-energy effective action [174, 175], and the known four-loop RG β-functions of some $N = (2,2)$ supersymmetric NLSM with torsion [176] as well.

Calculation of the background dependence of several hundred four-loop counterterms in the general supersymmetric NLSM with torsion is very tedious. Even if one restricts oneself to the graphs contributing to the RG β-functions (Fig. 3.4), only writing down the answer takes many pages [161]. Therefore, in this subsection, only those counterterms will be explicitly written down that do not contain the covariant derivatives of the curvature and torsion (Fig. 3.6). This means that the explicit results given below are only valid for NLSM manifolds $\mathcal{M}$ with covariantly constant curvature and torsion. In particular, they are enough to calculate the four-loop RG β-function of the supersymmetric WZNW model (see [161] for the general supersymmetric NLSM four-loop RG β-function). The relevant UV counterterms are all of $(D\phi)^2$-type, which considerably simplifies the calculation. Their contributions to the RG β-function according to the notation of Fig. 3.6 are (the common factor $\lambda^6 \zeta(3)/2^4(4\pi)^4$ is implicit):

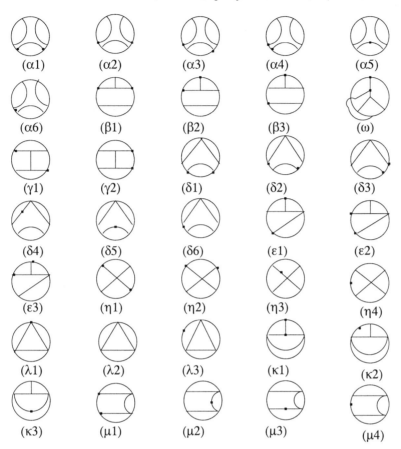

Fig. 3.6. The four-loop supergraphs contributing to the four-loop RG β-functions of the supersymmetric WZNW model. The dots stand for the $\Pi^{(3,1)}$, $\Pi^{(2,2)}$ and $\Pi^{(4,2)}$ vertices — see (1.21)

$$\alpha1 = -\widehat{R}_i{}^{krs}\widehat{R}_{jskp}(H^2)^p_t(H^2)^t_r ,$$

$$\alpha2 = -\tfrac{1}{2}\widehat{R}_i{}^{krs}\widehat{R}_j{}^{qph}H_{spt}H^t_{kh}(H^2)_{qr} ,$$

$$\alpha3 = -\widehat{R}_i{}^{krs}\widehat{R}_j{}^{qph}H_{skm}(H^2)^m_q H_{phr} ,$$

$$\alpha4 = -\widehat{R}_{is}{}^{rk}\widehat{R}_j{}^{spq}H_{pqh}(H^2)^{ht}H_{trk} ,$$

$$\beta1 = \tfrac{1}{3}\left[\widehat{R}_i{}^{krs}\widehat{R}_j{}^{pqh} - 2\widehat{R}_i{}^{krs}\widehat{R}_j{}^{qph}\right]H^t_{sp}H_{tkh}(H^2)_{rq} ,$$

$$\beta2 = 2\widehat{R}_i{}^{ksr}\widehat{R}_{js}{}^{pq}H^h_{kq}H^t_{hp}(H^2)_{tr} ,$$

$$\beta3 = -2\widehat{R}_i{}^{rks}\widehat{R}_j{}^{pqm}H^n_{ps}H^t_{nk}H^h_{tr}H_{hqm} ,$$

$$\omega = -2\widehat{R}_{iksr}\widehat{R}_{jqp}{}^{s}(H^4)^{kqrp} \ ,$$

$$\gamma 1 = -(2/3^2)\Big[(19\widehat{R}_i{}^{krs}+4\widehat{R}_i{}^{rks})\widehat{R}_j{}^{qpn}$$

$$+(16\widehat{R}_i{}^{krs}-17\widehat{R}_i{}^{rks})\widehat{R}_j{}^{pqn}\Big]H_{sn}^{m}H_{mk}^{h}H_{hq}^{t}H_{trp} \ ,$$

$$\gamma 2 = -(2/3^2)\Big[-10\widehat{R}_{isrk}-32\widehat{R}_{iksr}\Big]\widehat{R}_{jqp}{}^{k}(H^4)^{rspq} \ ,$$

$$\delta 1 = -(1/2^2)\widehat{R}_i{}^{krs}\widehat{R}_{jsk}{}^{p}\widehat{R}^{mnqh}H_{mnp}H_{qhr} \ ,$$

$$\delta 2 = -(1/2^3)\widehat{R}_i{}^{rks}\widehat{R}_j{}^{qph}R_{sk}{}^{nt}H_{ntq}H_{phr} \ ,$$

$$\delta 3 = (1/2^2)\widehat{R}_i{}^{krs}\widehat{R}_{jkpq}R^{pqmn}H_{mn}^{t}H_{trs} \ ,$$

$$\varepsilon 1 = \tfrac{2}{3}\Big[3\widehat{R}_i{}^{krs}\widehat{R}_j{}^{nmq}R_{spqk}+2\widehat{R}_i{}^{(kr)s}\widehat{R}_j{}^{mnq}R_{skpq}\Big]H_{m}^{ph}H_{nrh} \ ,$$

$$\varepsilon 2 = -2\widehat{R}_i{}^{(sr)k}\widehat{R}_{jpqk}R^{mpqn}H_{smh}h_{rn}^{H} \ ,$$

$$\varepsilon 3 = \tfrac{1}{3}\Big[\widehat{R}_i{}^{(kr)s}\widehat{R}_{js}{}^{pq}-\widehat{R}_i{}^{(sk)r}\widehat{R}_j{}^{qp}{}_{s}\Big]R^{mt}{}_{pn}H_{qk}^{n}H_{mtr} \ ,$$

$$\eta 1 = (1/3^3\cdot 2^2)\Big[(4\widehat{R}_i{}^{krs}-35\widehat{R}_i{}^{rks})\widehat{R}_j{}^{qmn}$$

$$+(-125\widehat{R}_i{}^{krs}+88\widehat{R}_i{}^{rks})\widehat{R}_j{}^{mqn}\Big]R_{kphq}H_{sn}^{h}H_{rm}^{p} \ ,$$

$$\eta 2 = \tfrac{2}{3}\Big[3(\widehat{R}_{irk}{}^{s}\widehat{R}_{jpsq}-\widehat{R}_{ikr}{}^{s}\widehat{R}_{jqsp})R^{knmq}-(5\widehat{R}_{irk}{}^{s}\widehat{R}_{jspq}$$

$$+5\widehat{R}_{ikr}{}^{s}\widehat{R}_{jspq}+3\widehat{R}_{ikr}{}^{s}\widehat{R}_{jqsp})R^{kqmn}\Big]H_{mh}^{r}H_{pn}^{h} \ ,$$

$$\lambda 1+\lambda 3+\kappa 1+\mu 1 = -2^2\widehat{R}_{ikhm}\widehat{R}_{jr}{}^{tm}\left(R^{k}{}_{qs}{}^{r}R_{t}{}^{qsh}+R^{k}{}_{qst}R^{hrsq}\right) \ ,$$

$$\alpha 5+\alpha 6+\delta 4+\eta 3 = -2^4\widehat{R}_i{}^{(kr)}{}_{j}\widehat{R}_k{}^{phq}(H^4)_{rpqh} \ ,$$

$$\delta 5+\eta 4 = -2^3\widehat{R}_i{}^{(kr)}{}_{j}R^{m(pn)q}H_{kq}^{t}H_{tn}^{h}H_{hm}^{s}H_{spr} \ ,$$

$$\delta 6 = -\widehat{R}_i{}^{(kr)}{}_{j}R^{mpqn}H_{pmk}(H^2)_{rs}H_{nq}^{s} \ ,$$

$$\kappa 3+\mu 2 = \tfrac{1}{3}\widehat{R}_i{}^{(kr)}{}_{j}R_k{}^{(pn)m}R_{r(ps)}{}^{h}H_{htm}H_{n}^{st} \ ,$$

$$\lambda 2 = (2^3/3)\Big[\widehat{R}_{ikhm}\widehat{R}^{m(tr)}{}_{j}R_{qtrs}R^{ghks}-\widehat{R}_{ikrm}\widehat{R}^{m(ht)}{}_{j}R^{k}{}_{qsh}R_{t}{}^{qsr}$$

$$-\widehat{R}_{ir}{}^{rm}\widehat{R}_m{}^{(ht)}{}_{j}R_{kqsh}R^{qrs}{}_{t}\Big] \ ,$$

$$\kappa 2 = (2^2/3)\widehat{R}_i{}^{(kr)}{}_{j}R_{q(sk)h}R^{q(mp)h}H_{rpt}H_{m}^{ts} \ ,$$

$$\mu 3 = \tfrac{1}{2}\widehat{R}_i{}^{(kr)}{}_{j}R_{pr}{}^{hq}R_{hq}{}^{mn}H_{mnt}H_{r}^{tp} \ , \tag{2.13}$$

$$\mu 4 = (1/2^3)\widehat{R}_i{}^{(kr)}{}_{j}R^{pmst}R_{st}{}^{hn}H_{pmk}H_{nhr} \ ,$$

where we have used the following restrictions on $\mathcal{M}$:

$$\widehat{\mathcal{D}}_k \widehat{R}_{mnpq} = \widehat{\mathcal{D}}_k H_{mnp} = H_{sm[n} H^s_{pq]} = 0 , \tag{2.14}$$

valid for any group manifold, and the notation

$$
\begin{aligned}
H^4_{kqtr} &= H^m_{kn} H^n_{qh} H^h_{tf} H^f_{rm} , \\
H^2_{kq} &= H^{mn}_k H_{qmn} , \\
H^2 &= H_{ijk} H^{ijk} .
\end{aligned}
\tag{2.15}
$$

In the purely metric case ($H_{ijk} = 0$), our results coincide with those of [71, 72]. An inspection of (2.13) shows that, even with non-vanishing torsion, all the four-loop countertems have at least one factor $\widehat{R}_{ijkl}$. It is enough to claim vanishing of the four-loop supersymmetric β-functions for the parallelized NLSM manifolds $\mathcal{M}$ with $\widehat{R}_{ijkl} = 0$. It also easily follows from (2.13) that the antisymmetric part of the four-loop β-functions vanishes under the conditions (2.14), e.g., on group manifolds.

As regards generic NLSM manifolds with torsion, it can be shown [161] that the sum of the $\widehat{R}H^6$- and $\widehat{R}RH^4$-terms (graphs $\delta 4, \delta 5, \delta 6$ and $\eta 3$ in Fig. 3.6, respectively) surprisingly combines into the $\widehat{R}^2 H^4$-terms. In addition, all the antisymmetric contributions to the four-loop β-function turn out to be proportional to $\widehat{R}^2$ at least: the $\widehat{R}RH^4$- and $\widehat{R}R^2 H^2$-terms together are unified into the $\widehat{R}^2 RH^2$-terms. The dependence of the general supersymmetric NLSM four-loop RG β-function upon the curvatures $\widehat{R}$ and R, and the torsion H, is symbolically given by [161]

$$
\begin{aligned}
\beta^{(g)}_{ij} \sim\ & \widehat{R}_i \widehat{R}_j (H^4) + \widehat{R}_i \widehat{R}_j R(H^2) + \widehat{R}_i R_j \widehat{R}(H^2) + \widehat{R}_{i\cdots j} \widehat{R}(H^4) \\
& + \widehat{R}_{i\cdots j} R(H^4) + \widehat{R}_{i\cdots j}(R^2)(H^2) + \widehat{R}_i \widehat{R} H_j (H^3) + (\widehat{R}^2) H_i H_j (H^2) \\
& + \widehat{R}_i \widehat{R} R H_j H + (\widehat{R}^2) R H_i H_j + \widehat{R}_i \widehat{R}_j (R^2) + \cdots ,
\end{aligned}
\tag{2.16a}
$$

$$
\begin{aligned}
\beta^{(h)}_{ij} \sim\ & \widehat{R}_i \widehat{R}_j (H^4) + \widehat{R}_i \widehat{R} H_j (H^3) + (\widehat{R}^2) H_i H_j (H^2) + \widehat{R}_i \widehat{R}_j R(H^2) \\
& + \widehat{R}_i \widehat{R} R H_j H + (\widehat{R}^2) R H_i H_j + \widehat{R}_i R_j \widehat{R}(H^2) + \widehat{R}_i \widehat{R}_j (R^2) + \cdots ,
\end{aligned}
\tag{2.16b}
$$

where all the contracted indices are suppressed.

Using (2.1.22), (2.1.23), (2.4.43) and (2.14), we can simplify (2.13) by rewriting it in the tangent space of the supersymmetric NLSM group manifold as

$$
\begin{aligned}
\alpha 1 &= \tfrac{1}{2}(1 - \eta^2)^2 \eta^4 Q^4 \delta_{ij} , & \alpha 2 &= \tfrac{1}{4}(\alpha 1) , \\
\alpha 3 &= \alpha 4 = -2(\alpha 1) , & \beta 1 &= \beta 2 = \beta 3 = \alpha 1 , \\
\omega 1 &= 4(\alpha 1) , & \gamma 1 &= \tfrac{412}{9}(\alpha 1) , \quad \gamma 2 = \tfrac{56}{9}(\alpha 1) ,
\end{aligned}
$$

$$\alpha 5 + \alpha 6 + \delta 4 + \eta 3 \;=\; 16(\alpha 1) \;,$$

$$\delta 1 \;=\; \tfrac{1}{8}(1-\eta^2)^2\eta^2 Q^4 \delta_{ij} \;,$$

$$\delta 3 \;=\; -(\delta 1) \;, \quad \delta 4 = 2(\delta 1) \;, \quad \varepsilon 1 = 18(\delta 1) \;, \quad \varepsilon 2 = 34(\delta 1) \;,$$

$$\varepsilon 3 \;=\; -\tfrac{4}{3}(\delta 1) \;, \quad \eta 1 = -\tfrac{19}{6}(\delta 1) \;, \quad \eta 2 = \tfrac{1}{2}(\delta 1) \;,$$

$$\mu 1 \;=\; (1-\eta^2)\eta^2 Q^4 \delta_{ij} \;,$$

$$\mu 1 + \kappa 3 \;=\; \kappa 2 = -\tfrac{2}{3}(\mu 3) \;, \quad \mu 4 = \tfrac{1}{4}(\mu 3) \;,$$

$$\delta 6 \;=\; 2(1-\eta^2)\eta^4 Q^4 \delta_{ij} \;, \quad \delta 5 + \eta 4 = 4(\delta 6) \;,$$

$$\lambda 2 \;=\; \tfrac{25}{4}(1-\eta^2)^2 Q^4 \delta_{ij} \;, \quad \lambda 1 + \lambda 3 + \kappa 1 + \mu 1 = -\tfrac{16}{25}(\lambda 2) \;.$$

(2.17)

The sum of these contributions determines the four-loop RG β-function (2.4.45) of the supersymmetric WZNW model [161, 177],

$$b_\lambda = -\frac{\lambda^4 Q}{2\pi}(1-\eta^2) - \frac{\lambda^{10}Q^4\zeta(3)}{2^7(4\pi)^4}(1-\eta^2)\left[q_0 + q_2\eta^2 + q_4\eta^4 + q_6\eta^6\right] \;, \quad (2.18)$$

where

$$q_0 = 18, \quad q_2 = 94/3 \;, \quad q_4 = 319 \;, \quad q_6 = -289 \;. \qquad (2.19)$$

The four-loop contribution to the Zamolodchikov action for the purely metric supersymmetric NLSM ($\lambda^2 = 1$) reads [175]

$$I_{\text{Zam.}} = \int \mathrm{d}^D\phi \sqrt{g}\left[-R + \frac{\zeta(3)}{(4\pi)^3}\left(R_{hmnk}R_p{}^{mn}{}_q R^{hrsp}R^q{}_{rs}{}^k\right.\right.$$

$$\left.\left.+\tfrac{1}{2}R_{hkmn}R_{pq}{}^{mn}R^{hrsp}R^q{}_{rs}{}^k\right)\right] \;. \qquad (2.20)$$

Equation (2.20) agrees with the results of superstring theory [174, 172].

A generalization of the action (2.20) to the NLSM with torsion is not known, to our knowledge. Despite having the proper supersymmetric NLSM four-loop RG β-functions whose vanishing is to be interpreted as the equations of motion of Zamolodchikov's action [161], we are unable to calculate this action explicitly, because of enormous technical problems. Some general remarks are, however, in order. First, assuming the equivalence with the low-energy superstring effective action, we may use the known fact [178, 179] that all the DH-dependent terms should enter the action in the minimal form, i.e. via the combination

$$R_{abcd} \to \tilde{R}_{abcd} \equiv R_{abcd} + D_c H_{abd} - D_d H_{abc} \;. \qquad (2.21)$$

Second, it is known [175] that the explicit covariant derivatives of the curvature can be avoided in the low-energy superstring effective action. Hence, after assuming that all the terms with the derivatives can be absorbed into the generalized curvature, the general structure of the allowed terms in the

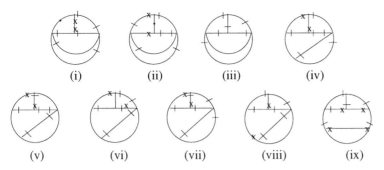

Fig. 3.7. The ill-defined four-loop supergraphs in the supersymmetric NLSM with non-vanishing torsion

Zamolodchikov action at the given order of quantum NLSM perturbation theory should be of the form

$$I_{\text{Zam.}} = \int d^D\phi \sqrt{g} \left\{ -R + \frac{\zeta(3)}{(4\pi)^3} \left[(\widehat{R})^4 + (\widehat{R})^3 (H)^2 \right] \right\} , \qquad (2.22)$$

where the $\widehat{R}^4$-terms are supposed to be obtained from (2.20) after the substitution $R_{abcd} \to \widehat{R}_{abcd}$. There exist nine independent terms having the structure $(\widehat{R})^3(H)^2$, while the $(\widehat{R})^2(H)^4$-type terms should be absent in the action (2.22) since there are no $\widehat{R}(H)^6$-type terms in the NLSM four-loop RG β-functions.

At the end of this subsection, we discuss the quantum ambiguities associated with the four-loop quantum calculations in superspace, and the related inconsistency of SDR [161]. Some of the four-loop contributions to the supersymmetric NLSM β-functions turn out to be dependent upon the order of calculations, i.e. upon the renormalization prescription used. We find two different types of quantum ambiguities in SDR and our renormalization scheme. The corresponding ill-defined four-loop supergraphs are depicted in Fig. 3.7.

The SDR-based non-minimal renormalization prescription used above for the calculation of UV divergences of multi-loop Feynman supergraphs has the following steps in the given order: (i) the D-algebra and θ-integration are performed in two dimensions, (ii) the resulting momentum integrals are analytically continued to d dimensions, (iii) the (minimal) subdivergence procedure only applies to the basic divergent integrals that are irreducible with respect to integration by parts of the partial derivatives (momenta) on the internal lines of a given supergraph. In accordance with the previous subsections, this procedure is equivalent to the standard renormalization scheme of minimal subtractions within the SDR up to the three-loop level.

At the four-loop level some new features arise. The first type of quantum ambiguity is related to the order of subdivergence subtractions that can be done in the momentum integrals either before or after integration by parts.

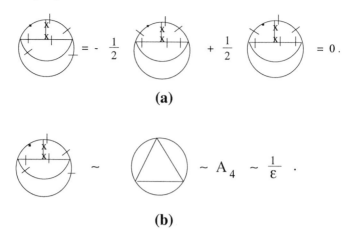

(a)

(b)

Fig. 3.8. The quantum ambiguity that can be removed by field redefinitions of the NLSM metric and torsion potential

The results for the NLSM β-functions may differ, while the corresponding ambiguities in the Zamolodchikov action are precisely those that are associated with redefinitions of the metric and torsion potential in string theory [85, 180, 181, 182, 183].

As an example, let us consider the supergraph $\kappa 2$ in Fig. 3.6, with the spinor derivatives on its internal lines as indicated in Fig. 3.7(i). After integration by parts, as shown in Fig. 3.8(a), we find no contribution to the RG β-functions of NLSM. Moreover, it is easy to check that any supergraph on the right-hand-side of Fig. 3.8(a) does not contain $1/\epsilon$-divergences. However, after the D-algebra and integration by parts with respect to the ∂-derivatives, one gets a non-vanishing contribution proportional to A_4, as shown in Fig. 3.8(b). This clearly demonstrates the dependence of the NLSM β-functions upon the renormalization prescription. The calculation according to the pattern (a) in Fig. 3.8 corresponds to the standard (*minimal*) renormalization prescription since we used integration by parts for the spinor derivatives only. The calculation according to the pattern (b) in Fig. 3.8 corresponds to our non-minimal renormalization prescription where integration by parts of the spacetime derivatives is also allowed.

The same conclusions equally apply to the supergraph (ii) in Fig. 3.7. Its background dependence has the structure $\widehat{R}_{iabj}\Omega^{ab}(R, H)$, while it can be removed from the β-functions by a field redefinition of the NLSM metric and torsion potential.

Quantum ambiguities of another type appear at the four-loop level in the supersymmetric NLSM with a non-vanishing torsion due to the presence of the γ_3-matrix that only exists in two dimensions (see the supergraphs (iii)–(ix) in Fig. 3.8). All these ambiguities disappear after the formal replacement $(\gamma_3)_\alpha{}^\beta \to \delta_\alpha{}^\beta$, while they also lead to a four-loop inconsistency of the SDR

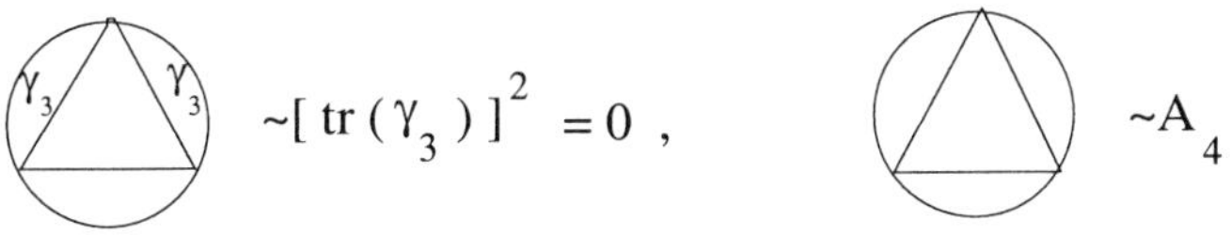

$$\sim [\, \mathrm{tr}\,(\gamma_3)\,]^2 = 0\ ,\qquad\qquad \sim A_4$$

Fig. 3.9. Quantum ambiguity due to γ_3 matrices

prescription [161]. Indeed, in the presence of γ_3, one can *either* perform the D-algebra without using any contraction of γ_3 matrices with each other, i.e. just by using the identities for the spinor derivatives in order to form $\mathrm{tr}[(\gamma_3)^2] = 2$ at the end of calculation, *or* one can integrate by parts the spinor derivatives in order to get $[\mathrm{tr}(\gamma_3)]^2 = 0$ first. The simple-pole UV divergence generically turn out to be different in both cases (see Fig. 3.9 for an example), which implies ambiguous RG β-function contributions.

For instance, let us consider the four-loop supergraph whose derivative structure is given by the first pictures in Figs. 3.10 and 3.11. On the one hand (Fig. 3.10), after integrating by parts the spinor derivatives and commuting the γ_3 matrix (on the left) to the γ_3 matrix (on the right), one arrives at the same graph but without both γ_3 matrices. The latter generically contributes to the supersymmetric NLSM β-functions according to Fig. 3.4.

On the other hand, when doing calculations with the same supergraph but in the order given by Fig. 3.11, one arrives at the vanishing factor $\mathrm{tr}(\gamma_3) = 0$ and, hence, no contribution to the RG β-functions. Note that the transformations in Fig. 3.10 used integrations by parts with respect to the spinor derivatives D_α and the γ-algebra in 2d, but no integration by parts with respect to the spacetime (vector) derivatives, which corresponds to the minimal subtraction procedure. Similar remarks apply to the other supergraphs in Fig. 3.7. We verified that the formal removal of all γ_3 matrices yields well-defined four-loop supergraphs. When considering the background dependence of the supergraphs depictured in Fig. 3.7, we find that their contributions to the RG β-functions can be altered to zero by the use of the superstring theory ambiguity given by the field redefinition freedom of the NLSM metric and torsion potential [161]. Though the metric and torsion field redefinitions are quite natural in D-dimensional string or superstring theory that is merely

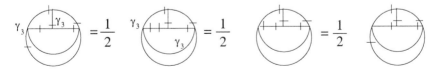

Fig. 3.10. The contraction of γ_3 matrices in a supergraph, which results in the non-vanishing trace $\mathrm{tr}[(\gamma_3)^2] = 2$

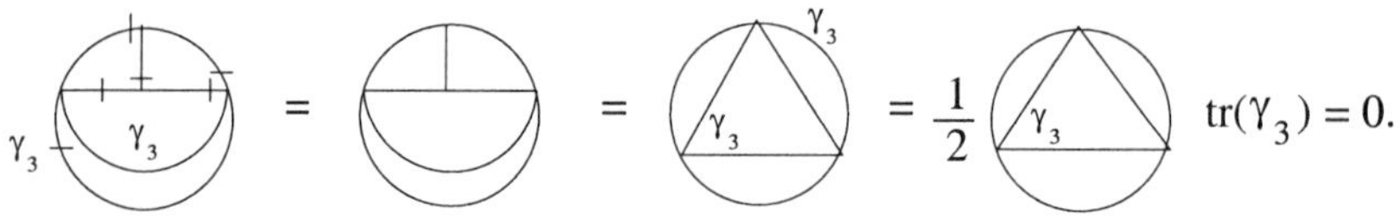

Fig. 3.11. The contraction of γ_3 matrices in the same supergraph (Fig. 3.10), which results in the vanishing trace $\text{tr}(\gamma_3) = 0$

defined on-shell via its S-matrix, there is apparently no such freedom in the 2d NLSM field theory.

We interpret the perturbative NLSM quantum inconsistencies discussed above as particular manifestations of a single fundamental inconsistency of the SDR in two-dimensional supersymmetric field theories. Similar remarks also apply to four-dimensional supersymmetric gauge field theories [184, 185, 186]. Strings and superstrings apparently avoid this problem, and we believe that this fact is deeply related to the intrinsically *on-shell* nature of any perturbative string theory.

3.3 NLSM on Homogeneous Spaces

The important class of 2d (bosonic or supersymmetric) NLSM is given by the NLSM whose target space is homogeneous, i.e. given by a *coset G/H* (see Sect. 5.3 or [45] for the related group theory and 2d CFT). The NLSM with homogeneous target space G/H describes the G symmetry spontaneously broken to the H symmetry [187]. The NLSM fields transform linearly with respect to H, but non-linearly with respect to G/H, while they can be considered as Goldstone fields [188].

The (super)conformally invariant 2d NLSM are relevant for (super)string compactification [170, 171, 172]. Superstrings are naturally defined in *ten* spacetime dimensions, so that six dimensions may have to be compactified on a (compact) manifold K, if one wants to relate the superstrings with particle phenomenology. The effective action of a test superstring in the compactified spacetime background takes the form of a NLSM over K (Sect. 6.1), which should be conformally invariant. In particular, the NLSM β-functions should vanish. A group manifold K is allowed since the WZNW model is conformally invariant in its fixed point. However, the superstring compactification on a group manifold is ruled out by a simple phenomenological argument: as is well known in string theory [189], the number of chiral families in the effective four-dimensional field theory to be obtained from the compactified superstrings is given by $\frac{1}{2}|\chi(K)|$ where $\chi(K)$ is the Euler characteristic of K, while the latter vanishes for any group manifold. As a better choice, one may try a coset manifold $K = G/H$, and then look at the vanishing conditions on the one-loop RG β-functions in the corresponding (bosonic or supersymmetric) 2d NLSM [190, 191].

Let Q_A, $A = (i, a) = 1, \ldots, \dim H, \ldots, \dim G$, be the generators of a compact and semisimple Lie group G. We divide them into two sets: the generators Q_i, $i = 1, \ldots, \dim H$, are supposed to be the generators of a subgroup H, whereas the rest of the generators, Q_a, $a = \dim H + 1, \ldots, \dim G$, are supposed to belong to the coset G/H. In string theory, it is usually assumed that $\operatorname{rank} G = \operatorname{rank} H$ that is the necessary condition for the existence of chiral fermions in a compactified string or superstring theory [192]. The anticommutation relations in terms of the structure constants $f^A{}_{BC}$ of G are given by

$$[Q_i, Q_j] = f^k{}_{ij} Q_k ,$$

$$[Q_i, Q_a] = f^b{}_{ia} Q_b , \qquad (3.1)$$

$$[Q_a, Q_b] = f^i{}_{ab} Q_i + f^c{}_{ab} Q_c .$$

The homogeneous space G/H is called symmetric if

$$f^c{}_{ab} = 0 , \qquad (3.2)$$

and non-symmetric otherwise [193].

We identify our latin indices with the flat indices of the tangent space of the NLSM manifold $\mathcal{M} = G/H$. There is a natural choice of metric and torsion for the homogeneous space G/H. The G group metric (2.1.21) induces the metric in the coset G/H, while the natural torsion three-form on G/H is given by

$$H_{abc} = \eta f_{abc} . \qquad (3.3)$$

The generalized curvature (2.2.23) then takes the form

$$\widehat{R}_{abcd} = f_{abi} f^i{}_{cd} + \tfrac{1}{2}(1+\eta) f_{abg} f^g{}_{cd} + \tfrac{1}{4}(1+\eta)^2 (f_{acg} f^g{}_{db} - f_{adg} f^g{}_{cb}) . \quad (3.4)$$

The one-loop RG β-functions of the bosonic and supersymmetric NLSM on G/H are proportional to the generalized Ricci tensor constructed from the curvature (3.4), see Subsect. 2.1. We find

$$\widehat{R}_{ab} = -f_{aci} f_b{}^{ci} - \tfrac{1}{4}(1-\eta^2) f_{acd} f_b{}^{cd} . \qquad (3.5)$$

Since the G algebra is semisimple, we have

$$2 f_{aci} f_b{}^{ci} + f_{acd} f_b{}^{cd} = c \delta_{ab} , \qquad (3.6)$$

where we have introduced the Casimir constant c in the adjoint of G. It follows from (3.5) and (3.6) that

$$\widehat{R}_{ab} = -\tfrac{1}{2} c \delta_{ab} + \tfrac{1}{4}(1+\eta^2) f_{acd} f_b{}^{cd} . \qquad (3.7)$$

We are now in a position to formulate the condition of the one-loop UV finiteness of the bosonic NLSM on G/H (it equally applies to the supersymmetric NLSM on G/H — see Subsect. 3.2.2), in the form

$$f_{acd}f_b{}^{cd} = \left[\frac{2c}{1+\eta^2}\right]\delta_{ab} . \tag{3.8}$$

It is obvious from (3.8) that the homogeneous space G/H cannot be a symmetric space since $c \neq 0$. Equations (3.6) and (3.8) imply

$$f_{aci}f_b{}^{ci} = \left[\frac{c(\eta^2-1)}{2(\eta^2+1)}\right]\delta_{ab} . \tag{3.9}$$

The particular case of $\eta^2 = 1$ is realized in the case of a group manifold G/H with parallelizing torsion.

The vanishing two-loop contribution to the NLSM β-functions implies extra restrictions on the coset [191]. According to Subsect. 2.4.2, they are

$$\widehat{R}_{d(ab)c}\widehat{R}_f{}^{(ab)c} + \tfrac{3}{2}\widehat{R}_{dabf}H^a_{gh}H^{bgh} = 0 , \tag{3.10}$$

where we have used (3.3) and (3.4). In the case of a homogeneous space G/H, there are some additional useful identities,

$$\widehat{R}_{[abc]d} = \widehat{R}_{a[bcd]} = \widehat{R}_{[abcd]} = f_{[ab}{}^i f_{cd]i} + \tfrac{1}{2}(1+\eta)(2+\eta)f_{[ab}{}^g f_{cd]g} . \tag{3.11}$$

It follows that

$$\widehat{R}_a{}^{(bc)}{}_d = -f^{(b}{}_{ai}f^{c)}{}_d{}^i + \frac{(\eta^2-1)}{4}f^{(b}{}_{ag}f^{c)}{}_d{}^g . \tag{3.12}$$

Equation (3.10) then amounts to the relation

$$f^{(a}{}_{di}f^{b)}{}_c{}^i f_{afj}f_b{}^{cj} - \frac{(\eta^2-1)}{2}f_a{}^{gc}f_{dg(d}f_{f)j}{}^{(a}f^{b)}{}_c{}^j$$
$$+\frac{(\eta^2-1)^2}{16}f_{adg}f_{bc}{}^g f^{(a}{}_{fh}f^{b)ch} = 0 . \tag{3.13}$$

The left-hand side of (3.13) vanishes for the group manifolds ($f^a{}_{bi} = 0$) with parallelizing torsion ($\eta^2 = 1$), while it does not seem to allow any more solutions. This agrees with the independent observation [194] that the bosonic NLSM on a homogeneous space G/H realizes a representation of the (Kac-Moody) current algebra only if G/H is a group manifold. The existence of 2d conformally invariant quantum field theories associated with the coset spaces (Sect. 5.3) does not contradict our conclusion. It implies, however, that a conformally invariant 2d NLSM with the coset target space has different metric and torsion.

Superstring compactification on a six-dimensional homogeneous space is very restrictive [190, 192]. Amongst the non-symmetric homogeneous spaces, only

$$\frac{G_2}{SU(3)} \; , \quad \frac{SU(3)}{U(1) \times U(1)} \; , \quad \frac{Sp(4)}{SU(2) \times U(1)} \; , \tag{3.14}$$

survive the three-loop UV finiteness test in the supersymmetric NLSM with $\eta^2 = 5$, while all of them have the most general $SO(6)$ holonomy group.

3.4 Gauging NLSM Isometries

In this section we discuss gauging isometries in 4d supersymmetric NLSM along the lines of [195, 114]. The NLSM gauging is closely related to spontaneous breaking of internal symmetry. All the considerations in this section are entirely classical, while they equally apply to (2,2) supersymmetric NLSM in 2d.

3.4.1 $N = 1$ NLSM in 4d, and Kähler Geometry

Scalar fields in 4d, $N = 1$ supersymmetry are described by complex chiral superfields Φ^i and their conjugates $\bar{\Phi}_i$, $i = 1, 2, \ldots, n$, satisfying the off-shell superspace constraints

$$\bar{D}_{\dot{\alpha}} \Phi^i = 0 \; , \quad D_\alpha \bar{\Phi}_i = 0 \; , \tag{4.1}$$

where we have introduced the chiral and antichiral (flat) superspace covaraint derivatives, D_α and $\bar{D}_{\dot{\alpha}}$, respectively. They satisfy the anticommutation relation of $N = 1$ supersymmetry in 4d,

$$\{D_\alpha, \bar{D}_{\dot{\alpha}}\} = i\partial_{\alpha\dot{\alpha}} \; , \quad \{D_\alpha, D_\beta\} = \{\bar{D}_{\dot{\alpha}}, \bar{D}_{\dot{\beta}}\} = 0 \; . \tag{4.2}$$

We use the two-component spinor notation that is standard in 4d supersymmetry [138, 139, 140, 141, 142, 143, 144, 145, 146, 148]. The ordinary field components of the chiral superfield Φ^i are

$$A^i = \Phi^i\big| \; , \quad \psi^i_\alpha = D_\alpha \Phi^i\big| \; , \quad F^i = \tfrac{1}{2} D^\alpha D_\alpha \Phi^i\big| \; , \tag{4.3}$$

where $|$ means taking the first (leading, or θ-independent) component of a superfield or an operator. The scalars A and the spinors ψ are the propagating fields, whereas the scalars F are the auxiliary fields.

A 4d NLSM has the $N = 1$ supersymmetric extension if and only if the NLSM target manifold $\mathcal{M}$ is Kähler [153]. It follows from the observation (based on dimensional reasons) that the most general 4d kinetic action for the $N = 1$ chiral superfields (without higher derivatives in components) reads in superspace as follows:

$$S = \int d^4x d^4\theta \, K(\Phi^i, \bar{\Phi}_j) = -\tfrac{1}{2} \int d^4x \; K_i{}^j\big| \partial^{\alpha\dot{\alpha}} A^i \partial_{\alpha\dot{\alpha}} \bar{A}_j + \ldots \; , \tag{4.4}$$

where we have explicitly written down the leading NLSM action for the physical scalars, while

$$K_i{}^j\big| \equiv \frac{\partial}{\partial A^i}\frac{\partial}{\partial \bar{A}_j}\,K\big|\;. \qquad (4.5)$$

We use the notation [114]

$$K^{j_1\cdots j_n}_{i_1\cdots i_m} = \frac{\partial}{\partial \Phi^{i_1}}\cdots\frac{\partial}{\partial \Phi^{i_m}}\frac{\partial}{\partial \bar{\Phi}_{j_1}}\cdots\frac{\partial}{\partial \bar{\Phi}_{j_n}}K. \qquad (4.6)$$

The right-hand side of (4.4) has the standard NLSM form (2.1.1), while it also implies a complex NLSM manifold $\mathcal{M}$ *and* a Kähler metric described by the line element

$$\mathrm{d}s^2 = 2\,K_i{}^j\big|\,\mathrm{d}A^i\mathrm{d}\bar{A}_j\;. \qquad (4.7)$$

A complex manifold whose metric can be written down in terms of a locally defined potential K, as in (4.7), is called Kähler manifold [196, 197, 198]. The line element (4.7) is preserved under Kähler gauge transformations of the Kähler potential K,

$$K(\Phi^i,\bar{\Phi}_j) \to K(\Phi^i,\bar{\Phi}_j) + \Lambda(\Phi^i) + \bar{\Lambda}(\bar{\Phi}_i)\;, \qquad (4.8)$$

where Λ is an arbitrary *holomorphic* function of Φ^i. The form of the superfield NLSM action (4.4) is preserved under arbitrary reparametrizations of Φ^i *and* $\bar{\Phi}_j$. The line element (4.7) is merely preserved under holomorphic reparametrizations of A^i and $\bar{A}_j$, $A^i \to f^i(A^j)$, while it can be extended to holomorphic transformations of the chiral superfields, $\Phi^i \to f^i(\Phi^j)$. All equations below, in this subsection, are equally valid either in terms of the chiral superfields or their leading bosonic components, as the arguments.

The only non-vanishing components of the connection are given by

$$\Gamma_{jk}{}^i = (K^{-1})^i{}_n K^n_{jk} \qquad (4.9)$$

and their complex conjugates $\bar{\Gamma}_{jk}{}^i$, where we have introduced the inverse of $K_i{}^j$. It follows that the covariant derivatives of covariant quantities (P) are

$$P_{i;j} = P_{i,j} - \Gamma_{ij}{}^k P_k = K_i{}^n\{P_k(K^{-1})_n{}^k\}_{,j}\;, \quad P_i{}^{;j} = P_i{}^{,j}\;, \qquad (4.10a)$$

whereas the covariant derivatives of contravariant quantities (Π) are

$$\Pi_{i;j} = \Pi_{i,j}\;, \quad \Pi_i{}^{;j} = \Pi_i{}^{,j} + \bar{\Gamma}_i{}^{jk}\Pi_k = (K^{-1})_i{}^k(\Pi_m K_k{}^m)^{,j}\;. \qquad (4.10b)$$

The only non-vanishing components of the Riemann tensor are

$$R_i{}^k{}_j{}^p = K^{kp}_{ij} - \Gamma_{ij}{}^m\bar{\Gamma}_n{}^{kp}K_m{}^n = K_m{}^k(\Gamma_{ij}{}^m)^{,p}\;. \qquad (4.11)$$

Accordingly, the contracted connection and the Ricci tensor are given by

$$\Gamma_i \equiv K_j{}^k\Gamma_{ik}{}^j = \{\ln\det(K_j{}^k)\}_{,i} \qquad (4.12)$$

and
$$R_i{}^j \equiv K_k{}^m R_i{}^k{}_m{}^j = \{\ln \det(K_k{}^m)\}_{,i}{}^{,j} \tag{4.13}$$

respectively. By using holomorphic coordinate transformations it is always possible to transform to *normal* coordinates at a given point of $\mathcal{M}$. The normal coordinates are characterized by the properties

$$K_i{}^j\big|_0 = \delta_j^i , \qquad K_{i_1\cdots i_m}^j\big|_0 = K_i^{j_1\cdots j_n}\big|_0 = 0 , \quad m, n > 1 , \tag{4.14}$$

valid at the origin associated with a given point. The connection also vanishes at this point, while the Riemann tensor in normal coordinates takes the simple form

$$R_i{}^k{}_m{}^j = K_{im}^{kj} . \tag{4.15}$$

The Kähler gauge invariance (4.8) can be fixed by imposing more restrictions,

$$K_{i_1\cdots i_m}\big|_0 = K^{j_1\cdots j_n}\big|_0 = 0 . \tag{4.16}$$

A description of isometries of a Kähler manifold $\mathcal{M}$ has some special features that are associated with the invariance of a given complex structure in $\mathcal{M}$ under holomorphic transformations that do not mix A^i and $\bar{A}_j$ or, equivalently, Φ^i and $\bar{\Phi}_j$. It is, therefore, quite natural to distinguish the *holomorphic* isometries that possess the same property. The Lie derivative associated with a holomorphic Killing vector (see below) leaves both the metric *and* the complex structure of $\mathcal{M}$ invariant. We follow [114, 195] here.

In adapted coordinates (Sect. 2.6) the action of the holomorphic isotropy subgroup in (2.6.5) takes the form

$$\delta A^i = i\lambda^i{}_j A^j , \quad \delta \bar{A}_i = -i\bar{A}_j \lambda^j{}_i , \quad \text{where} \quad \lambda^i{}_j = \lambda^X (T_X)^i{}_j , \tag{4.17}$$

with a constant Hermitian matrix parameter λ.

The Kähler potential is to be invariant under an isometry modulo a Kähler gauge transformation, i.e.

$$\delta K = K_i \delta A^i + K^i \delta \bar{A}_i = \eta(A) + \bar{\eta}(\bar{A}) . \tag{4.18}$$

In the isotropic case, one can always choose a *Kähler gauge* where η vanishes, so that

$$\delta K = i\lambda^i{}_j (K_i A^j - K^j \bar{A}_i) = 0 . \tag{4.19}$$

To describe general holomorphic isometries, one introduces Killing vectors k_B with holomorphic components $k_B^i(A)$ and their complex conjugates, $\bar{k}_{Bi}(\bar{A})$, so that

$$\begin{aligned}
\delta A^i &= \lambda^B k_B^i = [\lambda^B k_B^j \partial/\partial A^j, A^i] \equiv L_{\lambda \cdot k} A^i , \\
\delta \bar{A}_i &= \lambda^B \bar{k}_{Bi} = [\lambda^B \bar{k}_{Bj} \partial/\partial \bar{A}_j, \bar{A}_i] \equiv L_{\lambda \cdot k} \bar{A}_i .
\end{aligned} \tag{4.20a}$$

In adapted coordinates (4.20a) becomes (4.17). The finite form of the infinitesimal transformations (4.20a) is obviously given by

$$A'^i = \exp\{L_{\lambda \cdot k}\}A^i , \quad \bar{A}' = \exp\{L_{\lambda \cdot k}\}\bar{A}_i . \tag{4.20b}$$

The holomorphic and antiholomorphic components of the Killing vectors generate two separate isometry algebras,

$$k_{[A}{}^j k_{B],j}{}^i = f_{AB}{}^C k^i{}_C , \quad \bar{k}_{j[A}\bar{k}_{B]i}{}^{,j} = f_{AB}{}^C \bar{k}_{Ci} . \tag{4.21}$$

The condition (4.18) in general coordinates takes the form

$$\delta K \equiv \lambda^A \left(K_i k_A^i + K^i \bar{k}_{Ai} \right) = \lambda^A \left\{ \eta_A(\Phi) + \bar{\eta}(\bar{\Phi}) \right\} . \tag{4.22}$$

The holomorphic Killing vectors, $k_A^{i,j} = \bar{k}_{Ai,j} = 0$, for a Kähler manifold in general coordinates thus have to obey the Killing equations

$$K^i{}_j k^j_{A;k} + K^j{}_k \bar{k}_{Aj}{}^{;i} = 0 . \tag{4.23}$$

The form of (4.22) is rather suggestive for the introducion of a real function $X_A(\Phi, \bar{\Phi}) = \bar{X}_A$ as [195]

$$k_A^i K_i = iX_A + \eta_A , \quad \bar{k}_{Ai}K^i = -iX_A + \bar{\eta}_A . \tag{4.24}$$

The function X_A is defined modulo an additive real constant since η_A is only determined modulo an additive imaginary constant. Differentiating (4.24) once yields

$$k_A^i K_i^j = iX_A^j , \quad \bar{k}_{Ai}K^i{}_j = -iX_{Aj} . \tag{4.25}$$

In its turn, (4.25) implies

$$k^i{}_A X_{Bi} + \bar{k}_{Bi} X_A{}^i = 0 , \tag{4.26}$$

and, hence,

$$\delta X_A \equiv \lambda^B \left(\bar{k}_{Bi} X^i{}_A + k^i{}_B X_{Ai} \right) = \tfrac{1}{2}\lambda^B \left(\bar{k}_{i[B} X^i{}_{A]} + k^i{}_{[B} X_{A]i} \right) . \tag{4.27}$$

It is also straightforward to deduce from (4.21) and (4.22) that

$$k^i{}_{[A}\eta_{B]i} + \bar{k}_{i[A}\bar{\eta}^i{}_{B]} = f_{AB}{}^C (\eta_C + \bar{\eta}_C) . \tag{4.28}$$

The holomorphicity of the Killing vectors together with (4.28) now imply

$$k^i{}_{[A}\eta_{B]i} = f_{AB}{}^C \eta_C + ic_{AB} \quad \text{and} \quad \bar{k}_{i[A}\bar{\eta}_{B]}{}^i = f_{AB}{}^C \bar{\eta}_C - ic_{AB} , \tag{4.29}$$

where c_{AB} is a *constant* real antisymmetric matrix [195]. The Jacobi identity leads to yet another constraint

$$c_{A[B} f_{CD]}{}^A = 0 \; . \tag{4.30}$$

Since any non-vanishing constant c_{AB} appears to be a topological obstruction to the gauging [195, 114], it is worth investigating when c_{AB} can be removed altogether. If a solution to (4.30) takes the form $c_{AB} = f_{AB}{}^C \xi_C$ with some real constants ξ_C, then c_{AB} can indeed be eliminated by a shift $\eta_A + i\xi_A \to \eta_A$. This would simultaneously remove the ambiguity in η_A and X_A except for invariant Abelian subgroups. [2] If the isometry group is *semisimple*, it always works, while the Jacobi identity then implies

$$\xi_A = f_{AB}{}^D c_{DE} g^{BE} \; , \tag{4.31}$$

where g^{AB} is the inverse of the non-degenerate Killing metric defined by $g_{AB} = -f_{AC}{}^D f_{BD}{}^C$. If the Killing metric is degenerate, the c_{AB} may not be removable. Assuming, nevertheless, that $c_{AB} = 0$, allows one to put (4.29) into the form

$$k^i_{[A} \eta_{B]i} = f_{AB}{}^C \eta_C \quad \text{and} \quad \bar{k}_{i[A} \bar{\eta}_{B]}{}^i = f_{AB}{}^C \bar{\eta}_C \; . \tag{4.32}$$

Equations (4.21), (4.24), (4.27) and (4.32) together imply the following transformation law of the Killing potential X_A [195]:

$$\delta X_A = \lambda^B \left(\bar{k}_{Bi} X_A^i + k_B^i X_{Ai} \right) = -\lambda^B f_{AB}{}^C X_C \; . \tag{4.33}$$

In the case of a semisimple isometry group, (4.25) and (4.33) yield [114]

$$X_A = 2i f_{AB}{}^C k_D^i \bar{k}_{Cj} K_i^j g^{BD} \; . \tag{4.34}$$

3.4.2 Gauging Isometries of 4d, $N = 1$ NLSM

As was demonstrated in Sect. 2.6, it is possible to gauge any (continuous) subgroup of the isometry group in a 4d bosonic NLSM. In this subsection, it is generalized to the case of the 4d, $N = 1$ NLSM with a Kählerian target space $\mathcal{M}$ [195, 199]. We follow [114, 195] here.

It is straightforward to gauge a subgroup of the isotropy group. The Kähler potential can be chosen to be globally invariant as in (4.18), so that the (infinitesimal) symmetry transformation laws in adapted coordinates take the 'flat space' form (4.17). In order to promote this rigid symmetry to a local (Yang-Mills) symmetry, its constant real parameters λ^X are to be promoted to $N = 1$ *chiral* superfields Λ^A. Since the $N = 1$ superfield gauge parameters are necessarily complex, $\Lambda \neq \bar{\Lambda}$, the original symmetry group is going to act on the chiral NLSM superfields $\Phi^i, \bar{\Phi}_i$ via its complexification. For example, finite gauge transformations of the NLSM chiral superfields Φ^i read

---

[2] The last ambiguity precisely corresponds to the possibility of adding a Fayet-Iliopoulos term to the action for each Abelian factor of the gauged isometry group [114].

$$\Phi'^i = (\mathrm{e}^{\mathrm{i}\Lambda})^i{}_j \Phi^j \ , \qquad \bar{\Phi}'_i = \bar{\Phi}_j (\mathrm{e}^{-\mathrm{i}\bar{\Lambda}})^j{}_i \ , \tag{4.35}$$

where $\Lambda^i{}_j = \Lambda^A (T_A)^i{}_j$, with $(T_A)^i{}_j$ being the representation matrices of the gauge group generators, $[T_A, T_B] = \mathrm{i} f_{AB}{}^C T_C$. The gauge (Yang-Mills) $N = 1$ superfield potential V^A transforms in the standard (in 4d supersymmetry) way [138, 140, 141, 142, 143, 144, 146, 148],

$$\mathrm{e}^{V'} = \mathrm{e}^{\mathrm{i}\bar{\Lambda}} \mathrm{e}^{V} \mathrm{e}^{-\mathrm{i}\Lambda} \ , \qquad V = V^A T_A, \quad \Lambda = \Lambda^A T_A \ , \tag{4.36}$$

where T_A are the gauge group generators in the adjoint representation. The gauge potential superfield acts like the 'bridge' converting the gauge Λ-transformations into the gauge $\bar{\Lambda}$-transformations. It is therefore useful to introduce the new superfields $\tilde{\Phi}_i$, in terms of $\bar{\Phi}_i$ and the gauge potential V, as follows [195, 114]:

$$\tilde{\Phi}_i = \bar{\Phi}_j (\mathrm{e}^V)^j{}_i \ , \qquad V^j{}_i = V^A (T_A)^j{}_i \ . \tag{4.37}$$

The latter transform under the gauge Λ-transformations only,

$$\tilde{\Phi}'_i = \tilde{\Phi}_j (\mathrm{e}^{-\mathrm{i}\Lambda})^j{}_i \ . \tag{4.38}$$

Since $\tilde{\Phi}$ and Φ transform under local transformations just as $\bar{\Phi}$ and Φ transform under global transformations, and since the Kähler potential $K(\Phi, \bar{\Phi})$ does not contain any derivatives of the superfields, the substitution $\bar{\Phi} \to \tilde{\Phi}$ converts any globally invariant Kähler potential $K(\Phi, \bar{\Phi})$ into a local invariant $K(\Phi, \tilde{\Phi})$. A globally invariant superpotential $P(\Phi)$, if any, is automatically invariant under local transformations. [3]

The gauging procedure just described is only valid in special (adapted) coordinates at a given point of $\mathcal{M}$. In a generic coordinate system, we have to deal with the geometrical description of the global symmetries in terms of holomorphic Killing vectors (Subsect. 3.4.1). Since gauging the global symmetries means promoting constants λ to chiral superfields Λ and $\bar{\Lambda}$, we again need a superfield $\tilde{\Phi}$ that transforms with Λ rather than $\bar{\Lambda}$. Equations (4.46) and (4.49) tell us that $\tilde{\Phi}$ was earlier defined in terms of $\bar{\Phi}$ by making a 'gauge transformation' with parameter $\bar{\Lambda}^A \to \mathrm{i} V^A$. Hence, in general coordinates, (4.38) should be generalized to [114]

$$\tilde{\Phi}_i = \mathrm{e}^L \bar{\Phi}_i \ , \qquad \text{where} \quad L \equiv L_{\mathrm{i} V \cdot \bar{k}} = \mathrm{i} L_{V \cdot \bar{k}} \ . \tag{4.39}$$

The gauging in general coordinates then amounts to a substitution $\bar{\Phi}_i \to \tilde{\Phi}_i$. The conditions for the invariance of the superpotential, if any, are given by

$$\delta P = P_i k_A^i \Lambda_A = 0 \ . \tag{4.40}$$

---

[3] We do not consider here $N = 1$ supersymmetric gauge kinetic terms that may also be added to the $N = 1$ gauged NLSM action, see [195, 114].

The simple gauging prescription given above applies to globally invariant Kähler potentials. In general, a NLSM Kähler potential may be merely invariant under the isometry transformations (4.20) up to a Kähler gauge transformation (4.18). It happens, for example, when some of the isometries of $\mathcal{M}$ do not leave points of $\mathcal{M}$ invariant, i.e. when they are non-linearly realized (or spontaneously broken). Though the extra term $\int \mathrm{d}^4 x \mathrm{d}^4 \theta \, \lambda^A \bar{\eta}_A$ vanishes in the rigid variation of the NLSM action, the *local* variation of its naively gauged $N = 1$ NLSM counterpart yields a non-vanishing result, $\int \mathrm{d}^4 x \mathrm{d}^4 \theta \, \Lambda^A \bar{\eta}_A$. A solution to this problem amounts to the introduction of yet another $N = 1$ chiral superfield ζ with the local transformation law [4]

$$\delta \zeta = \eta_A(\Phi) \lambda^A , \quad \delta \bar{\zeta} = \bar{\eta}_A(\bar{\Phi}) \Lambda^A , \tag{4.41}$$

and the subsequent modification of the original $N = 1$ NLSM Lagrangian,

$$K(\Phi, \bar{\Phi}) \rightarrow K(\Phi, \bar{\Phi}) - \zeta - \bar{\zeta} , \tag{4.42}$$

which is invariant under the rigid isometry transformations [114]. It is worth mentioning that the modified (ungauged) NLSM action does not depend upon ζ, like the final gauged $N = 1$ NLSM action in eq. (4.46) below. The ζ-transformations are, however, to be included into the new Killing vectors

$$k'_A \equiv k_A^i \partial/\partial \Phi^i + \eta_A \partial/\partial \zeta , \quad \bar{k}'_A \equiv \bar{k}_{Ai} \partial/\partial \bar{\Phi}_i + \bar{\eta}_A \partial/\partial \bar{\zeta} , \tag{4.43}$$

since the corresponding isometries leave the new Kähler potential (4.42) invariant. Since the transformation laws (4.41) should be a realization of the symmetry group, equations (4.32) have to be satisfied, i.e. the obstruction constants c_{AB} are to be removable, at least for the isometry subgroup that is going to be gauged.

The $N = 1$ supersymmetric gauging now amounts to the substitutions

$$\bar{\Phi}_i \rightarrow \tilde{\Phi}_i \quad \text{and} \quad \bar{\zeta} \rightarrow \tilde{\zeta} , \tag{4.44}$$

where $\tilde{\zeta}$ is supposed to be defined in analogy with (4.39), i.e.

$$\tilde{\zeta} = \mathrm{e}^{L'} \bar{\zeta} = \left(1 + \frac{\mathrm{e}^{L'} - 1}{L'} L' \right) \bar{\zeta} = \bar{\zeta} + \mathrm{i} \frac{\mathrm{e}^{L'} - 1}{L'} \bar{\eta}_A V^A$$
$$= \bar{\zeta} + \mathrm{i} \frac{\mathrm{e}^L - 1}{L} \bar{\eta}_A V^A , \tag{4.45}$$

where we have used the notaton $L' \equiv L_{\mathrm{i} V^A \bar{k}'_A}$ and $L \equiv L_{\mathrm{i} V^A \bar{k}_A}$. Hence, after dropping the irrelevant ζ-dependent terms, the locally invariant and $N = 1$ supersymmetric (gauged) NLSM action reads [114]

---

[4] The $\zeta, \bar{\zeta}$ may be interpreted as the coordinates of the enlarged NLSM target space.

$$S = \int \mathrm{d}^4 x \mathrm{d}^4 \theta \left[K(\Phi^i, \tilde{\Phi}_j) - \mathrm{i} \frac{\mathrm{e}^L - 1}{L} \bar{\eta}_A V^A \right] . \tag{4.46}$$

It is straightforward to rewrite the action (4.46) in terms of X_A by using (4.25) and the identity $\mathrm{e}^L \Phi^i = \Phi^i$. One finds

$$K(\Phi^i, \tilde{\Phi}_j) = \mathrm{e}^L K(\Phi^i, \bar{\Phi}_j) = K + \frac{\mathrm{e}^L - 1}{L} K^i \bar{k}_{Ai}(iV^A) , \tag{4.47}$$

which implies [199]

$$S = \int \mathrm{d}^4 x \mathrm{d}^4 \theta \left[K(\Phi^i, \bar{\Phi}_j) + \frac{\mathrm{e}^L - 1}{L} X_A V^A \right] . \tag{4.48}$$

A Fayet-Iliopoulos term for any Abelian factor of the gauge group may also be introduced into the action (4.48), via the freedom of adding a real constant to X_A [195, 114]. Given a non-vanishing vacuum expectation value, $\langle \Phi^i \rangle = \Phi_0^i \neq 0$, the gauge symmetry $\mathcal{G}$ of the gauged $N = 1$ supersymmetric NLSM is spontaneously broken to the isotropy subgroup $\mathcal{H}_{\Phi_0}$. The 4d, $N = 1$ supersymmetry is also spontaneously broken in this case.

The standard example of a Kähler manifold is given by the complex projective space

$$\mathcal{M} = CP^n = \frac{SU(n+1)}{S(U(1) \otimes U(n))} , \tag{4.49}$$

with the Kähler potential

$$K = \ln \left(1 + A^i \bar{A}_i \right) , \tag{4.50}$$

and the isometry group $SU(n+1)$ whose isotropy subgroup is $S(U(1) \otimes U(n))$. The Killing vectors for the linearly realized isotropy subgroup are [114]

$$k_X^i = \mathrm{i}(T_X)^i{}_j A^j , \quad k_{n^2}^i = \mathrm{i} A^i , \tag{4.51}$$

where $(T_X)^i{}_j$, $X = 1, 2, \ldots, n^2 - 1$, are the Hermitian traceless generators of $SU(n)$, and $k_{n^2}^i$ is the $U(1)$ Killing vector. In addition, there are $2n$ holomorphic Killing vectors associated with the non-linearly realized isometries [114],

$$k_p^i = \delta_p^i + \delta_j^p A^j A^i \quad \text{and} \quad k_{p'}^i = \mathrm{i}(\delta_{p'}^i - \delta_j^{p'} A^j A^i) , \tag{4.52}$$

where $p, p' = 1, 2, \ldots, n$. Gauging isometries in the $N = 1$ supersymmetric CP^n model (4.49) and (4.50) yields $c_{pq'} = -2d_{pq'}$ as the only non-vanishing constants amongst c_{AB}. Since the structure constants are given by

$$f_{pq'}{}^{n^2} = -2\delta_{pq'} , \quad \text{while} \quad f_{AB}{}^{n^2} = 0 \quad \text{when} \quad AB \neq pq' , \tag{4.53}$$

one easily finds that $c_{AB} = f_{AB}{}^C \beta_C$ with $\beta_A = \delta_A^{n^2}$ [114]. All the non-vanishing c_{AB} are thus removable in this case, so that there are no obstructions to the gauging procedure described above.

4. NLSM and Extended Superspace

Given a 2d NLSM action of the form (Chap. 2)

$$S_{\text{bosonic}}[\phi] = \tfrac{1}{2} \int \mathrm{d}^2 x \left\{ g_{ij}(\phi)\partial^\mu \phi^i \partial_\mu \phi^j + \varepsilon^{\mu\nu} h_{ij}(\phi)\partial_\mu \phi^i \partial_\nu \phi^j \right\} , \qquad (0.1)$$

it can be supersymmetrized by using the *Noether* (trial and error) procedure. The Noether procedure amounts to defining fermionic partners λ^i of ϕ^i by the supersymmetry (susy) transformation law, $\delta\phi^i = \bar\varepsilon\lambda^i$, where the susy parameter ε and the fields $\lambda^i(x)$ may be either chiral or non-chiral, and demanding the action (0.1) to be invariant under susy by modifying it with fermionic-dependent terms. For dimensional reasons, the extra fermionic terms can be either quadratic or quartic in λ. The susy algebra requires the susy transformations to be closed to translations on all fields subject to their equations of motion. The outcome of the (non-chiral) Noether (1,1) supersymmetrization of the action (0.1) reads [149, 200]

$$S_{\text{susy}}[\phi, \lambda] = S_{\text{bosonic}} + \tfrac{1}{2} \int \mathrm{d}^2 x \left\{ \mathrm{i} g_{ij} \bar\lambda_+^i \gamma^\mu (D_\mu \lambda_+)^j + \mathrm{i} g_{ij} \bar\lambda_-^i \gamma^\mu (D_\mu \lambda_-)^j \right.$$

$$\left. + \tfrac{1}{4} R_{ijkl}^+ (\bar\lambda_+^i \gamma^\mu \lambda_+^j)(\bar\lambda_-^k \gamma_\mu \lambda_-^l) \right\} ,$$

$$(0.2)$$

where we have introduced the notation

$$(D_\mu \lambda_\pm)^k = \partial_\mu \lambda_\pm^k + \Gamma_{\pm jl}^k \lambda_\pm^j \partial_\mu \phi^l , \qquad \lambda_\pm^k = \tfrac{1}{2}(1 \pm \gamma_3)\lambda^k , \qquad (0.3)$$

and

$$\Gamma_{\pm jk}^i = \Gamma_{jk}^i \pm H_{jk}^i , \qquad H_{ijk} = \tfrac{3}{2} h_{[ij,k]} , \qquad R_{ijkl}^\pm = R_{ijkl}(\Gamma_\pm) . \qquad (0.4)$$

The action (0.2) is invariant under rigid $N = 1$ susy transformations with the infinitesimal spinor parameter $\varepsilon = \varepsilon_+ + \varepsilon_-$,

$$\delta\phi^i = \bar\varepsilon_+ \lambda_-^i + \bar\varepsilon_- \lambda_+^i \equiv \delta_+ \phi^i + \delta_- \phi^i ,$$

$$\delta\lambda_\pm^i = -\mathrm{i}\gamma^\mu \partial_\mu \phi^i \varepsilon_\mp - \Gamma_{\pm jk}^i \lambda_\pm^j (\delta_\pm \phi^k) , \qquad (0.5)$$

by construction, while the algebra of these transformations closes *on-shell*, i.e. on the field equations of motion.

It can also be investigated whether the 2d, $N = 1$ supersymmetric NLSM action (0.2) has, in fact, more supersymmetries beyond the one that we have imposed. For dimensional reasons, the extra susy transformation law should be of the form $\delta\phi^i = \bar{\varepsilon}_2 J^i{}_j \lambda^j$, in terms of some functions $J^i{}_j(\phi)$. When imposing the invariance of the action (0.2) under those transformations, one finds that J has to be covariantly constant with respect to the covariant derivative in the NLSM target space $\mathcal{M}$ (with torsion), whereas imposing the extended susy algebra implies that J is an almost complex structure (Subsect. 4.1.1). For example, in the case of (non-chiral) 2d, $N = (2, 2)$ susy without torsion ($h_{ij} = 0$), one finds that the almost complex structure J is actually integrable and covariantly constant, which amounts to the Kähler geometry of $\mathcal{M}$ (cf. Subsect. 3.4.1). Similarly, in the case of (non-chiral) 2d, $N = 4$ susy of the action (0.2) without torsion, one gets three almost complex structures $J^{(A)}$, $A = 1, 2, 3$, that are integrable, covariantly constant and satisfy a quaternionic algebra, which amounts to the hyper-Kähler geometry of $\mathcal{M}$ (Subsect. 4.1.1).

The component result (0.2) is equivalent to (3.1.18b) derived in Subsect. 3.3.1 from the NLSM superspace action (3.1.18a). The form of the (manifestly supersymmetric) superfield action (3.1.18a) is easily guessed from the bosonic NLSM action (0.1) and dimensional reasons, while its susy transformations close *off-shell*, i.e. they are model-independent. This shows the power of superspace and its natural advantanges against the component (Noether) approach in constructing $N = 1$ supersymmetric NLSM actions.

Superspace becomes especially useful for constructing 2d NLSM with N-extended supersymmetry. As was just outlined above, according to the Noether procedure, extended susy poses certain restrictions on the NLSM metric, which are closely related to complex geometry [196, 197, 198]. The extended superspace formulations of NLSM automatically incorporate those restrictions. The simplest example is provided by the purely metric NLSM with $N = 2$ extended susy in 2d. The non-chiral $N = 2$ susy in 2d has four supercharges, and in 2d it corresponds to *simple* $N = 1$ susy having the same number of supercharges. The basic (scalar) $N = 1$ matter supermultiplet in 4d is described by a complex chiral superfield Φ satisfying the chirality condition $\bar{D}_{\underset{\alpha}{\bullet}}\Phi = 0$ (Subsect. 3.4.1). The most general 4d, $N = 1$ superspace action in terms of a set of chiral superfields $\{\Phi^i\}$, with merely the second-order (NLSM) kinetic terms for its physical bosonic components, can be easily fixed for dimensional reasons (Subsect. 3.4.1),

$$S[\Phi] = \int \mathrm{d}^4 x \mathrm{d}^4 \theta \, K(\Phi, \bar{\Phi}) \ , \tag{0.6}$$

in terms of a single function $K(\Phi, \bar{\Phi})$ called the Kähler potential [152]. It is straightforward to dimensionally reduce the 4d action (0.6) down to 2d, in order to get a NLSM of the form (0.2) with the Kähler metric [153]

$$g_{i\bar{j}} = \partial_i \partial_{\bar{j}} K \ . \tag{0.7}$$

We thus conclude that imposing $N = 2$ extended supersymmetry on the NLSM action (0.2) with vanishing torsion ($h_{ij} = 0$) requires the complex Kähler geometry of the NLSM target manifold $\mathcal{M}$. A single potential also exists in the more general case of 2d, $N = 2$ NLSM with a non-vanishing torsion. [1] Generally speaking, any *extended* supersymmetry is associated with *complex* geometry of the NLSM target space (Subsect. 4.1.1).

Imposing a non-chiral $N = 4$ susy (having eight supercharges) in 2d requires a hyper-Kähler metric in the purely metric 2d NLSM, and vice versa [149, 201]. A 2d, $N = 4$ supersymmetric NLSM with a non-vanishing torsion gives rise to a quaternionic NLSM metric. A non-trivial 2d NLSM of the form (0.2), invariant under more than eight susy charges, does not exist [149]. [2] The most natural classification of complex Riemannian manifolds (of real dimension $D = 2n$) is based on their holonomy group $\mathrm{Hol}(\mathcal{M})$, whose generators are the $O(2n)$-valued curvature components [196, 197, 198]. A complex manifold $\mathcal{M}$ of $\mathrm{Hol}(\mathcal{M}) \subset U(n) \subset O(2n)$ is just a Kähler manifold. If the holonomy of $\mathcal{M}$ is a subgroup of $Sp(n/2)$, the manifold is called hyper-Kähler. Its real dimension D must be a multiple of four, $D = 4m$. The holonomy of a quaternionic manifold is a subgroup of $Sp(n/2) \times Sp(1)$. The hyper-Kähler NLSM Lagrangian in components is similar to (0.2),

$$\mathcal{L}_{\text{hyper}-\text{Kahler}} = \tfrac{1}{2}g_{ij}\partial^\mu \phi^i \partial_\mu \phi^j + \tfrac{i}{2}\bar{\lambda}_a \gamma^\mu \nabla_\mu \lambda^a$$
$$+ \tfrac{1}{4}\Omega_{abcd}(\bar{\lambda}^a \gamma^\mu \lambda^b)(\bar{\lambda}^c \gamma_\mu \lambda^d) , \tag{0.8}$$

where $i, j = 1, 2, \ldots, 4m$, and $a, b, c, \ldots = 1, 2, \ldots, 2m$, in terms of the Dirac spinors λ^a, the $Sp(m)$ covariant derivatives ∇_μ and the hyper-Kähler curvature Ω_{abcd}.

In this chapter we extensively use both the conventional ($N = 1$ and $N = 2$) superspace and its refined (projective and harmonic) versions, in order to construct large classes of supersymmetric NLSM with *maximal* supersymmetry. In addition to remarkable geometrical properties, all the 2d NLSM with maximal $N = 4$ susy turn out to be UV-*finite* as quantum field theories, to all orders of quantum perturbation theory. Their 4d, $N = 2$ supersymmetric NLSM counterparts naturally appear as the low-energy effective actions for (charged) hypermultiplet matter in quantum 4d, $N = 2$ supersymmetric (non-Abelian) gauge field theories (Chap. 8).

4.1 Complex Geometry and NLSM

In this section we mostly discuss the 4d, $N = 2$ susy NLSM with hyper-Kähler metrics in the context of 4d, N=1 superspace [114]. This section serves as a

---

[1] See Subsect. 6.3.2 about *local* $N = 2$ supersymmetry in 2d NLSM with torsion.

[2] See, however, [202] for a discussion of 2d WZNW models with $N = 7$ and $N = 8$ (exceptional) non-linear supersymmetry based on octonions.

prerequisite for the next sections where several different versions of extended superspace will be introduced and the maximally supersymmetric NLSM will be constructed in terms of extended superfields.

Basic facts about complex geometry are summarized in Subsect. 4.1.1. In Subsects. 4.1.2 and 4.1.3 a close relation between $N = 2$ susy and hyper-Kähler geometry in 4d NLSM is established in $N = 1$ superspace. Gauging isometries of the hyper-Kähler NLSM in $N = 1$ superspace is discussed in Subsect. 4.1.4.

4.1.1 ABC of Complex Geometry

A full account of complex geometry in available in the books [196, 197, 198]. A short presentation for physicists can be found e.g., in [114, 201, 203].

Let $\mathcal{M}$ be a manifold of real dimension $2n$, covered by a system of co-ordinate neighbourhoods (charts) $\{x^i\}$. In an intersection of two charts the associated coordinates are supposed to be related by smooth and locally in-vertible *transition functions*.

A mixed second-rank tensor $J_i{}^j(x)$ with real components is called an almost complex structure if it satisfies the relation

$$J_i{}^j J_j{}^k = -\delta_i{}^k \ . \tag{1.1}$$

The manifold $\mathcal{M}$ equipped with an almost complex structure is called an almost complex manifold of complex dimension n.

It is clear that the almost complex structure defines multiplication by i on vectors in each coordinate chart. The projectors

$$P_\pm = \tfrac{1}{2}(1 \pm \mathrm{i}J) \tag{1.2}$$

can be used to split any vector V^i into two projections $V^i_\pm$. In particular, the basis 1-forms $\mathrm{d}x^j$ can be split into

$$\omega^i_\pm = P_{\pm j}{}^i \mathrm{d}x^j \ . \tag{1.3}$$

The almost complex structure is said to be integrable if

$$(\mathrm{d}\omega_\pm)^k_\mp = P_{\mp i}{}^k \mathrm{d}P_{\pm j}{}^i \mathrm{d}x^j = 0 \ . \tag{1.4}$$

In this case J is called a complex structure , while $\mathcal{M}$ is then called a complex manifold. The complex manifold is characterized by *holomorphic* transition functions. Equation (1.4) is equivalent to the vanishing torsion on $\mathcal{M}$,

$$N_{ij}{}^k \equiv J_{[i|}{}^n \partial_n J_{|j]}{}^k + J_n{}^k \partial_{[j} J_{i]}{}^n = 0 \ , \tag{1.5}$$

where $N_{ij}{}^k$ is called the Nijenhuis tensor.

Equations (1.4) and (1.5) allow us to put the complex structure J into canonical form, after introducing holomorphic and antiholomorphic coordinates, z^i and $\bar{z}^i$,

$$J_j{}^i = \begin{pmatrix} i\delta_i^j & 0 \\ 0 & -i\delta_{\bar{i}}^{\bar{j}} \end{pmatrix} \quad . \tag{1.6}$$

In fact, the existence of a system of complex coordinate charts, in which the almost complex structure takes the form (1.6), is equivalent to the integrability condition (1.4) [196, 197, 198]. The sufficiency of (1.6) is obvious.

Given a Riemannian manifold $\mathcal{M}$ with a metric g_{ij} and an (almost) complex structure $J_i{}^j$, the invariance of the metric with respect to the complex structure means

$$J_i{}^k J_j{}^m g_{km} = g_{ij} \ , \quad \text{or, equivalently,} \quad J_{ij} \equiv g_{jk} J_i{}^k = -J_{ji} \ . \tag{1.7}$$

The metric satisfying (1.7) is called Hermitian. Given an arbitrary metric $\tilde{g}_{ij}$ on $\mathcal{M}$, it can always be extended to a Hermitian metric g_{ij},

$$g_{ij} = \tilde{g}_{ij} + J_i{}^m J_j{}^k \tilde{g}_{km} \ . \tag{1.8}$$

An (almost) complex manifold equipped with a Hermitian metric is called an (almost) Hermitian manifold. The Hermitian manifold thus possesses the fundamental two-form

$$\omega \equiv J_{ij} \, dx^i \wedge dx^j \ . \tag{1.9}$$

In the special coordinates (where the complex structure J is of canonical form), it reads

$$\omega = 2ig_{i\bar{j}} \, dz^i \wedge d\bar{z}^{\bar{j}} \ . \tag{1.10}$$

An (almost) Kähler manifold is an (almost) Hermitian manifold with closed fundamental two-form, i.e.

$$d\omega = 0 \quad \text{or, equivalently,} \quad J_{[ij,k]} = 0 \ . \tag{1.11}$$

As is clear from (1.10) and (1.11), the metric of a Kähler manifold can be (locally) expressed in terms of a Kähler potential, as in (0.7). The conditions (1.5) and (1.11) together are equivalent to the covariant constancy of the complex structure,

$$\nabla_i J_j{}^k = 0 \ . \tag{1.12}$$

An (almost) quaternionic structure is a set of three linearly independent (almost) complex structures $J_i{}^{(A)j}$, $A = 1, 2, 3$, satisfying the $su(2)$ algebra,

$$J_i{}^{(A)k} J_k{}^{(B)j} = -\delta^{AB} \delta_i^j + \varepsilon^{ABC} J_i{}^{(C)j} \ . \tag{1.13}$$

A manifold with the quaternionic structure is called a quaternionic manifold.

If a quaternionic manifold possesses a metric that is Hermitian with respect to all three covariantly constant complex structures, then the manifold is hyper-Kähler. In other words, the hyper-Kähler manifold has three linearly independent complex structures satisfying the quaternionic algebra, while there exists a coordinate system for a given complex structure where the latter takes the canonical form (1.6).

Any hyper-Kähler manifold is Ricci-flat. In four dimensions the hyper-Kähler condition is equivalent to Ricci flatness and the Kähler condition together. In fact, a four-dimensional hyper-Kähler manifold is always (anti)self-dual and vice versa [204]. See e.g., the book [205] for more details about hyper-Kähler manifolds.

4.1.2 4d, $N = 2$ NLSM in $N = 1$ Superspace

Having constructed the most general 4d, $N = 1$ supersymmetric NLSM action (0.6) in terms of a Kähler potential $K(\Phi, \bar{\Phi})$ depending upon the $N = 1$ chiral superfields Φ^i and $\bar{\Phi}_i$, one can impose extra non-manifest supersymmetry on the action (0.6) in order to get 4d, $N = 2$ NLSM with a hyper-Kähler potential K [140, 114]. Of course, in the absence of $N = 2$ auxiliary fields, the extended supersymmetry algebra is expected to be closed only on-shell. We follow [114] here, and use the notation introduced in Sect. 3.4.

The most general form of the extra rigid susy law is given by

$$\delta\Phi^i = \bar{D}^2(\bar{\varepsilon}\bar{\Omega}^i) , \quad \delta\bar{\Phi}_i = D^2(\varepsilon\Omega_i) , \tag{1.14}$$

where ε is a constant chiral superfield parameter, $\bar{D}_{\dot\alpha}\varepsilon = D^2\varepsilon = \partial_\mu\varepsilon = 0$, and $\bar{\Omega}$ is a function of Φ and $\bar{\Phi}$ (modulo an additive chiral term). The closure of the algebra of the transformations (1.14) implies the relations [114]

$$\Omega_{i,j}\bar{\Omega}^{j,k} = \Omega_{j,i}\bar{\Omega}^{k,j} = -\delta_i^k ,$$

$$\bar{\Omega}^{j,[m|}\bar{\Omega}^{i,|k]}{}_j = 0 , \tag{1.15}$$

$$\bar{D}^2\bar{\Omega}^i = \bar{\Omega}^{i,j}\bar{D}^2\bar{\Phi}_j + \tfrac{1}{2}\bar{\Omega}^{i,jk}\bar{D}_{\dot\alpha}\bar{\Phi}_j\bar{D}^{\dot\alpha}\bar{\Phi}_k = 0 ,$$

as well as their complex conjugates.

The $N = 1$ NLSM action (0.6) is invariant under the transformations (1.14) provided that [114]

$$\bar{\omega}^{jm} \equiv K_i{}^j\Omega^{i,m} = -\bar{\omega}^{mj} ,$$

$$K_i{}^j\bar{\Omega}^{i,mk} + K_i{}^{mk}\bar{\Omega}^{i,j} = 0 , \tag{1.16}$$

$$K_i{}^j\bar{\Omega}^{i,m}{}_k + K_{ik}{}^j\bar{\Omega}^{i,m} = 0 .$$

This is to be compared to the field equations following from the action (0.6),

$$\bar{D}^2 K_i = K_i{}^j\bar{D}^2\bar{\Phi}_j + \tfrac{1}{2}K_i{}^{jk}\bar{D}_{\dot\alpha}\bar{\Phi}_j\bar{D}^{\dot\alpha}\bar{\Phi}_k = 0 . \tag{1.17}$$

The first two equations (1.16) now imply that the third equation (1.15) is equivalent to the equation of motion (1.17). Hence, the $N = 2$ susy algebra closes only on-shell, as expected.

It is now straightforward to check that (1.15) and (1.16) amount to hyper-Kähler geometry (see the previous subsection). In particular, the quaternionic structure (1.13) comprises the canonical complex structure $J^{(3)}$ of (1.6) and two other (almost) complex structures

$$J_i^{(1)j} = \begin{pmatrix} 0 & \Omega_{j,i} \\ \bar{\Omega}^{j,i} & 0 \end{pmatrix} , \quad J_i^{(2)j} = \begin{pmatrix} 0 & i\Omega_{j,i} \\ -i\bar{\Omega}^{j,i} & 0 \end{pmatrix} , \tag{1.18}$$

with mixed (one covariant and one contravariant) indices, because of the first equation (1.15). Both $J^{(1)}$ and $J^{(2)}$ are integrable due to the second equation (1.15), while they are covariantly constant due to the second and third equations (1.16). Finally, the NLSM metric is Hermitian with respect to all three complex structures due to the first equation (1.16). According to Subsect. 4.1.1, this precisely amounts to the hyper-Kähler structure.

The canonical complex structure, $J^{(3)}$, is obviously related to the manifest Kähler structure of the $N = 1$ NLSM that we started with. The coordinate system, where the metric takes the Kähler form with respect to a *non-canonical* complex structure, is related to the canonical one by a *non-holomorphic* coordinate transformation. All non-canonical complex structures on the top of the canonical one form a two-sphere. Hence, any 'democratic' treatment of all complex structures requires the use of *twistors* (cf. [206]) parametrizing the two-sphere. The twistors will be introduced in Sects. 4.3 and 4.4, where a construction of NLSM with hyper-Kähler metrics in twistor (projective or harmonic) superspace is discussed.

Equation (1.18) gives the complex structures in terms of the derivatives of the non-holomorphic functions Ω_i and $\bar{\Omega}^i$ entering the extra susy transformation laws (1.14). It is also possible to reconstruct these functions from the given Kähler potential and one of the non-holomorphic complex structures associated with a hyper-Kähler NLSM [114]. Since a non-canonical complex structure J anticommutes with the canonical one, the former can be written down in the form

$$J_i^j = \begin{pmatrix} 0 & \Omega_{ji} \\ \bar{\Omega}^{ji} & 0 \end{pmatrix} \tag{1.19}$$

with some matrix Ω and its complex conjugate $\bar{\Omega}$. Hence, J^{ij} is block-diagonal,

$$J^{ij} = \begin{pmatrix} \bar{\gamma}^{ij} & 0 \\ 0 & \gamma_{ij} \end{pmatrix} , \quad \text{where} \quad \bar{\gamma}^{ij} \equiv (K^{-1})_k^i \bar{\Omega}^{kj} . \tag{1.20}$$

The covariant constancy of J now implies that $\bar{\gamma}^{jk}$ is holomorphic whereas γ_{jk} is antiholomorphic, $\bar{\partial}^i \bar{\gamma}^{jk} = \partial_i \gamma_{jk} = 0$. Similarly one finds that $\bar{\omega}^{ij}$ introduced in the first equation (1.16) is antiholomorphic whereas $\omega_{ij} \equiv K_i^k \Omega_{kj}$ is holomorphic.

It is now straightforward to verify that the functions

$$\bar{\Omega}^i \equiv \bar{\gamma}^{ij} K_j \tag{1.21}$$

obey the desired relation

$$\bar{\Omega}^{i,j} = \bar{\Omega}^{ij} \tag{1.22}$$

and satisfy all the equations (1.15) valid for the hyper-Kähler manifolds. Equations (1.20) and (1.21) further imply that

$$K_i \bar{\Omega}^i = 0 \quad \text{and} \quad \nabla_i \bar{\Omega}^i = 0 . \tag{1.23}$$

The quaternionic (or hyper-Kähler) structure thus plays a fundamental role in the hyper-Kähler geometry. In particular, the Christoffel connection (3.4.9) can also be expressed in terms of the hyper-Kähler structure,

$$\Gamma_{jk}{}^m = -\bar{\Omega}^{m,i}\Omega_{i,jk} , \tag{1.24}$$

where (1.16) has been used.

4.1.3 Triholomorphic Isometries of $N = 2$ NLSM

The importance of holomorphic isometries in the context of the 4d, $N = 1$ supersymmetric NLSM on Kähler manifolds was emphasized in Sect. 3.4. A holomorphic isometry of the 4d, $N = 2$ supersymmetric NLSM on a hyper-Kähler manifold can be holomorphic either with respect to just one of the hyper-Kähler complex structures, or with respect to all (three) of them. In the latter case, the holomorphic isometry and the corresponding Killing vector are called *triholomorphic* [114] or *translational* [204]. The triholomorphic isometries commute with $N = 2$ extended supersymmetry, so that they can be consistently gauged, i.e. in the $N = 2$ supersymmetric way.

By definition, a triholomorphic Killing vector k^m satisfies the relation

$$P_{\pm i}{}^j \nabla_j \left(P_{\mp m}^n k^m \right) = 0 , \tag{1.25}$$

where we have used the projection operators $P_\pm$ of (1.2), for each of the complex structures (or, equivalently, for any two linearly independent complex structures). Equation (1.25) is equivalent to the vanishing Lie derivative of all complex structures,

$$L_k J_i{}^j \equiv k^m J_i{}^j{}_{,m} - k^j{}_{,m} J_i{}^m + k^m{}_{,i} J_m{}^j = 0 . \tag{1.26}$$

In the special coordinate system, where the complex structures take the form (1.18) and the Killing vector k^m is manifestly holomorphic, the triholomorphicity condition reads

$$\bar{\Omega}^{ij}\bar{k}_j{}^{;m} - \bar{\Omega}^{jm}\bar{k}^i{}_{;j} = 0 , \tag{1.27}$$

or, equivalently,

$$\bar{\omega}^{j[i}\bar{k}_j{}^{;m]} = 0 . \tag{1.28}$$

We follow [114] here. Equations (3.4.23) and (1.27) further imply that

$$k^i_{;i} = \bar{k}_i^{;i} = 0 \ . \tag{1.29}$$

The triholomorphic condition (1.25) or (1.26) can be considered as the integrability condition for the existence of a real Killing potential $X^{(J)}$ associated with the complex structure J and satisfying a differential equation (cf. Sect. 3.4)

$$k^i J_{ij} = -X^{(J)}_{,j} \ . \tag{1.30}$$

In the special coordinate system specified by (1.6) and (1.18) the existence of the real Killing potential $X^{(J^A)}$ amounts to the existence of a holomorphic Killing potential P and an antiholomorphic Killing potential $\bar{P}$ to be defined with respect to $J^{(1)} \mp i J^{(2)}$, respectively. It follows from (1.30) that

$$k^i \omega_{ij} = -P_{,j} \ , \quad \bar{k}_i \bar{\omega}^{ij} = -\bar{P}^{,j} \ . \tag{1.31}$$

To compute the potentials, one integrates an equation [114]

$$(\partial_i + i\Omega_{ji}\bar{\partial}^j)(\bar{P}_A - P_A - X_A) = 0 \ , \tag{1.32}$$

which is a consequence of (3.4.25), (1.16) and (1.31). One finds

$$X_A = \bar{P}_A - P_A + \chi_A \ , \tag{1.33}$$

where χ_A satisfies the equation

$$(\partial_i + i\Omega_{ji}\bar{\partial}^j)\chi_A = 0 \ . \tag{1.34}$$

Similarly, when using another operator $\partial_i - i\Omega_{ji}\bar{\partial}^j = -i\Omega_{mi}(\bar{\partial}^m - i\bar{\Omega}^{jm}\partial_j)$, one gets [114]

$$X_A = P_A - \bar{P}_A + \bar{\chi}_A \quad \text{and} \quad (\bar{\partial}^i - i\bar{\Omega}^{ji}\partial_j)\bar{\chi}_A = 0 \ . \tag{1.35}$$

Taken together, (1.33) and (1.35) imply

$$2X_A = \chi_A + \bar{\chi}_A \quad \text{and} \quad 2(P_A - \bar{P}_A) = \chi_A - \bar{\chi}_A \ . \tag{1.36}$$

Equations (1.34) and (1.35) tell us that χ (and $\bar{\chi}$) are holomorphic (and antiholomorphic) with respect to a non-canonical complex structure, whereas the first equation (1.36) can be used to fix them up to an imaginary constant (cf. (3.4.22)). The second equation (1.36) can then be used to determine P and $\bar{P}$, again up to yet another real constant. The total ambiguity in P is thus given by a complex constant.

The gauge transformations of χ_A and P_A follow from (3.4.29) and (3.4.33),

$$\delta P_A = \lambda^B k^i_B P_{A,i} = -\lambda^B(f_{AB}{}^C P_C + \tilde{c}_{AB}) \ , \tag{1.37}$$

where $\tilde{c}_{AB}$ is the complex analogue of c_{AB} introduced in (3.4.29), $\tilde{c}_{AB} = -\tilde{c}_{BA}$. Like in Sect. 3.4, if the constants $\tilde{c}_{AB}$ cannot be removed, they represent an obstruction to $N = 2$ supersymmetric gauging. [3] As in Sect. 3.4,

---

[3] The obstructions $c_{AB} \neq 0$ and $\tilde{c}_{AB} \neq 0$ are independent!

whenever the isometry (sub)group to be gauged is semisimple, the obstruction is removable, while (1.37) can be solved for P_A as follows [114]:

$$P_A = f_{AB}{}^C k_D^i k_C^j \omega_{ji} g^{BD} . \tag{1.38}$$

As an example [114], let us consider the case of flat four-dimensional space that is obviously hyper-Kähler, with the Kähler potential

$$K = \bar{\Phi}_i \Phi^i , \qquad i = 1, 2 . \tag{1.39}$$

The translational isometries are generated by

$$k_A = \delta_A^i \left(\partial/\partial \Phi^i + \partial/\partial \bar{\Phi}_i \right) , \qquad A = 1, 2 . \tag{1.40}$$

The chiral functions η_A and the real functions X_A defined by (3.4.22) and (3.4.24), respectively, are easily calculated in this case,

$$\eta_A = \delta_A^i \Phi^i , \quad X_A = i\delta_A^i (\Phi^i - \bar{\Phi}_i) , \tag{1.41}$$

while (1.41) implies $c_{AB} = 0$ (cf. (3.4.40)).

Choosing complex structures as

$$\Omega_{ij} = \begin{pmatrix} 0 & i \\ -i & 0 \end{pmatrix} \tag{1.42}$$

gives rise to the operator (1.32) in the form

$$\partial_i + i\Omega_{ji}\bar{\partial}^j = \begin{pmatrix} \partial_1 + \bar{\partial}^2 \\ \partial_2 - \bar{\partial}^1 \end{pmatrix} . \tag{1.43}$$

A general solution to (1.34) in this case is given by $\chi(\Phi^1 - \bar{\Phi}_2, \Phi^2 + \bar{\Phi}_1)$. The first equation (1.36) and (1.41) now imply

$$\begin{aligned} \chi_1 &= i(\Phi^1 - \bar{\Phi}_2 - (\Phi^2 + \bar{\Phi}_1)) , \\ \chi_2 &= i(\Phi^1 - \bar{\Phi}_2 + (\Phi^2 + \bar{\Phi}_1)) . \end{aligned} \tag{1.44}$$

The second equation (1.36) then yields

$$P_1 = -2i\Phi^2 \quad \text{and} \quad P_2 = 2i\Phi^1 , \tag{1.45}$$

or, equivalently, $P_A = -2\delta_A^i \Omega_{ij}\Phi^j$. The constants $\tilde{c}_{AB}$ can now be computed from (1.37), with the result [114]

$$\tilde{c}_{12} = -k_1 P_2 = k_2 P_1 = -2i . \tag{1.46}$$

Since the group is Abelian, there is no way to remove this $\tilde{c}$. One concludes that only the isometry generated by one linear combination of k_1 and k_2 can be gauged in the $N = 2$ supersymmetric way [114]. We shall return to this issue in the framework of a manifestly $N = 2$ supersymmetric (harmonic superspace) approach in Subsect. 4.4.4.

4.1.4 Gauging $N = 2$ NLSM Isometries in $N = 1$ Superspace

Gauging isometries of the 4d, $N = 1$ NLSM on Kähler manifolds in $N = 1$ superspace was discussed in Sect. 3.4. In the case of a hyper-Kähler manifold the 4d, $N = 1$ NLSM possesses extra supersymmetry (Subsect. 4.1.2). To be consistent with $N = 2$ extended (non-manifest and on-shell) supersymmetry, the Killing vector of the isometry to be gauged should be triholomorphic (Subsect. 4.1.3). The $N = 2$ supersymmetric gauging procedure then goes along the lines of Sect. 3.4. We continue to follow [114] in this subsection, by using the notation of Sect. 3.4.

Let us consider a 4d, $N = 1$ NLSM action (0.6) that is invariant under the extra susy transformations (1.14) and possesses isometries associated with the triholomorphic Killing vectors k_A^i. To gauge the triholomorphic isometries in a (non-manifestly) $N = 2$ supersymmetric way, one introduces $N = 2$ gauge vector multiplets (V^A, S^A) comprising the $N = 1$ vector multiplets, V^A, and the $N = 1$ chiral multiplets, S^A, in $N = 1$ superspace. The second susy transformations are given by

$$
\begin{aligned}
\delta \Phi^i &= \bar{D}^2 \left(\bar{\varepsilon} \bar{\Omega}^i(\Phi, e^L \bar{\Phi}) \right) , \\
\delta \bar{\Phi}_i &= D^2 \left(\epsilon \Omega_i(\bar{\Phi}, e^{\bar{L}} \Phi) \right) , \\
\delta e^V &= \varepsilon \bar{S} e^V + e^V S \bar{\varepsilon} , \\
\delta S &= -i W^\alpha D_\alpha \varepsilon = \bar{D}^2 \left((e^{-V} D^\alpha e^V) D_\alpha \varepsilon \right) ,
\end{aligned}
\tag{1.47}
$$

where the naive covariantization rule, $\Phi \to \tilde{\Phi}$ according to (3.4.39) has been used. The gauge transformations are given by

$$
\begin{aligned}
\delta \Phi^i &= \Lambda^A k_A^i , \\
\delta \bar{\Phi}_i &= \bar{\Lambda}^A \bar{k}_{Ai} , \\
\delta e^V &= i \left(\bar{\Lambda} e^V - e^V \Lambda \right) , \\
\delta S &= i[\Lambda, S] , \quad \Lambda \equiv \Lambda^A T_A .
\end{aligned}
\tag{1.48}
$$

The NLSM action invariant under the transformations (1.48) is the $N = 1$ covariantized action (3.4.48) with the additional S-dependent terms dictated by the extra supersymmetry [114],

$$
\int d^4 x d^4 \theta \left[K(\Phi^i, \bar{\Phi}_j) + \frac{e^L - 1}{L} X_A V^A + g_{AB} S^A (e^{-V} \bar{S} e^V)^B \right] , \tag{1.49}
$$

where the potentials P_A have been introduced in (1.31).

The $N = 2$ supersymmetry of the action (1.49) can be checked in the Wess-Zumino gauge where $V^3 = 0$ [114]. This is enough to ensure the $N = 2$ invariance in any gauge because of the gauge invariance of the action (1.49) by construction. As was shown in the preceeding subsection (see Sect. 3.4 also), the obstruction coefficients c and $\tilde{c}$ should vanish, while this is always the case for a semisimple gauge group. As far as Abelian factors, if any, of the gauge group are concerned, their obstruction coefficients cannot be removed and thus, if non-vanishing, they are the true obstruction against $N = 2$ supersymmetric gauging of the $N = 2$ NLSM isometries. In the Abelian case, P_A is determined up to an additive complex constant for each Abelian factor. Together with the freedeom of additng a real constant to X_A this precisely corresponds to the possibility of adding an (electric) $N = 2$ Fayet-Iliopoulos term to the action (1.49),[4]

$$\sum_{\substack{\text{Abelian} \\ \text{factors}}} \left\{ \int \mathrm{d}^4 x \mathrm{d}^4 \theta \, E_{3A} V^A + \left(\int \mathrm{d}^4 x \mathrm{d}^4 \theta \, \mathrm{i} E_A S^A + \text{h.c.} \right) \right\} \ . \qquad (1.50)$$

As an example, let us consider gauging of $U(2)$ isometries in the $N = 2$ NLSM on the Eguchi-Hanson (EH) (hyper-Kähler) manifold [114]. The EH Kähler potential reads [207, 208]

$$K(\eta, \zeta) = \sqrt{1 + \rho^4} - \ln \frac{1 + \sqrt{1 + \rho^4}}{\rho^2} \ , \quad \rho^2 \equiv |\eta|^2 + |\zeta|^2 \ , \qquad (1.51)$$

where η and ζ are the complex coordinates ($N = 1$ chiral superfields) on the EH manifold. The holomorphic components of the Killing vectors (in the basis $\partial_\eta, \partial_\zeta$) are given by

$$k_1 = \mathrm{i}(\eta, \zeta) \ , \quad k_1 = \mathrm{i}(\eta, -\zeta) \ , \quad k_3 = \mathrm{i}(\zeta, \eta) \ , \quad k_4 = (-\zeta, \eta) \ . \qquad (1.52)$$

The first vector k_1 is the generator of the invariant $U(1)$ subgroup, whereas (k_2, k_3, k_4) generate an $SU(2)$ isometry group. All Killing vectors (1.52) are holomorphic in the given coordinate system. The corresponding Killing potentials are

$$X_1 = \sqrt{1 + \rho^4} \ ,$$

$$X_2 = \frac{\sqrt{1 + \rho^4}}{\rho^2} \left(|\eta|^2 - |\zeta|^2 \right) \ ,$$

$$X_3 = \frac{\sqrt{1 + \rho^4}}{\rho^2} \left(\zeta \bar{\eta} + \eta \bar{\zeta} \right) \ , \qquad (1.53)$$

$$X_4 = \mathrm{i} \frac{\sqrt{1 + \rho^4}}{\rho^2} \left(\zeta \bar{\eta} - \eta \bar{\zeta} \right) \ .$$

Since the $SU(2)$ group is semisimple, while the c-obstructions for the $U(1)$ factor vanish in this case, the whole $U(2)$ isometry group can be gauged

---

[4] See Sects. 4.4 and 8.3 for more about gauging NLSM isometries and FI terms in $N = 2$ superspace.

consistently with $N = 1$ supersymmetry according to Sect. 3.4, thus leading to the $N = 1$ supersymmetric gauged NLSM action given by (3.4.48).

The quaternionic structures of the EH manifold are given by (1.18), with $\bar{\Omega}^{i,j}$ being obtained from the potential $\bar{\Omega}^i$ according to (1.16) and (1.23). One finds in the EH case that

$$\bar{\Omega}^\eta = K_\zeta = \left(\frac{\mathrm{d}K}{\mathrm{d}\rho^2}\right)\bar{\zeta} ,$$

$$\bar{\Omega}^\zeta = -K_\eta = -\left(\frac{\mathrm{d}K}{\mathrm{d}\rho^2}\right)\bar{\eta} .$$

(1.54)

It follows now from (1.25) and (1.54) that the generators (k_2, k_3, k_4) are all triholomorphic, whereas k_1 is not. The Killing vector k_1 is thus not holomorphic in the coordinate system where another complex structure takes the canonical form. This implies that the $U(1)$ factor cannot be gauged without breaking $N = 2$ supersymmetry, whereas the $SU(2)$ symmetry can. The potentials P_A in the latter case are easily calculated from (1.38),

$$\{P_1, P_2, P_3\} = \left\{ \mathrm{i}\eta\zeta, \tfrac{1}{2}\mathrm{i}(\zeta^2 - \eta^2), -\tfrac{1}{2}(\zeta^2 + \eta^2) \right\} .$$

(1.55)

The gauged action is given by (1.49) after substituting K, X_A and P_A from above. The Lagrangian turns out to be polynomial in the Wess-Zumino gauge $(V^3 = 0)$, while its component form can be found in [114].

4.2 Special Geometry and Renormalization

In this section we consider the particular class of $N = 4$ supersymmetric 2d NLSM with the so-called *special* geometry, which can be obtained by dimensional reduction from the Seiberg-Witten type 4d action (see also Sect. 8.2). Our analysis [137] is based on extended superspace (see also [209, 210, 211, 212, 213, 214] for details about irreducible extended superfields in 4d, superspace construction of invariant Lagrangians, constraints and superprojectors, etc.). Our presentation of the relevant superfields and their components, the dimensional reduction from 4d to 2d, the special NLSM geometry and the superspace Feynman rules, as well as our proof of the UV-finiteness of the $N = 4$ NLSM in 2d, are all self-contained and very explicit.

4.2.1 $N = 2$ Susy and Restricted Chiral Superfields in 4d

The $N = 2$ superalgebra $SUSY_4^2$ in 4d is a graded extension of the 4d Poincaré algebra. In addition to the generators of the Poincaré algebra $(P_\mu, M_{\lambda\rho})$, the Lie superalgebra $SUSY_4^2$ contains Majorana spinor generators $(Q_\alpha^i, \bar{Q}_i^{\dot\alpha})$ and generators of the $U(2)$ automorphism group transformations $A^i{}_j$, which satisfy the following (anti)commutation relations:

$$\frac{1}{i}[M_{\mu\nu}, M_{\rho\lambda}] = \eta_{\nu\rho}M_{\mu\lambda} - \eta_{\nu\lambda}M_{\mu\rho} + \eta_{\mu\lambda}M_{\nu\rho} - \eta_{\mu\rho}M_{\nu\lambda} ,$$

$$\frac{1}{i}[P_\mu, M_{\nu\lambda}] = \eta_{\mu\nu}P_\lambda - \eta_{\mu\lambda}P_\nu , \quad [P_\mu, P_\nu] = 0 ,$$

$$[Q^i_\alpha, M_{\mu\nu}] = \frac{i}{4}(\sigma_{\mu\nu}Q^i)_\alpha , \quad [\bar{Q}^{\dot\alpha}_j, M_{\mu\nu}] = \frac{i}{4}(\tilde{\sigma}_{\mu\nu}\bar{Q}_j)^{\dot\alpha} ,$$

$$[P_\mu, Q^i_\alpha] = [P_\mu, \bar{Q}^{\dot\alpha}_i] = \{Q^i_\alpha, Q^j_\beta\} = \{\bar{Q}^{\dot\alpha}_i, \bar{Q}^{\dot\beta}_j\} = 0 ,$$

$$\{Q^i_\alpha, \bar{Q}_{j\dot\alpha}\} = \delta^i{}_j\sigma^\mu_{\alpha\dot\alpha}P_\mu ,$$

$$[A^i{}_j, A^l{}_m] = \delta^l{}_j A^i{}_m - \delta^i{}_m A^l{}_j ,$$

$$[A^i{}_j, Q^l_\alpha] = \delta^l{}_j Q^i_\alpha, \quad [A^i{}_j, \bar{Q}^{\dot\alpha}_l] = -\delta^i{}_l \bar{Q}^{\dot\alpha}_j ,$$

$$[A^i{}_j, M_{\mu\nu}] = [A^i{}_j, P_\mu] = 0 . \tag{2.1}$$

In particular, the $U(1)$ generator $B \equiv A^i{}_i$ has the commutation relations

$$[B, Q^i_\alpha] = Q^i_\alpha , \quad [B, \bar{Q}^{\dot\alpha}_i] = -\bar{Q}^{\dot\alpha}_i . \tag{2.2}$$

We use middle Greek letters to denote (flat) spacetime vector indices, $\mu = 0, 1, 2, 3$, early Latin indices to denote (flat) space indices, $\mu = (0, a)$, $a = 1, 2, 3$, early Greek letters to denote spacetime spinor indices, $\alpha = 1, 2$, and middle Latin letters to denote the indices associated with the $SU(2)$ internal symmetry in the fundamental representation, $i = 1, 2$.

In flat $N = 2$ superspace with coordinates $(x^\mu, \theta^\alpha_i, \bar{\theta}^i_{\dot\alpha})$ one can introduce the covariant derivatives $(D^i_\alpha, \bar{D}^{\dot\alpha}_j)$ anticommuting with all the $N = 2$ susy charges and satisfying the algebra

$$\{D^i_\alpha, \bar{D}_{j\dot\alpha}\} = \delta^i{}_j\sigma^\mu_{\alpha\dot\alpha}P_\mu , \quad \{D^i_\alpha, D^j_\beta\} = \{\bar{D}^{\dot\alpha}_i, \bar{D}^{\dot\beta}_j\} = 0 . \tag{2.3}$$

The explicit realization of the $N = 2$ superspace covariant derivatives reads

$$D^i_\alpha = \frac{\partial}{\partial\theta^\alpha_i} - \frac{i}{2}\partial_{\alpha\dot\alpha}\bar{\theta}^{\dot\alpha i} , \quad \bar{D}^{\dot\alpha}_j = \frac{\partial}{\partial\bar{\theta}^j_{\dot\alpha}} - \frac{i}{2}\partial^{\dot\alpha\alpha}\theta_{\alpha j} . \tag{2.4}$$

A general $N = 2$ superfield is the irreducible representation of the enlarged superalgebra $SUSY^{2,D}_4$ defined by adding the covariant derivatives $(D^i_\alpha, \bar{D}^{\dot\alpha}_j)$ to the generators of $SUSY^2_4$. The same superfield is, however, reducible with respect to the susy algebra $SUSY^2_4$, while its irreducible constituents can be defined either by imposing certain superspace constraints or by using superprojectors.

We use the following notation for 4d spacetime:

$$\eta_{\mu\nu} = \text{diag}(+,-,-,-) , \quad \varepsilon_{0123} = 1 . \tag{2.5}$$

The 4×4 Dirac gamma matrices and the charge conjugation matrix C in 4d satisfy the defining relations

$$\{\Gamma_\mu, \Gamma_\nu\} = 2\eta_{\mu\nu} , \quad \Gamma_5 \equiv \Gamma_0\Gamma_1\Gamma_2\Gamma_3 , \quad \Gamma_5^2 = -1 ,$$

$$\Sigma_{\mu\nu} \equiv \frac{i}{2}[\Gamma_\mu, \Gamma_\nu] , \quad C^{-1}\Gamma_\mu C = -\Gamma_\mu^T , \quad C^T = -C . \tag{2.6}$$

A convenient realization of the Dirac gamma matrices, which is useful for an introduction of the 2-component formalism for spinors, is given by

$$\Gamma_0 = \begin{pmatrix} 0 & I_2 \\ I_2 & 0 \end{pmatrix} , \quad \Gamma_a = \begin{pmatrix} 0 & \sigma_a \\ -\sigma_a & 0 \end{pmatrix} , \quad i\Gamma_5 = \begin{pmatrix} I_2 & 0 \\ 0 & -I_2 \end{pmatrix} , \tag{2.7}$$

where σ_a are Pauli matrices.

The 2-component spinor formalism is formally introduced as follows:

$$\Gamma_\mu = \begin{pmatrix} 0 & \sigma_\mu \\ \tilde{\sigma}_\mu & 0 \end{pmatrix} , \quad \psi = \begin{pmatrix} \psi_\alpha \\ \bar{\psi}^{\dot{\alpha}} \end{pmatrix} , \quad \bar{\psi} = (\psi^\alpha, \bar{\psi}_{\dot{\alpha}}) , \tag{2.8}$$

where ψ is a four-component 4d Majorana spinor, $\psi = C\bar{\psi}^T$.

We easily find that

$$\sigma_\mu\tilde{\sigma}_\nu + \sigma_\nu\tilde{\sigma}_\mu = 2\eta_{\mu\nu} , \quad \tilde{\sigma}_\mu\sigma_\nu + \tilde{\sigma}_\nu\sigma_\mu = 2\eta_{\mu\nu} , \tag{2.9}$$

where we have introduced the notation

$$\sigma_{\mu\alpha\dot{\beta}} = (1,\boldsymbol{\sigma})_{\alpha\dot{\beta}} , \quad \tilde{\sigma}_\mu^{\dot{\alpha}\beta} = (1,-\boldsymbol{\sigma})^{\dot{\alpha}\beta} ; \quad \alpha = 1,2, \quad \dot{\alpha}=\dot{1},\dot{2} . \tag{2.10}$$

Similarly, let us define the spacetime derivatives

$$\partial_{\alpha\dot{\alpha}} = (\sigma^\mu\partial_\mu)_{\alpha\dot{\alpha}} \equiv \partial\!\!\!/_{\alpha\dot{\alpha}} , \quad \partial^{\dot{\alpha}\alpha} = (\tilde{\sigma}^\mu\partial_\mu)^{\dot{\alpha}\alpha} \equiv \partial\!\!\!/^{\dot{\alpha}\alpha} . \tag{2.11}$$

Our conventions for 2-component spinors are

$$\psi_\alpha = \varepsilon_{\alpha\beta}\psi^\beta , \quad \psi^\beta = \varepsilon^{\beta\gamma}\psi_\gamma ; \quad \bar{\psi}_{\dot{\alpha}} = \varepsilon_{\dot{\alpha}\dot{\beta}}\bar{\psi}^{\dot{\beta}} , \quad \bar{\psi}^{\dot{\beta}} = \varepsilon^{\dot{\beta}\dot{\alpha}}\bar{\psi}_{\dot{\alpha}} , \tag{2.12}$$

where the 2×2 antisymmetric Levi-Civita symbols

$$\varepsilon_{12} = \varepsilon^{21} = \varepsilon^{\dot{2}\dot{1}} = \varepsilon_{\dot{1}\dot{2}} = -1 , \quad \varepsilon_{\alpha\beta}\varepsilon^{\beta\gamma} = \delta_\alpha^\gamma \tag{2.13}$$

have been introduced. The 4×4 matrices $\Sigma_{\mu\nu}$ and C take the diagonalized form,

$$\Sigma_{\mu\nu} = \frac{1}{2}\begin{pmatrix} \sigma_{\mu\nu} & 0 \\ 0 & \tilde{\sigma}_{\mu\nu} \end{pmatrix} , \quad C = \begin{pmatrix} \varepsilon_{\alpha\beta} & 0 \\ 0 & \varepsilon^{\dot{\alpha}\dot{\beta}} \end{pmatrix} , \tag{2.14}$$

where we have introduced the notation

$$\sigma_{\mu\nu} = \sigma_\mu\tilde\sigma_\nu - \sigma_\nu\tilde\sigma_\mu \ , \quad \tilde\sigma_{\mu\nu} = \tilde\sigma_\mu\sigma_\nu - \tilde\sigma_\nu\sigma_\mu \ . \tag{2.15}$$

Here are some useful identities for the newly introduced derivatives and σ-matrices:

$$\partial_{\alpha\dot\beta}\partial^{\dot\beta\gamma} = \delta_\alpha^\gamma\Box \ , \quad \partial^{\dot\beta\alpha}\partial_{\alpha\dot\gamma} = \delta_{\dot\gamma}^{\dot\beta}\Box \ ,$$

$$\tilde\sigma_\mu^{\dot\alpha\beta} = \varepsilon^{\dot\alpha\dot\beta}\varepsilon^{\beta\alpha}\sigma_{\mu\alpha\dot\beta} = \sigma_\mu^{\dot\beta\alpha} \ , \quad \sigma_{\mu\alpha\dot\beta} = \varepsilon_{\alpha\beta}\varepsilon_{\dot\beta\dot\alpha}\tilde\sigma_\mu^{\dot\alpha\beta} = \tilde\sigma_{\mu\dot\beta\alpha} \ ,$$

$$\sigma^\mu_{\alpha\dot\alpha}\sigma_{\mu\beta\dot\beta} = 2\varepsilon_{\alpha\beta}\varepsilon_{\dot\alpha\dot\beta} \ , \quad \tfrac{1}{2}\varepsilon^{\mu\nu\rho\lambda}\sigma_{\mu\nu} = \mathrm{i}\sigma^{\rho\lambda} \ , \tag{2.16}$$

as well as for their products:

$$\tilde\sigma_\mu\sigma_\nu\tilde\sigma_\lambda = \eta_{\mu\nu}\tilde\sigma_\lambda - \eta_{\mu\lambda}\tilde\sigma_\nu + \eta_{\nu\lambda}\tilde\sigma_\mu - \mathrm{i}\varepsilon_{\mu\nu\lambda\rho}\tilde\sigma^\rho \ ,$$

$$\sigma_\mu\tilde\sigma_\nu\sigma_\lambda = \eta_{\mu\nu}\sigma_\lambda - \eta_{\mu\lambda}\sigma_\nu + \eta_{\nu\lambda}\sigma_\mu + \mathrm{i}\varepsilon_{\mu\nu\lambda\rho}\sigma^\rho \ ,$$

$$\sigma_{\mu\nu}\sigma_\lambda\tilde\sigma_\epsilon = 2(\eta_{\nu\lambda}\sigma_\mu\tilde\sigma_\epsilon - \eta_{\mu\lambda}\sigma_\nu\tilde\sigma_\epsilon) - 2\mathrm{i}\varepsilon_{\mu\nu\lambda\rho}\sigma^\rho\tilde\sigma_\epsilon \ ,$$

$$\tilde\sigma_{\mu\nu}\tilde\sigma_\lambda = 2(\eta_{\nu\lambda}\tilde\sigma_\mu - \eta_{\mu\lambda}\tilde\sigma_\nu) - 2\mathrm{i}\varepsilon_{\mu\nu\lambda\rho}\tilde\sigma^\rho \ ,$$

$$\sigma_{\mu\nu}\sigma_\lambda = 2(\eta_{\nu\lambda}\sigma_\mu - \eta_{\mu\lambda}\sigma_\nu) + 2\mathrm{i}\varepsilon_{\mu\nu\lambda\rho}\sigma^\rho \ ,$$

$$\sigma_{\mu\nu}\sigma_{\lambda\epsilon} = 2(\eta_{\nu\lambda}\sigma_\mu\tilde\sigma_\epsilon - \eta_{\mu\lambda}\sigma_\nu\tilde\sigma_\epsilon - \eta_{\nu\epsilon}\sigma_\mu\tilde\sigma_\lambda + \eta_{\mu\epsilon}\sigma_\nu\tilde\sigma_\lambda)$$
$$+ 2\mathrm{i}(\varepsilon_{\mu\nu\epsilon\rho}\sigma^\rho\tilde\sigma_\lambda - \varepsilon_{\mu\nu\lambda\rho}\sigma^\rho\tilde\sigma_\epsilon) \ ,$$

$$\{\sigma_{\mu\nu}, \sigma_{\lambda\epsilon}\} = 4(\eta_{\nu\lambda}\eta_{\epsilon\mu} - \eta_{\mu\lambda}\eta_{\epsilon\nu} - \eta_{\nu\epsilon}\eta_{\lambda\mu} + \eta_{\mu\epsilon}\eta_{\lambda\nu})$$
$$+ 2\mathrm{i}(\varepsilon_{\mu\nu\epsilon\rho}\sigma^\rho\tilde\sigma_\lambda - \varepsilon_{\mu\nu\lambda\rho}\sigma^\rho\tilde\sigma_\epsilon - \varepsilon_{\lambda\epsilon\mu\rho}\sigma^\rho\tilde\sigma_\nu + \varepsilon_{\lambda\epsilon\nu\rho}\sigma^\rho\tilde\sigma_\mu) \ ,$$

$$\{\sigma_{\mu\nu}, \sigma_{\lambda\epsilon}\}\partial^\nu\partial^\epsilon = 8(\partial_\mu\partial_\lambda - \eta_{\mu\lambda}\Box) \ . \tag{2.17}$$

Some useful traces of the products of σ-matrices are given by

$$\mathrm{tr}(\sigma_\mu\tilde\sigma_\nu) = 2\eta_{\mu\nu} \ ,$$

$$\mathrm{tr}(\sigma_\rho\tilde\sigma_\lambda\sigma_\mu\tilde\sigma_\nu) = 2(\eta_{\rho\lambda}\eta_{\mu\nu} - \eta_{\rho\mu}\eta_{\lambda\nu} + \eta_{\rho\nu}\eta_{\lambda\mu}) + 2\mathrm{i}\varepsilon_{\rho\lambda\mu\nu} \ ,$$

$$\mathrm{tr}(\sigma_{\mu\nu}\sigma_{\lambda\rho}) = 8(\eta_{\mu\rho}\eta_{\nu\lambda} - \eta_{\mu\lambda}\eta_{\nu\rho}) + 8\mathrm{i}\varepsilon_{\mu\nu\lambda\rho} \ ,$$

$$\mathrm{tr}(\tilde\sigma_{\mu\nu}\tilde\sigma_{\lambda\rho}) = 8(\eta_{\mu\rho}\eta_{\nu\lambda} - \eta_{\mu\lambda}\eta_{\nu\rho}) - 8\mathrm{i}\varepsilon_{\mu\nu\lambda\rho} \ . \tag{2.18}$$

In order to simplify our notation, we sometimes suppress spinorial indices in their 'natural' position,

$$(\theta\sigma_\mu\bar\eta) \equiv \theta^\alpha\sigma_{\mu\alpha\dot\beta}\bar\eta^{\dot\beta} = -(\bar\eta\tilde\sigma_\mu\theta) \equiv -\bar\eta_{\dot\alpha}\tilde\sigma_\mu^{\dot\alpha\beta}\theta_\beta \ ,$$

$$(\theta\sigma_{\mu\nu}\eta) \equiv \theta^\alpha(\sigma_{\mu\nu})_\alpha{}^\beta\eta_\alpha = -(\eta\sigma_{\mu\nu}\theta) \ . \tag{2.19}$$

As regards the internal $SU(2)$ indices, we have

$$\psi^i = \varepsilon^{ij}\psi_j \; , \quad \varepsilon^{ij}\varepsilon_{jk} = \delta^i_k \; , \quad \varepsilon_{12} = \varepsilon^{21} = -1 \; . \tag{2.20}$$

We raise and lower spinor $SL(2,\mathbf{C})$ indices by using the charge conjugation matrix $\varepsilon^{\alpha\beta}$, $\varepsilon_{\alpha\beta}$ and $\varepsilon^{\dot{\alpha}\dot{\beta}}$, $\varepsilon_{\dot{\alpha}\dot{\beta}}$, whereas fundamental $SU(2)$ indices are raised and lowered by the use of ε^{ij} and ε_{ij}. The notation $\boldsymbol{\tau} \equiv \{\tau_m\}$, $m = 1,2,3$, is used to denote an extra set of Pauli matrices, in order to distinguish them from the $SL(2,\mathbf{C})$-related Pauli matrices $\{\sigma_a\}$ introduced earlier. We have

$$A^i{}_j = (\tau_m)^i{}_j A_m \equiv \boldsymbol{\tau}^i{}_j \cdot \boldsymbol{A} \; , \quad A_m = \tfrac{1}{2}(\tau_m)^i{}_j A^j{}_i \equiv \tfrac{1}{2}\mathrm{tr}(\tau_m A) \; , \tag{2.21}$$

for any $SU(2)$ vector $\boldsymbol{A}$, and the identity

$$\mathrm{tr}(\tau_m \tau_k) = 2\delta_{mk} \; . \tag{2.22}$$

The non-vanishing products of the $N = 2$ superspace Grassmann anticommuting coordinates can be conveniently arranged into the following irreducible combinations:

$$\theta_{ij} = \theta^\alpha_i \theta_{\alpha j} \; , \quad \bar{\theta}_{ij} = \bar{\theta}_{\dot{\alpha}i} \bar{\theta}^{\dot{\alpha}}_j \; ,$$

$$\theta_{\alpha\beta} = \theta_{\alpha i}\theta^i_\beta \; , \quad \bar{\theta}_{\dot{\alpha}\dot{\beta}} = \bar{\theta}_{\dot{\alpha}i}\bar{\theta}^i_{\dot{\beta}} \; ,$$

$$(\theta^3)^i_\alpha = \frac{\partial}{\partial\theta^\alpha_i}\theta^4 \; , \quad (\bar{\theta}^3)^{\dot{\alpha}}_i = \frac{\partial}{\partial\bar{\theta}^i_{\dot{\alpha}}}\bar{\theta}^4 \; , \tag{2.23}$$

where we have introduced the maximal non-vanishing product of four θ's as $\theta^4 = \theta^1_1\theta^1_2\theta^2_1\theta^2_2$, and similarly for $\bar{\theta}^4$. Here are some useful identities:

$$\theta^\alpha_i(\theta^3)^j_\beta = \delta^j_i \delta^\alpha_\beta \theta^4 \; , \quad \bar{\theta}^j_{\dot{\beta}}(\bar{\theta}^3)^{\dot{\alpha}}_i = \delta^j_i \delta^{\dot{\alpha}}_{\dot{\beta}}\bar{\theta}^4 \; ,$$

$$\theta^{\alpha\beta}\theta^{\gamma\epsilon} = -2\theta^4(\varepsilon^{\alpha\gamma}\varepsilon^{\beta\epsilon} + \varepsilon^{\beta\gamma}\varepsilon^{\alpha\epsilon}) \; , \quad \bar{\theta}_{\dot{\alpha}\dot{\beta}}\bar{\theta}_{\dot{\gamma}\dot{\epsilon}} = -2\bar{\theta}^4(\varepsilon_{\dot{\alpha}\dot{\gamma}}\varepsilon_{\dot{\beta}\dot{\epsilon}} + \varepsilon_{\dot{\beta}\dot{\gamma}}\varepsilon_{\dot{\alpha}\dot{\epsilon}}) \; ,$$

$$\theta_{ij}\theta_{kl} = 2\theta^4(\varepsilon_{ik}\varepsilon_{jl} + \varepsilon_{il}\varepsilon_{jk}) \; , \quad \bar{\theta}^{ij}\bar{\theta}^{kl} = 2\bar{\theta}^4(\varepsilon^{ik}\varepsilon^{jl} + \varepsilon^{il}\varepsilon^{jk}) \; ,$$

$$\theta_{\alpha i}\theta_{\beta j} = \tfrac{1}{2}\varepsilon_{\alpha\beta}\theta_{ij} - \tfrac{1}{2}\varepsilon_{ij}\theta_{\alpha\beta} \; , \quad \bar{\theta}^i_{\dot{\alpha}}\bar{\theta}^j_{\dot{\beta}} = -\tfrac{1}{2}\varepsilon_{\dot{\alpha}\dot{\beta}}\bar{\theta}^{ij} - \tfrac{1}{2}\varepsilon^{ij}\bar{\theta}_{\dot{\alpha}\dot{\beta}} \; ,$$

$$\theta^\alpha_i \theta^\beta_j \theta_{kl} = -\varepsilon^{\alpha\beta}(\varepsilon_{ik}\varepsilon_{jl} + \varepsilon_{il}\varepsilon_{jk})\theta^4 \; ,$$

$$\theta^\alpha_i \theta^\beta_j \theta^{\gamma\epsilon} = \varepsilon_{ij}(\varepsilon^{\alpha\gamma}\varepsilon^{\beta\epsilon} + \varepsilon^{\beta\gamma}\varepsilon^{\alpha\epsilon})\theta^4 \; ,$$

$$\theta^{ij}\theta_{j\alpha} = 3(\theta^3)^i_\alpha \; , \quad \bar{\theta}_{ij}\bar{\theta}^{\dot{\alpha}j} = 3(\bar{\theta}^3)^{\dot{\alpha}}_i \; ,$$

$$\theta_{\alpha i}\theta_{\beta\gamma} = -\varepsilon_{\alpha\beta}(\theta^3)_{i\gamma} , \quad \theta^{\alpha\beta}\theta_{ij} = \bar{\theta}^{\dot\alpha\dot\beta}\bar{\theta}_{ij} = 0 ,$$

$$\theta_{\alpha i}\theta_{kl} = -\varepsilon_{ik}(\theta^3)_{\alpha l} - \varepsilon_{il}(\theta^3)_{\alpha k} ,$$

$$\theta_i^\alpha\theta_j^\beta\theta_k^\gamma\theta_l^\epsilon = -\tfrac{1}{2}\varepsilon_{ij}\varepsilon_{kl}(\varepsilon^{\alpha\gamma}\varepsilon^{\beta\epsilon} + \varepsilon^{\beta\gamma}\varepsilon^{\alpha\epsilon})\theta^4 + \tfrac{1}{2}\varepsilon^{\alpha\beta}\varepsilon^{\gamma\epsilon}(\varepsilon_{ik}\varepsilon_{jl} + \varepsilon_{il}\varepsilon_{jk})\theta^4 ,$$

$$\bar\theta^i_{\dot\alpha}\bar\theta^j_{\dot\beta}\bar\theta^k_{\dot\gamma}\bar\theta^l_{\dot\epsilon} = -\tfrac{1}{2}\varepsilon^{ij}\varepsilon^{kl}(\varepsilon_{\dot\alpha\dot\gamma}\varepsilon_{\dot\beta\dot\epsilon} + \varepsilon_{\dot\beta\dot\gamma}\varepsilon_{\dot\alpha\dot\epsilon})\bar\theta^4 + \tfrac{1}{2}\varepsilon_{\dot\alpha\dot\beta}\varepsilon_{\dot\gamma\dot\epsilon}(\varepsilon^{ik}\varepsilon^{jl} + \varepsilon^{il}\varepsilon^{jk})\bar\theta^4 ,$$

$$\theta^4 = \tfrac{1}{12}\theta_{ij}\theta^{ij} = -\tfrac{1}{12}\theta_{\alpha\beta}\theta^{\alpha\beta} , \quad \bar\theta^4 = \tfrac{1}{12}\bar\theta_{ij}\bar\theta^{ij} = -\tfrac{1}{12}\bar\theta_{\dot\alpha\dot\beta}\bar\theta^{\dot\alpha\dot\beta} . \qquad (2.24)$$

The same notation and identities are valid for the products of the superspace covariant derivatives D_α^i and $\bar{D}_i^{\dot\alpha}$.

A normalization of the (Berezin) integration measure over the anticommuting $N = 2$ superspace coordinates is fixed by the conditions

$$\int \mathrm{d}^4\theta\, \theta^4 = 1 , \quad \int \mathrm{d}^4\bar\theta\, \bar\theta^4 = 1 . \qquad (2.25)$$

All our (anti)symmetrizations are defined with unit weight, as usual. The Grassmann derivatives are all chosen to be the left ones.

We are now in a position to consider a 4d chiral superfield Φ in $N = 2$ extended superspace, which satisfies the off-shell constraints

$$\bar{D}_i^{\dot\alpha}\Phi = 0 , \qquad D_\alpha^i\bar\Phi = 0 , \qquad (2.26)$$

where the bar means complex conjugation. Unlike its $N = 1$ superspace chiral counterpart, the complex $N = 2$ chiral representation $(\Phi, \bar\Phi)$ is reducible [215, 216], since there exists a projection that makes this supermultiplet real in a certain sense [210]. The generalized (off-shell) reality condition in $N = 2$ extended superspace reads

$$D^4\Phi = \Box\bar\Phi . \qquad (2.27)$$

Solving the constraints (2.26) and (2.27) is fully straightforward [215, 216], and it results in the irreducible *restricted chiral* $N = 2$ superfield,

$$\Phi = \exp\left\{-\tfrac{i}{2}\theta_i\partial\bar\theta^i\right\}\left[A + \theta_i^\alpha\psi_\alpha^i - \tfrac{1}{2}\theta_i^\alpha(\tau_m)^i{}_j\theta_\alpha^j C_m + \tfrac{1}{8}\theta_i^\alpha(\sigma_{\mu\nu})_\alpha{}^\beta\theta_\beta^i F^{\mu\nu}\right.$$

$$\left. -\mathrm{i}(\theta^3)^{i\alpha}\partial_{\alpha\dot\beta}\bar\psi_i^{\dot\beta} + \theta^4\Box\bar{A}\right] ,$$

$$(2.28)$$

whose components are given by

$$\left\{A, \quad \psi_\alpha^i, \quad C_m, \quad \widetilde{F}_{\mu\nu}, \quad \bar\psi_i^{\dot\alpha}, \quad \bar{A}\right\} , \qquad (2.29)$$

where A is a complex scalar, ψ^i is a Majorana spinor isodoublet, $\widetilde{F}_{\mu\nu}$ is an antisymmetric tensor satisfying the constraint

$$\partial^\mu \widetilde{F}_{\mu\nu} = 0 \ , \tag{2.30a}$$

and C_m is a real auxiliary isovector. The constraint $(2.30a)$ can be rewritten in terms of the dual tensor $F_{\mu\nu}$ as

$$\varepsilon^{\mu\nu\lambda\rho} \partial_\nu F_{\lambda\rho} = 0 \ , \tag{2.30b}$$

while the latter is usually interpreted as the 'Bianchi identity' for the Abelian vector field strength $F_{\mu\nu} = \partial_\mu A_\nu - \partial_\nu A_\mu$. This observation allows us to interpret the restricted chiral $N = 2$ multiplet as an $N = 2$ *vector* supermultiplet. We prefer to deal with the 4d constraint (2.30), and solve it (in terms of a pseudo-scalar — see (2.36) below) *after* dimensional reduction to 2d.

The $N = 2$ susy transformation properties of the components (2.29) can be most easily deduced from the chiral $N = 2$ superspace (z^μ, θ_i^α), where they are induced by a shift of the coordinates,

$$\delta\theta_i^\alpha = \varepsilon_i^\alpha \ , \qquad \delta z^\mu = -\mathrm{i}\theta_i^\alpha (\sigma^\mu)_{\alpha\dot\beta} \bar\varepsilon^{\dot\beta i} \ , \tag{2.31}$$

with infinitesimal $N = 2$ susy parameters ε_i^α and $\bar\varepsilon^{\dot\alpha i}$. We find

$$\delta A = \varepsilon_i \psi^i \ ,$$

$$\delta\psi_\alpha^i = -(\tau_m)^i{}_j C_m \varepsilon_\alpha^j - \mathrm{i}\partial_{\alpha\dot\beta} A \bar\varepsilon^{\dot\beta i} + \tfrac{1}{4}(\sigma_{\mu\nu}\varepsilon^i)_\alpha \widetilde{F}^{\mu\nu} \ ,$$

$$\delta C_m = -\tfrac{\mathrm{i}}{2}(\bar\varepsilon_i \partial \tau_{mj}^i \psi^j) - \tfrac{\mathrm{i}}{2}(\varepsilon_i \partial \tau_{mj}^i \bar\psi^j) \ ,$$

$$\delta\widetilde{F}^{\mu\nu} = -\tfrac{1}{2}\varepsilon^{\mu\nu\lambda\rho}\partial_\lambda \left[(\varepsilon_i \sigma_\rho \bar\psi^i) + (\bar\varepsilon_i \tilde\sigma_\rho \psi^i) \right] \ , \tag{2.32}$$

$$\delta\bar\psi^{\dot\beta i} = (\tau_m)^i{}_j C_m \bar\varepsilon^{\dot\beta j} + \mathrm{i}\partial^{\dot\beta\alpha}\bar A \varepsilon_\alpha^i + \tfrac{1}{4}(\tilde\sigma_{\mu\nu}\bar\varepsilon^i)^{\dot\beta} \widetilde{F}^{\mu\nu} \ ,$$

$$\delta\bar A = \bar\varepsilon_i \bar\psi^i \ .$$

4.2.2 Dimensional Reduction and 2d, $N = 4$ Special NLSM

A plain Dimensional Reduction (DR) to 2d means that all 4d fields are supposed to be only dependent upon two spacetime coordinates, i.e. $\partial_2 = \partial_3 = 0$. The 4d gamma matrices can be decomposed as

$$\Gamma_\mu = \gamma_\mu \otimes I_2 \ , \qquad \Gamma_{2,3} = \gamma_3 \otimes \mathrm{i}\tau_{1,2} \ , \tag{2.33}$$

where the index μ now takes only two values $\mu = 0, 1$, and γ_μ are the 2d gamma matrices, $\gamma_3 = \gamma_0\gamma_1$ and $\gamma_3^2 = 1$. Similarly, the 4d charge conjugation matrix C_4 can be written down in terms of the 2d charge conjugation matrix C_2 as

$$C_4 = C_2 \otimes \tau_1 \ , \qquad C_2\gamma^\mu C_2^{-1} = -\gamma_\mu^{\mathrm{T}} \ . \tag{2.34}$$

A 4d isospinor Ψ^i takes the form

$$\Psi^i = \begin{pmatrix} \psi^i \\ \tilde{\psi}^i \end{pmatrix} , \quad \text{where} \quad \tilde{\psi}^i = \varepsilon^{ij} C_2 \bar{\psi}_j^{\mathrm{T}} . \tag{2.35}$$

A general solution to the constraint (2.30) after DR to 2d is given by

$$\widetilde{F}_{\mu 2} = \frac{1}{2}\varepsilon_{\mu\nu}\partial^\nu(B + \bar{B}) , \quad \widetilde{F}_{\mu 3} = \frac{1}{2i}\varepsilon_{\mu\nu}\partial^\nu(B - \bar{B}) ,$$

$$\widetilde{F}_{01} = m = \text{const.} , \quad \widetilde{F}_{23} \equiv D , \tag{2.36}$$

in terms of the propagating complex (pseudo-)scalar field B, the auxiliary real scalar field D, and the dimensionful constant m. The restricted chiral superfield Φ after DR to 2d reads

$$\begin{aligned}
\Phi = \exp\left\{-\tfrac{i}{2}\bar{\theta}_i \partial\!\!\!/\, \theta^i\right\} &\left(A + \bar{\theta}_i \tfrac{1}{2}(1 + \gamma_3)\psi^i + \bar{\psi}^i \tfrac{1}{2}(1 + \gamma_3)\theta_i \right. \\
&- \bar{\theta}_i \tau^i_{mj} \tfrac{1}{2}(1 + \gamma_3)\theta^j C_m - \tfrac{1}{2}\bar{\theta}_i \gamma_3 \theta^i m - \tfrac{1}{2}\bar{\theta}_i \gamma_3 \theta^i D \\
&+ \tfrac{i}{2}\bar{\theta}_i \partial\!\!\!/\, \tfrac{1}{2}(1 + \gamma_3)\theta^i B + \tfrac{i}{2}\bar{\theta}_i \partial\!\!\!/\, \tfrac{1}{2}(1 - \gamma_3)\tilde{\theta}^i \bar{B} \\
&- \tfrac{1}{3}\left\{\bar{\theta}^i \tfrac{1}{2}(1 + \gamma_3)\theta^j + (i \leftrightarrow j)\right\}\left[\bar{\theta}_j \tfrac{1}{2}(1 + \gamma_3)i\partial\!\!\!/\, \psi_i \right. \\
&\left. - \partial_\mu \bar{\psi}_i \tfrac{1}{2}(1 - \gamma_3)i\gamma^\mu \theta_j\right] + \tfrac{1}{12}\left\{\bar{\theta}_i \tfrac{1}{2}(1 + \gamma_3)\theta_j + (i \leftrightarrow j)\right\} \\
&\left. \times \left\{\bar{\theta}^i \tfrac{1}{2}(1 + \gamma_3)\theta^j + (i \leftrightarrow j)\right\} \Box\bar{A}\right) .
\end{aligned} \tag{2.37}$$

The $N = 4$ susy transformation laws of its 2d components are given by

$$\begin{aligned}
\delta A &= \bar{\varepsilon}_i \tfrac{1}{2}(1 + \gamma_3)\psi^i + \bar{\psi}^i \tfrac{1}{2}(1 + \gamma_3)\varepsilon_i , \\
\delta\psi^i &= -\tau^i_{mj} C_m \gamma_3 \varepsilon^j - \tfrac{1}{2}(1 + \gamma_3)i\partial\!\!\!/\, A\varepsilon^i + \tfrac{1}{2}(1 - \gamma_3)i\partial\!\!\!/\, \bar{A}\varepsilon^i \\
&\quad - iD\varepsilon^i - 2m\gamma_3 \varepsilon^i + i\partial\!\!\!/\, \tilde{\varepsilon}^i \bar{B} , \\
\delta C_m &= -\tfrac{1}{2}\bar{\varepsilon}_i \partial\!\!\!/\,(\tau_m)^i_{\ j}\psi^j + \text{h.c.} , \\
\delta B &= -\bar{\varepsilon}_i \gamma_3 \tilde{\psi}^i , \quad \delta\bar{B} = -\bar{\tilde{\psi}}^i \gamma_3 \varepsilon_i , \\
\delta D &= -\tfrac{1}{2}\bar{\varepsilon}_i \gamma_3 \partial\!\!\!/\, \psi^i + \text{h.c.}
\end{aligned} \tag{2.38}$$

Hence, the dimensionally reduced restricted chiral $N = 4$ miltiplet in 2d has the following $8 + 8$ off-shell components:

$$\left\{A , \quad B , \quad \psi^i , \quad C_m , \quad D\right) , \tag{2.39}$$

where A and B are 2d scalars, ψ^i is a 2d Dirac spinor isodoublet, C_m is a real auxiliary isovector, and D is a real auxiliary scalar. The dimensional constant m induces spontaneous supersymmetry breaking that leads to the appearance of massive fields and Goldstone fermions in 2d field theory (see below).

The most general (of the second-order in spacetime derivatives) invariant action of the restricted chiral superfields [5] $\{\Phi^a\}$ in 4d reads [217]

$$S = \int d^4x d^4\theta \, V(\Phi) + \text{h.c.} , \qquad (2.40)$$

and it is fully determined by a single holomorphic function $V(\Phi)$. In particular, a free action is given by

$$S_0 = -\tfrac{1}{2} \int d^4x d^4\theta \, \Phi^2 . \qquad (2.41)$$

Being (canonically) dimensional in 4d, the superfield Φ implies dimensional coupling constants in the 4d action (2.40), which is ultimately responsible for its non-renormalizability (Subsect. 4.2.3). After DR the 2d superfield Φ gets a vanishing (canonical) dimension, whereas the action (2.40) takes the form

$$S = \int d^2x d^2\theta_R d^2\tilde{\theta}_L \, V(\Phi) + \text{h.c.} , \qquad (2.42)$$

where both the function $V(\Phi)$ and the 2d full superspace measure are dimensionless.

It is straightforward to deduce the 2d component Lagrangian out of the action (2.42). We find [217]

$$
\begin{aligned}
\mathcal{L} \;=\; & g_{ab}\partial_\mu A^a \partial^\mu \bar{A}^b - g_{ab}\bar{\psi}^{ia}\tfrac{1}{2}(1+\gamma_3)\mathrm{i}\partial\!\!\!/\psi_i^b - g_{ab}\bar{\psi}^{ia}\tfrac{1}{2}(1-\gamma_3)\mathrm{i}\overleftarrow{\partial\!\!\!/}\,\psi_i^b \\
& + g_{ab}C_m^a C_m^b - g_{ab}m^a m^b + g_{ab}D^a D^b + g_{ab}\partial_\mu B^a \partial^\mu \bar{B}^b + 2\mathrm{i}g_{ab}m^a D^b \\
& + g_{ab}\varepsilon^{\mu\nu}\partial_\mu \bar{B}^a \partial_\nu B^b - g_{abc}\bar{\psi}^a\tfrac{1}{2}(1+\gamma_3)\tau_m\psi^b C_m^c \\
& + 2g_{abc}\bar{\psi}_i^a\tfrac{1}{2}(1+\gamma_3)\psi^{bi}m^c + +2\mathrm{i}g_{abc}\bar{\psi}_i^a\tfrac{1}{2}(1+\gamma_3)\psi^{bi}D^c \\
& - \tfrac{1}{2}\mathrm{i}g_{abc}\left[\bar{\psi}_i^a\tfrac{1}{2}(1+\gamma_3)\gamma^\mu\tilde{\psi}^{bi}\partial_\mu \bar{B}^c + \bar{\tilde{\psi}}_i^a\tfrac{1}{2}(1-\gamma_3)\gamma^\mu\psi^{bi}\partial_\mu B^c\right] \\
& - \tfrac{1}{24}\left\{(\bar{\psi}^{ia}\psi^{ib})(\bar{\psi}_i^c\psi_j^d + \bar{\psi}_j^c\psi_i^d) + (\bar{\psi}^{ia}\gamma_3\psi^{ib})(\bar{\psi}_i^c\gamma_3\psi_j^d + i \leftrightarrow j)\right\} \\
& - \tfrac{1}{12}g_{abcd}(\bar{\psi}^{ia}\gamma_3\psi^{jb})(\bar{\psi}_i^c\psi_j^d + i \leftrightarrow j) + \text{h.c.} ,
\end{aligned}
$$
$$\qquad (2.43)$$

where we have introduced the notation

$$g_{a_1\cdots a_n}(A) = \left.\frac{\partial^n V(\Phi)}{\partial\Phi^{a_1}\cdots\partial\Phi^{a_n}}\right|_{\Phi=A} . \qquad (2.44)$$

The auxiliary fields C_m and D can be removed from (2.43) by using their algebraic equations of motions, whose solutions are given by

---

[5] The early Latin letters numerate different (super)fields in what follows.

$$C_m^c = \tfrac{1}{2}[(g+\bar{g})^{-1}]^{cb}\left\{\tfrac{1}{2}(g_{bad}+\bar{g}_{bad})(\bar{\psi}_i^a \tau_m \psi^{di})\right.$$

$$\left.+\tfrac{1}{2}(g_{bad}-\bar{g}_{bad})(\bar{\psi}_i^a \tau_m \gamma_3 \psi^{di})\right\}\ ,$$

$$D^c = -\tfrac{1}{2}[(g+\bar{g})^{-1}]^{cb}\left\{2\mathrm{i}(g_{bd}-\bar{g}_{bd})m^d + \mathrm{i}(g_{bda}+\bar{g}_{bda})\tfrac{1}{2}(\bar{\psi}_i^d \psi^{ai})\right.$$

$$\left.+\tfrac{1}{2}\mathrm{i}(g_{bda}-\bar{g}_{bda})\bar{\psi}_i^d \gamma_3 \psi^{ai}\right\}\ . \tag{2.45}$$

The complex scalars (A^a, B^a) and the 2d Dirac spinor isodoublets ψ_i^a are the physical fields. The bosonic sector $(\psi = 0)$ of the final 2d NLSM is

$$\mathcal{L}_{\text{bos.}} = 2\frac{\partial^2 H}{\partial N^a \partial N^b}\left\{\partial_\mu N^a \partial^\mu N^b + \partial_\mu M^a \partial^\mu M^b + \partial_\mu Q^a \partial^\mu Q^b + \partial_\mu P^a \partial^\mu P^b\right\}$$

$$+4\frac{\partial^2 H}{\partial N^a \partial M^b}\varepsilon^{\mu\nu}\partial_\mu Q^a \partial_\nu P^b - 4\frac{\partial^2 H}{\partial N^a \partial M^b}m^a m^b$$

$$-4m^a\frac{\partial^2 H}{\partial M^a \partial M^b}\left[\frac{\partial^2 H}{\partial N \partial N}\right]^{-1bc}\frac{\partial^2 H}{\partial M^c \partial N^d}m^d\ , \tag{2.46}$$

where we have introduced the notation

$$A \equiv \frac{1}{\sqrt{2}}(M + \mathrm{i}N)\ , \quad B \equiv \frac{1}{\sqrt{2}}(P + \mathrm{i}Q)\ , \quad H(M,N) \equiv \mathrm{Re}\,V(A)\ , \tag{2.47}$$

and used the identities

$$\frac{\partial^2 H}{\partial N^a \partial M^b} = \frac{\partial^2 H}{\partial N^b \partial M^a}\ , \quad \frac{\partial^2 H}{\partial N^a \partial N^b} = -\frac{\partial^2 H}{\partial M^a \partial M^b}\ . \tag{2.48}$$

A free theory corresponds to the function $H = \tfrac{1}{8}(N^2 - M^2)$.

Equation (2.46) is a particular 2d NLSM Lagrangian of the type

$$\mathcal{L}(\phi, E) = \tfrac{1}{2}g_{ab}(\phi)\partial_\mu \phi^a \partial^\mu \phi^b + \tfrac{1}{2}g_{ab}(\phi)\partial_\mu E^a \partial^\mu E^b$$

$$+\tfrac{1}{2}f_{ab}(\phi)\varepsilon^{\mu\nu}\partial_\mu E^a \partial_\nu E^b + W(\phi)\ , \tag{2.49}$$

which has a scalar potential $W(\phi)$ in addition to the NLSM kinetic terms, if $m \neq 0$. The scalar potential can be easily deduced from (2.46), whereas the geometry of the 2d, $N = 2$ NLSM (2.49) is described by the quadratic form $S = g_{ab}\mathrm{d}\phi^a \mathrm{d}\phi^b$ and the external two-form $T = f_{ab}\mathrm{d}E^a \wedge \mathrm{d}E^b$. In their turn, they are fully determined by the single 'pre-potential' V.

It is easy to verify that the Ricci tensor of the special 2d NLSM metric does not vanish,

$$R_{ab} \propto \left(\frac{\partial^2 H}{\partial N^2}\right)^{-2}\left\{\left(\frac{\partial^3 H}{\partial N^3}\right)^2 + \left(\frac{\partial^3 H}{\partial M^2 \partial N}\right)^2\right\} \neq 0\ . \tag{2.50}$$

In particular, this immediately implies that the special Kähler geometry is *not* hyper-Kähler because the latter requires a vanishing Ricci tensor (Subsect. 4.1.1).

4.2.3 UV-Finiteness Versus Non-Renormalizability

To derive the manifestly $N = 4$ supersymmetric Feynman rules in 2d, we have to construct the generating functional of the Green functions in extended superspace. We assume for notational simplicity that $m = 0$ in what follows.

The restricted chiral superfield Φ and its conjugate antichiral superfield $\overline{\Phi}$ are related via the generalized 'reality' constraint (2.27). It is, therefore, possible to treat the superfield Φ as a 'real' general superfield in the chiral $N = 2$ extended subsuperspace. Since the $N = 2$ vector multiplet action (2.40) depends upon the gauge vector A_μ only via its field strength $F_{\mu\nu}$, no gauge fixing is necessary. The constraint (2.30) is automatically preserved in $N = 2$ superfield perturbation theory. The variational derivative $\delta\Phi(1)/\delta\Phi(2)$ reads

$$\frac{\delta\Phi(1)}{\delta\Phi(2)} = \exp\left(-\tfrac{i}{2}\theta_1\partial_1\bar{\theta}_1 - \tfrac{i}{2}\theta_2\partial_2\bar{\theta}_2\right)\delta^4(\theta_1 - \theta_2)\delta^2(x_1 - x_2) , \qquad (2.51a)$$

which is quite similar to the standard rule for the (unextended) $N = 1$ chiral superfields, see e.g., [135, 136, 137]. In our $N = 2$ case, we have in addition

$$\frac{\delta\overline{\Phi}(1)}{\delta\Phi(2)} = \frac{D_1^4}{\Box_1}\frac{\delta\Phi(1)}{\delta\Phi(2)} = \exp\left(+\tfrac{i}{2}\theta_1\partial_1\bar{\theta}_1 - \tfrac{i}{2}\theta_2\partial_2\bar{\theta}_2 - i\theta_2\partial_1\bar{\theta}_1\right)\delta^2(x_1 - x_2) \qquad (2.51b)$$

and

$$\frac{\delta\overline{\Phi}(1)}{\delta\overline{\Phi}(2)} = \exp\left(+\tfrac{i}{2}\theta_1\partial_1\bar{\theta}_1 + \tfrac{i}{2}\theta_2\partial_2\bar{\theta}_2\right)\delta^4(\bar{\theta}_1 - \bar{\theta}_2)\delta^2(x_1 - x_2) . \qquad (2.51c)$$

Let us now introduce a source J, which is also a restricted chiral $N = 2$ superfield, and write down a free action with the source term,

$$S[\Phi, J] = \int d^2x\, d^4\theta \left(-\tfrac{1}{2}\Phi^2 + \Phi J\right) . \qquad (2.52)$$

It is not difficult to verify that the corresponding equations of motion are

$$\Phi = J , \qquad (2.53)$$

while they are obviously consistent with the fact that J is the restricted chiral superfield like Φ. It is now clear how to define the quantum generating functional in question, after replacing the free action (2.41) by the interacting action (2.40),

$$Z[J] = \exp(iW[J]) = \exp\left\{i\left[\sum_{n=3}^{+\infty}\frac{g_n}{n!}\int d^2x\, d^4\theta \left(i\frac{\delta}{\delta J}\right)^n \right.\right.$$

$$\left.\left. + \sum_{n=3}^{+\infty}\frac{g_n}{n!}\int d^2x\, d^4\bar{\theta} \left(i\frac{D^4}{\Box}\frac{\delta}{\delta J}\right)^n\right]\right\}\exp\left(\frac{i}{2}\int d^2x\, d^4\theta\, J^2\right) , \qquad (2.54)$$

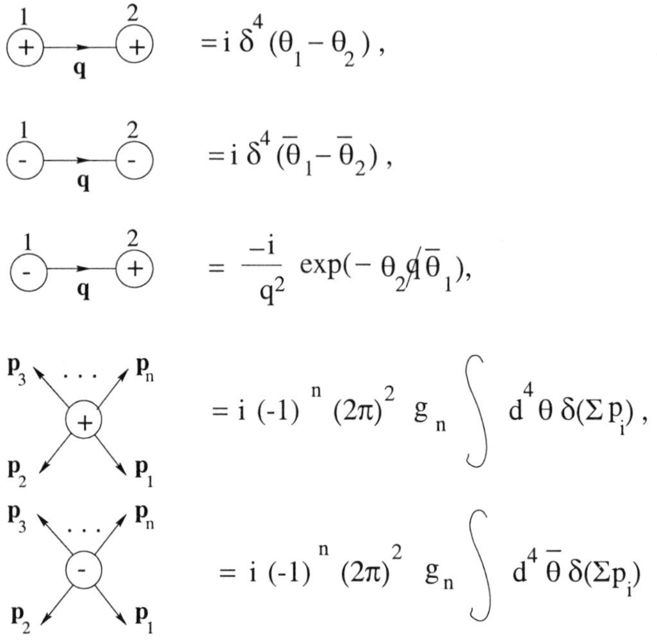

$$\text{\raisebox{0pt}{$\overset{1}{\oplus}$}}\!\!-\!\!\xrightarrow{q}\!\!\text{\raisebox{0pt}{$\overset{2}{\oplus}$}} \quad = i\,\delta^4(\theta_1 - \theta_2),$$

$$\text{\raisebox{0pt}{$\overset{1}{\ominus}$}}\!\!-\!\!\xrightarrow{q}\!\!\text{\raisebox{0pt}{$\overset{2}{\ominus}$}} \quad = i\,\delta^4(\bar\theta_1 - \bar\theta_2),$$

$$\text{\raisebox{0pt}{$\overset{1}{\ominus}$}}\!\!-\!\!\xrightarrow{q}\!\!\text{\raisebox{0pt}{$\overset{2}{\oplus}$}} \quad = \frac{-i}{q^2}\,\exp(-\theta_2\,\slashed{q}\,\bar\theta_1),$$

$$= i\,(-1)^n\,(2\pi)^2\,g_n\int d^4\theta\,\delta(\Sigma p_i),$$

$$= i\,(-1)^n\,(2\pi)^2\,g_n\int d^4\bar\theta\,\delta(\Sigma p_i)$$

Fig. 4.1. Feynman rules for extended supergraphs in the $N = 4$ NLSM with special geometry in 2d

where we have introduced the coupling constants

$$g_n \equiv \left.\frac{\partial^n V(\Phi)}{\partial \Phi^n}\right|_{\Phi=0}. \tag{2.55}$$

The superspace variational derivatives with respect to the source J are defined similarly to (2.51a).

Expanding (2.54) in a perturbation series results in the $N = 4$ extended superfield Feynman rules of our 2d NLSM. When using those rules in calculations of extended Feynman supergraphs, one notices a cancellation of all the exponential factors associated with the chirality shifts at vertices, which is quite similar to the well-known situation with simple supergraphs [135, 136, 137]. This observation allows us to introduce the simplified Feynman rules (in momentum space) for our $N = 4$ extended supergraphs, as shown in Fig. 4.1.

As regards the UV divergences, it is enough to consider the supergraph loops in which the chiralities (denoted by signs) of vertices alternate (this is obvious from Fig. 4.1). Note that a supergraph, all of whose vertices have the same chirality, identically vanishes. Let us consider first a cycle that may be either part of a multi-loop graph or the one-loop graph itself (Fig. 4.2).

The relevant analytic expression corresponding to Fig. 4.2 reads

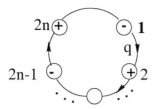

Fig. 4.2. A one-loop supergraph with vertices of alternating chirality

$$\int \frac{\mathrm{d}^2 q}{(q^2)^{2n}} \exp\left[(\theta_{2n} - \theta_2)q\bar{\theta}_1 + (\theta_2 - \theta_4)q\bar{\theta}_3 + \ldots + (\theta_{2n-2} - \theta_{2n})q\bar{\theta}_{2n-1}\right] \;,$$

$$(2.56)$$

where we have only taken into account the dependence upon the internal (loop) momenta. The number of terms in the exponential (2.56) can be easily reduced by one after a simple regrouping. Hence, to estimate the divergence index of the supergraph in Fig. 4.2, we can write

$$\int \frac{\mathrm{d}^2 q}{(q^2)^{2n}} (q^2)^{2(n-1)} \; < \; \infty \tag{2.57}$$

at large loop momenta. This means that our NLSM does not have one-loop UV divergences at all, since the loop integrals at external momenta are automatically UV-finite by naive dimensional reasons.

This remarkable observation can be generalized to all loop orders of perturbation theory. Let us consider the divergence index ω of an arbitrary supergraph with L loops, I internal lines and $V_{\pm}$ vertices of chirality $(\pm)$. We have

$$\omega = 2L - 2I + 4\mu \;, \tag{2.58}$$

where μ denotes the *effective* number of exponents with internal momenta. Taking $V_{-} \leq V_{+}$ for definiteness, let us now estimate μ. The exponential factor generically reads as

$$\sum_{i=1}^{V_{-}} \left(\sum_{j=1}^{V_{+}} \theta_{2j}q_{ji} \right) \bar{\theta}_{2i-1} \quad . \tag{2.59}$$

Hence, $\mu \leq V_{-}$. Moreover, the momentum conservation at the vertices,

$$\sum_{i=1}^{V_{-}} q_{ji} = 0 \;, \tag{2.60}$$

implies a linear dependence of the coefficients in (2.59),

$$\sum_{i=1}^{V_{-}} \left(\sum_{j=1}^{V_{+}} \theta_{2j}q_{ji} \right) = 0 \;, \tag{2.61}$$

and, therefore, $\mu \leq V_- - 1$. By using the topological relations

$$L - I = 1 - V , \qquad V \equiv V_- + V_+ , \tag{2.62}$$

we finally get

$$\omega \leq 2 - 2(V_- + V_+) + 4\left[\min(V_+, V_-) - 1\right] \leq -2 . \tag{2.63}$$

This means that $N = 4$ supersymmetry of the 2d NLSM under consideration results in the exact cancellation of all the UV divergences to all orders
of quantum perturbation theory! To appreciate this fact, one should compare the simple analysis given above with the conventional component approach, and verify the eventual cancellation of the UV divergences between
the bosonic and fermionic loops. Though this is hardly possible in any loop
order, we checked the actual divergence cancellation in components up to
three loops [218].

Taking into account the non-vanishing parameter m does not change the
main result. The IR convergence can always be achieved by adding a (soft)
mass term.

The UV-finiteness of *any* $N = 4$ supersymmetric NLSM in 2d can also
be proved by other (indirect) methods [219, 220, 221, 222, 223, 224]. More
examples can be found in Sect. 4.3. Here we would like to emphasize the
unique and remarkable role of extended superspace in extracting the geometrical data and proving the UV-finiteness in a very clear and straightforward
way.

Let us now consider the UV properties of the original 4d theory (2.40) to
be rewritten in the form

$$S = -\frac{1}{2} \int \mathrm{d}^4 x \mathrm{d}^4 \theta \, \Phi^2 + \int \mathrm{d}^4 x \mathrm{d}^4 \theta \mathrm{d}^4 \bar{\theta} \left\{ \delta^4(\bar{\theta}) V_{\text{int}}(\Phi) + \delta^4(\theta) V_{\text{int}}(\bar{\Phi}) \right\} . \tag{2.64}$$

In the simplest non-trivial case, we have an interaction

$$V_{\text{int}}(\Phi) = \frac{1}{3!} g \Phi^3 . \tag{2.65}$$

The corresponding component Lagrangian was calculated in [215], but we
do not need it for our purposes here. The 4d, $N = 2$ supersymmetric model
(2.64) with the self-interaction (2.65) can be formally considered as the $N = 2$
supersymmetric generalization of the standard $N = 1$ supersymmetric Wess-
Zumino model [125] in 4d.

The free equations of motion are given by

$$D_{ij} \Phi = 0 . \tag{2.66}$$

Let us introduce the $N = 2$ superfield source $C_{ij} = C_{ji}$ as

$$D_{ij} \Phi = C_{ij} . \tag{2.67}$$

The superspace constraints (2.26) and (2.27) on the superfield Φ imply certain constraints on the superfield C_{ij}, because of (2.67). We find

$$D_\alpha^{(i} C^{jk)} = \overline{D}_{\stackrel{.}{\alpha}}^{(i} C^{jk)} = 0 \quad \text{and} \quad C^{ij} = \varepsilon^{ik}\varepsilon^{jl}\overline{C}_{kl} \; . \tag{2.68}$$

The off-shell extended superspace constraints (2.68) define a 4d real *tensor* $N = 2$ superfield C^{ij} (see Sect. 4.3 for more details).

A solution to (2.67) reads

$$\Phi = \frac{1}{12\Box}\overline{D}^{ij}C_{ij} \; . \tag{2.69}$$

Hence, the generating functional of the connected Green functions of the free theory with the source C_{ij} takes the form

$$W_0(C_{ij}) = \int \mathrm{d}^4 x \mathrm{d}^4\theta \mathrm{d}^4\bar\theta\, C_{ij} \left(\frac{D^{ij}\overline{D}^{kl}}{288\Box^3} \right) C_{kl} \; . \tag{2.70}$$

After a redefinition of the superfield argument according to the relation

$$\frac{1}{12\Box}\overline{D}^{ij}C_{ij} \equiv J \; , \tag{2.71}$$

we are in a position to write down the 4d generating functional of the interacting QFT (2.64) with the potential (2.65) in the form that is quite similar to the two-dimensional equation (2.54), namely,

$$Z[J] = \exp(\mathrm{i}W[J]) = \exp\left\{ \frac{\mathrm{i}g}{3!}\int \mathrm{d}^4x\mathrm{d}^4\theta \left(\mathrm{i}\frac{\delta}{\delta J} \right)^3 \right.$$

$$\left. + \int \mathrm{d}^4x\mathrm{d}^4\bar\theta \left(\mathrm{i}\frac{D^4}{\Box}\frac{\delta}{\delta J} \right)^3 \right\} \exp\left(\frac{\mathrm{i}}{2}\int \mathrm{d}^4x\mathrm{d}^4\theta\, J^2 \right) \; . \tag{2.72}$$

The 4d variational derivative $\delta/\delta J$ with respect to the restricted chiral superfield J is also quite similar to (2.51a),

$$\frac{\delta J(1)}{\delta J(2)} = \exp\left(-\tfrac{\mathrm{i}}{2}\theta_1\slashed{\partial}_1\bar\theta_1 - \tfrac{\mathrm{i}}{2}\theta_2\slashed{\partial}_2\bar\theta_2 \right) \delta^4(\theta_1 - \theta_2)\delta^4(x_1 - x_2) \; . \tag{2.73}$$

The Feynman rules for the 4d extended supergraphs follow from the perturbative expansion of (2.72), as usual. There are two types of vertices according to their chirality, as in the 2d case considered above. In the one-loop approximation, all the relevant supergraphs are schematically pictured in Fig. 4.3.

It is straightforward to check that the supergraphs (b), (c), (d) and (e) identically vanish, whereas the supergraphs (a), (f) and (g) are UV divergent. Since the UV counterterms to the divergent graphs do not have the structure of the original action (in fact, their component expressions are of higher order in derivatives), the 4d super-Φ^3 theory is non-renormalizable [225]. The non-renormalizability proof in the case of a general function $V(\Phi)$ goes along similar lines.

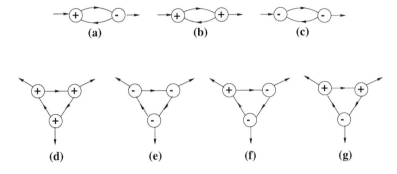

Fig. 4.3. One-loop supergraphs in the 4d super-Φ^3 theory

4.3 NLSM in Projective Superspace

The NLSM considered in the preceeding section represent the very special class of $N = 4$ supersymmetric 2d NLSM. The most general 2d, $N = 4$ NLSM can be formulated in (a 2d version of) harmonic (or twistor) superspace [226].

Being applied to hyper-Kähler geometry, the basic idea of twistor construction is to extend a hyper-Kähler manifold $\mathcal{M}$ by the two-sphere (S^2), representing all non-canonical complex structures of $\mathcal{M}$ (Subsect. 4.1.1), to a larger complex manifold $\mathcal{M} \times S^2$ [201]. Twistors form a local basis on the Riemann sphere $S^2 = SO(3)/SO(2) \cong CP^1$. Similarly, having identified the $SU(2)$ automorphisms of $N = 2$ susy algebra in 4d with (the covering group of) the rotational $SO(3)$ symmetry of the Riemann sphere, one may naturally arrive at the idea of extending $N = 2$ superspace to its product with the Riemann sphere S^2 or with the projective complex line CP^1. Both $N = 2$ susy and its $SU(2)$ automorphisms can be made manifest after introducing the twistors $u_i^\pm \in SU(2)$, called *harmonics*. In the original Harmonic Superspace (HSS) approach [220], one considers the product of $N = 2$ extended superspace with the group $SU(2)$ instead of the coset $SU(2)/U(1)$, by assigning all harmonic superfields to be *equivariant* with respect to the $U(1)$ subgroup (Subsect. 4.4.2). [6]

The Riemann sphere S^2 considered as the complex projective line CP^1 is obtained by patching together two copies of the complex plane with coordinates ξ and $\tilde{\xi}$ related by a conformal transformation $\tilde{\xi} = \xi^{-1}$ in the overlapping region. This parametrization is used in the Projective Superspace (PSS) extension of the conventional 4d, $N = 2$ superspace [229, 230]. As is clear from this definition of PSS, it may not be sufficient to merely assume a polynomial dependence of the PSS superfields upon ξ, which essentially amounts to the assumption about a finite number of the auxiliary

---

[6] In the mathematical literature, this construction is known as a *flag* manifold. We refer the reader to [227, 228] for a discussion of HSS from the viewpoint of flag manifolds.

fields (Subsect. 4.3.1). Allowing a generic (holomorphic) dependence of the PSS superfields upon ξ is essentially equivalent to introducing HSS.

The HSS method provides a universal approach to deal with 2d, $N = 4$ and 4d, $N = 2$ supersymmetric field theories. The HSS generically implies an infinite number of auxiliary fields that make a derivation of component results from HSS highly non-trivial (cf. the twistor geometry versus the standard geometry). The PSS approach can be formulated with a finite number of auxiliary fields that are associated with certian (non-universal) off-shell tensor versions of a *hypermultiplet* (Subsect. 4.3.1). The 4d, $N = 2$ PSS is also close to the usual 4d, $N = 1$ superspace (Sect. 4.1). We adopt a balanced approach in this book, by introducing both methods and emphasizing their advantages and disadvantages in various applications to NLSM (see Chap. 8 also).

In this section we introduce the PSS action describing a universal self-interaction of 4d, $N = 2$ tensor multiplets and their higher-isospin generalizations. We also consider some general geometrical properties of the associated 4d NLSM obtained via the (generalized) Legendre transform in components or in superfields, give some explicit examples, and calculate Feynman rules for extended supergraphs. The PSS is used as a tool for extracting information about geometry and UV renormalization properties of a large class of maximally supersymmetric NLSM. The use of the HSS approach for a derivation of four-dimensional hyper-Kähler metrics is considered in Sect. 4.4.

4.3.1 $N = 2$ Tensor Multiplet and Its Generalizations in 4d

Let us introduce 4d, $N = 2$ superfields $L^{i_1 \cdots i_n}$ that are totally symmetric with respect to their $SU(2)_{\mathrm{R}}$ internal indices and satisfy the following constraints [229, 230, 231, 232, 233]:

$$D_\alpha^{(k} L^{i_1 \cdots i_n)} = \overline{D}_{\dot{\alpha}}^{(k} L^{i_1 \cdots i_n)} = 0 \,, \qquad n \geq 2 \,. \tag{3.1}$$

In the case of an even number of indices, $n = 2p$, the superfields (3.1) are supposed to satisfy the reality condition

$$\overline{L}_{i_1 \cdots i_{2p}} \equiv (L^{i_1 \cdots i_{2p}})^* = \varepsilon_{i_1 j_1} \cdots \varepsilon_{i_{2p} j_{2p}} L^{i_1 \cdots i_{2p}} \,. \tag{3.2}$$

The off-shell superspace constraints (3.1) and (3.2) at $n = 2$ define an $N = 2$ tensor multiplet in 4d, in precise agreement with (2.68). For higher n, (3.1) and (3.2) define the $O(n)$ *projective* (or generalized tensor) $N = 2$ multiplets [230, 231, 232, 233]. They are all irreducible off-shell representations of $N = 2$ extended 4d susy, with superspin $Y = 0$ and superisospin $I = (n - 2)/2$. The list of their $8(n - 1) + 8(n - 1)$ off-shell field components is most conveniently represented in terms of the $SU(2)$ Young tableaux

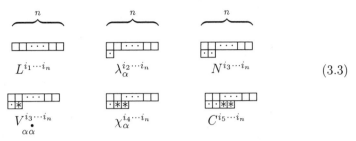

$$(3.3)$$

where the boxes with dots and stars denote the $N = 2$ superspace covariant derivatives D_α^i and $\bar{D}_i^{\dot\alpha}$, respectively. It follows from matching the numbers of the bosonic and fermionic degrees of freedom in (3.3) that the vector $V_{\alpha\dot\alpha}$ of an $N = 2$ tensor multiplet ($n = 2$) is *conserved*, $\partial^{\alpha\dot\alpha}V_{\alpha\dot\alpha} = 0$. The vector $V^{i_3\cdots i_n}_{\alpha\dot\alpha}$ of any projective $N = 2$ multiplet with $n > 2$ turns out to be an *unconstrained* (general) vector field.

Choosing $n = 1$ in (3.1) results in the special (on-shell) case defining the Fayet-Sohnius (FS) hypermultiplet (without a central charge) [234, 235]. It is not difficult to verify that the constraints (3.1) at $n = 1$ imply free equations of motion, $\Box L^i = 0$ [236]. We exclude the FS hypermultiplet from our considerations until Sect. 4.4.

The constraints (3.1) and (3.2) can be generalized even further, in order to incorporate various types of the so-called *relaxed* hypermultiplets, with the simplest example being provided by the constraints [237]

$$D_\alpha^{(i}L^{jk)} = D_{\alpha l}L^{ijkl} , \qquad D_\alpha^{(i}L^{jklm)} = 0 , \qquad (3.4)$$

and their conjugates, subject to the reality conditions (3.2). Unlike the $N = 2$ tensor multiplet itself ($n = 2$), the relaxed hypermultiplet (3.4) does not contain a conserved vector, so that it can be minimally coupled to $N = 2$ gauge fields in $N = 2$ superspace [237, 238]. Equation (3.4) is just the simplest pattern of relaxation, $2 \to 4$. It is not difficult to define other types of relaxation, e.g. $n \to (n+2) \to (n+4) \to \ldots \to (n+2k)$ with any positive integer k. The *infinitely* relaxed FS hypermultiplet ($n = 1$), according to the scheme $1 \to 3 \to 5 \to \ldots$ up to $k = \infty$, in fact, exactly amounts to an off-shell realization of a FS hypermultiplet in harmonic superspace [239, 240, 241].

An $N = 2$ supersymmetric self-interaction of the generalized (and relaxed) $N = 2$ tensor multiplets in 4d can be defined by applying the $N = 2$ PSS method [229, 230, 231, 232, 233]. Let us introduce a function $G(L^{i_1\cdots i_n}; \xi, \eta)$ satisfying the linear differential equations

$$\nabla_\alpha G \equiv (D_\alpha^1 + \xi D_\alpha^2)G = 0 , \qquad \Delta_{\dot\alpha} G \equiv (\bar{D}_{\dot\alpha}^1 + \eta \bar{D}_{\dot\alpha}^2)G = 0 . \qquad (3.5)$$

It is straightforward to verify that a general solution to (3.5) reads

$$G = G(Q_{(n)}(\xi); \xi) , \qquad \eta = \xi , \qquad Q_{(n)}(\xi) \equiv \xi_{i_1} \cdots \xi_{i_n} L^{i_1\cdots i_n} , \qquad \xi_i \equiv (1, \xi) , \qquad (3.6)$$

in terms of an *arbitrary* function $G(Q(\xi), \xi)$ on the right-hand side of this equation.

Since the function G does not depend upon some (a half) of the Grassmann coordinates of $N = 2$ superspace by its definition (3.5), its integration over the rest of the $N = 2$ superspace coordinates is invariant under $N = 2$ supersymmetry. This leads to the following $N = 2$ invariant action:

$$S[L] = \int d^4x \, \frac{1}{2\pi i} \oint_C d\xi \, (1 + \xi^2)^{-4} \widetilde{\nabla}^2 \widetilde{\Delta}^2 G(Q_A, \xi) + \text{h.c.} , \qquad (3.7)$$

where we have introduced the new superspace derivatives,

$$\widetilde{\nabla}_\alpha \equiv \xi D_\alpha^1 - D_\alpha^2 , \quad \widetilde{\Delta}_{\dot\alpha} \equiv \xi \overline{D}_{\dot\alpha}^1 - \overline{D}_{\dot\alpha}^2 , \qquad (3.8)$$

in the directions orthogonal to the 'vanishing' directions defined by (3.5). The integration contour C in the complex ξ-plane is supposed to be chosen to make the action (3.7) non-trivial (i.e. not equal to zero). The points $\xi_\pm = \pm i$, where the linear independence of the derivatives (3.5) and (3.8) breaks down, should be outside of the contour C.

The form of the PSS action (3.7) is universal, i.e. it applies to any set of (generalized and relaxed) $N = 2$ tensor multiplets $L_{(n)}$ that may enter the action via the corresponding function $Q_{(n)}(\xi)$ defined by (3.6), while the whole action is governed by a holomorphic potential G. The existence of a single potential in the maximally supersymmetric NLSM was also noticed in other approaches [239, 242, 243]. Our construction is easily generalizable to the case of relaxed hypermultiplets too. For example, in the case of the relaxed Howe-Stelle-Townsend (HST) hypermultiplet (3.4), we find

$$Q_{(2),\text{rel.}} = Q_{(2)} - \frac{5}{4} \frac{\partial Q_{(4)}}{\partial \xi} . \qquad (3.9)$$

Some comments are in order. In the odd case of $n = 2p + 1$ the conjugated superfields $\overline{L}^{i_1 \cdots i_n}$ may also enter the action (3.7) via the corresponding polynomial $\overline{Q}_{(2p+1)}(\xi)$. The factor $(1 + \xi^2)^{-4}$ in (3.7) was introduced to simplify the transformation properties of the integrand under the internal $SU(2)$ automorphisms of the $N = 2$ susy algebra. It is worth mentioning that the PSS construction of the action (3.7) is in general not invariant under the internal $SU(2)$ rotations of two supersymmetries. The variable ξ should be considered as a CP^1 (projective) coordinate having the rational transformation law

$$\xi' = \frac{\bar{a}\xi - \bar{b}}{a + b\xi} , \qquad (3.10)$$

whose complex $SU(2)$-transformation parameters (a, b) are constrained by the condition $|a|^2 + |b|^2 = 1$. The action (3.7) is, nevertheless, $SU(2)$ invariant if the function G transforms as

$$G(Q', \xi') = \frac{1}{(a + b\xi)^2} G(Q, \xi) \qquad (3.11)$$

under the projective transformations (3.10), modulo an additive total derivative. Because of the transformation law

$$Q'_{(n)}(\xi') = \frac{1}{(a + b\xi)^n} Q_{(n)}(\xi) , \qquad (3.12)$$

equation (3.11) is a severe restriction on the choice of G. For example, in the case of an $O(2)$ tensor multiplet ($n = 2$), the $SU(2)$ invariant PSS Lagrangian should be essentially linear in Q_2 outside the origin of the complex ξ-plane (in fact, up to the $\log Q_2$ factor transforming a line integral into a contour integral). This observation is closely related to the existence of the *improved* (i.e. $N = 2$ superconformally invariant) action of the $N = 2$ tensor multiplet (Subsects. 4.3.3 and 8.3.2). Similarly, as far as an $O(4)$ projective multiplet ($n = 4$) is concerned, the $SU(2)$-invariant (improved) PSS potential $G(Q_4)$ should be proportional to $\sqrt{Q_4}$ (Subsect. 4.4.3).

The restrictions (3.5) can be interpreted as *analyticity* conditions in PSS (cf. Subsect. 4.4.2). The projective $O(n)$ multiplets ($n \geq 2$) can be equally defined in HSS as *analytic* harmonic superfields subject to certain off-shell HSS constraints (Subsect. 8.3.2) whose precise form can be deduced from (3.1) and (3.2) by contracting their free $SU(2)_R$ indices with harmonics [239, 240] — see Subsects. 4.4.3, 8.3.2 and 8.3.3 too.

4.3.2 Reduction to 4d, $N = 1$ Superspace and Components

In this subsection we deduce the component results that follow from the general $N = 2$ superspace equations of the preceeding subsection in some particular cases, and also rewrite them in terms of the conventional 4d, $N = 1$ superfields [231, 232, 233, 244].

Let us consider first the standard $N = 2$ tensor superfield L^{ij}. Its components (3.3) can be identified with the leading (i.e. θ-independent) parts of the independent (secondary) $N = 2$ superfields,

$$L^{ij} , \qquad \lambda_\alpha^j = D_{\alpha k} L^{jk} , \qquad \bar{\lambda}^{\cdot j}_{\ \alpha} = \bar{D}_{\dot\alpha k} L^{jk} ,$$

$$M = -2D_{ij} L^{ij} , \qquad V_{\alpha\dot\alpha} = i[\bar{D}_{\dot\alpha j}, D_{\alpha i}] L^{ij} , \qquad (3.13)$$

which are obtained from the defining constraints (3.1) and (3.2) with $n = 2$ by applying the $N = 2$ superspace covariant derivatives. It is now straightforward to calculate the 4d invariant action (3.7) with $Q = \xi_i \xi_j L^{ij}$ by using the definitions given above. We find

$$S = \int d^4x \frac{1}{2\pi i} \oint_C \left\{ \frac{\partial^2 G}{\partial Q^2} \left[2\partial_\mu L^{ij} \partial_\mu L_{ij} - \tfrac{1}{9} V_\mu^2 - \tfrac{4}{3}(1 + \xi^2)^{-1} \right.\right.$$

$$\left. \times \left(V_\mu \partial^\mu L^{\mathrm{T}} + \tfrac{4}{3}\bar{\lambda}_{\dot\alpha} i\bar\sigma_\mu^{\dot\alpha\alpha}\partial^\mu \lambda_\alpha^{\mathrm{T}} - \tfrac{4}{3}\lambda^\alpha i\sigma^\mu_{\alpha\dot\alpha}\partial_\mu \bar\lambda^{\dot\alpha\mathrm{T}} \right) + \tfrac{1}{36}\overline{M}M \right]$$

$$+ \frac{\partial^3 G}{\partial Q^3}\left[-\tfrac{2}{27}(M\bar\lambda^2 + \overline{M}\lambda^2) + \tfrac{8i}{27} V^\mu \lambda^\alpha \sigma_{\mu\alpha\dot\alpha}\bar\lambda^{\dot\alpha} + \tfrac{16}{9}\lambda^\alpha \frac{i\not{\partial}_{\alpha\dot\alpha} L^{\mathrm{T}}}{1 + \xi^2}\bar\lambda^{\dot\alpha} \right]$$

$$+ \tfrac{16}{81}\frac{\partial^4 G}{\partial Q^4}\lambda^2\bar\lambda^2 \bigg\} d\xi + \text{h.c.} \,,$$

$$(3.14)$$

where we have introduced the notation

$$\lambda_\alpha(\xi) \equiv \xi_i \lambda_\alpha^i \,, \quad \lambda_\alpha^{\mathrm{T}}(\xi) \equiv \xi_i \lambda_{i\alpha} \,, \quad \lambda^2 \equiv \lambda^\alpha(\xi)\lambda_\alpha(\xi) \,,$$

$$L^{\mathrm{T}} \equiv \xi_m \xi_n L_m{}^n \,, \quad V_{\alpha\dot\alpha} \equiv \sigma^\mu_{\alpha\dot\alpha} V_\mu \,, \quad \partial_\mu V^\mu = 0 \,. \qquad (3.15)$$

The constraint $\partial_\mu V^\mu = 0$ can be solved in terms of a gauge antisymmetric tensor $E_{\mu\nu}$ as

$$V^\mu = \tfrac{1}{2}\varepsilon^{\mu\nu\lambda\rho}\partial_\nu E_{\lambda\rho} \,, \qquad (3.16)$$

subject to the gauge invariance

$$\delta E_{\lambda\rho} = \partial_\lambda \zeta_\rho - \partial_\rho \zeta_\lambda \,, \qquad (3.17)$$

with the vector gauge parameter $\zeta_\lambda(x)$. However, we have not arrived at a supersymmetric NLSM yet, which is supposed to be entirely formulated in terms of scalars and spinors. A simple way to solve this problem in components is given by the *Legendre transform*, by trading the conserved vector V_μ for a scalar Lagrange multiplier in 4d [245]. Though being fully consistent with the physical on-shell content of the field theory under investigation, the component Legendre transform is, however, inconsistent with the original (off-shell and linearly realized) $N = 2$ supersymmetry. It is worth noticing that there is no problem *after* Dimensional Reduction (DR) to 2d because a 2d transverse vector is equivalent to a pseudo-scalar (Sect. 4.2). The supersymmetric treatment in 4d uses duality transformations either in $N = 1$ superspace [246] (see below) or in $N = 2$ harmonic superspace (Sect. 8.3).

In the component approach, the constraint $\partial_\mu V^\mu = 0$ is managed by adding the extra term $-R\partial_\mu V^\mu$ with the Lagrange multiplier R to the action (3.14). It is not difficult to calculate the *non-linear* $N = 2$ susy transformation law of the scalar field R from requiring the invariance of the whole action. We find

$$\delta R = B_{ijmn}\delta_m{}^{(i} L_k{}^n \varepsilon^{\alpha k)}\lambda_\alpha^j + \text{h.c.} \,, \qquad (3.18)$$

where we have introduced the notation

$$B_{ijmn}(L) \equiv \frac{1}{2\pi i}\oint d\xi \frac{4}{9}\frac{\partial^3 Q}{\partial Q^3}\frac{\xi_i\xi_j\xi_m\xi_n}{1 + \xi^2} \,. \qquad (3.19)$$

It is now straightforward to eliminate the auxiliary fields (M, V_μ) from the action, by using their algebraic equations of motion. This results in a NLSM Lagrangian in terms of the scalars (L^{ij}, R) and the spinors $(\lambda_\alpha^i, \overline{\lambda}_j^{\dot\alpha})$ only, with a hyper-Kähler NLSM metric by $N = 2$ susy of the construction. The $N = 2$ susy (scalar) transformation laws given by (3.18) and

$$\delta L^{ij} = -\tfrac{1}{2}\varepsilon^{(i\alpha}\lambda_\alpha^{j)} + \text{h.c.} \tag{3.20}$$

determine the hyper-Kähler complex structures according to the general rules in components (Sect. 4.1). The bosonic part of the NLSM Lagrangian reads

$$\mathcal{L}_{\text{bos.}} = \frac{1}{a}\big\{(a^2 + 4b^2)\partial_\mu T \partial^\mu T + a^2\partial_\mu S \partial^\mu S + 4(a^2 + c^2)\partial_\mu\phi\partial^\mu\phi$$
$$+\tfrac{9}{4}\partial_\mu R\partial^\mu R + 8bc\partial_\mu\phi\partial^\mu T - 6b\partial_\mu T\partial^\mu R - 6c\partial_\mu\phi\partial^\mu R\big\} \tag{3.21}$$
$$\equiv \tfrac{1}{2}g_{AB}(X)\partial_\mu X^A\partial^\mu X^B \ ,$$

where we have introduced the notation

$$a \equiv 2\,\text{Re}\sum\text{Res}\frac{\partial^2 G}{\partial Q^2} \ , \quad b \equiv 2\,\text{Im}\sum\text{Res}\frac{\partial^2 G}{\partial Q^2}\frac{\xi}{1 + \xi^2} \ ,$$

$$c \equiv 2\,\text{Im}\sum\text{Res}\frac{\partial^2 G}{\partial Q^2}\frac{\xi^2 - 1}{\xi^2 + 1} \ , \quad L^{11} - L^{22} \equiv -iT \ ,$$

$$L^{11} + L^{22} \equiv S \ , \quad L^{12} \equiv -i\phi \ . \tag{3.22}$$

In particular, the determinant of the NLSM metric is

$$\det g_{AB} = (12a)^2 \ . \tag{3.23}$$

It is also worth mentioning that the hyper-Kähler metric g_{AB} does not depend upon R, i.e. it has a (translational) isometry. The presence of triholomorphic isometries is, in fact, a general feature of all hyper-Kähler metrics to be obtained from the self-interaction of $N = 2$ tensor multiplets ($n = 2$) in PSS. Since a generic hyper-Kähler metric does not have any isometries, its superspace derivation may require higher $O(k)$ projective multiplets in PSS or analytic superfields in HSS.

In two dimensions (2d), any NLSM metric having an isometry can be dualized (sect. 2.5). This is obvious in our case after DR to 2d: the constraint $\partial_\mu V^\mu = 0$ on the dimensionally reduced 2d vector V^μ can be solved in terms of a pseudo-scalar B as $V^\mu = \varepsilon^{\mu\nu}\partial_\nu B$, which results in the *dual* 2d NLSM with torsion and linearly realized $N = 4$ supersymmetry.

The 4d, $N = 2$ invariant action (3.7) of an $N = 2$ tensor multiplet can be easily rewritten in 4d, $N = 1$ superspace [244],

$$S = \int \mathrm{d}^4x\mathrm{d}^4\theta\frac{1}{2\pi\mathrm{i}}\oint_C \mathrm{d}\xi\xi^{-2}\,G(\chi - \mathrm{i}\xi g + \xi^2\bar{\chi}, \xi) + \text{h.c.} \ , \tag{3.24}$$

in terms of the $N = 1$ complex chiral superfield $\chi = L^{11}|$ and the $N = 1$ real linear superfield $g = L^{12}|$, where $|$ denotes the $(\bar{\theta}_2, \theta_1)$-independent part of a superfield or an operator. The $N = 1$ multiplets χ and g together constitute an $N = 2$ tensor multiplet in 4d. The $N = 1$ superspace covariant derivatives are given by $D = D^2|$ and $\overline{D} = \overline{D}^1|$, whereas the $N = 1$ superfields χ and g satisfy the constraints

$$\overline{D}^{\dot{\alpha}}\chi = D_\alpha \overline{\chi} = 0 , \tag{3.25}$$

and

$$\overline{D}_{\dot{\alpha}}\overline{D}^{\dot{\alpha}}g = D^\alpha D_\alpha g = 0 , \tag{3.26}$$

respectively. The Legendre transform in 4d, $N = 1$ superspace allows one [246] to trade the $N = 1$ linear superfield g for yet another $N = 1$ chiral superfield ψ. Being applied to the action (3.24), this leads to the 4d, $N = 1$ NLSM action

$$S = \int d^4x d^4\theta \, K(\psi + \overline{\psi}, \chi, \overline{\chi}) , \tag{3.27}$$

whose hyper-Kähler NLSM metric has the Kähler potential

$$K = \left[\frac{1}{2\pi i} \oint_C d\xi \xi^{-2} G(\chi - i\xi H + \xi^2 \overline{\chi}, \xi) + \text{h.c.}\right] + (\psi + \overline{\psi})H , \tag{3.28}$$

where the function $H(\chi, \overline{\chi}, \psi + \overline{\psi})$ is a solution to the algebraic equation

$$\psi + \overline{\psi} = \frac{1}{2\pi} \oint_C d\xi \xi^{-1} \frac{\partial G}{\partial Q}(Q, \xi) + \text{h.c.} , \qquad Q = \chi - i\xi H + \xi^2 \overline{\chi} . \tag{3.29}$$

For instance, the choice of $n = 2$ and $G(Q, \xi) = F/\xi$ above, with an arbitrary *holomorphic* function $F(Q^A)$ of $N = 2$ tensor multiplets, $A = 1, 2, \ldots, m$, and the contour C encircling the origin, was considered in [247]. Equation (3.24) yields in this case the $N = 2$ NLSM whose $N = 1$ superspace action is

$$
\begin{aligned}
S[\chi, \bar{\chi}; H] &= \int d^4x d^4\theta \left\{ F_A(\chi)\bar{\chi}^A - \tfrac{1}{2}F_{AB}(\chi)H^A H^B + \text{h.c.} \right\} \\
&= \int d^4x d^4\theta \left\{ K(\chi, \bar{\chi}) - \tfrac{1}{2}g_{AB}(\chi, \bar{\chi})H^A H^B \right\} ,
\end{aligned}
\tag{3.30}
$$

where we have used the notation [247]

$$F_A = \frac{\partial F}{\partial Q^A} , \qquad F_{AB} = \frac{\partial^2 F}{\partial Q^A \partial Q^B} , \tag{3.31}$$

and

$$K(\chi, \bar{\chi}) = F_A \bar{\chi}^A + \bar{F}_A \chi^A , \qquad g_{AB}(\chi, \bar{\chi}) = F_{AB}(\chi) + \bar{F}_{AB}(\bar{\chi}) . \tag{3.32}$$

The $N = 1$ superspace Legendre transform eliminates all the $N = 1$ linear superfields H^A in favour of some new $N = 1$ chiral superfields ψ_A according to the algebraic equation (3.29),

$$\psi_A + \bar{\psi}_A = g_{AB}(\chi, \bar{\chi}) H^B \tag{3.33}$$

whose solution is given by

$$H^A = g^{AB}(\chi, \bar{\chi})(\psi_A + \bar{\psi}_A) \tag{3.34}$$

in terms of the inverse matrix g^{AB}. Substituting (3.34) back into (3.30) yields a Kähler potential in $N = 1$ superspace,

$$K(\chi, \bar{\chi}, \psi, \bar{\psi}) = K(\chi, \bar{\chi}) + \tfrac{1}{2} g^{AB}(\chi, \bar{\chi})(\psi_A + \bar{\psi}_A)(\psi_B + \bar{\psi}_B) . \tag{3.35}$$

By construction, the Kähler potential (3.35) is parametrized by a single holomorphic potential $F(\chi)$, while it must give rise to a hyper-Kähler NLSM metric by $N = 2$ extended supersymmetry. This construction is known as the c-map [248] between the special Kähler geometry defined in terms of the Kähler potential $K(\chi, \bar{\chi})$ of (3.32) and of complex dimension m, on the one side, and the hyper-Kähler geometry defined by (3.35) and having the double complex dimension $2m$, on the other. As is clear from the above construction, the $N = 1$ chiral superfields ψ are the covectors accociated with the special Kähler manifold [248].

It is straightforward to verify that the metric associated with the Kähler potential (3.35) is hyper-Kähler, which amounts to checking its Ricci flatness. In Kähler geometry, the Ricci flatness means that the metric determinant is constant. This implies a Monge-Ampère (MA) non-linear partial differential equation on the Kähler potential. In the case (3.35), the determinant of the metric is indeed a constant. The hyper-Kähler metric of the Kähler potential (3.35) has $2n$ isometries generated by $\delta\psi_A = F_{AB}(\chi)k^B + i\eta_A$, where (k^B, η_A) are real parameters [248].

It is straightforward to generalize the $N = 2$ tensor multiplet self-interaction ($n = 2$) to $n > 2$, which is based on the same formula (3.7). For example, the components of the projective $O(4)$ superfield L^{ijkl} defined by the constraints (3.1) and (3.2) with $n = 4$ are given by

$$L^{ijkl} \; ; \qquad \lambda_\alpha{}^{ijk} = D_{\alpha l} L^{ijkl} \, , \qquad \bar{\lambda}_{\dot{\alpha}}{}^{ijk} = \bar{D}_{\dot{\alpha} l} L^{ijkl} \; ;$$

$$M^{ij} = -2D_{kl} L^{ijkl} \, , \qquad \overline{M}^{ij} = -2\overline{D}_{kl} L^{ijkl} \; ; \qquad V^{ij}_{\alpha\dot{\alpha}} = i[\overline{D}_{\dot{\alpha}k}, D_{\alpha l}] L^{ijkl} \; ;$$

$$\chi_{\alpha k} = D_\alpha^l M_{kl} \, , \qquad \bar{\chi}_{\dot{\alpha}}{}^k = \overline{D}_{\dot{\alpha} l} \overline{M}^{kl} \; ; \qquad C = -2\overline{D}_{ij} D^{kl} L^{ij}{}_{kl} . \tag{3.36}$$

The $N = 2$ invariant PSS action (3.7) in terms of the L^{ijkl} components (3.36) reads

$$S = \int \mathrm{d}^4 x \frac{1}{2\pi i} \oint_C \mathrm{d}\xi \left\{ \frac{256}{625} \frac{\partial^4 G}{\partial Q^4} \overline{\lambda}^2 \lambda^2 + \frac{192}{125} \frac{\partial^3 G}{\partial Q^3} \lambda^\alpha i V_{\alpha\dot\alpha} \overline{\lambda}^{\dot\alpha} \right.$$

$$- \frac{24}{125} \frac{\partial^3 G}{\partial Q^3} (M\overline{\lambda}^2 + \overline{M}\lambda^2) + \frac{64}{25} \frac{\partial^3 G}{\partial Q^3} \lambda^\alpha \frac{i\slashed\partial_{\alpha\dot\alpha} L_4^{\mathrm{T}}}{1+\xi^2} \overline{\lambda}^{\dot\alpha} - \frac{8}{25} \frac{\partial^2 G}{\partial Q^2} (\overline{\chi}\,\overline{\lambda} - \lambda\chi)$$

$$+ \frac{64}{25} \frac{\partial^2 G}{\partial Q^2} (1+\xi^2)^{-1} \left(-\overline{\lambda}_{\dot\alpha} i \slashed\partial^{\dot\alpha\alpha} \lambda_\alpha^{\mathrm{T}} + \lambda^\alpha i \slashed\partial_{\alpha\dot\alpha} \overline{\lambda}^{\dot\alpha\mathrm{T}} \right)$$

$$+ \frac{9}{100} \frac{\partial^2 G}{\partial Q^2} M\overline{M} - \frac{18}{25} \frac{\partial^2 G}{\partial Q^2} V^{\alpha\dot\alpha} V_{\alpha\dot\alpha} - \frac{12}{5} \frac{\partial^2 G}{\partial Q^2} \frac{V^{\alpha\dot\alpha} \slashed\partial_{\alpha\dot\alpha} L_4^{\mathrm{T}}}{1+\xi^2}$$

$$- \frac{4}{(1+\xi^2)^2} \frac{\partial^2 G}{\partial Q^2} (\partial_\mu L_4^{\mathrm{T}})^2 - \frac{1}{10} \frac{\partial G}{\partial Q} C - \frac{8}{5}(1+\xi^2)^{-1} \frac{\partial G}{\partial Q} \slashed\partial^{\alpha\dot\alpha} V_{\alpha\dot\alpha}^{\mathrm{T}}$$

$$\left. -4(1+\xi^2)^{-2} \frac{\partial G}{\partial Q} \partial^2 L_4^{\mathrm{TT}} \right\} + \text{h.c.} ,$$

$$(3.37)$$

where we have introduced the notation

$$L_4^{\mathrm{T}}(\xi) = \xi_i \xi_j \xi_k \xi_l L^{ijk}{}_l , \qquad L_4^{\mathrm{TT}}(\xi) = \xi_i \xi_j \xi_k \xi_l L^{ij}{}_{kl} ,$$

$$\lambda_\alpha(\xi) = \xi_i \xi_j \xi_k \psi_\alpha^{ijk} , \qquad \lambda_\alpha^{\mathrm{T}}(\xi) = \xi_i \xi_j \xi_k \psi^{ij}{}_{\alpha k} , \quad \text{and similarly for } \overline{\lambda}_{\dot\alpha} ,$$

$$M(\xi) = \xi_i \xi_j M^{ij} , \quad \text{and similarly for } \overline{M} ,$$

$$V_{\alpha\dot\alpha}(\xi) = \xi_i \xi_j V_{\alpha\dot\alpha}^{ij} , \qquad V_{\alpha\dot\alpha}^{\mathrm{T}}(\xi) = \xi_i \xi_j V^i{}_{\alpha\dot\alpha j} . \qquad (3.38)$$

The fields $(L^{ijkl}, \lambda_\alpha^{ijk})$ are physical (i.e. propagating), the fields $(M^{ij}, V_{\alpha\dot\alpha}^{ij})$ are auxiliary, while the fields (C, χ_α^i) play the role of Lagrange multipliers. Varying the action (2.37) with respect to C yields the algebraic constraint

$$\text{Re} \oint_C \mathrm{d}\xi \frac{\partial G}{\partial Q} = 0 \qquad (3.39)$$

that reduces the number of independent bosonic degrees of freedom on-shell. Indeed, the scalars L^{ijkl} comprise five real bosonic components, while (2.39) reduces this number by one, in agreement with the known fact that the dimension of any hyper-Kähler manifold is always a multiple to four.

The off-shell $N = 2$ susy transformation laws of the $O(4)$ projective supermultiplet components in 4d follow from the general rule

$$\delta X = \varepsilon_i^\alpha D_\alpha^i X| + \bar\varepsilon_{\dot\alpha}^{\,i} \bar{D}_i^{\,\dot\alpha} X| , \qquad (3.40)$$

where $|$ denotes the projection on the leading component of a superfield. We find

$$D_\alpha^i L^{jklm} = \tfrac{4}{5}\varepsilon^{i(j}\lambda_\alpha^{klm)} \, ,$$

$$D_{\alpha n}\lambda_\beta^{ijk} = -\tfrac{3}{16}\varepsilon_{\alpha\beta}\delta_n^{(i}M^{jk)} \, ,$$

$$\overline{D}_{\dot\alpha l}\lambda_\alpha^{ijk} = -\tfrac{i}{4}\left(\delta_l^i V_{\alpha\dot\alpha}^{jk} + \delta_l^j V_{\alpha\dot\alpha}^{ki} + \delta_l^k V_{\alpha\dot\alpha}^{ij}\right) - \tfrac{5i}{4}\partial_{\alpha\dot\alpha}L^{ijk}{}_l \, ,$$

$$D_\alpha^i M^{jk} = 0 \, , \quad \overline{D}_{\dot\alpha}^i M^{jk} = \tfrac{2}{3}\varepsilon^{i(j}\overline\chi_{\dot\alpha}^{k)} - \tfrac{16}{3}i(\partial\lambda^{ijk})_{\dot\alpha} \, ,$$

$$D^i{}_\alpha V_{\beta\dot\beta}^{jk} = \tfrac{2}{3}\varepsilon_{\alpha\beta}\varepsilon^{i(j}\overline\chi_{\dot\beta}^{k)} - \tfrac{5}{3}\varepsilon_{\alpha\beta}(\partial\lambda^{ijk})_{\dot\beta} - 2\partial_{\dot\beta(\alpha}\lambda_{\beta)}^{ijk} \, ,$$

$$D^i{}_\alpha \overline\chi_{\dot\alpha}^j = i\partial_{\alpha\dot\alpha}M^{ij} \, , \quad \overline{D}_{\dot\alpha}^i\overline\chi_{\dot\beta}^j = -\tfrac{1}{4}\varepsilon_{\dot\alpha\dot\beta}\varepsilon^{ij}C - 2\varepsilon_{\dot\alpha\dot\beta}(\partial\cdot V^{ij}) + 2\partial^\alpha{}_{(\dot\alpha}V_{\dot\beta)\alpha}^{ij} \, ,$$

$$D_\alpha^i C = -2i(\partial\overline\chi^i)_\alpha \, , \quad \overline{D}_i^{\dot\alpha}C = -2i(\partial\chi_i)^{\dot\alpha} \, .$$

$$(3.41)$$

The 4d, $N = 1$ superspace reformulation of the interacting theory (3.7) in terms of the projective $O(4)$ multiplet is obtained after a decomposition of the $N = 2$ extended superfield L^{ijkl} in terms of its $N = 1$ superfield constituents, according to the truncation rules

$$L^{1111}\big| = \chi \, , \quad L^{2222}\big| = \overline\chi \, , \quad 4\,L^{1112}\big| = W \, , \quad 6\,L^{1122}\big| = V \, , \quad (3.42)$$

where χ is the $N = 1$ complex chiral superfield satisfying the constraints (3.25), W is the $N = 1$ *complex* linear superfield satisfying the constraint [249]

$$\overline{D}_{\dot\alpha}\overline{D}^{\dot\alpha}W = 0 \, , \qquad (3.43)$$

and V is the general $N = 1$ *real* scalar superfield. This gives rise to the following $N = 1$ superspace form of the action (3.7):

$$S_1 = \int \mathrm{d}^4x\mathrm{d}^4\theta \frac{1}{2\pi i}\oint_C \frac{\mathrm{d}\xi}{\xi^2}G(\chi + \xi W + \xi^2 V - \xi^3\overline{W} + \xi^4\overline\chi, \xi) + \text{h.c.} \quad (3.44)$$

The $N = 1$ complex linear multiplet W is dual to an $N = 1$ chiral multiplet ψ, since the constraint (3.43) can be accounted for by introducing the $N = 1$ chiral Lagrange multiplier ψ into the action (3.44). This yields the master action

$$S = S_1 + \int \mathrm{d}^4x\mathrm{d}^4\theta \left(\psi\overline{W} + \overline\psi\overline{W}\right) \, , \qquad (3.45)$$

where W is now a general complex scalar $N = 1$ superfield. Varying (3.45) with respect to ψ yields back the constraint (3.43) and the action S_1. Varying (3.45) with respect to W instead (i.e. performing a Legendre transform) yields an algebraic equation for W, which can (at least, in principle) be solved in terms of the other superfields. After being substituted back into the action (3.45), it gives rise to the dual $N = 1$ supersymmetric action,

$$S_{\text{dual}} = \int \mathrm{d}^4x\mathrm{d}^4\theta \, K(\chi, \overline\chi, \psi, \overline\psi; V) \, , \qquad (3.46)$$

with a certain function K. The general superfield V can now be determined from (3.46), by the use of its algebraic equation of motion. Substituting the result back into (3.46) results in the hyper-Kähler potential $K_{\text{hyper-K.}}(\chi, \psi; \overline{\chi}, \overline{\psi})$ that is only dependent upon the $N = 1$ chiral superfields and their conjugates.

As a simple illustration, consider a free theory defined by the function

$$G(Q_4, \xi) \propto \frac{Q_4^2}{\xi^3} , \quad \text{where} \quad Q_4 = \chi + \xi W + \xi^2 V - \xi^3 \overline{W} + \xi^4 \overline{\chi} , \quad (3.47)$$

with the contour C encircling the origin in the complex ξ-plane. The corresponding $N = 1$ superspace action reads

$$S_1 = \int d^4x d^4\theta \left(\chi \overline{\chi} - W \overline{W} + V^2 \right) , \quad (3.48)$$

while its dual chiral (on-shell equivalent) action is given by

$$S_{\text{dual}} = \int d^4x d^4\theta \left(\chi \overline{\chi} + \psi \overline{\psi} \right) . \quad (3.49)$$

Another example is provided by an extension of the c-map to the $O(4)$ tensor multiplet PSS self-interaction parametrized by a holomorphic potential [247],

$$G(Q_{(4)}^A; \xi) = \frac{F(Q_{(4)}^A)}{\xi^3} , \quad (3.50)$$

where $F(Q)$ is a holomorphic function of $Q_{(4)}^A(\xi)$. Equations (3.7) and (3.50) lead to the $N = 1$ action

$$
\begin{aligned}
S_1 = \int d^4x d^4\theta & \left\{ K(\chi, \overline{\chi}) + g_{AB}(\chi, \overline{\chi}) \left[\tfrac{1}{2} V^A V^B - W^A \overline{W}^B \right] \right. \\
& + \tfrac{1}{2} \left[F_{ABC}(\chi) W^B W^C + \text{h.c.} \right] V^A \\
& \left. + \tfrac{1}{4!} \left[F_{ABCD}(\chi) W^A W^B W^C W^D + \text{h.c.} \right] \right\} ,
\end{aligned}
\quad (3.51)
$$

where we have used the notation (3.31) and (3.32). The action (3.51) is quadratic in the general V-superfields which can be eliminated according to their algebraic equations of motion,

$$V^D = -\tfrac{1}{2} g^{AD}(\chi, \overline{\chi}) \left[F_{ABC}(\chi) W^B W^C + \text{h.c.} \right] . \quad (3.52)$$

After being substituted back into the action (3.51), this gives rise to the action [247]

$$
\begin{aligned}
S = \int d^4x d^4\theta & \left\{ K(\chi, \overline{\chi}) - g_{AB}(\chi, \overline{\chi}) W^A \overline{W}^B \right. \\
& - \tfrac{1}{4} g^{EF}(\chi, \overline{\chi}) F_{ABE}(\chi) \overline{F}_{CDF}(\overline{\chi}) W^A W^B \overline{W}^C \overline{W}^D \\
& \left. + \tfrac{1}{4!} \left[\mathcal{F}_{ABCD}(\chi, \overline{\chi}) W^A W^B W^C W^D + \text{h.c.} \right] \right\} ,
\end{aligned}
\quad (3.53)
$$

where we have used the notation [247]

$$\mathcal{F}_{ABCD}(\chi, \bar{\chi}) = F_{ABCD}(\chi) - 3F_{ABE}(\chi)g^{EF}(\chi, \bar{\chi})F_{CDF}(\chi) . \tag{3.54}$$

Unfortunately, dualizing the complex linear superfields W^A in favour of the $N = 1$ chiral superfields by the $N = 1$ superfield Legendre transform cannot be explicitly performed in (3.53). A highly non-trivial example of the $N = 2$ PSS action (3.7) in terms of the $O(4)$ projective $N = 2$ multiplet, which leads to the dual $N = 2$ NLSM with the Atiyah-Hitchin metric, is discussed in Subsect. 4.3.3.

It is straightforward to investigate any $N = 2$ invariant PSS action (3.7) with any kind and number of off-shell $O(n)$ projective multiplets and their relaxed versions. In the 'bad' case of the 'relaxed' FS hypermultiplet $(1 \rightarrow 3)$, one finds [240] that the free theory to be constructed by using (3.7) and the *Ansatz* $Q_{1,\text{rel.}} = Q_1 - \frac{4}{3}\partial Q_3/\partial \xi$ has a *doubled* number of on-shell degrees of freedom compared to that of the standard FS hypermultiplet. Therefore, the term 'relaxed' is not proper in this particular context.

4.3.3 Quaternionic Geometry and Extended Supersymmetry

As regards all four-dimensional hyper-Kähler metrics with a triholomorphic (or translational) isometry, they are known to be governed by a real three-dimensional harmonic potential, whereas all four-dimensional hyper-Kähler metrics with a non-triholomorphic (or rotational) isometry are also known to be governed by another real potential obeying the non-linear Toda equation again in three dimensions (see Subsect. 4.4.1). This implies that the *c-map* with its holomorphic potential may be of limited use for a derivation of hyper-Kähler metrics from PSS since, a harmonic function may be identified with the imaginary part of a holomorphic function only in two dimensions. Assuming any of the known 3d potentials of a 4d hyper-Kähler metric to be independent of one more coordinate obviously implies yet another isometry of the metric.

In this subsection we (partially) calculate geometrical data (metric and torsion) of the 4d, $N = 2$ NLSM and their dimensionally reduced 2d, $N = 4$ counterparts from the 4d, $N = 2$ self-interaction (3.7) of the $O(2)$ and $O(4)$ projective $N = 2$ multiplets in PSS.

Our first example deals with the 4d, $N = 2$ tensor multiplet self-interaction (3.7) given by the sum of the naive (quadratic) free action and the so-called *improved* (scale-invariant and non-polynomial) free action of the same $O(2)$ multiplet [245, 246, 229, 250, 251],

$$\oint G = \oint_{C_1} \frac{Q_{(2)}^2}{2\xi} + m \oint_{C_2} Q_{(2)} \left(\ln Q_{(2)} - 1 \right) , \tag{3.55}$$

where the contour C_1 is a loop encircling the origin, whereas the contour C_2 encircles the roots of the quadratic equation $Q_{(2)}(\xi)\big| = 0$, see Fig. 4.4.

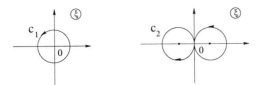

Fig. 4.4. The integration contours C_1 and C_2 in (3.55)

The real constant m represents the only (mass) parameter. Equation (3.55) describes an interacting theory even though each *separate* term on the right-hand-side of (3.55) actually amounts to a free theory. The self-interaction (3.55) is also formally $SU(2)$ invariant, in accordance with (3.10), (3.11) and (3.12). The improved (scale-invariant) action of an $O(2)$ tensor multiplet was originally invented in $N = 2$ supergravity, where the $O(2)$ tensor multiplet can be used as an $N = 2$ superconformal compensator [245]. It is straightforward to calculate the component action defined by (3.55) in PSS. The alternative formulation of the same theory also exists in the conventional 4d, $N = 2$ extended superspace that we now describe.

Both the improved and unimproved 4d, $N = 2$ invariant actions of an $O(2)$ tensor multiplet can be formulated in the conventional $N = 2$ superspace, by using the $N = 2$ supersymmetric duality relation (Subsect. 4.2.3) between the $N = 2$ tensor superfield L^{ij} and the $N = 2$ restricted chiral superfield Φ in 4d. In particular, the free (unimproved) action of L^{ij} can be written down in the chiral $N = 2$ superspace,

$$
\begin{aligned}
S_{\text{unimpr.}} &= \int \mathrm{d}^4x \mathrm{d}^4\theta (\overline{D}^{ij} L_{ij}) \frac{1}{576\Box} (\overline{D}^{kl} L_{kl}) \\
&= \frac{1}{4} \int \mathrm{d}^4x \mathrm{d}^4\theta \left(\Psi \Box \Psi + \Psi \overline{D}^4 \overline{\Psi} \right) + \text{h.c.} ,
\end{aligned}
\tag{3.56}
$$

where we have introduced the chiral $N = 2$ superfield pre-potential Ψ of L^{ij} according to the relation

$$
L^{ij} = D^{ij}\Psi + \overline{D}^{ij}\overline{\Psi}
\tag{3.57}
$$

that solves the defining constraints (3.1) and (3.2) of the $O(2)$ tensor multiplet [246].

The improved ($N = 2$ superconformally invariant) $N = 2$ tensor multiplet action can also be written down in the $N = 2$ chiral superspace form

$$
S_{\text{impr.}} = \int \mathrm{d}^4x \mathrm{d}^4\theta \, \Phi(L) \frac{1}{12\Box} \overline{D}^{ij} L_{ij} = \int \mathrm{d}^4x \mathrm{d}^4\theta \, \Psi \Phi(L(\Psi, \overline{\Psi})) + \text{h.c.} ,
\tag{3.58}
$$

where the composite $N = 2$ chiral superfield [245, 233]

$$
\Phi(L) \equiv \tfrac{1}{12}(\overline{D}^{ij} L_{ij})L^{-1} - \tfrac{1}{9}(\overline{D}_{\dot\alpha}{}^k L_{ki})(\overline{D}^{l\dot\alpha} L_{lj})L^{ij} L^{-3}
\tag{3.59}
$$

has been introduced, as well as the notation

$$L \equiv \sqrt{L^{ij} L_{ij}} \ . \tag{3.60}$$

It is straightforward to verify that the right-hand-side of (3.59) satisfies the constraints (2.26) and (2.27) defining the $N = 2$ restricted chiral superfield.

It is the sum of (3.56) and (3.58) that is equivalent to (3.55), while the linearly realized $N = 2$ susy is manifest in both cases. However, there is still a problem with the NLSM interpretation of this $N = 2$ invariant action, since the $N = 2$ tensor multiplet contains a conserved vector V_μ amongst its field components. Solving the constraint in terms of a gauge antisymmetric tensor $E_{\mu\nu}$, as in (3.16), gives rise to the components $(L^{ij}, \lambda^i, M, E_{\mu\nu})$ and the non-vanishing central charge associated with the gauge invariance (3.17). The bosonic part of the component action calculated from (3.56), (3.58) and (3.59) is given by

$$\begin{aligned}
\mathcal{L}_{\text{bosonic}} = & \left(\tfrac{1}{2} + mL^{-1} \right) \left\{ -\tfrac{1}{2} |\partial_\mu L_{ij}|^2 - \bar{\lambda}^i \gamma^\mu \overset{\leftrightarrow}{\partial}_\mu \lambda_i + |M|^2 + V^\mu V_\mu \right\} \\
& + m \left[-|\bar{\lambda}_i \lambda_j|^2 L^{-3} + 3 |\bar{\lambda}_i \lambda_j L^{ij}|^2 L^{-5} - \bar{\lambda}^k L_{ik} \gamma^\mu \overset{\leftrightarrow}{\partial}_\mu L^{ij} \lambda_j L^{-3} \right. \\
& - 2 L_{ik} \bar{\lambda}^k \gamma^\mu G_\mu \lambda^i L^{-3} - \tfrac{1}{2} \varepsilon^{\mu\nu\rho\sigma} \partial_\mu L_{ik} \partial_\nu L_{jl} E_{\rho\sigma} L^{kl} \varepsilon^{ij} L^{-3} \\
& \left. - M \bar{\lambda}_i \lambda_j L^{ij} L^{-3} - \bar{M} \bar{\lambda}^i \lambda^j L_{ij} L^{-3} \right] \ .
\end{aligned} \tag{3.61}$$

In order to get rid of the antisymmetric tensor gauge field $E_{\rho\sigma}$ by dualizing it in favour of a pseudo-scalar, we first have to rewrite (3.61) in another form where $E_{\rho\sigma}$ only enters via its field strength V'_μ. It then allows us to apply the Legendre transform in components, and thus finally represent all the bosonic degrees of freedom in terms of scalars. An inspection of (3.61) shows that $E_{\rho\sigma}$ explicitly enters the action only once, namely, in the combination

$$\tfrac{1}{2} \varepsilon^{\mu\nu\rho\sigma} E_{\rho\sigma} F_{\mu\nu}(L) \ , \tag{3.62}$$

where the antisymmetric tensor

$$F_{\mu\nu}(L) \equiv (\partial_\mu L \times \partial_\nu L) \cdot \frac{L}{|L|^3} \ , \quad L \equiv \tfrac{1}{2} \tau_{ij} L^{ij} \ , \tag{3.63}$$

has been introduced [245]. Equation (3.63) is formally identical to the electromagnetic field strength of a magnetic monopole. Therefore, there exists a (locally defined) vector potential $A_\mu(L)$ such that

$$F_{\mu\nu}(L) = \partial_\mu A_\nu - \partial_\nu A_\mu \ . \tag{3.64}$$

An explicit magnetic monopole solution for the locally defined vector potential $A_\mu(L)$ cannot be invariant under the 'rotational' $SO(3) = SU(2)_R/Z_2$ symmetry. [7] Nevertheless, it can be easily written down as a function of the

---

[7] It also implies that the $SU(2)_R$ symmetry of the underlying $N = 2$ NLSM cannot be gauged (cf. Subsect. 4.4.4).

$SO(2)$-irreducible L^{ij}-components defined by the decomposition

$$L^{ij} = \delta^{ij}S + P^{(ij)}_{\text{traceless}} \; . \tag{3.65}$$

After integrating by parts and introducing the Lagrange multiplier R as

$$^*EF = {}^*E\mathrm{d}A \to -\mathrm{d}^*EA = -V'_\mu A^\mu \to -V_\mu A^\mu - V^\mu \partial_\mu R \; , \tag{3.66}$$

one can eliminate V_μ via its algebraic equation of motion. This results in the dual 4d NLSM action with four real scalars representing the bosonic NLSM degrees of freedom.

The 4d Legendre transform can also be performed in terms of $N = 1$ superfields, after rewriting the action (3.55) in $N = 1$ superspace, where it takes the form [246]

$$S_1 = \int \mathrm{d}^4x\mathrm{d}^4\theta \left[-\tfrac{1}{2}G^2 + \bar{\chi}\chi + m\sqrt{G^2 + 4\bar{\chi}\chi} \right. \\ \left. -mG\ln\left(G + \sqrt{G^2 + 4\bar{\chi}\chi}\right) \right] \; , \tag{3.67}$$

in terms of the $N = 1$ chiral superfield χ and the $N = 1$ real linear superfield G. Dualizing G in favour of another $N = 1$ chiral superfield ϕ yields the NLSM action

$$S_{\text{dual}} = \int \mathrm{d}^4x\mathrm{d}^4\theta \left[\bar{\chi}\chi(1 + 2F^2) + 2m\sqrt{\bar{\chi}\chi + \bar{\chi}\chi F^2} \right] \; , \tag{3.68}$$

where $F(a,b)$ is a solution to the algebraic equation

$$\sinh(aF + b) + F = 0 \; , \tag{3.69}$$

whose $N = 1$ superfield-dependent coefficients are given by

$$a = \frac{2}{m}\sqrt{\bar{\chi}\chi} \; , \quad b = \bar{\phi} + \phi \; . \tag{3.70}$$

Unfortunately, it is not easy to find an explicit solution to the algebraic equation (3.69). This clearly shows the important role of a 'smart' parametrization in identifying the NLSM superfield coordinates that make possible an explicit derivation of the NLSM metric out of extended superspace.

After DR of the 4d theory (3.55) down to 2d we get a 2d, $N = 4$ supersymmetric field theory whose bosonic part can be most easily deduced from (3.61),

$$\mathcal{L}_{\text{bosonic}} = \left(1 + \frac{\gamma}{\sqrt{2A^2 + P^2}}\right)\left\{-\tfrac{1}{2}(\partial_\mu A)^2 - \tfrac{1}{4}(\partial_\mu P_{ij})^2 + \tfrac{1}{2}V_\mu^2\right\} \\ - \frac{\gamma A}{2P^2\sqrt{2A^2 + P^2}}P_{ij}\overleftrightarrow{\partial}_\mu P_{jk}\varepsilon^{ik}V^\mu \; , \tag{3.71}$$

where we have used (3.65), introduced a conserved 2d vector V^μ, and defined $\gamma \equiv m/2$ and $P^2 \equiv (P_{ij})^2$. The complex scalar field $V_2 + iV_3$ is auxiliary in 2d.

The constraint $\partial_\mu V^\mu = 0$ can be easily solved, $V^\mu = \varepsilon^{\mu\nu}\partial_\nu B$, in terms of a pseudo-scalar B. Then we arrive at a quaternionic 2d NLSM with torsion (by construction), whose bosonic Lagrangian is of the type

$$\mathcal{L}_{\text{bosonic}} = -\tfrac{1}{2}g_{ab}(X)\partial_\mu X_a \partial^\mu X_b - \tfrac{1}{2}h_{ab}(X)\varepsilon^{\mu\nu}\partial_\mu X_a \partial_\nu X_b , \tag{3.72}$$

while the metric and torsion potential are given by

$$\begin{aligned} g_{ab} &= \text{diag}\left(1 + \frac{\gamma}{X_1\sqrt{2}}\right) , \qquad a,b = 0,1,2,3 , \\ h_{20} &= \frac{\gamma X_1 X_3}{\sqrt{2}(X_2^2 + X_3^2)S} , \qquad h_{30} = \frac{\gamma X_1 X_2}{\sqrt{2}(X_2^2 + X_3^2)S} , \end{aligned} \tag{3.73}$$

and all the other components of h_{ab} vanish. In (3.72) and (3.73) we have used the notation

$$X_0 \equiv B , \quad X_1 \equiv A , \quad S^2 \equiv X_1^2 + X_2^2 + X_3^2 , \tag{3.74}$$

and

$$P_{ij} \equiv \begin{pmatrix} X_2 & X_3 \\ X_3 & -X_2 \end{pmatrix} . \tag{3.75}$$

The presence of NLSM torsion in the bosonic NLSM action (3.72) appears to be the price one has to pay for the linearly realized $N = 4$ supersymmetry in the full 2d, $N = 4$ supersymmetric NLSM action (with all fermionic and auxiliary fields) to be constructed via DR.

As an alternative approach, one can go back to (3.71) and then perform the Legendre transform in components, by introducing a Lagrange multiplier D for the constraint $\partial^\mu V_\mu = 0$. Dualizing the vector in favour of the Lagrange multiplier leads to the dual 2d, $N = 4$ supersymmetric NLSM without torsion and, hence, with a hyper-Kähler metric. The bosonic NLSM terms are

$$\begin{aligned} \mathcal{L}_{\text{bosonic, dual}} &= \left(1 + \frac{\gamma}{\sqrt{2A^2 + P^2}}\right)\left\{-\tfrac{1}{2}(\partial_\mu A)^2 - \tfrac{1}{4}(\partial_\mu P_{ij})^2\right\} \\ &\quad - \frac{1}{2}\left(\frac{\sqrt{2A^2 + P^2}}{\gamma + \sqrt{2A^2 + P^2}}\right)\left(\partial_\mu D + \frac{\gamma A P_{ij}\overleftrightarrow{\partial}_\mu P_{jk}\varepsilon^{ik}}{P^2\sqrt{4A^2 + 2P^2}}\right)^2 \\ &\equiv -\tfrac{1}{2}\hat{g}_{ab}(\hat{X})\partial_\mu \hat{X}_a \partial^\mu \hat{X}_b , \qquad a,b = 1,2,3,4 , \end{aligned} \tag{3.76}$$

where we have introduced the notation

$$\hat{X}_1 \equiv A , \quad \hat{X}_2 \equiv P_{11} , \quad \hat{X}_3 \equiv P_{12} , \quad \hat{X}_4 \equiv D . \tag{3.77}$$

The explicit hyper-Kähler metric $\hat{g}_{ab}(\hat{X})$ can be read off from (3.77). We find

$$\hat{g}_{11} = 1 + \frac{\gamma}{S} \ , \quad \hat{g}_{22} = \hat{g}_{11} + \frac{\gamma^2 \hat{X}_1^2 \hat{X}_3^2}{(\gamma + S)S(\hat{X}_2^2 + \hat{X}_3^2)^2} \ ,$$

$$\hat{g}_{33} = \hat{g}_{11} + \frac{\gamma^2 \hat{X}_1^2 \hat{X}_2^2}{(\gamma + S)S(\hat{X}_2^2 + \hat{X}_3^2)^2} \ , \quad \hat{g}_{44} = \frac{S}{\gamma + S} \ ,$$

$$\hat{g}_{23} = \frac{-\gamma^2 \hat{X}_1^2 \hat{X}_2 \hat{X}_3}{(\gamma + S)S(\hat{X}_2^2 + \hat{X}_3^2)^2} \ , \quad \hat{g}_{24} = \frac{-\gamma \hat{X}_1 \hat{X}_3}{(\gamma + S)(\hat{X}_2^2 + \hat{X}_3^2)} \ ,$$

$$\hat{g}_{34} = \frac{\gamma \hat{X}_1 \hat{X}_2}{(\gamma + S)(\hat{X}_2^2 + \hat{X}_3^2)} \ , \quad \hat{g}_{12} = \hat{g}_{13} = \hat{g}_{14} = 0 \ . \tag{3.78}$$

It is also worth noticing that

$$\det \hat{g}_{ab} = \hat{g}_{11}^2 = (1 + \gamma/S)^2 \ . \tag{3.79}$$

The analytical computer program REDUCE 3.0 was used to directly check the hyper-Kähler properties of the metric (3.78), namely, the self-duality of the curvature and Ricci flatness, as well as to verify its equivalence to the Taub-NUT metric (Sect. 4.4).

We now describe a derivation of the Atiyah-Hitchin (AH) metric [204] [8] from PSS [252]. In order to fix the corresponding PSS potential $G(Q, \xi)$ in (3.7), it is useful to recall the following basic properties of the AH metric [204]:

- the AH metric has a twistor description, while the corresponding twistor space has a holomorphic projection on the holomorphic $O(4)$ bundle;
- the AH metric possesses a rotational $SO(3)$ isometry;
- the AH metric asymptotically approaches the Taub-NUT metric with a negative mass parameter, $m = -1/2$ (in proper normalization).

The first property implies that the PSS construction (3.7) of the AH metric is possible in terms of a single $O(4)$ projective multiplet, i.e. $G(Q, \xi) = G(Q_{(4)}(\xi), \xi)$. The contour C in the complex ξ-plane either encircles the origin (denoted by C_1) or the roots of the quartic polynomial of $Q_{(4)}(\xi)| = 0$ (denoted by C_2). The second property together with (3.11) and (3.12) further imply that outside of the origin the potential $G(Q_{(4)}, \zeta)$ should be proportional to $\sqrt{Q_{(4)}}$. Finally, the third property of the AH metric implies that $Q_{(4)}$ should asymptotically approach $Q_{(2)}^2$ in order to reproduce the Taub-NUT potential in (3.55) with $m = -1/2$. Taken together, this leads to the following PSS description [252] of the $N = 2$ NLSM with the Atiyah-Hitchin metric (see [251] for the equivalent HSS description):

---

[8] See Sect. 4.4 for more about the AH metric.

$$\oint G = -\frac{1}{2\pi i} \oint_{C_1} \frac{Q_{(4)}}{\xi} + \oint_{C_2} \sqrt{Q_{(4)}} . \tag{3.80}$$

To extract the component NLSM metric out of (3.7) and (3.80), it is convenient to rewrite the PSS action in the $N = 1$ superspace form (3.44), solve the algebraic constraint (3.39) that amounts to the relation

$$\oint_{C_2} \frac{d\xi}{\sqrt{Q_{(4)}}} = 1 \tag{3.81}$$

on the $N = 1$ superfield components (3.42) of the $O(4)$ projective $N = 2$ superfield $Q_{(4)}$ (or, equivalently, it determines the $N = 1$ Lagrange multiplier V in terms of the rest of the $N = 1$ superfields), and, finally, dualize the complex chiral $N = 1$ superfield W by the $N = 1$ superfield Legendre transform.

The quartic polynomial $Q_4(\xi)$ of (3.47) can be rewritten in the form

$$Q_4(\xi) = c(\xi - \alpha)(\xi - \beta)(\bar\alpha\xi + 1)(\bar\beta\xi + 1) , \tag{3.82}$$

where the reality condition (3.2) has been used, in terms of a real scale factor c and two complex roots, α and β. The branch cuts of the root in (3.80) and (3.81) can be chosen to run from α to $-1/\bar\beta$ and from β to $-1/\bar\alpha$. The contour integration over C_2 reduces to the integration over the brunch cuts, while the constraint (3.81) takes the form of a complete elliptic integral of the first kind (in Legendre normal form) [252, 253, 254],

$$\frac{4}{\sqrt{c(1 + \bar\alpha\alpha)(1 + \bar\beta\beta)}} \int_0^1 \frac{d\gamma}{\sqrt{(1 - \gamma^2)(1 - k^2\gamma^2)}} = 1 , \tag{3.83}$$

whose modulus k is given by

$$k^2 = \frac{(1 + \alpha\bar\beta)(1 + \beta\bar\alpha)}{(1 + \alpha\bar\alpha)(1 + \beta\bar\beta)} . \tag{3.84}$$

Though it seems to be difficult to solve (3.83) for V, one can easily determine the scale factor c in terms of the roots,

$$c = \frac{16K^2(k)}{(1 + \alpha\bar\alpha)(1 + \beta\bar\beta)} , \tag{3.85}$$

where $K(k)$ is the Legendre complete integral of the first kind. However, in order to make contact with the explicit AH metric (see Sect. 4.4), it turns out to be more useful to use yet another parametrization that solves the constraint (3.81) and has *manifest* rotational $SO(3)$ symmetry [252]:

$$\chi = 2e^{2i\varphi} \left[\cos(2\psi)(1 + \cos^2 \vartheta) + 2i\sin(2\psi)\cos\vartheta + (2k^2 - 1)\sin^2 \vartheta\right] K^2(k),$$

$$W = 8e^{i\varphi}\sin\vartheta \left[\sin(2\psi) - i\cos(2\psi)\cos\vartheta + i(2k^2 - 1)\cos\vartheta\right] K^2(k) ,$$

$$V = 4\left[-3\cos(2\psi)\sin^2\vartheta + (2k^2 - 1)(1 - 3\cos^2\vartheta)\right] K^2(k) .$$

$$\tag{3.86}$$

It is straightforward to verify that the last equation (3.86) solves the constraint (3.85), after calculating the roots of the quartic polynomial $Q_4(\xi)$ with the coefficients given by (3.86). The functions (3.86) form (in fact, by construction [252]) a five-dimensional representation of the $SU(2)_R$ symmetry rotating the complex structures and acting on ξ in the form of the fractional transformations (3.10), whose generators have the standard form

$$K_3 = -i\frac{\partial}{\partial\varphi} ,$$

$$K_+ = e^{i\varphi}\left(i\frac{\partial}{\partial\vartheta} + \frac{1}{\sin\vartheta}\frac{\partial}{\partial\psi} - \cot\vartheta\frac{\partial}{\partial\varphi}\right) , \qquad (3.87)$$

$$K_- = e^{-i\varphi}\left(-i\frac{\partial}{\partial\vartheta} + \frac{1}{\sin\vartheta}\frac{\partial}{\partial\psi} - \cot\vartheta\frac{\partial}{\partial\varphi}\right) .$$

The variables $(k, \vartheta, \psi, \varphi)$ can be identified with the standard parameters of the AH metric (Subsect. 4.4.1). A direct comparison of the metric to be obtained by the Legendre transform with the standard form of the AH metric may be difficult in this approach [252]. The most compelling argument is based on computing the complex structures (or Kähler forms) of the metric in the basis $(dk, d\vartheta, d\psi, d\varphi)$, and comparing them to the standard (AH) ones (Subsect. 4.4.1). The results apparently agree [252], while the same hyper-Kähler structures imply the same hyper-Kähler metrics.

Other important examples are provided by hyper-Kähler Asymptotically Locally Euclidean (ALE) metrics that approach $\mathbf{R}^4/\Gamma$ at infinity, where Γ is a finite subgroup of $SU(2)$. They admit an $A - D - E$ classification [255]. The A_k series are described by the PSS potential [252]

$$G(Q_2, \xi) = \sum_{i=1}^{k+1}(Q_2 - a_i)\left[\ln(Q_2 - a_i) - 1\right] , \qquad (3.88)$$

where, in accordance with (3.24), $Q_2 = \chi - i\xi g + \xi^2\bar{\chi}$. In addition, (3.88) is dependent upon some fixed real sections of the $O(2)$ bundle, $a_i(\xi) = p_i - i\xi q_i + \xi^2\bar{p}_i$, which parametrize the moduli space of A_k metrics. The PSS integration contours encircle the roots of $(Q_2 - a_i)(\xi) = 0$. In particular, the Eguchi-Hanson (EH) metric is associated with the ALE A_1 space. The general ALE multi-Eguchi-Hanson metric is described by the PSS potential (3.88), whose moduli are given by the differences $(p_i - p_j)$ and $(q_i - q_j)$.

Similarly, the PSS *Ansatz* in terms of an $O(4)$ projective multiplet [252],

$$G(Q_4, \xi) = \sum_{i=1}^{k}\left\{(\sqrt{Q_4} - a_i)\ln(\sqrt{Q_4} - a_i) + (\sqrt{Q_4} + a_i)\ln(\sqrt{Q_4} + a_i)\right\} ,$$

$$(3.89)$$

gives rise to the D_k series of ALE metrics [255]. The PSS potentials for Asymptotically Locally Flat (ALF) metrics are obtained by adding either a bilinear (kinetic) term for the $O(2)$ multiplet to (3.88) in the A_k (multi-Taub-NUT) case, or a linear term for the $O(4)$ multipet to (3.89) in the D_k case. See [256, 257, 258] for the proofs and more examples by combining (3.55), (3.80), (3.88) and (3.89). The HSS construction of the (ALE- and ALF-type) series of hyper-Kähler metrics can be found in [251].

4.3.4 Perturbation Theory in Extended Superspace

To define quantum perturbation theory with the $N = 2$ superspace action (3.7), we need $N = 2$ superfield propagators of the projective $O(n)$ multiplets in 4d. They are given by [232] [9]

$$\left\langle L^{i_1 \cdots i_n}(1) L_{j_1 \cdots j_n}(2) \right\rangle \propto p^{-2} D^{(i_1 i_2}(1) \overline{D}_{(j_1 j_2}(2) \delta^{i_3 \cdots i_n)}_{j_3 \cdots j_n)} \delta^4_{\text{ch.}}(\theta_1 - \theta_2) , \quad (3.90)$$

where we have introduced the $N = 2$ chiral delta-function

$$\delta^4_{\text{ch.}}(\theta_1 - \theta_2) \equiv \exp\left(-\tfrac{1}{2}\theta_1 \not{p} \bar{\theta}_1 - \tfrac{1}{2}\theta_2 \not{p} \bar{\theta}_2\right) \delta^4(\theta_1 - \theta_2) . \quad (3.91)$$

It is not difficult to verify that the $N = 2$ extended superpropagator (3.90) gives rise to the standard scalar propagators of the leading components of the projective $O(n)$ supermultiplet [232]. The rest of the component propagators are dictated by $N = 2$ supersymmetry, and, therefore, it should be automatically correct too. [10] The simple way to obtain (3.90) is to use the duality between the projective $O(n)$ superfields and the restricted chiral superfields, since the superpropagators of the latter are most easily constructed in chiral $N = 2$ superspace. The $N = 2$ supersymmetric duality relations are given by

$$L^{i_1 \cdots i_n} = D^{(i_1 i_2} \Phi^{i_3 \cdots i_n)} , \quad \Phi^{i_3 \cdots i_n} \propto \Box^{-1} \overline{D}_{i_1 i_2} L^{i_1 \cdots i_n} . \quad (3.92)$$

It is now straightforward to check that the constraints (3.1) and (3.2) on $L^{i_1 \cdots i_n}$ are equivalent to that in (2.26) and (2.27) on $\Phi^{i_3 \cdots i_n}$. In the case of an even number of external $SU(2)$ indices, $n = 2p$, (3.90) can be put into the equivalent form

$$\left\langle L^{i_1 \cdots i_{2p}}(1) L_{j_1 \cdots j_{2p}}(2) \right\rangle \propto p^{-4} D^{(i_1 i_2}(1) \overline{D}_{(j_1 j_2}(2) \delta^{i_3 \cdots i_{2p})}_{j_3 \cdots j_{2p})}$$
$$\times \exp\left(-\tfrac{1}{2}\theta_1 \not{p} \bar{\theta}_1 + \tfrac{1}{2}\theta_2 \not{p} \bar{\theta}_2 - \theta_1 \not{p} \bar{\theta}_2\right) . \quad (3.93)$$

The whole theory (3.7) can also be rewritten in the $N = 2$ chiral superspace form [261, 232],

$$S = \int d^4x d^4\theta \frac{1}{2\pi i} \oint_C d\xi \left\{ \Phi_A \frac{\partial H_A}{\partial Q_B} \Box \Phi_B + \Phi_A \frac{\Box H_A}{\partial Q_B \partial Q_C} [(\nabla^\alpha \partial_\mu \Phi_B) \right.$$
$$\left. \times (\nabla_\alpha \partial^\mu \Phi_C) + \tfrac{1}{2}(\nabla^\alpha \partial^\mu \Phi_B)(\sigma_{\mu\nu})_\alpha{}^\beta (\nabla_\beta \partial^\nu \Phi_C)] \right\} + \text{h.c.} , \quad (3.94)$$

---

[9] We ignore overall combinatorial factors in front of the superpropagators.

[10] See [259, 260] for more details, as well as a generalization to the massive case.

where we have introduced the notation

$$G(Q,\xi) \equiv \frac{1}{16} \sum_A H_A Q_A , \quad Q_A = \nabla^2 \Phi_A ,$$

$$\Phi_A = \xi_{i_1} \cdots \xi_{i_{2p-2}} \Phi_A^{i_1 \cdots i_{2p-2}} , \quad L_A^{i_1 \cdots i_{2p}} = D^{(i_1 i_2} \Phi_A^{i_3 \cdots i_{2p})} , \qquad (3.95)$$

in terms of the $N = 2$ restricted chiral superfields Φ_A satisfying the constraints

$$\overline{D}{}^i_{\overset{.}{\alpha}} \Phi_A = 0 , \quad D^4 \Phi_A = \Box \bar{\Phi}_A . \qquad (3.96)$$

Given a function $G(Q)$, the functions $H_A(Q)$ are defined by the first line of (3.95).

Calculation of the divergence index ω of an arbitrary extended PSS supergraph, after dimensional reduction of the chiral theory (3.94) down to 2d, is very similar to that of Subsect. 4.2.3. The effective number μ of the exponential factors can be estimated by $4\mu \leq 2V - 4$, where V is the number of vertices in the supergraph. For dimensional and topological reasons, this implies that $\omega \leq -2$ in 2d, thus leading to the UV-finiteness of the corresponding 2d, $N = 4$ supersymmetric NLSM by the divergence index.

At the end of this subsection, we consider UV properties of the four-dimensional (4d) theory (3.7) or (3.94), and demonstrate its on-shell non-renormalizability for any (non-flat) choice of the NLSM geometry [262]. The non-existence of a non-trivial renormalizable self-interaction of $N = 2$ hypermultiplets in 4d was also proved in the component approach [263], under the assumption of the unbroken $SU(2)_R$ symmetry rotating two supersymmetries, and in the $N = 2$ HSS approach as well [221]. In the PSS approach, we merely assume the unbroken $U(1)$ subgroup of $SU(2)_R$, and avoid the infinite number of auxiliary fields present in HSS. Moreover, the general HSS results of [221] only apply to the formally non-renormalizable one-loop UV divergences of $N = 2$ NLSM in 4d, while they do not address the geometrical functions in front of those divergences (the former may vanish for particular NLSM geometries). We demonstrate that those functions only vanish for a flat NLSM metric, in agreement with the general theorem in $N = 1$ superspace [264]. We restrict ourselves to $N = 2$ superfields L^{ijkl} for definitness and simplicity of our notation, though our results apply equally to the other $N = 2$ projective superfields satisfying (3.1) and (3.2).

We use a *linear* background-quantum field splitting here (cf. Chaps. 2 and 3),

$$L^{ijkl} \rightarrow L^{ijkl} + l^{ijkl} , \quad Q \rightarrow Q + q , \qquad (3.97)$$

where capital letters are used to denote classical background fields, whereas lower-case letters are used to denote quantum fields. The 4d, $N = 2$ super-propagator reads

$$\langle l^{ijkl}(1) l_{mnpq}(2) \rangle = - \frac{\mathrm{i}}{p^4} D^{(ij}(1) \overline{D}_{(mn}(2) \delta^{kl)}_{pq)}$$

$$\times \exp\left(-\tfrac{1}{2}\theta_1 p\bar{\theta}_1 + \tfrac{1}{2}\theta_2 p\bar{\theta}_2 - \theta_1 p\bar{\theta}_2\right) , \qquad (3.98)$$

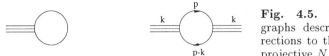

Fig. 4.5. The one-loop PSS graphs describing quantum corrections to the self-interaction of projective $N = 2$ multiplets in 4d

in agreement with (3.93). Equation (3.98) can be rewritten in the $N = 2$ chiral form as follows :

$$\langle q_A(1,\xi)q_B(1,\zeta)\rangle = \frac{-i\delta_{AB}}{p^4}4!(\xi_i\zeta^i)^2\nabla^2(1)\Delta^2(2)$$
$$\times \exp\left(-\tfrac{1}{2}\theta_1\slashed{p}\bar\theta_1 + \tfrac{1}{2}\theta_2\slashed{p}\bar\theta_2 - \theta_1\slashed{p}\bar\theta_2\right) \ . \tag{3.99}$$

The one-loop effective action is determined by the second-order term in the expansion of the action $S(Q + q)$ in powers of the quantum superfields q (Chap. 2). We have

$$S_{(2)} = \int \mathrm{d}^4x \frac{1}{2\pi i}\oint_C \frac{\mathrm{d}\xi}{(1+\xi^2)^4}\widetilde\nabla^2\widetilde\Delta^2\tfrac{1}{2}qG''(L)q + \text{h.c.} \ , \tag{3.100}$$

where primes denote differentiation with respect to Q. We assume that $S_{(2)}$ only contains the interacting piece. To disentangle the UV and IR divergences, we introduce a mass regulator and use SDR (Sect. 3.1).

The relevant one-loop supergraphs are schematically pictured in Fig. 4.5. The first graph in Fig. 4.5 vanishes since

$$\langle q(1,\xi)q(1,\xi)\rangle = 0 \ , \tag{3.101}$$

where we have used (3.99) and the identity $\xi_i\xi^i \equiv \varepsilon^{ij}\xi_i\xi_j = 0$. The analytic contribution to the quantum effective action due to the second supergraph in Fig. 4.5 is given by

$$\frac{1}{4}(4!)^2\frac{1}{(2\pi i)^2}\int \frac{\mathrm{d}^4k}{(2\pi)^4}\oint_{C_1}\frac{\mathrm{d}\xi}{(1+\xi^2)^4}\oint_{C_2}\frac{\mathrm{d}\zeta}{(1+\zeta^2)^4}(\xi_i\zeta^i)^4\widetilde\nabla^2(-ik,\theta,\xi)$$

$$\times\widetilde\Delta^2(-ik,\theta,\xi)\widetilde\nabla^2(ik,\eta,\zeta)\widetilde\Delta^2(ik,\eta,\zeta)\int\frac{\mathrm{d}^dp}{(2\pi)^d}\left[\nabla^2(-ip,\theta,\xi)\right.$$

$$\times\Delta^2(ip,\eta,\zeta)\exp\left(-\tfrac{1}{2}\theta\slashed{p}\bar\theta + \tfrac{1}{2}\eta\slashed{p}\bar\eta - \theta\slashed{p}\bar\eta\right)]\frac{1}{p^2(k-p)^2}$$

$$\times G''(k,\theta,\xi)G''(-k,\eta,\zeta)\left[\nabla^2(-ik+ip,\theta,\xi)\Delta^2(ik-ip,\eta,\zeta)\right.$$

$$\times \exp\left(-\tfrac{1}{2}\theta(\slashed{k}-\slashed{p})\bar\theta + \tfrac{1}{2}\eta(\slashed{k}-\slashed{p})\bar\eta - \theta(\slashed{k}-\slashed{p})\bar\eta\right)] \ , \tag{3.102}$$

where the $N = 2$ superspace covariant derivatives in momentum space are

$$\widetilde\nabla_\alpha(-ik,\theta,\xi) = \xi_i D_{i\alpha}(-ik,\theta) = \xi_i\left(\frac{\partial}{\partial\theta^{\alpha i}} - \tfrac{1}{2}k_{\alpha\dot\alpha}\bar\theta_i^{\dot\alpha}\right) \ ,$$

$$\widetilde\Delta^{\dot\alpha}(-ik,\theta,\xi) = \xi_i\overline{D}_i^{\dot\alpha}(-ik,\theta) = -\xi_i\left(\frac{\partial}{\partial\bar\theta_{\dot\alpha}^i} - \tfrac{1}{2}k^{\alpha\dot\alpha}\theta_{\alpha i}\right) \ , \tag{3.103}$$

and similarly for ∇_α and $\Delta^{\dot\alpha}$.

In computing the leading (logarithmic) UV-divergence of (3.102), we can restrict ourselves to the first component of $\nabla^2 \Delta^2 \exp(-\theta\not p\bar\eta)$ in its expansion over the anticommuting superspace coordinates θ and η, on dimensional grounds. Then we easily find that the UV-divergent piece of (3.102) is given by

$$\tfrac{1}{4}(4!)^2 \int \frac{d^4k}{(2\pi)^4} \frac{1}{(2\pi i)^2} \oint_{C_1} \frac{d\xi}{(1+\xi^2)^4} \oint_{C_2} \frac{d\zeta}{(1+\zeta^2)^4} (\xi_i \zeta^i)^4$$

$$\times \widetilde\nabla^2(-ik,\theta,\xi)\widetilde\Delta^2(-ik,\theta,\xi)\widetilde\nabla^2(ik,\eta,\zeta)\widetilde\Delta^2(ik,\eta,\zeta)$$

$$\times G''(k,\theta,\xi)G''(-k,\eta,\zeta) \int \frac{d^d p}{(2\pi)^d} \frac{1}{p^2(k-p)^2} \quad . \tag{3.104}$$

The superspace derivatives $\widetilde\nabla$ and $\widetilde\Delta$ together form the full $N = 2$ superspace measure $d^8\theta$, so that the UV divergence (3.104) can be rewritten in the form

$$\int \frac{d^4k}{(2\pi)^4} F_{ijklmnpq}(k) F^{ijklmnpq}(-k) \int \frac{d^d p}{(2\pi)^d} \frac{1}{p^2(k-p)^2} \quad , \tag{3.105}$$

where we have introduced the notation

$$F_{ijklmnpq}(k) \equiv \tfrac{1}{2}4! \frac{1}{2\pi i} \oint_C \frac{d\xi}{(1+\xi^2)^4} \xi_i \xi_j \xi_k \xi_l \xi_m \xi_n \xi_p \xi_q$$
$$\times \widetilde\nabla^2(-ik,\theta,\xi)\widetilde\Delta^2(-ik,\theta,\xi)G''(k,\theta,\xi) \quad . \tag{3.106}$$

As is clear from (3.105), the non-renormalizable one-loop UV divergence of the self-interacting $N = 2$ tensor multiplet L^{ijkl} vanishes if and only if the tensor F of (3.106) vanishes, i.e. if the theory is free. The UV divergence (3.105) is non-renormalizable since it leads to higher derivative terms in components, after integrating over all eight anticommuting (Grassmann) coordinates of $N = 2$ superspace.

4.4 Hyper-Kähler Metrics and Harmonic Superspace

In this section we introduce a Harmonic Superspace (HSS) [220], and discuss some of its applications in the theory of 4d, $N = 2$ supersymmetric NLSM. Our primary goal is to demonstrate the power of HSS in (i) constructing most general $N = 2$ matter couplings and gauging their isometries in 4d, and (ii) generating explicit hyper-Kähler metrics and scalar potentials that are consistent with extended supersymmetry. In Subsect. 4.4.1 we discuss four-dimensional hyper-Kähler metrics, and review the $SO(3)$-invariant metrics of Taub-NUT, Eguchi-Hanson and Atiyah-Hitchin. Subsection 4.4.2 is devoted to HSS. In Subsect. 4.4.3 the most general 4d, $N = 2$ NLSM [265] is formulated, and the $SO(3)$ invariant four-dimensional hyper-Kähler metrics are derived from HSS. Gauging isometries of 4d, $N = 2$ supersymmetric NLSM in HSS [266] is discussed in Subsect. 4.4.4.

4.4.1 Four-Dimensional Hyper-Kähler Metrics

Four-dimensional hyper-Kähler metrics attracted much attention both in theoretical high-energy physics and in general relativity, most notably in connection with the theory of (BPS) magnetic monopoles, whose moduli space is a hyper-Kähler manifold [204], the theory of gravitational instantons [267, 268], and integrable systems [269]. More recently, it was found that some 4d, $N = 2$ NLSM with hyper-Kähler metrics arise as the hypermultiplet low-energy effective actions in 4d, $N = 2$ supersymmetric quantum gauge field theories (Sect. 8.3). The four-dimensional hyper-Kähler metrics also naturally appear in the context of M-theory, e.g., in the form of Kaluza-Klein monopoles (Sect. 8.4).

The hyper-Kähler condition (Subsect. 4.1.1) on a four-dimensional (Euclidean) metric is equivalent to the self-duality (or anti-self-duality) of the Riemann curvature. In turn, it is equivalent to the Kähler condition and Ricci-flatness together. A classification of all (regular and complete) hyper-Kähler 4-manifolds (without singularities and with complete geodesic lines) is still an open problem. Moreover, an explicit metric is not available even for some well-known four-dimensional hyper-Kähler manifolds, with the only compact and simply connected K3-space being the most famous example [270, 268]. All other hyper-Kähler 4-manifolds are not compact, while there exist infinitely many of them. The boundary conditions play a very important role in any classification of hyper-Kähler 4-manifolds. Amongst the four-dimensional (Euclidean) non-compact hyper-Kähler spaces one usually distinguishes the Asymptotically Locally Euclidean (ALE) spaces approaching R^4/Γ at infinity, and the Asymptotically Locally Flat (ALF) spaces approaching $(S^1 \times \mathbf{R}^3)/\Gamma$ at infinity, where Γ is a finite subgroup of $SU(2)$ [255].

In the absence of full classification of hyper-Kähler 4-manifolds, we would like to concentrate on those having isometries. If the action of the Killing vector associated with an isometry is holomorphic with respect to all three complex structures, the Killing vector is called triholomorphic or translational (Subsect. 4.1.3). There may be isometries that do not preserve the complex structures and thus cannot be gauged in rigid $N = 2$ susy — see Subsects. 4.1.4 and 4.4.4. The non-triholomorphic isometries are also known as rotational [271, 272]. In real coordinates $\{\phi^\mu\}$ of $\mathcal{M}$, $\mu = 1, 2, 3, 4$, a Killing vector field $K_\mu(\phi)$ (by definition) satisfies the Killing equation, $K_{(\mu;\nu)} = 0$. The triholomorphic (or translational) condition (Subsect. 4.1.3) on the Killing vector in real coordinates puts a restriction on the antisymmetric part of $K_{\mu;\nu}$ [271, 273],

$$K_{\mu;\nu} = \pm \tfrac{1}{2} \sqrt{g}\, \varepsilon_{\mu\nu}{}^{\lambda\rho} K_{\lambda;\rho} \,, \tag{4.1}$$

where the metric $g_{\mu\nu}$ of $\mathcal{M}$ and the four-dimensional Levi-Civita symbol have been introduced, $g = \det(g_{\mu\nu})$. The sign ambiguity in (4.1) is related to the self-dual or anti-self-dual nature of $\mathcal{M}$.

All hyper-Kähler 4-manifolds having at least one triholomorphic $U(1)$ isometry (or, equivalently, a translational Killing vector) have been classified [273]. The corresponding localized solutions are given by (multi)-Taub-NUT [208] and (multi)-Eguchi-Hanson multi-centre metrics [273]. Indeed, let us consider a general 4-metric having an isometry. In adapted real coordinates where the isometry is manifest, the metric reads

$$ds^2 = H^{-1}(d\tau + C_i d\phi^i)^2 + H\gamma_{ij}d\phi^i d\phi^j , \tag{4.2}$$

where the functions H, C_i and γ_{ij} are all independent of τ, so that $K_\mu = (1,0,0,0)$ is the Killing vector associated with the isometry, $i,j = 1,2,3$. If K_μ is translational, the adapted coordinate system can always be specified even further, so that

$$\nabla H = \pm \nabla \times C \quad \text{and} \quad \gamma_{ij} = \delta_{ij} , \tag{4.3}$$

where $\nabla = \{\partial_i\}$. The self-duality condition is then equivalent to the three-dimensional Laplace equation [208],

$$\Delta H = 0 , \tag{4.4}$$

with respect to the flat 3-metric δ_{ij}. The localized solutions to (4.4),

$$H(\phi) = \lambda + \sum_{s=1}^{n} \frac{m}{|\phi - \phi_{0,s}|} , \tag{4.5}$$

describe the regular multi-centre metrics parametrized by $(\lambda, m, \phi_{0,s})$ provided that τ is periodic (to avoid conical singularities). The ALE metrics are obtained in the case of $\lambda = 0$, whereas the ALF (multi-Taub-NUT) metrics arise when $\lambda = 1$.

According to [271], the rotational isometries (they do not obey (4.1)) either form an $so(3)$ algebra or there is only one of them, if any. Given the existence of a rotational isometry to be represented in adapted coordinates (4.2) by the Killing vector $K = \partial/\partial\tau$, it is useful to introduce the so-called nut potential b_{nut} defined by the equation [269, 272] [11]

$$\partial_i b_{\mathrm{nut}} = H^{-2}\sqrt{\det\gamma}\, \varepsilon_i{}^{jk}\partial_j C_k , \tag{4.6}$$

and the related quantity [12]

$$S_\pm \equiv b_{\mathrm{nut}} \pm H^{-1} , \tag{4.7}$$

whose partial derivatives measure the 'non-triholomorphicity' of the isometry: $S_\pm$ is constant only for trihilomorphic isometries [271], whereas it can always

---

[11] A solution to (4.6) always exists for any Ricci-flat metric (4.2).

[12] Only S_+ is considered below.

be identified with one of the coordinates (say, ϕ^3), up to an overall normalization given by $1/\sqrt{\gamma^{ij}(\partial_i S_\pm)(\partial_j S_\pm)} \equiv 1/(\Delta S_\pm)$. In the so-called Toda frame associated with the coordinates $\phi^1 = x$, $\phi^2 = y$ and $\phi^3 = z = S_+/(\Delta S_+)$, one can reduce the metric (4.2) without loss of generality to the form

$$H = \partial_z \Psi , \quad C_1 = \mp \partial_y \Psi , \quad C_2 = \pm \partial_x \Psi , \quad C_3 = 0 , \tag{4.8}$$

with diagonal metric γ_{ij},

$$\gamma_{11} = \gamma_{22} = e^\Psi , \quad \gamma_{33} = 1 . \tag{4.9}$$

In the self-dual (i.e. hyper-Kähler) geometry, the scalar potential $\Psi(x, y, z)$ satisfies the three-dimensional continuous *Toda equation* [271, 274]

$$\left(\partial_x^2 + \partial_y^2\right) \Psi + \partial_z^2 e^\Psi = 0 . \tag{4.10}$$

The construction of four-dimensional hyper-Kähler metrics with a rotational isometry thus boils down to finding solutions of the non-linear Toda equation (4.10). In the Toda frame one uses local coordinates adapted to the rotational isometry given by τ-shifts. Being not triholomorphic, the τ-shifts do not leave invariant any of the Kähler forms on $\mathcal{M}$. Three independent Kähler forms of $\mathcal{M}$ can be explicitly written down [272] in the Toda frame, where they naturally form an $SO(2)$-doublet

$$\begin{pmatrix} \omega_1 \\ \omega_2 \end{pmatrix} = \exp(\tfrac{1}{2}\Psi) \begin{pmatrix} \cos\frac{\tau}{2} & \sin\frac{\tau}{2} \\ \sin\frac{\tau}{2} & -\cos\frac{\tau}{2} \end{pmatrix} \begin{pmatrix} f_1 \\ f_2 \end{pmatrix} , \tag{4.11}$$

with

$$\begin{aligned} f_1 &= (\mathrm{d}\tau + C_2 \mathrm{d}y) \wedge \mathrm{d}x - H \mathrm{d}z \wedge \mathrm{d}y , \\ f_2 &= (\mathrm{d}\tau + C_1 \mathrm{d}x) \wedge \mathrm{d}y + H \mathrm{d}z \wedge \mathrm{d}x , \end{aligned} \tag{4.12}$$

and an $SO(2)$-singlet

$$\omega_3 = (\mathrm{d}\tau + C_1 \mathrm{d}x + C_2 \mathrm{d}y) \wedge \mathrm{d}z + H e^\Psi \mathrm{d}x \wedge \mathrm{d}y . \tag{4.13}$$

Under the $SO(3)$ symmetry transformations, the Kähler forms (4.11) and (4.12) either belong to a triplet (and then all the $SO(3)$ generators are rotational, i.e. non-triholomorphic), or they are all singlets (and then the $SO(3)$ generators are all translational or triholomorphic).

Regular hyper-Kähler 4-manifolds with only one rotational isometry (and no translational isometries) apparently exist, though in a rather implicit form (as algebraic curves) [275, 276]. In our studies of 4d, $N = 2$ supersymmetric NLSM, we are mostly interested in a derivation of *all* known four-dimensional hyper-Kähler metrics having non-Abelian $SO(3)$ isometries, either rotational or translational, from extended superspace (PSS or HSS). There exist only three (complete and regular) metrics having an $SO(3)$ isometry, namely, (i) the Taub-NUT metric, (ii) the Eguchi-Hanson metric, and (iii) the Atiyah-Hitchin metric [273]. They are also the most important hyper-Kähler metrics in various applications. We now describe them in some detail [272].

The line element of any four-dimensional (Euclidean) metric with $SO(3)$ symmetry can be written down in the Bianchi IX formalism [204, 267] as

$$ds^2 = f^2(t)dt^2 + A^2(t)\sigma_1^2 + B^2(t)\sigma_2^2 + C^2(t)\sigma_3^2 , \qquad (4.14)$$

in terms of the $SO(3)$-invariant 1-forms

$$
\begin{aligned}
\sigma_1 &= +\tfrac{1}{2}\left(\sin\psi d\vartheta - \sin\vartheta\cos\psi d\varphi\right) , \\
\sigma_2 &= -\tfrac{1}{2}\left(\cos\psi d\vartheta + \sin\vartheta\sin\psi d\varphi\right) , \\
\sigma_3 &= +\tfrac{1}{2}\left(d\psi + \cos\vartheta d\varphi\right) ,
\end{aligned}
\qquad (4.15)
$$

where the Euler angles $(\vartheta, \psi, \varphi)$ have been introduced. The 1-forms (4.15) satisfy

$$\sigma_i \wedge \sigma_j = \tfrac{1}{2}\varepsilon_{ijk} d\sigma_k , \quad i,j,k = 1,2,3 . \qquad (4.16)$$

The coordinate t in (4.14) can be chosen to satisfy the relation [269, 272]

$$f(t) = \tfrac{1}{2}ABC , \qquad (4.17)$$

after a suitable reparametrization of t.

The four-dimensional self-duality conditions imposed on the Riemann curvature of the metric (4.14) give rise to second-order differential equations in t, which can be integrated once to yield the following *first-order* system of equations [269]:

$$
\begin{aligned}
2\frac{A'}{A} &= B^2 + C^2 - 2\lambda_1 BC - A^2 , \\
2\frac{B'}{B} &= C^2 + A^2 - 2\lambda_2 CA - B^2 , \\
2\frac{C'}{C} &= A^2 + B^2 - 2\lambda_3 AB - C^2 ,
\end{aligned}
\qquad (4.18)
$$

where the parameters $(\lambda_1, \lambda_2, \lambda_3)$ are all integration constants, and the primes denote the derivatives with respect to the variable t chosen in accordance with (4.17). The sign ambiguity associated with the self-dual or anti-self-dual curvature is associated with the transformation $t \to -t$.

Taub-NUT Metric. The Taub-NUT metric [268] is a solution to (4.18) with

$$\lambda_1 = \lambda_2 = \lambda_3 = 1 \quad \text{and} \quad A = B , \qquad (4.19)$$

having the form

$$A^2 = B^2 = \frac{1 + 4m^2 t}{4m^2 t^2} , \quad C^2 = \frac{4m^2}{1 + 4m^2 t} , \qquad (4.20)$$

where m is the Taub-NUT parameter. In the limit $m \to \infty$ we obtain the flat metric

$$ds^2 = dR^2 + R^2(\sigma_1^2 + \sigma_2^2 + \sigma_3^2) \tag{4.21}$$

with $R^2 = 1/t$. The standard form of the Taub-NUT metric,

$$ds^2_{\text{Taub-NUT}} = \frac{1}{4}\left(\frac{r+m}{r-m}\right) dr^2 + (r^2 - m^2)(\sigma_1^2 + \sigma_2^2) + 4m^2\left(\frac{r-m}{r+m}\right)\sigma_3^2 , \tag{4.22}$$

is obtained from (4.20) after the reparametrization

$$r = m + \frac{1}{2mt} . \tag{4.23}$$

In addition to the $SO(3)$ symmetry of the *Ansatz* (4.14), the Taub-NUT metric possesses yet additional $O(2)$ isometry generated by the Killing vector field $\partial/\partial\psi$ that commutes with the $SO(3)$ isometries due to the second relation (4.19). The $SO(3)$ symmetry of the Taub-NUT metric turns out to be rotational with respect to *all* of its generators, whereas the additional isometry generated by $\partial/\partial\psi$ turns out to be translational or triholomorphic. The full (covering) isometry group of the Taub-NUT metris is thus given by $U(2) = SU(2) \times U(1)$. [13]

To verify that the Killing vector $\partial/\partial\psi$ is translational, it is certainly enough to submit the explicit coordinate transformation from the frame (4.2) to the special frame where (4.3) and (4.4) are satisfied [272],

$$\phi = (r - m)(\sin\vartheta\cos\varphi, \sin\vartheta\sin\varphi, \cos\vartheta) , \quad \tau = \tfrac{m}{2}\psi , \tag{4.24}$$

where

$$H = \frac{1}{4}\left(1 + \frac{2m}{|\phi|}\right) , \quad C = \frac{m\phi_3}{2|\phi|(\phi_1^2 + \phi_2^2)}(-\phi_2, \phi_1, 0) . \tag{4.25}$$

To verify that the isometry generated by $\partial/\partial\varphi$ is not trihilomorphic, one may calculate the nut potential from (4.6). One finds [272]

$$b_{\text{nut}} = \frac{C}{4}\left(2A - C - (A - C)\sin^2\vartheta\right) , \quad H^{-1} = \frac{1}{4}\left(C^2 + (A^2 - C^2)\sin^2\vartheta\right) . \tag{4.26}$$

This implies, in accordance with (4.7), that

$$(\Delta S_+)^2 \equiv \gamma^{ij}(\partial_i S_+)(\partial_j S_+) = 4 > 0 . \tag{4.27}$$

The explicit coordinate transformation to the Toda frame is given by [272]

$$x = \psi , \quad y = -\frac{1}{4m^2t}\cos\vartheta + \log\left(\tan\frac{\vartheta}{2}\right) ,$$

---

[13] Since our considerations are purely local, we do not distinguish between $SU(2)$ and $SO(3)$ symmetries here.

$$z = \frac{1}{4t}\left(1 + \frac{1}{8m^2 t}\sin^2 \vartheta\right) \ , \quad \tau = 2\varphi \ . \tag{4.28}$$

This leads to the Toda potential [272]

$$e^\Psi = \frac{1}{16}A^2 C^2 \sin^2 \vartheta \equiv \frac{1}{16t^2}\sin^2 \vartheta \ . \tag{4.29}$$

The Taub-NUT metric can be physically interpreted either as a gravitational instanton [268], or as the metric on the moduli space of a BPS $SU(2)$ monopole [204]. More recently (Sect. 8.4), it also appeared in the context of M-theory, as the metric describing a Kaluza-Klein monopole in 11d.

Eguchi-Hanson Metric. The Eguchi-Hanson (EH) metric [268] is a solution to (4.18) with

$$\lambda_1 = \lambda_2 = \lambda_3 = 0 \quad \text{and} \quad A = B \ , \tag{4.30}$$

having the form (after transforming t to $-t$)

$$A^2 = B^2 = m^2 \coth(m^2 t) \ , \quad C^2 = \frac{2m^2}{\sinh(2m^2 t)} \quad , \tag{4.31}$$

where m is the EH parameter. The standard description of the EH metric,

$$ds_{\mathrm{EH}}^2 = \frac{dr^2}{1 - \left(\frac{m}{r}\right)^4} + r^2\left[\sigma_1^2 + \sigma_2^2 + \left(1 - \left(\frac{m}{r}\right)^4\right)\sigma_3^2\right] \ , \tag{4.32}$$

is recovered from (4.14) and (4.31) after a reparametrization

$$r^2 = m^2 \coth(m^2 t) \ . \tag{4.33}$$

Taking the limit $m \to 0$ in (4.32) results in the flat metric (4.21) with $R = r$.

The $SO(3)$ isometry of the EH metric is translational or triholomorphic, whereas the extra isometry generated by $\partial/\partial\psi$ turns out to be rotational, i.e. not triholomorphic (see below). The full isometry group $U(2)$ of the EH metric is therefore the same as that of the Taub-NUT, though its origin is different.

The nut potential associated with the EH Killing vector $\partial/\partial\psi$ can be easily calculated from the defining equation (4.6),

$$b_{\mathrm{nut}} = \frac{r^2}{4}\left(1 + \frac{m^4}{r^4}\right) \ . \tag{4.34}$$

This is to be compared with the EH harmonic potential H,

$$H^{-1} = \frac{r^2}{4}\left(1 - \frac{m^4}{r^4}\right) \ , \tag{4.35}$$

which implies $S_+ = b_{\text{nut}} + H^{-1} = r^2/2 \neq$ const. Hence, the Killing vector $\partial/\partial\psi$ is truly rotational. Similarly, in the case of the Killing vector $\partial/\partial\varphi$, one easily finds that $S_+ = 0$, which means its triholomorphicity.

The coordinate transformation to the Toda frame for the EH metric reads [272]

$$x = 2\sqrt{2}\,\cos\varphi\tan\frac{\vartheta}{2}\ , \quad y = 2\sqrt{2}\,\sin\varphi\tan\frac{\vartheta}{2}\ ,$$

$$z = \tfrac{1}{4}r^2 \quad \text{and} \quad \tau = 2(\psi + \varphi)\ . \tag{4.36}$$

The corresponding Toda potential is given by [272]

$$e^\Psi = \frac{z^2 - \alpha^2}{2\left(1 + \tfrac{1}{8}(x^2 + y^2)\right)^2}\ , \quad z^2 \geq \alpha^2\ , \tag{4.37}$$

where $4\alpha = m^2$.

Atiyah-Hitchin Metric. The Atiyah-Hitchin (AH) metric can be introduced either as the metric on the centred moduli space of two BPS $SU(2)$ monopoles [204], or as a regular (ALF) solution to (4.18) with purely rotational $SO(3)$ symmetry [277, 272]. The AH metric components are usually given in terms of complete elliptic integrals of the first and second kind,

$$K(k) = \int_0^{\pi/2} \frac{d\gamma}{\sqrt{1 - k^2\sin^2\gamma}}\ , \quad E(k) = \int_0^{\pi/2} d\gamma\,\sqrt{1 - k^2\sin^2\gamma}\ , \tag{4.38}$$

respectively. In the standard parametrization [204]

$$ds_{\text{AH}}^2 = \frac{1}{4}A^2B^2C^2\left(\frac{dk}{kk'^2K^2}\right)^2 + A^2(k)\sigma_1^2 + B^2(k)\sigma_2^2 + C^2(k)\sigma_3^2\ , \tag{4.39}$$

the AH solution reads [204]

$$AB = -K(k)\left(E(k) - K(k)\right)\ ,$$

$$BC = -K(k)\left(E(k) - k'^2K(k)\right)\ , \tag{4.40}$$

$$AC = -K(k)E(k)\ ,$$

where k is the AH modulus, $0 < k < 1$, and k' is the complementary modulus, $k'^2 = 1 - k^2$.

Since (4.39) has the same form as (4.14), the AH metric has an $SO(3)$ symmetry, while the coordinate t satisfying (4.17) is given by

$$t = -\frac{2K(k')}{\pi K(k)}\ , \tag{4.41}$$

up to an additive constant. The AH metric then satisfies (4.18) with $\lambda_1 = \lambda_2 = \lambda_3$, as the result of certain differential identities [253, 254] for the elliptic functions (4.38). In the limit $k \to 1$, the AH metric becomes exponentially close to the Taub-NUT metric [204, 277]. This easily follows after a redefinition

$$k' = \sqrt{1 - k^2} \cong 4 \exp\left(\frac{1}{\gamma}\right) , \qquad (4.42)$$

when considering the limit $\gamma \to 0^-$. By using the standard expansions of the elliptic integrals [253, 254] one arrives at the relations (in the vicinity of 0^-)

$$A^2 \cong B^2 \cong \frac{1 + \gamma}{\gamma^2} , \quad C^2 \cong \frac{1}{1 + \gamma} , \qquad (4.43)$$

which reproduce the Taub-NUT configuration (4.20) with $t = \gamma$ and $m^2 = 1/4$. A closer investigation, based on another useful parametrization in terms of the coordinate $r = 2K(k)$ with $\pi < r < \infty$, where the Taub-NUT limit is approached asymptotically as $r \to \infty$, shows that the corresponding Taub-NUT mass parameter is negative, $m = -1/2$ [204].

In monopole physics, the AH metric describes the scattering of two BPS $SU(2)$ monopoles. When the monopoles are separated, their moduli space is well approximated by the Taub-NUT metric, as it should. Another interesting limit, $k \to 0$, in the parametrization (4.39), leads to a coordinate bolt-type singularity [204], which corresponds to the coincidence limit of two monopoles. The $SO(3)$ symmetry of the AH metric appears to be purely rotational, while there are no other isometries [273]. Being compared to the Taub-NUT metric having extra translational isometry because of $a^2 = b^2$ in (4.19), the absence of this symmetry in the AH metric has profound consequences in monopole physics, where it leads to the generation of an electric charge for monopoles, thus converting them into dyons [278].

The transformation to the Toda frame in the AH case is known [277],

$$y + \mathrm{i}x = K(k)\sqrt{1 + k'^2 \sinh^2 \nu} \left(\cos \vartheta + \frac{\tanh \nu}{K(k)} \int_0^{\pi/2} \mathrm{d}\gamma \, \frac{\sqrt{1 - k^2 \sin^2 \gamma}}{1 - k^2 \tanh^2 \nu \sin^2 \gamma} \right)$$

and

$$z = \frac{1}{8} K^2(k) \left(k^2 \sin^2 \vartheta + k'^2 (1 + \sin^2 \vartheta \sin^2 \psi) - \frac{2E(k)}{K(k)} \right) , \qquad (4.44)$$

where

$$\nu \equiv \log\left(\tan \frac{\vartheta}{2}\right) + \mathrm{i}\psi , \quad \tau = 2\left(\varphi + \arg(1 + k'^2 \sinh^2 \nu)\right) . \qquad (4.45)$$

The corresponding Toda potential is given by [272]

$$e^\Psi = \frac{1}{16} K^2(k) \sin^2 \vartheta \left| 1 + k'^2 \sinh^2 \nu \right| . \qquad (4.46)$$

The AH Toda potential (4.46) reduces to the Taub-NUT Toda potential (4.29) in the limit $k \to 1$, after using the Taub-NUT identity $A^2 C^2 = A(A - C)$ valid at $m^2 = 1/4$.

4.4.2 Harmonic Superspace

As was already discussed in the preceeding sections, four-dimensional (4d) field theories with $N = 2$ extended supersymmetry can be formulated in the conventional $N = 2$ extended superspace parametrized by the coordinates $Z^M = (x^\mu, \theta_i^\alpha, \bar{\theta}^{\dot{\alpha}i})$, $\mu = 0, 1, 2, 3$, $\alpha = 1, 2$, $i = 1, 2$, and $\overline{\theta_i^\alpha} = \bar{\theta}^{\dot{\alpha}i}$, in terms of *constrained* $N = 2$ superfields. However, the standard $N = 2$ superspace constraints [279],

$$\{\mathcal{D}_\alpha^i, \bar{\mathcal{D}}_{\dot{\alpha}j}\} = -2i\delta_j^i \mathcal{D}_{\alpha\dot{\alpha}} \,, \quad \{\mathcal{D}_\alpha^i, \mathcal{D}_\beta^j\} = -2\varepsilon_{\alpha\beta}\varepsilon^{ij}\bar{W} \,, \tag{4.47}$$

and

$$\mathcal{D}_\alpha^{(i}q^{j)} = \bar{\mathcal{D}}_{\dot{\alpha}}^{(i}q^{j)} = 0 \,, \tag{4.48}$$

defining a (non-Abelian) $N = 2$ vector multiplet and a FS hypermultiplet, respectively, in terms of the gauge- and supercovariant (Lie algebra-valued) derivatives

$$\mathcal{D}_M \equiv (\mathcal{D}_\mu, \mathcal{D}_\alpha^i, \bar{\mathcal{D}}_{\dot{\alpha}i}) = D_M + \mathcal{A}_M \,, \tag{4.49}$$

do not have a manifestly holomorphic (or analytic) structure. Accordingly, they also do not have a simple solution in terms of *unconstrained* $N = 2$ superfields needed for a manifestly $N = 2$ supersymmetric quantization. The situation is even more dramatic for the FS hypermultiplet, since its defining equations (4.48) (in the absence of $N = 2$ central charges) are merely *on-shell* constraints, whereas the known off-shell formulations of a hypermultiplet in the conventional $N = 2$ superspace (Sect. 4.3) are either not universal (like an $N = 2$ tensor multiplet) or very inconvenient (like the relaxed hypermultiplet), so that their potential applications are limited.

In the HSS formalism [220] the standard 4d, $N = 2$ superspace is extended by adding extra bosonic variables (or 'zweibein') $u^{\pm i}$ belonging to the group $SU(2)$ and charged with respect to $U(1)$. One can then make manifest the hidden analyticity structure of the standard $N = 2$ superspace constraints (4.47) and (4.48), and find their manifestly $N = 2$ supersymmetric solutions in terms of unconstrained (analytic) $N = 2$ superfields. By definition of harmonics,

$$\begin{pmatrix} u^{+i} \\ u^{-i} \end{pmatrix} \in SU(2) \,, \tag{4.50}$$

they obey the unimodularity constraints,

$$u^{+i}u_i^- = 1 \,, \quad u^{+i}u_i^+ = u^{-i}u_i^- = 0 \,, \quad \text{and} \quad \overline{u^{i+}} = u_i^- \,. \tag{4.51}$$

Instead of using an explicit parametrization of the twistor sphere S^2, it is more convenient to deal with equivariant ($u_i^\pm$-dependent) functions of definite $U(1)$ charge U defined by $U(u_i^\pm) = \pm 1$, with the integration rules

$$\int du = 1 \quad \text{and} \quad \int du \, u^{+(i_1}\cdots u^{+i_m}u^{-j_1}\cdots u^{-j_n)} = 0 \quad \text{otherwise} \,, \tag{4.52}$$

so that any integral over a $U(1)$-charged quantity vanishes.

The usual complex conjugation does not preserve analyticity. However, it does, after being combined with another (star) conjugation acting on $U(1)$ indices as $(u_i^+)^* = u_i^-$ and $(u_i^-)^* = -u_i^+$. One easily finds that

$$\overset{*}{u^{\pm i}} = -u_i^\pm \, , \qquad \overset{*}{u_i^\pm} = u^{\pm i} \, . \tag{4.53}$$

The covariant harmonic derivatives, preserving the defining equations (4.50) and (4.51), are given by

$$D^{++} = u^{+i}\frac{\partial}{\partial u^{-i}} \, , \quad D^{--} = u^{-i}\frac{\partial}{\partial u^{+i}} \, , \quad D^0 = u^{+i}\frac{\partial}{\partial u^{+i}} - u^{-i}\frac{\partial}{\partial u^{-i}} \, . \tag{4.54}$$

It is easy to check that they satisfy an $su(2)$ algebra,

$$[D^{++}, D^{--}] = D^0 \, , \quad [D^0, D^{\pm\pm}] = \pm 2D^{\pm\pm} \, , \tag{4.55}$$

and commute with the $N = 2$ superspace derivatives (4.49). Equation (4.55) is supposed to be added to the constraints (4.47) and (4.48).

The key feature of $N = 2$ HSS is the existence of the so-called *analytic* subspace parametrized by the coordinates

$$(\zeta; u) \equiv \left\{ x_A^\mu = x^\mu - 2i\theta^{(i}\sigma^\mu\bar\theta^{j)}u_i^+ u_j^- \, , \quad \theta_\alpha^+ = \theta_\alpha^i u_i^+ \, , \quad \bar\theta_{\dot\alpha}^+ = \bar\theta_{\dot\alpha}^i u_i^+ \, ; \quad u_i^\pm \right\} \tag{4.56}$$

and invariant under $N = 2$ supersymmetry. It is also closed under the combined conjugation (4.53) [220]. This allows one to define *analytic* superfields of any non-negative and integer $U(1)$ charge q by the analyticity conditions

$$D_\alpha^+ \phi^{(q)} = \bar D_{\dot\alpha}^+ \phi^{(q)} = 0 \, , \quad \text{where} \quad D_\alpha^+ = D_\alpha^i u_i^+ \quad \text{and} \quad \bar D_{\dot\alpha}^+ = \bar D_{\dot\alpha}^i u_i^+ \, . \tag{4.57}$$

The analytic measure reads $d\zeta^{(-4)}du \equiv d^4 x_A d^2\theta^+ d^2\bar\theta^+ du$. It carries the $U(1)$ charge (-4), whereas the full neutral measure of $N = 2$ HSS is given by

$$d^4 x d^4\theta d^4\bar\theta du = d\zeta^{(-4)}du(D^+)^4 \, , \tag{4.58}$$

where

$$(D^+)^4 = (D^+)^2(\bar D^+)^2 = \frac{1}{16}(D^{+\alpha}D_\alpha^+)(\bar D_{\dot\alpha}^+\bar D^{+\dot\alpha}) \, . \tag{4.59}$$

In the analytic subspace, the harmonic derivative D^{++} reads

$$D^{++}_{\text{analytic}} = D^{++} - 2i\theta^+\sigma^\mu\bar\theta^+\partial_\mu \, , \tag{4.60}$$

it preserves analyticity, and it allows one to integrate by parts. The original (central) basis and the analytic basis can be used on an equal footing in HSS. In what follows we omit the subscript 'analytic' on the covariant derivatives in the analytic basis, in order to simplify our notation.

It is an advantage of the analytic $N = 2$ HSS compared to the ordinary (or chiral) $N = 2$ superspace that both an off-shell $N = 2$ vector multiplet and an off-shell (FS) hypermultiplet can be introduced there on an equal footing. There exist two basic off-shell hypermultiplet versions in HSS, which are dual to each other. The so-called Fayet-Sohnius (FS) type hypermultiplet is defined as an unconstrained complex analytic superfield q^+ of $U(1)$-charge $(+1)$, whereas its dual, called the Howe-Stelle-Townsend (HST) type hypermultiplet, is a real unconstrained analytic superfield ω with the vanishing $U(1)$-charge. It is worth mentioning that both FS and HST multiplets were originally introduced in the *conventional* $N = 2$ superspace [234, 235, 238], whereas we use the same names to denote the $N = 2$ *analytic* (harmonic) superfields that are very different off-shell, being equivalent to the FS and HST multiplets on-shell. The on-shell physical components of the FS hypermultiplet comprise an $SU(2)$ doublet of complex scalars and a Dirac spinor that is an $SU(2)$ singlet. The on-shell physical components of the HST hypermultiplet comprise a real singlet and a real triplet of scalars, and a doublet of chiral spinors. The FS hypermultiplets are natural in describing charged $N = 2$ matter, whereas the HST hypermultiplets are appropriate in describing neutral $N = 2$ matter.

Similarly, an $N = 2$ vector multiplet is described by an unconstrained analytic superfield V^{++} of $U(1)$-charge $(+2)$. The V^{++} is real in the sense $\overline{V^{++}}^* = V^{++}$, while it can be naturally introduced as a connection to the harmonic derivative D^{++}.

A free FS hypermultiplet HSS action is given by (in canonical normalization) [220]

$$S[q] = - \int d\zeta^{(-4)} du \, \overset{*}{q}{}^+ D^{++} q^+ , \qquad (4.61)$$

whereas its minimal coupling to an Abelian $N = 2$ gauge superfield reads [220]

$$S[q, V] = - \int d\zeta^{(-4)} du \, \overset{*}{q}{}^+ (D^{++} + iV^{++}) q^+ . \qquad (4.62)$$

It is not difficult to verify now that the free FS hypermultiplet equations of motion, $D^{++}q^+ = 0$, imply $q^+ = q^i(Z)u_i^+$ and the (on-shell) Fayet-Sohnius constraints (4.48) in the conventional $N = 2$ superspace,

$$D_\alpha^{(i} q^{j)}(Z) = D_{\overset{\cdot}{\alpha}}^{(i} q^{j)}(Z) = 0 . \qquad (4.63)$$

Similarly, a free HSS action of the HST hypermultiplet is given by [220]

$$S[\omega] = -\tfrac{1}{2} \int d\zeta^{(-4)} du \, (D^{++}\omega)^2 , \qquad (4.64)$$

and it is equivalent (dual) to the free $N = 2$ tensor multiplet action (Sect. 4.3).

The constraints (4.47) defining the $N = 2$ Super-Yang-Mills (SYM) theory in the conventional $N = 2$ superspace imply the existence of a (covariantly) chiral [14] and gauge-covariant $N = 2$ SYM field strength W satisfying the reality condition (the $N = 2$ Bianchi 'identity')

$$\mathcal{D}^\alpha_{(i} \mathcal{D}_{j)\alpha} W = \bar{\mathcal{D}}_{\dot{\alpha}(i} \bar{\mathcal{D}}^{\dot\alpha}_{j)} \bar{W} . \tag{4.65}$$

An $N = 2$ supersymmetric solution to the non-Abelian $N = 2$ SYM constraints (4.47) in the ordinary $N = 2$ superspace is not known in analytic form (see, however, [280] for some partial results). It is the $N = 2$ HSS reformulation of the $N = 2$ SYM theory that makes it possible. The exact non-Abelian (highly non-linear and complicated) relation between the constrained, harmonic-independent $N = 2$ superfield strength W and the unconstrained analytic (harmonic-dependent) $N = 2$ HSS superfield V^{++} is given in [220]. An Abelian relation is much simpler,

$$W = \{\bar{\mathcal{D}}^+_{\dot\alpha}, \bar{\mathcal{D}}^{-\dot\alpha}\} = -(\bar{D}^+)^2 A^{--} , \tag{4.66}$$

where the non-analytic harmonic superfield connection $A^{--}(Z, u)$ to the derivative D^{--} has been introduced, $\mathcal{D}^{--} = D^{--} + iA^{--}$.

It follows from the $N = 2$ HSS Abelian constraint $[\mathcal{D}^{++}, \mathcal{D}^{--}] = \mathcal{D}^0 = D^0$ that the connection A^{--} satisfies the relation

$$D^{++} A^{--} = D^{--} V^{++} , \tag{4.67}$$

whereas (4.65) can be rewritten in the form

$$(D^+)^2 W = (\bar{D}^+)^2 \bar{W} . \tag{4.68}$$

The solution to A^{--} in terms of the analytic unconstrained superfield V^{++} easily follows from (4.67) when using the identity [220]

$$D^{++}_1 (u^+_1 u^+_2)^{-2} = D^{--}_1 \delta^{(2,-2)}(u_1, u_2) , \tag{4.69}$$

where we have introduced the harmonic delta-function $\delta^{(2,-2)}(u_1, u_2)$ and the harmonic distribution $(u^+_1 u^+_2)^{-2}$ according to their definitions [221], hopefully, in self-explanatory notation. [15] One finds [281]

$$A^{--}(z, u) = \int dv \, \frac{V^{++}(z, v)}{(u^+ v^+)^2} , \tag{4.70}$$

and

---

[14] A covariantly-chiral superfield can be transformed into a chiral superfield by field redefinition.

[15] The explicit harmonic series, defining the harmonic delta-function and the harmonic distributions $(u^+_1 u^+_2)^{-n}$, are not needed for our calculations (see, however, [221]).

$$W(z) = -\int du (\bar{D}^-)^2 V^{++}(z,u) , \quad \bar{W}(z) = -\int du (D^-)^2 V^{++}(z,u) ,$$
(4.71)

by using yet another identity,

$$u_i^+ = v_i^+(v^- u^+) - v_i^-(u^+ v^+) ,$$
(4.72)

which is a simple consequence of the definitions (4.51).

The free equations of motion of the Abelian $N = 2$ vector multiplet are given by the vanishing analytic superfield

$$(D^+)^4 A^{--}(Z,u) = 0 ,$$
(4.73)

while the corresponding $N = 2$ Abelian action is

$$S[V] = -\frac{1}{2e^2} \int d^4x d^4\theta \, W^2 = -\frac{1}{2e^2} \int d^4x d^4\theta d^4\bar{\theta} du \, V^{++}(Z,u) A^{--}(Z,u)$$

$$= -\frac{1}{2e^2} \int d^4x d^4\theta d^4\bar{\theta} du_1 du_2 \, \frac{V^{++}(Z,u_1)V^{++}(Z,u_2)}{(u_1^+ u_2^+)^2} ,$$
(4.74)

where we have introduced an electromagnetic coupling constant e.

The Abelian analytic HSS pre-potential V^{++} in a WZ-like gauge reads [220]

$$V^{++}(x_A, \theta^+, \bar{\theta}^+, u) = \bar{\theta}^+ \bar{\theta}^+ a(x_A) + \bar{a}(x_A)\theta^+\theta^+ - 2i\theta^+ \sigma^\mu \bar{\theta}^+ V_\mu(x_A)$$

$$+ \bar{\theta}^+ \bar{\theta}^+ \theta^{\alpha+} \psi_\alpha^i(x_A) u_i^- + \theta^+ \theta^+ \bar{\theta}_{\dot\alpha}^+ \bar{\psi}^{\dot\alpha i}(x_A) u_i^- \quad (4.75)$$

$$+ \theta^+ \theta^+ \bar{\theta}^+ \bar{\theta}^+ D^{(ij)}(x_A) u_i^- u_j^- ,$$

where $(a, \psi_\alpha^i, V_\mu, D^{ij})$ are the usual 4d, $N = 2$ vector multiplet components (Sect. 4.2).

In the non-Abelian case, the gauge Lie algebra-valued $N = 2$ analytic superfield $V^{++} = V^{a++} t_a$, $\text{tr}(t_a t_b) = 2\delta_{ab}$, has the gauge transformation law

$$V^{++\prime} = e^\Lambda V^{++} e^{-\Lambda} - e^\Lambda D^{++} e^{-\Lambda} ,$$
(4.76)

whose gauge parameter $\Lambda = \Lambda^a t_a$ is the $N = 2$ analytic superfield too. The form of (4.76) is dictated by the gauge covariance of the minimally extended harmonic derivative $\nabla^{++} = D^{++} - V^{++}$. The pure $N = 2$ SYM action in HSS reads [281]

$$S_{N=2 \text{ SYM}} = \frac{1}{4g^2} \text{tr} \int d^4x d^4\theta \, W^2$$

$$= \frac{1}{2g^2} \text{tr} \int d^4x d^4\theta d^4\bar{\theta} \sum_{n=2}^{\infty} \frac{(-i)^n}{n} \int du_1 du_2 \cdots du_n \quad (4.77)$$

$$\times \frac{V^{++}(Z,u_1)V^{++}(Z,u_2) \cdots V^{++}(Z,u_n)}{(u_1^+ u_2^+)(u_2^+ u_3^+) \cdots (u_n^+ u_1^+)} ,$$

where g is the YM coupling constant.

A hypermultiplet (BPS) mass can only come from the central charges in $N = 2$ susy algebra since, otherwise, the number of massive hypermultiplet components has to be increased. The most natural way to introduce central charges $(Z, \bar{Z})$ is to identify them with spontaneously broken $U(1)$ generators of dimensional reduction from six dimensions via the Scherk-Schwarz mechanism [282]. After being rewritten in six dimensions and then 'compactified' down to four dimensions, the harmonic derivative (4.60) receives an extra 'connection' term in 4d,

$$\mathcal{D}^{++} = D^{++} + v^{++} , \quad \text{where} \quad v^{++} = \mathrm{i}(\theta^+\theta^+)\bar{Z} + \mathrm{i}(\bar{\theta}^+\bar{\theta}^+)Z . \quad (4.78)$$

Comparing (4.78) with (4.62), (4.71) and (4.75) gives rise to the conclusion that the $N = 2$ central charges can be equivalently treated as the Abelian $N = 2$ vector superfield background with the covariantly constant $N = 2$ chiral superfield strength [282, 283]

$$\langle W \rangle = \langle a \rangle = Z . \quad (4.79)$$

It is worth mentioning that introducing central charges into the algebra (4.47) of the $N = 2$ superspace covariant derivatives also implies corresponding changes in the $N = 2$ susy algebra and, hence, in the $N = 2$ susy transformation laws of $N = 2$ superfields and their components. The HSS formalism automatically incorporates those changes. The non-vanishing $N = 2$ central charges also break the rigid R-symmetry,

$$\theta^i_\alpha \to \mathrm{e}^{-\mathrm{i}\gamma}\theta^i_\alpha , \quad \bar{\theta}^{\dot{\alpha}i} \to \mathrm{e}^{+\mathrm{i}\gamma}\bar{\theta}^{\dot{\alpha}i} , \quad (4.80)$$

of a massless $N = 2$ supersymmetric field theory. This fact alone is responsible for the presence of anomalous (holomorphic) terms in the $N = 2$ gauge low-energy effective action (Chap. 8).

4.4.3 Hyper-Kähler NLSM in HSS

$N = 1$ supersymmetric matter in 4d is described by the chiral $N = 1$ multiplets and the linear $N = 1$ multiplets dual to the chiral ones (Sect. 4.3). The $N = 1$ chiral superfields Φ may have a chiral scalar superpotential described by a holomorphic function $W(\Phi)$. As regards fundamental QFT actions, the function $W(\Phi)$ should be restricted to a cubic polynomial by renormalizability [138, 140, 141], while there is no such restriction for the Low-Energy Effective Action (LEEA) of a quantum 4d, $N = 1$ field theory. $N = 2$ supersymmetric matter in 4d is described by FS and HST hypermultiplets dual to each other (Sect. 4.3). The $N = 2$ NLSM in 4d may also have a scalar potential, in the presence of central charges [282].

A duality transformation between the FS and HST analytic superfields is given by [220]

$$q_a^+ = u_a^+ \omega + u_a^- f^{++} , \qquad (4.81)$$

where we have defined $q_a^+ = (q^+, \overset{*}{\bar{q}}{}^+)$, $a = 1, 2$. The auxiliary analytic complex superfield f^{++} in (4.81) is a Lagrange multiplier. Inverting (4.81) yields

$$\omega = u_a^- q^{a+} \quad \text{and} \quad f^{++} = -u_a^+ q^{a+} . \qquad (4.82)$$

Unlike $N = 1$ susy models, no obvious $N = 2$ susy invariant exists that would generate a non-trivial scalar potential for the scalar components of a hypermultiplet (beyond the BPS mass term generated by $N = 2$ central charges), either in a fundamental $N = 2$ susy field theory action or in the corresponding LEEA, provided that $N = 2$ susy is not broken. In fact, at the fundamental level, any non-trival hypermultiplet potential is indeed forbidden by renormalizability and unitarity. However, contrary to naive expectations, a non-trivial scalar potential may appear in the hypermultiplet LEEA provided that $N = 2$ central charges do not vanish [282, 284]. Some explicit examples are given below. In this subsection, we also describe the $N = 2$ supersymmetric NLSM with the four-dimensional hyper-Kähler $SO(3)$-invariant metrics (Subsect. 4.4.1) in HSS.

The Ricci flatness of a Kähler manifold $\mathcal{M}$ implies the non-linear (MA) partial differential equation on its Kähler potential K (Subsect. 4.3.2). The $N = 2$ HSS (Subsect. 4.4.2) offers a formal solution to this equation in the form of the most general 4d, $N = 2$ supersymmetric NLSM action [265],

$$S[q, \overset{*}{\bar{q}}] = \frac{1}{\kappa^2} \int d\zeta^{(-4)} du \, \mathcal{L}^{(+4)}(q^+, \overset{*}{\bar{q}}{}^+, D^{++}q^+, D^{++}\overset{*}{\bar{q}}{}^+, \dots; u_i^\pm), \quad (4.83)$$

where the dots stand for possible higher harmonic derivatives of q^+. [16] The HSS Lagrangian $\mathcal{L}^{(+4)}$ has to be of $U(1)$ charge $(+4)$, in order to compensate the opposite $U(1)$ charge (-4) of the analytic HSS measure.

Due to its manifest $N = 2$ susy, the equations of motion for the HSS action (4.83) determine (at least, in principle) the component hyper-Kähler NLSM metric in terms of a single analytic function $\mathcal{L}^{(+4)}$. An explicit form of this relation is, however, not known because (4.83) contains infinitely many auxiliary field components whose elimination requires solving inifinitely many linear differential equations on the sphere. This highly non-trivial procedure in HSS was actually done in some special cases of $N = 2$ NLSM with four-dimensional hyper-Kähler target spaces [285, 286, 287]. Yet another caveat related to the infinite number of auxiliary fields in HSS is the considerable redundancy in the HSS description of $N = 2$ NLSM, which exhibits itself in the existence of several, apparently different, analytic HSS Lagrangians leading to the same hyper-Kähler metric in components (some examples are given below). Nevertheless, HSS offers the unique opportunity of classifying all hyper-Kähler metrics in terms of their potentials, similarly to Kähler metrics.

---

[16] The HSS harmonic derivative D^{++} is dimensionless.

We only consider those analytic HSS Lagrangians $\mathcal{L}^{(+4)}$ that have a well-defined kinetic term,

$$\mathcal{L}^{(+4)} = - \overset{*}{\bar{q}}{}^+ D^{++} q^+ + \mathcal{K}^{(+4)}(\overset{*}{\bar{q}}{}^+, q^+; u^\pm) . \tag{4.84}$$

An analytic function $\mathcal{K}^{+(4)}$ is called the *hyper-Kähler potential* [265, 285, 286].

The NLSM (4.84) naturally arises as the exact hypermultiplet LEEA in 4d quantized $N = 2$ supersymmetric gauge field theories (Sect. 8.3). Any explicit dependence of the function $\mathcal{K}$ upon harmonics signals the breaking of the internal $SU(2)_{\mathrm{R}}$ symmetry rotating spinor charges of $N = 2$ susy. Since the duality relation (4.81) between the FS and HST hypermultiplets involves harmonics, it may sometimes be useful to re-introduce both superfields q and ω into (4.84) if it results in the absence of an explicit dependence upon harmonics. This is relevant e.g., in the context of the hypermultiplet LEEA, since the latter is normally dependent upon the dynamically generated real scale Λ that can be interpreted as the vacuum expectaion value of a real Higgs hypermultiplet ω, i.e. $\Lambda = \langle \omega \rangle = \mathrm{const.} > 0$ (Sect. 8).

As is clear from (4.83) and (4.84), a general hyper-Kähler metric does not have any isometries. The absence of isometries makes it very difficult to explicitly construct the metric. Though the most general $N = 2$ NLSM action with hyper-Kähler geometry and no isometries can be formulated in HSS, the elimination of the HSS auxiliary fields, which is necessary to recover a component metric of the $N = 2$ NLSM, represents the fundamental technical problem in the HSS approach.

We restrict ourselves to the four-dimensional (Euclidean) hyper-Kähler NLSM target spaces having at least one isometry. Having assumed this isometry to be trihilomorphic (or translational), we are led to the multi-Taub-NUT metrics as the only ALF solutions (Subsect. 4.4.1). An explicit relation between the ALF harmonic potential H defined by (4.2) and (4.4), and the hyper-Kähler potential $\mathcal{K}$ of the corresponding $N = 2$ NLSM in HSS was established in [287]. One needs just a single q-hypermultiplet in order to parametrize a four-dimensional hyper-Kähler NLSM target space, while a triholomorphic isometry can always be represented (in adapted coordinates) by a rigid $U(1)$ symmetry of the hyper-Kähler potential with respect to the hypermultiplet rotations

$$q^+ \to e^{i\alpha} q^+ , \quad \overset{*}{\bar{q}}{}^+ \to e^{-i\alpha} \overset{*}{\bar{q}}{}^+ . \tag{4.85}$$

This implies that the hyper-Kähler potential of $U(1)$ charge $(+4)$ is an analytic function of the invariant product $(q \overset{*}{\bar{q}})$ of $U(1)$ charge $(+2)$, i.e. $\mathcal{K} = \mathcal{K}(q\overset{*}{\bar{q}}, u)$. Hence, one has [287]

$$\mathcal{K}^{(+4)} = \sum_{l=0}^{\infty} \xi^{(-2l)} \frac{(\overset{*}{\bar{q}}{}^+ q^+)^{l+2}}{l + 2} , \tag{4.86}$$

where the harmonic-dependent 'coefficients'

$$\xi^{(-2l)} = \xi^{(i_1 \cdots i_{2l})} u_{i_1}^- \cdots u_{i_{2l}}^- , \qquad l = 1, 2, \dots , \tag{4.87}$$

have been introduced. They are subject to the reality condition

$$\overset{*}{\xi}{}^{(-2l)} = (-1)^l \xi^{(-2l)} . \tag{4.88}$$

A general solution to (4.4) in the three-dimensional space R^3 parametrized by $\{y\}$ reads (outside of the origin)

$$H = \frac{\text{const.}}{2r} + \frac{U(y)}{2} , \quad r = \sqrt{y^2} , \tag{4.89}$$

where the function $U(y)$ is supposed to be non-singular. The latter can be decomposed in terms of the standard momentum eigenfunctions $Y_l^m(\vartheta, \varphi)$ depending upon the spherical coordinates (r, ϑ, φ),

$$U(y) = \sum_{l=0}^{+\infty} \sum_{m=-l}^{m=+l} c_{lm} r^l Y_l^m(\vartheta, \varphi) . \tag{4.90}$$

The one-to-one correspondence between the integration constants c_{lm} of (4.90) and the hyper-Kähler potential coefficients of (4.87) is given by [287]

$$\xi^{i_1=1,\dots,i_{l-m}=1,i_{l-m+1}=2,\dots,i_{2l}=2} = \frac{c_{lm}}{C} \frac{(2l+1)}{(l+1)} , \tag{4.91}$$

where C is a normalization constant.

The physical meaning of the harmonic potential H is transparent in the case of the multi-centre regular ALF metrics specified by (4.5) whose moduli are usually identified with charges and locations of monopoles. Though the HSS moduli $\xi^{(i_1 \cdots i_{2l})}$ of the same multi-monopole configuration have no direct physical interpretation, and they are not even independent, the HSS description in terms of the analytic hyper-Kähler potential (4.86) is manifestly non-singular. The regular description of (Kaluza-Klein) monopoles is useful in M-theory and brane technology (Chap. 8).

Taub-NUT NLSM in HSS with Central Charges. The $U(2)$ isometry of the Taub-NUT metric (Subsect. 4.4.1) leads to the simple HSS *Ansatz* for the corresponding 4d, $N = 2$ NLSM, when one identifies the $SU(2)$ part of the isometry with the $SU(2)_{\text{R}}$ automorphisms of $N = 2$ susy, and one associates the remaining $U(1)$ isometry factor with the charge carried by the q^+ superfield. The corresponding $N = 2$ HSS Lagrangian cannot explicitly depend upon harmonics, while there is only one candidate for the hyper-Kähler potential of $U(1)$ charge $(+4)$ [285, 284],

$$S[q] = -\int d\zeta^{(-4)} du \left\{ \overset{*}{q}{}^+ D^{++} q^+ + \frac{\lambda}{2} (q^+)^2 (\overset{*}{q}{}^+)^2 \right\} , \tag{4.92}$$

where the covariant derivative D^{++} has been introduced in the analytic basis with non-vanishing central charges Z and $\overline{Z}$,

$$D^{++} = \partial^{++} - 2\mathrm{i}\theta^+ \sigma^m \bar\theta^+ \partial_m + \mathrm{i}\theta^+ \theta^+ \overline{Z} + \mathrm{i}\bar\theta^+ \bar\theta^+ Z , \qquad (4.93)$$

while the parameter λ is supposed to be related to the Taub-NUT mass. The q^+ superfields have dimension of mass, whereas the coupling constant λ has dimension of length squared.

It is straightforward to calculate the bosonic equations of motion (ignoring fermionic contributions) out of the equations of motion for the NLSM action (4.92) in components, and then eliminate all the auxiliary fields (see [285, 284] for details). As a result, one arrives at the NLSM with the Taub-NUT metric in the form (4.22). In addition, one finds a non-trivial scalar potential [282]

$$V(r) = \frac{Z\overline{Z}\rho^2}{1 + \lambda\rho^2} , \qquad (4.94)$$

where $\rho^2 = 2(r - m)m$ and $r \geq m = \frac{1}{2}\lambda^{-1/2}$ (see Subsect. 8.3.1 for more).

Eguchi-Hanson NLSM in HSS with Central Charges. The (EH) 4d, $N = 2$ NLSM in HSS takes the most elegant form in terms a single dimensionless HST ω-hypermultiplet [286, 284],

$$S_{\mathrm{EH}}[\omega] = -\frac{1}{2\kappa^2} \int \mathrm{d}\zeta^{(-4)} \mathrm{d}u \left\{ (D^{++}\omega)^2 - \frac{(\xi^{++})^2}{\omega^2} \right\} , \qquad (4.95)$$

where $\xi^{++} = u_i^+ u_j^+ \xi^{(ij)}$ with some constants $\xi^{(ij)}$, and κ is the coupling constant of dimension of length. After changing the variables according to (4.81), and eliminating the Lagrange multiplier f^{++} from the action via its algebraic equation of motion, one can rewrite (4.106) in the equivalent form, in terms of the FS q^+-hypermultiplet [286],

$$S_{\mathrm{EH}}[q] = -\frac{1}{2\kappa^2} \int \mathrm{d}\zeta^{(-4)} \mathrm{d}u \left\{ q^{a+} D^{++} q_a^+ - \frac{(\xi^{++})^2}{(q^{a+} u_a^-)^2} \right\} , \qquad (4.96)$$

where we have used the notation $q_a^+ = (\overset{*}{\bar q}{}^+, q^+)$ and $q^{a+} = \varepsilon^{ab} q_b^+$. In turn, (4.96) is classically equivalent to the following gauge-invariant action in terms of *two* FS hypermultiplets q_{aA}^+ ($A = 1, 2$) and the auxiliary real analytic $N = 2$ (Abelian) vector superfield V^{++} [286]:

$$S_{\mathrm{EH}}[q, V] = -\frac{1}{2\kappa^2} \int \mathrm{d}\zeta^{(-4)} \mathrm{d}u \left\{ q_A^{a+} D^{++} q_{aA}^+ + V^{++} \left(\tfrac{1}{2}\varepsilon^{AB} q_A^{a+} q_{Ba}^+ + \xi^{++} \right) \right\} . \qquad (4.97)$$

The bosonic part of the equations of motion for the NLSM (4.97) in components can be most easily calculated in the WZ-gauge for V^{++} [286, 284]. The corresponding NLSM metric appears to be equivalent to the Eguchi-Hanson metric up to field redefinition [288, 289]. Because of the $U(2)$ symmetry of the metric, which is manifest in (4.97), and the S^3/Z^2 topology of the boundary, [17] the correct identification follows from the known classification of the ALE spaces (Subsect. 4.4.1).

A non-trivial scalar potential associated with the Eguchi-Hanson $N = 2$ NLSM in 4d was calculated in [284]. In terms of the (independent) NLSM field coordinates (m, n, θ, ϕ) of [284] and $\xi^2 = -4$, it reads

$$
V = \frac{|Z|^2 \sin^2(\theta + \phi)}{m^2 + n^2} \left[\frac{4(m^2 - n^2)^2}{1 + (m^2 + n^2)^2 \sin^2(\theta + \phi)} \right.
$$
$$
\left. + \frac{1 + (m^2 + n^2)^2 \sin^2(\theta + \phi)}{\sin^4(\theta + \phi)} \right] .
$$
(4.98)

The potential V is positive definite, while it does not vanish due to the non-vanishing central charge $|Z|$. This apparently signals spontaneous (dynamical) breaking of $N = 2$ susy in the $N = 2$ NLSM under consideration.

Atiyah-Hitchin NLSM in HSS. The general $N = 2$ NLSM Lagrangian in HSS is given by (4.84), whose component metric is not known. A simplification arises when the hyper-Kähler potential $\mathcal{K}^{(+4)}$ is independent of harmonics, which immediately leads to a general solution in the form [285, 290]

$$
\mathcal{K}^{(+4)} = \tfrac{\lambda}{2} (\overset{*}{q}{}^+)^2 (q^+)^2 + \left[\gamma \, \overline{(q^+)}^{\,4} + \beta \, \overline{(q^+)}^{\,3} q^+ + \text{h.c.} \right] ,
$$
(4.99)

with one real (λ) and two complex (β, γ) parameters. The $SU(2)_{\text{PG}}$ transformations of q_a^+ leave the form of (4.99) invariant but not the coefficients. Since $SU(2)_{\text{PG}}$ is the invariance of the free hypermultiplet action, it may be used to reduce the number of coupling constants in the family of hyper-Kähler metrics associated with the hyper-Kähler potential (4.99) from five to two. Equation (4.99) also implies the conservation law [285]

$$
D^{++} \mathcal{K}^{(+4)} = 0
$$
(4.100)

on the HSS equations of motion given by $D^{++} \overset{*}{q}{}^+ = \partial \mathcal{K}^{(+4)} / \partial q^+$ and $D^{++} q^+ = -\partial \mathcal{K}^{(+4)} / \partial \overset{*}{q}{}^+$.

Equation (4.99) describes the two-parameter family of four-dimensional hyper-Kähler metrics, with the standard AH metric (Subsect. 4.4.1) being the only *regular* member of the family [204]. The relation to the PSS description of those metrics (Sect. 4.3) is provided by the substitution

$$
\mathcal{K}^{(+4)}(q, \overset{*}{q}) \equiv \tfrac{\lambda}{2} (\overset{*}{q}{}^+)^2 (q^+)^2 + \left[\gamma \, \overline{(q^+)}^{\,4} + \beta \, \overline{(q^+)}^{\,3} q^+ + \text{h.c.} \right] = L^{++++}(\zeta, u) ,
$$
(4.101)

---

[17] The topology is determined by a (non-compact) algebraic constraint arising from varying the action (4.97) with respect to the $N = 2$ gauge superfield V^{++}.

where the real analytic superfield L^{++++} satisfies the conservation law (4.100), i.e. $D^{++}L^{++++} = 0$. This can be recognized as the *off-shell* $N = 2$ superspace constraints (3.1) and the reality condition (3.2) defining an $O(4)$ projective supermultiplet. Unlike the $O(2)$ tensor supermultiplet, the $O(4)$ projective multiplet does not have a conserved vector (or a gauge antisymmetric tensor) amongst its field components (Subsect. 4.3), which implies the absence of an Abelian triholomorphic isometry in the $N = 2$ NLSM to be constructed in terms of L^{++++} (see Sect. 8.3 for more).

4.4.4 Gauging $N = 2$ NLSM Isometries in HSS

General invariant couplings of susy matter to susy gauge fields are most naturally formulated in superspace. As regards the $N = 2$ NLSM describing self-coupling of $N = 2$ matter, its general coupling to $N = 2$ SYM fields in 4d (and, in fact, to $N = 2$ supergravity too) can be most naturally described in terms of unconstrained (analytic) HSS superfields [220]. Other superfield methods use either $N = 1$ superfields [114] or $N = 2$ constrained superfields [291]. The HSS uses an infinite number of auxiliary field components, which makes its relation to the component (or $N = 1$ superfield) description of $N = 2$ susy field theories highly non-trivial. The HSS action of a given $N = 2$ NLSM with the hyper-Kähler metric is not unique, being dependent upon the choice of the HSS superfields describing hypermultiplets. The different choices are related by $N = 2$ duality transformations, and they lead to the same on-shell physics. Of course, the above-mentioned applies to all 4d, $N = 2$ susy field theories, not just the 4d, $N = 2$ NLSM. For example, the $N = 2$ SYM fields are described in HSS by (semisimple) Lie algebra-valued analytic superfields $V^{++}(\zeta; u)$ of $U(1)$-charge $(+2)$ (cf. Subsect. 4.4.2). Under infinitesimal $N = 2$ gauge transformations with analytic HSS superfield parameters $\varepsilon(\zeta; u)$, they transform as

$$\delta_{\text{gauge}}V^{++} = D^{++}\varepsilon + [V^{++}, \varepsilon] \ . \tag{4.102}$$

An arbitrary four-dimensional hyper-Kähler manifold $\mathcal{M}$ has a holonomy group that is a subgroup of $Sp(1)$, whereas a generic hyper-Kähler manifold of complex dimension $2n$ has a holonomy that is a subgroup of $Sp(n)$. Let us now assign extra $Sp(n)$ indices to the complex HSS superfields,

$$q^{a+} \rightarrow q^{A+} \ , \quad A = 1, 2, \ldots, 2n \ , \tag{4.103}$$

where the multi-index A can be raised and lowered by using the symplectic metric $\Omega_{AB} = \begin{pmatrix} O & I_n \\ -I_n & 0 \end{pmatrix}$. In order to fix the most general form of NLSM action in HSS, let us apply an arbitrary reparametrization of $\mathcal{M}$,

$$q^{A+} \rightarrow q^{A+\prime} = f^{A+}(q, u) \ , \quad \text{with} \quad u^{\pm i} \text{ inert} \ , \tag{4.104}$$

to the free HSS action

$$S_{\text{free}}[q^A] = \frac{1}{2\kappa^2} \int_{\text{analytic}} q_A^+ D^{++} q^{A+} . \qquad (4.105)$$

One easily finds that the free FS action (4.105) gets transformed into

$$S[q^A] = \frac{1}{2\kappa^2} \int_{\text{analytic}} \left\{ F_A^+(q, u) D^{++} q^{A+} + G^{(+4)}(q, u) \right\} , \qquad (4.106)$$

with the *particular* functions F_A^+ and $G^{(+4)}$ given by

$$F_A^+ = f_B^+ \frac{\partial f^{B+}}{\partial q^{A+}} , \quad G^{(+4)} = f_B^+ \partial^{++} f^{B+} . \qquad (4.107)$$

It is equation (4.106), with *arbitrary* complex functions $F_A^+(q, u)$ and a real function $G^{(+4)}(q, u)$, that represents the most general $N = 2$ NLSM in HSS, by dimensionality and $U(1)$ charge conservation alone. The general action (4.106) is invariant under the infinitesimal field reparametrizations,

$$\delta q^{A+} = \rho^{A+}(q, u) , \quad \delta u_i^\pm = 0 , \qquad (4.108)$$

provided that

$$\delta F_A^+ = F_B^+ \frac{\partial \rho^{B+}}{\partial q^{A+}} , \quad \delta G^{(+4)} = F_A^+ \partial^{++} \rho^{A+} . \qquad (4.109)$$

The 'vielbeine' F_A are pure gauge fields with respect to HSS reparametrizations, so that they can be 'gauge-fixed' to the 'canonical' form $F_A^+ = q_A^+$ in adapted NLSM field coordinates, where the NLSM action (4.106) takes the standard HSS form

$$S_{\text{NLSM}}[q] = \frac{1}{2\kappa^2} \int_{\text{analytic}} \left\{ q_A^+ D^{++} q^{A+} + \mathcal{K}^{(+4)}(q, u) \right\} . \qquad (4.110)$$

The $N = 2$ NLSM may have isometries (forming the group G) that are the symmetries of the NLSM action (4.106). A NLSM isometry can be realized either linearly (in adapted field parametrization) or non-linearly (in arbitrary field parametrization). The general procedure of gauging isometries of the 4d, $N = 2$ NLSM in HSS was described in [266]. [18]

The NLSM isometries generically do *not* commute with $N = 2$ susy and its $SU(2)_R$ automorphisms. If this is the case, their gauging is impossible within rigid $N = 2$ susy. In particular, the $SU(2)_R$ internal rotations generically cannot be gauged within rigid $N = 2$ susy since commuting a local $SU(2)_R$ rotation with a rigid $N = 2$ susy transformation yields a local $N = 2$

---

[18] See also [114, 291] for related work, in terms of $N = 1$ superfields or constrained $N = 2$ superfields, respectively.

susy transformation. We confine ourselves to the gauging of triholomorphic isometries, $H \subset G$, which commute with $N = 2$ susy and its $SU(2)_R$ automorphisms.

Since isometries are the symmetries of the HSS action, not of the HSS Lagrangian, the latter may vary in a total harmonic derivative, $\delta \mathcal{L}^{(+4)} = D^{++} \Lambda^{++}(q, u)$, because of the identity

$$\int_{\text{analytic}} D^{++} \Lambda^{++} \equiv \int_{\text{analytic}} \left[\partial^{++} \Lambda^{++} + \frac{\partial \Lambda^{++}}{\partial q^{A+}} D^{++} q^{A+} \right] = 0 . \quad (4.111)$$

The action (4.106) is invariant under the infinitesimal isometry transformation

$$\delta q^{A+} = \varepsilon^X \rho^{X A+} \quad (4.112)$$

with some constant parameters ε^X, $X = 1, 2, \ldots, \dim H$, and the Killing vectors $\rho^{X A+}(q, u)$, provided that [266]

$$\left(\frac{\partial F_A^+}{\partial q^{B+}} + \frac{\partial F_B^+}{\partial q^{A+}} \right) \rho^{X B+} = \frac{\partial \Lambda^{X++}}{\partial q^{A+}} ,$$

$$\left(\frac{\partial V^{(+4)}}{\partial q^{A+}} - \partial^{++} F_A^+ \right) \rho^{X A+} = \partial^{++} \Lambda^{X++} . \quad (4.113)$$

In adapted coordinates (4.113) simplifies to

$$\frac{\partial \Lambda^{X++}}{\partial q^{A+}} = -2\rho_A^{X+} , \quad -2\partial^{++} \Lambda^{X++} = \frac{\partial \mathcal{K}^{(+4)}}{\partial q_A^+} \frac{\partial \Lambda^{X++}}{\partial q^{A+}} . \quad (4.114)$$

By analogy with the $N = 1$ superspace description of $N = 2$ NLSM isometries in [114], the analytic HSS superfield Λ^{X++} is called the *HSS Killing potential* of the HSS Killing vector $\rho^{X A+}$ [266]. The HSS Killing potential superfield Λ^{X++} has a harmonic decomposition

$$\Lambda^{X++}(\zeta, u) = \Lambda^{X(ij)}(Z) u_i^+ u_j^+ + \ldots , \quad (4.115)$$

where the dots stand for higher-order (auxiliary terms) in $u^{\pm i}$, while the ordinary $N = 2$ superfields $\Lambda^{X(ij)}(Z)$ can be identified with the constrained 4d, $N = 2$ superfields considered in [291].

If the NLSM action in HSS still has the form (4.110) in adapted coordinates where the isometry is linearly realized, then (4.112) takes the form

$$\delta q^{A+} = i \varepsilon^X (T_X)^A{}_B q^{B+} , \quad (4.116)$$

where T_X are the generators of H. The corresponding Killing vectors are linear in q, whereas their Killing potentials are quadratic [291],

$$\Lambda^{X++} = -i q_A^+ (T_X)^A{}_B q^{B+} . \quad (4.117)$$

Gauging the NLSM symmetries (4.116) amounts to a mimimal extension of the derivative D^{++} to a gauge-covariant HSS derivative ∇^{++} according to the standard rule [220],

$$\nabla^{++}q^{+A} = D^{++}q^{A+} + iV^{X++}(T_X)^A{}_B q^{B+} . \tag{4.118}$$

Hence, in adapted coordinates, the gauged $N = 2$ NLSM action reads [266]

$$S_{\text{gauged}}[q, V] = \tfrac{1}{2} \int_{\text{analytic}} \left\{ q_A^+ D^{++} q^{A+} + \mathcal{K}^{(+4)} - \Lambda^{X++}V^{X++} \right\} , \tag{4.119}$$

where we have used (4.117) and (4.118). The gauged NLSM action in the form (4.119) equally applies to non-linearly realized isometries too, provided that the Killing potential Λ transforms in the adjoint representation of the gauge group, $\delta\Lambda^{++} = [\varepsilon, \Lambda^{++}]$.

The ambiguities of triholomorphic Killing potentials, whose origin is related to Abelian factors in the isometry group H [114], also have a natural description in HSS, where they appear due to the ambiguity in the solution to (4.114) for the HSS Killing potential Λ^{++} [266],

$$\Lambda^{++} \rightarrow \Lambda^{++} + c^{(ij)}u_i^+ u_j^+ , \tag{4.120}$$

with some arbitrary symmetric constants $c^{(ij)}$. It is the ambiguity (4.120) that allows one to add an $N = 2$ supersymmetric and gauge-invariant Fayet-Iliopolulos (FI) term,

$$\tfrac{1}{2} \int_{\text{analytic}} c^{++}V^{++} , \tag{4.121}$$

to the gauged $N = 2$ NLSM action (4.119).

The local obstructions [19] against gauging the whole isometry group G of the $N = 2$ NLSM in rigid $N = 2$ HSS can be overcomed in local (curved) $N = 2$ HSS (i.e. in the presence of $N = 2$ supergravity), where the $SU(2)_R$ symmetry may also be gauged [266]. However, in the context of $N = 2$ supergravity, the $N = 2$ NLSM metric is no longer hyper-Kähler, and it has to be replaced by a *quaternionic* metric whose holonomy is a subgroup of $Sp(n) \times Sp(1)$ with a non-trivial $Sp(1)$ curvature proportional to Newton's constant [292]. The associated almost quaternionic structure in the NLSM target manifold $\mathcal{M}$ is not globally defined (i.e. not integrable) or, equivalently, it is not covariantly constant. In fact, as one goes from one point of the NLSM target space to another, the almost complex structures rotate into each other under the $Sp(1)$ transformations induced by gravity.

---

[19] There may be, in addition, some topological (global) obstructions against gauging, which are however beyond the scope of our (local) analysis.

5. NLSM and 2d Conformal Field Theory

In this chapter the 2d Wess-Zumino-Novikov-Witten (WZNW) models (as the NLSM in field theory) are discussed from the viewpoint of 2d Conformal Field Theory (CFT) [45]. We also consider supersymmetric generalizations of the WZNW models and gauging of their symmetries.

5.1 WZNW Models and Their Symmetries

Here we introduce 2d CFT (Subsects. 5.1.1 and 5.1.2), and discuss the 2d bosonic WZNW models (Subsect. 5.1.3) as examples. The so-called rational CFT (with a rational central charge) arise after introducing a Lie group structure into general CFT. The associated conserved (chiral) currents define an Affine Kač-Moody (AKM) algebra. A WZNW model can be thought of as the field-theoretical realization of the rational CFT whose stress-energy tensor is of the Sugawara-Sommerfeld form (Subsect. 5.1.2).

5.1.1 2d Conformal Invariance

By definition, the conformal transformations (in d dimensions) are the restricted general coordinate transformations, $x \to \tilde{x}$, that leave the metric invariant up to a scale factor,

$$g_{\mu\nu}(x) \to \tilde{g}_{\mu\nu}(\tilde{x}) = \Omega(x)g_{\mu\nu} , \quad \Omega(x) \equiv e^{\omega(x)} . \tag{1.1}$$

The conformal transformations form a group known as the conformal group. If spacetime is flat, the reference metric can always be chosen to be flat also, $g_{\mu\nu} = \eta_{\mu\nu}$, where $\eta_{\mu\nu}$ is the Minkowski metric in d spacetime dimensions.

As is easily seen from their definition, the conformal transformations preserve the angle $A \cdot B / \sqrt{A^2 B^2}$ between any two vectors A^μ and B^μ, where $A \cdot B = g_{\mu\nu} A^\mu B^\nu$. That is why these transformations are called conformal. They also contain the Poincaré transformations (translations and rotations of flat spacetime) as a subgroup, since the latter also satisfy (1.1) with $g_{\mu\nu} = \eta_{\mu\nu}$ and $\Omega = 1$. To determine all conformal transformations in flat spacetime, let us consider the infinitesimal coordinate transformations $x^\mu \to x^\mu + \varepsilon^\mu(x)$ with respect to the flat metric $g_{\mu\nu} = \eta_{\mu\nu}$. One easily finds that

$$\eta_{\mu\nu} + \partial_\mu \varepsilon_\nu + \partial_\nu \varepsilon_\mu = \Omega \eta_{\mu\nu} = [1 + \omega(x)]\eta_{\mu\nu} , \quad \omega(x) = \frac{2}{d}\partial^\mu \varepsilon_\mu . \tag{1.2}$$

Hence, we arrive at the equation

$$\partial_\mu \varepsilon_\nu + \partial_\nu \varepsilon_\mu = \frac{2}{d}(\partial \cdot \varepsilon)\eta_{\mu\nu} . \tag{1.3}$$

The simplest way to find a general solution to (1.3) is to consider its corollary,

$$[\eta_{\mu\nu}\Box + (d-2)\partial_\mu\partial_\nu]\partial \cdot \varepsilon = 0 , \tag{1.4}$$

which follows after two differentiations and a contraction of indices from (1.3). It is then easy to conclude that $\varepsilon(x)$ is at most quadratic in x when $d > 2$. After substituting a general second-order polynomial (in x) into (1.3) and fixing the coefficients, one finds that the conformal algebra consists of the ordinary translations ($\varepsilon^\mu = a^\mu$) and rotations ($\varepsilon^\mu = \omega^\mu{}_\nu x^\nu, \omega_{\mu\nu} = -\omega_{\nu\mu}$), scale transformations ($\varepsilon^\mu = \lambda x^\mu$) and the special conformal transformations ($\varepsilon^\mu = b^\mu x^2 - 2x^\mu b \cdot x$). A special conformal transformation can be recognized as the composition of an inversion and a translation: $\tilde{x}^\mu/\tilde{x}^2 = x^\mu/x^2 + b^\mu$. The whole algebra is locally isomorphic to $so(2, d)$.

The conformal group comprises finite conformal transformations. They include the Poincaré group

$$x^\mu \to \tilde{x}^\mu = x^\mu + a^\mu ,$$

$$x^\mu \to \tilde{x}^\mu = \Lambda^\mu{}_\nu x^\nu , \quad \Lambda^\mu{}_\nu \in SO(1, d-1) , \tag{1.5}$$

the scale transformations (dilatations) and the special conformal transformations, respectively,

$$x^\mu \to \tilde{x}^\mu = \lambda x^\mu , \quad \omega = -2\ln\lambda ,$$

$$x^\mu \to \tilde{x}^\mu = \frac{x^\mu + b^\mu x^2}{1 + 2b \cdot x + b^2 x^2} , \quad \omega = 2\ln[1 + 2b \cdot x + b^2 x^2] . \tag{1.6}$$

The d-dimensional conformal group is finite dimensional for $d > 2$, while all its transformations are globally defined. It has been known for a long time that the conformal group is a symmetry of massless fields with dimensionless coupling constants. The conformal invariance is the maximal kinematical extension of relativistic invariance [293].

Let us now consider some consequences of the global scaling invariance, $x^\mu \to \lambda x^\mu$, which is a part of the global conformal group. The associated current is given by

$$j_\mu = x^\nu T_{\nu\mu} . \tag{1.7}$$

A conservation of the current (1.7) is equivalent to a tracelessness condition for the stress-energy tensor, $T^\mu{}_\mu = 0$. The latter is supposed to be defined by a local (Euclidean) action S via the relation

$$T^{\mu\nu}(x) = -\frac{2}{\sqrt{g}} \frac{\delta S}{\delta g_{\mu\nu}(x)} \quad , \tag{1.8}$$

so that it describes the reaction of the field theory to a metric deformation. The condition $T^{\mu}{}_{\mu} = 0$ is known as the *scale invariance* condition. It is easy to see that any current of the form $f^{\lambda}(x)T_{\lambda\nu}$ will also be conserved provided that

$$\partial^{\mu} f^{\nu} + \partial^{\nu} f^{\mu} - \varphi(x)\eta^{\mu\nu} = 0 , \tag{1.9}$$

where $\varphi(x)$ is an arbitrary function. In the case of special conformal transformations, the associated current reads $f_{\mu}{}^{\lambda}(x)T_{\lambda\nu} = x_{\mu}x^{\lambda}T_{\lambda\nu} - x^2 T_{\mu\nu}$. Therefore, the global scale invariance *implies* global conformal invariance in any *local* field theory.

The case of two Euclidean dimensions is special, as one can already see from (1.4). Given $g_{\mu\nu} = \delta_{\mu\nu}$ and $d = 2$, equation (1.3) takes the form of the Cauchy-Riemann equation,

$$\partial_1 \varepsilon_1 = \partial_2 \varepsilon_2 , \quad \partial_1 \varepsilon_2 = -\partial_2 \varepsilon_1 . \tag{1.10}$$

In terms of complex coordinates and fields,

$$z = x^1 + \mathrm{i}x^2 , \quad \bar{z} = x^1 - \mathrm{i}x^2 ,$$

$$\varepsilon^z(z,\bar{z}) = \varepsilon^1(z,\bar{z}) + \mathrm{i}\varepsilon^2(z,\bar{z}) , \quad \bar{\varepsilon}^{\bar{z}}(z,\bar{z}) = \varepsilon^1(z,\bar{z}) - \mathrm{i}\varepsilon^2(z,\bar{z}) , \tag{1.11}$$

(1.10) implies a holomorphic dependence, $\varepsilon^z = \varepsilon^z(z)$ and $\bar{\varepsilon}^{\bar{z}} = \bar{\varepsilon}^{\bar{z}}(\bar{z})$. The 2d conformal transformations can, therefore, be identified with *analytic* coordinate transformations,

$$z \to f(z) \quad \bar{z} \to \bar{f}(\bar{z}) , \quad f'(z) \neq 0 . \tag{1.12}$$

This can be seen in yet another way, after rewriting the line element in terms of complex coordinates on a plane, $\mathrm{d}s^2 = \mathrm{d}z\mathrm{d}\bar{z}$. Under the holomorphic transformations (1.12) we get

$$\mathrm{d}z\mathrm{d}\bar{z} \to \left|\frac{\partial f}{\partial z}\right|^2 \mathrm{d}z\mathrm{d}\bar{z} , \tag{1.13}$$

which just means that we have a conformal transformation, in accordance with the definition (1.1), and $\omega(x) = 2\ln|\partial f/\partial z|$.

A useful basis for the infinitesimal conformal transformations, $z \to \tilde{z} = z + \varepsilon(z)$ and $\bar{z} \to \tilde{\bar{z}} = \bar{z} + \bar{\varepsilon}(\bar{z})$, is given by

$$z \to z - a_n z^{n+1} , \quad \bar{z} \to \bar{z} - \bar{a}_n \bar{z}^{n+1} , \quad n \in \mathbf{Z} . \tag{1.14}$$

This basis is generated by the operators

$$l_n = -z^{n+1}\frac{\mathrm{d}}{\mathrm{d}z} , \quad \bar{l}_n = -\bar{z}^{n+1}\frac{\mathrm{d}}{\mathrm{d}z} , \quad n \in \mathbf{Z} , \tag{1.15}$$

that satisfy an algebra comprising two commuting pieces,

$$[l_n, l_m] = (n - m)l_{n+m} , \quad [\bar{l}_n, \bar{l}_m] = (n - m)\bar{l}_{n+m} , \quad [l_n, \bar{l}_m] = 0 , \qquad (1.16)$$

each one being known as a 2d local *conformal algebra*.

The independence of two algebras $\{l_n\}$ and $\{\bar{l}_n\}$ justifies the use of z and $\bar{z}$ as independent coordinates. This means a complexification of the initial 2d Euclidean space, $\mathbf{C} \to \mathbf{C}^2$. This gives us freedom of choosing various reality conditions by making 'sections' in the complexified space. In particular, the section defined by $\bar{z} = z^*$ recovers the Euclidean plane. The Minkowski plane can be recovered by the section $z^* = -z$, which implies

$$(z, \bar{z}) = \mathrm{i}(\tilde{\tau} + \tilde{\sigma}, \tilde{\tau} - \tilde{\sigma}) , \quad \mathrm{d}s^2 = -\mathrm{d}\tilde{\tau}^2 + \mathrm{d}\tilde{\sigma}^2 . \qquad (1.17)$$

These are, however, not the coordinates that we would like to associate with our 2d Minkowski spacetime. First, we make a conformal transformation $z = \mathrm{e}^{\zeta}$, $\bar{z} = \mathrm{e}^{\bar{\zeta}}$, with $\zeta = \tau + \mathrm{i}\sigma$, $\bar{\zeta} = \tau - \mathrm{i}\sigma$, from the z-plane to a cylinder, $-\infty < \tau < +\infty$, $0 \leq \sigma \leq 2\pi$. By definition, the Minkowski spacetime formulation of a field theory is obtained from its Euclidean formulation on the cylinder by a Wick rotation, $\zeta = \tau + \mathrm{i}\sigma \to \mathrm{i}(\tau + \sigma) \equiv \mathrm{i}\zeta^+$, $\bar{\zeta} = \tau - \mathrm{i}\sigma \to \mathrm{i}(\tau - \sigma) \equiv \mathrm{i}\zeta^-$, where the Minkowski *light-cone* coordinates $\zeta^\pm = \tau \pm \sigma$ have been introduced. In terms of those coordinates, the line element takes the form $\mathrm{d}s_\mathrm{M}^2 = \mathrm{d}\zeta^+\mathrm{d}\zeta^-$, while the conformal transformations take the form of reparametrizations of ζ^+ and ζ^-, i.e. $\zeta'^+ = f(\zeta^+)$, $\zeta'^- = g(\zeta^-)$, which leave the light-cone invariant. The line element $\mathrm{d}s_\mathrm{M}^2$ is preserved by these transformations up to the scale factor $\Omega = f'(\zeta^+)g'(\zeta^-)$.

The number of generators in the 2d local conformal algebra (1.16) is *infinite*. The infinite number of symmetries implies severe restrictions on the conformally invariant field theories in two dimensions. It is also worth mentioning that the general 2d conformal transformations are neither globally well defined nor invertible, even on the Riemann sphere $S^2 = \mathbf{C} \cup \infty$. This is because the vector fields

$$\mathcal{V}(z) = -\sum_n a_n l_n = \sum_n a_n z^{n+1} \frac{\mathrm{d}}{\mathrm{d}z} , \qquad (1.18)$$

generating holomorphic transformations, are globally defined only if $a_n = 0$ for $n < -1$ *and* $n > 1$. This is necessary for the absence of singularities of $\mathcal{V}(z)$ at $z \to 0$ and $z \to \infty$ (use the conformal transformation $z = -1/w$ in the latter case!). A global group of the well defined and invertible conformal transformations on the Riemann sphere is generated by $\{l_{-1}, l_0, l_1\} \cup \{\bar{l}_{-1}, \bar{l}_0, \bar{l}_1\}$ only. Equation (1.15) tells us that l_{-1} and $\bar{l}_{-1}$ can be identified with the generators of translations, $(l_0 + \bar{l}_0)$ and $\mathrm{i}(l_0 - \bar{l}_0)$ are the generators of dilatations and rotations, respectively, while l_1 and $\bar{l}_1$ are the generators of special conformal transformations. The corresponding finite transformations form the so-called complex Möbius group,

$$z \to \tilde{z} = \frac{az + b}{cz + d} \, , \tag{1.19}$$

where a, b, c, d are complex parameters and $ad - bc = 1$. The group of transformations (1.19) is parametrized by six real parameters, being isomorphic to $SL(2, \mathbf{C})/\mathbf{Z}_2 \cong SO(1,3)$. The need for the quotient $\mathbf{Z}_2$ is caused by the fact that the transformation (1.19) is not sensitive to a simultaneous change of sign for all parameters a, b, c, d.

Representations of the global conformal algebra (after quantization) assign quantum numbers to physical states. It is quite natural to assume the existence of the vacuum state $|0\rangle$ among the physical states, which is invariant under the Möbius transformations and has vanishing quantum numbers. The eigenvalues h and $\bar{h}$ of the operators l_0 and $\bar{l}_0$, respectively, are known as *conformal weights* of a physical state. Given conformal weights of a state, its scaling dimension Δ and spin s are given by $\Delta = h + \bar{h}$ and $s = h - \bar{h}$, in accordance with the similar assignment for the generators of dilatations and rotations.

Instead of the Riemann sphere, which is a *closed* Riemann surface of genus zero, the simplest *open* Riemann surface given by an upper half-plane (with infinity attached) may also be considered. In this case, the parameters a, b, c, d in (1.19) have to be real, thus leading to the *real* Möbius group $SL(2, \mathbf{R})/\mathbf{Z}_2$. There is an apparent similarity between the 'open' case and the holomorphic part of the 'closed' case (it extends to a similar relation between open and closed strings, see Sect. 6.4).

On the Euclidean plane parametrized by complex coordinates $z, \bar{z}$ with the line element $ds^2 = dz d\bar{z}$, the conservation of the stress-energy tensor, $\partial^\mu T_{\mu\nu} = 0$, takes the form

$$\partial_{\bar{z}} T_{zz} + \partial_z T_{\bar{z}z} = 0 \, , \quad \partial_z T_{\bar{z}\bar{z}} + \partial_{\bar{z}} T_{z\bar{z}} = 0 \, , \tag{1.20}$$

while the scale invariance condition $T^\mu{}_\mu = 0$ yields

$$T_{z\bar{z}} = T_{\bar{z}z} = 0 \, , \tag{1.21}$$

where the corresponding metric and stress-energy tensor components have been introduced:

$$g_{z\bar{z}} = g_{\bar{z}z} = \tfrac{1}{2} \, , \quad g_{zz} = g_{\bar{z}\bar{z}} = 0 \, ,$$

$$T_{zz} = \tfrac{1}{4}(T_{11} + 2iT_{12} - T_{22}) \, , \quad T_{\bar{z}\bar{z}} = \tfrac{1}{4}(T_{11} - 2iT_{12} - T_{22}) \, ,$$

$$T_{z\bar{z}} = T_{\bar{z}z} = \tfrac{1}{4}(T_{11} + T_{22}) \equiv \tfrac{1}{4} T^\mu{}_\mu \, . \tag{1.22}$$

The stress-energy tensor in any 2d CFT can, therefore, be split into a holomorphic part and an antiholomorphic part, [1]

---

[1] In what follows, we refer to the 2d stress-energy tensor as the stress tensor for brevity.

$$T_{zz} \equiv T(z) , \qquad T_{\bar{z}\bar{z}} \equiv \bar{T}(\bar{z}) . \tag{1.23}$$

The holomorphic and antiholomorphic parts of a field in Euclidean space are related by complex conjugation, whereas in Minkowski space they correspond to the Left-Moving (LM) and Right-Moving (RM) modes that are truly independent. The relation between the two formulations is provided via the complexification that allows us to consider z and $\bar{z}$ as independent variables.

The generators of infinitesimal conformal transformations can be defined in terms of $T(z)$ as

$$L_n = \oint_C \frac{\mathrm{d}z}{2\pi \mathrm{i}} z^{n+1} T(z) , \tag{1.24}$$

with the contour C encircling the origin. The formal operatorial equation (1.24) makes sense when acting on fields whose arguments are inside the integration contour. The contour shape is irrelevant because of Cauchy's theorem. The same theorem yields the statement

$$T(z) = \sum_{n \in \mathbf{Z}} L_n z^{-n-2} , \tag{1.25}$$

and similarly for $\bar{L}_n$ and $\bar{T}$. We now have the full power of complex calculus at our disposal.

To incorporate the standard machinery of canonical quantization into CFT, it is convenient to use *radial quantization* techniques [45] on the plane. This uses the following parameterization of $\mathbf{C}^2$:

$$z = \mathrm{e}^\zeta , \quad \zeta = \tau + \mathrm{i}\sigma , \tag{1.26}$$

in terms of the Euclidean time and space coordinates, $\tau \in \mathbf{R}$ and $0 < \sigma \leq 2\pi$, respectively. In string theory (Chap. 6), the latter are interpreted as the coordinates of a (Euclidean) closed string *world-sheet*.

One interprets (1.26) as the conformal map of a cylinder to a plane. Infinite past and future, $\tau = \mp\infty$, on the cylinder are mapped into the points $z = 0, \infty$, respectively, on the plane. Equal-time slices are circles of fixed radius on the plane, whereas equal-space slices are lines radiating from the origin. The time translations $\tau \to \tau + \lambda$ are the dilatations on $\mathbf{C}$: $z \to \mathrm{e}^\lambda z = z + \lambda z + \ldots$, whereas the space translations $\sigma \to \sigma + \theta$ are the rotations on $\mathbf{C}$: $z \to \mathrm{e}^{\mathrm{i}\theta} z$. Therefore, the Hamiltonian of the system can be identified with the dilatation generator on the plane, while the Hilbert space of states comprises surfaces of constant radius. The stress tensor components $T(z)$ and $\bar{T}(\bar{z})$ in (1.23) are identified with the generators of local conformal transformations on the z-plane. In the radial quantization, an 'equal-time' surface becomes a contour on the z-plane surrounding the origin.

For dimensional reasons and analyticity, the Operator Product Expansion (OPE) for the product of a stress tensor $T(z)$ with itself reads

$$T(z)T(w) = \frac{c/2}{(z-w)^4} + \frac{2}{(z-w)^2}T(w) + \frac{1}{z-w}\partial_w T(w) + \ldots , \tag{1.27}$$

where the constant c has been introduced, and the dots stand for finite terms. The simplest way to isolate the constant c is to consider the two-point correlation function of two T's,

$$\langle T(z)T(0)\rangle = \frac{c/2}{z^4} . \tag{1.28}$$

Equation (1.28) can be justified as a consequence of global conformal invariance. The constant c depends upon which CFT the T is computed for, and it is called the *central charge* or the conformal anomaly.

A non-vanishing central charge is a quantum effect. An anomalous QFT may generically have internal inconsistencies, since the anomaly often violates some of the basic symmetries used to build up the theory. However, the conformal symmetry violation due to a non-vanishing central charge merely implies the conformal anomaly 'sitting' at one singular point (say, the origin), which is harmless in CFT.

The additional and, in principle, independent constant $\bar{c}$ appears in similar antiholomorphic counterparts to (1.27) and (1.28). The difference $(c - \bar{c})$ is known as the local gravitational anomaly (see Subsect. 7.1.3).

The equivalent useful representation of the OPE (1.27) is given by the commutation relations between the operatorial modes $\hat{L}_n$ defined by the Laurent expansion of $T(z)$ in (1.25). They satisfy the *Virasoro algebra*

$$[L_n, L_m] = (n - m)L_{n+m} + \frac{c}{12}n(n^2 - 1)\delta_{n+m,0} , \tag{1.29}$$

which can be recognized as the central extension of the 2d conformal algebra (1.16). The uniqueness of the central extension, fixed by the only constant c, can also be seen in the following way [294]. Quantum mechanically, after taking into account the conformal anomalies, the l_n's of (1.16) become operators L_n in a centrally extended conformal algebra of the general form

$$[L_n, L_m] = (n - m)L_{n+m} + c_{n,m} . \tag{1.30}$$

The antisymmetric coefficients $c_{n,m}$ defined by this equation are constrained by the requirement of the algebra (1.30) to obey the Jacobi identity for the commutators. The only way to obey the Jacobi identity is to choose

$$c_{n,m} = \frac{c}{12}n(n^2 - 1)\delta_{n+m,0} , \tag{1.31}$$

modulo additive constant redefinitions of the generators. The coefficient c appearing in this equation is just the central charge. In the normalization (1.27) or (1.31), one finds $c = 1$ for a real scalar, and $c = 1/2$ for a 2d Majorana-Weyl (MW) fermion. The vanishing central terms for $m = 0, \pm 1$ in (1.31) reflect the $SL(2, \mathbf{C})$ invariance of the ground state. Therefore, the global conformal group $SL(2, \mathbf{C})$ remains the exact symmetry of 2d CFT, despite the central extension.

5.1.2 AKM Symmetries and SS Construction

In the previous subsection we considered only those CFT whose maximal symmetry algebra was the direct sum of two Virasoro algebras, Vir $\oplus$ $\overline{\text{Vir}}$. In general, the symmetry of a CFT is given by the algebra $\mathcal{A} = \mathcal{A}_{\text{L}} \oplus \mathcal{A}_{\text{R}}$ generated by holomorphic and antiholomorphic tensors of the type $(p, 0)$ and $(0, \bar{p})$, respectively, which contains the Virasoro algebra Vir $\oplus$ $\overline{\text{Vir}}$. Any $(p, 0)$ tensor or p-differential $S(z)$ obeys the conservation law

$$\bar{\partial} S(z) = 0 \tag{1.32}$$

which obviously implies an infinite number of conserved quantities,

$$\bar{\partial} \left(z^n S(z) \right) = 0 . \tag{1.33}$$

The conserved quantities can be identified with the coefficients in the Laurent series for $S(z)$,

$$S(z) = \sum_{n \in \mathbf{Z}} S_n z^{-n-p} . \tag{1.34}$$

The Virasoro algebra Vir is the particular case of (1.34), with $p = 2$ and $S(z) = T(z)$. Of particular importance is the case when the *chiral algebra* $\mathcal{A}_{\text{L}}$ is generated by the $(1, 0)$ conformal fields, called conformal currents. A free scalar field theory gives us the simplest example with the $U(1)$ current $j(z) = i\partial\phi(z)$ and the OPE

$$\partial j(z)\partial j(w) \sim \frac{1}{(z - w)^2} . \tag{1.35}$$

Equation (1.35) can be easily verified by using the mode expansion in terms of the bosonic oscillators $\{\alpha_n\}$,

$$j(z) = i\partial\phi(z) = \sum_{n \in \mathbf{Z}} \alpha_n z^{-n-1} , \tag{1.36}$$

and the commutation relations

$$[\alpha_n, \alpha_m] = n\delta_{n+m,0} . \tag{1.37}$$

Dimensional analysis constrains the OPE of $(1, 0)$ conformal currents $J^a(z)$ to the form generalizing that of (1.35) as follows:

$$J^a(z)J^b(w) \sim \frac{\tilde{k}^{ab}}{(z - w)^2} + \frac{i f^{abc}}{z - w} J^c(w) , \tag{1.38}$$

where the 'structure constants' f^{abc} are antisymmetric in a and b. The associativity of the operator product now implies that f^{abc} satisfy the Jacobi identity. This means that f^{abc} are just the structure constants of a Lie algebra

$\mathcal{G}$. It is usually assumed that the corresponding Lie group G is simple and compact, in order to have a positive definite Cartan metric. In addition, this allows one to choose a basis where the central extension takes the diagonal form, $\tilde{k}^{ab} = \tilde{k}\delta^{ab}$. Expanding $J^a(z)$ in a way similar to that of (1.36),

$$J^a(z) = \sum_{n \in \mathbf{Z}} J^a_n z^{-n-1} , \qquad (1.39)$$

we deduce from (1.38) the commutation relations

$$[J^a_m, J^b_n] = \mathrm{i}f^{abc} J^c_{m+n} + \tilde{k}m\delta^{ab}\delta_{m+n,0} \qquad (1.40)$$

that generalize (1.37) and have $m, n \in \mathbf{Z}$ and $a, b = 1, \ldots, \dim G$.

Equations (1.38) and (1.40) define the (untwisted) Affine Kač-Moody (AKM) algebra $\hat{\mathcal{G}}$ with a central extension. The subalgebra of zero modes J^a_0 is the ordinary Lie algebra $\mathcal{G}$ called the *horizontal* Lie subalgebra of the AKM algebra,

$$[J^a_0, J^b_0] = \mathrm{i}f^{abc} J^c_0 . \qquad (1.41)$$

The procedure $\mathcal{G} \to \hat{\mathcal{G}}$ is known as the (infinite-dimensional) *affinization* of a finite-dimensional Lie algebra $\mathcal{G}$.

After pulling $J(z)$ back to a cylinder,

$$J^a_{\mathrm{cyl}}(w) = \sum_n J^a_n \mathrm{e}^{-nw} , \qquad (1.42)$$

with real w, the modes J^a_n can be recognized as the infinitesimal generators of the group of gauge transformations $g(\sigma): S^1 \to G$ on a circle S^1. The Virasoro generators also define the projective representation of the conformal algebra. The mode operators J^a_n are associated with the loop algebra of G, $\mathcal{L}(z) = \sum_{a,n} \mathcal{L}^a_n t^a_{(r)} z^n$, where $\mathcal{L}^a_n$ are the loop algebra generators and $\{t^a_{(r)}\}$ are the generators of a matrix representation (r) of $\mathcal{G}$.

The representation theory of AKM algebras has many features similar to that of the Virasoro algebra. The AKM highest-weight representations are characterized by the highest weights $\hat{\Lambda}$ [45]. When $\hat{\Lambda}$ is an integral dominant weight, such representations are called integrable. According to the Gepner-Witten theorem [295], only integrable highest-weight representations can appear in the WZNW spectrum (Subsect. 5.1.3).

Given an AKM algebra, one can always build the associated Virasoro algebra. Indeed, AKM currents have conformal dimension one, while a stress tensor is of dimension two. It is therefore natural to use a bilinear in terms of the AKM currents to form the stress tensor known as the Sugawara-Sommerfeld (SS) construction [296, 297]. More insights are again provided by the theory of a single boson ϕ, where the $U(1)$ current is known, $J(z) = \mathrm{i}\partial\phi$, while the stress tensor is $T(z) = -\frac{1}{2} : \partial\phi(z)\partial\phi(z) : = \frac{1}{2} : J(z)J(z) :$. This prescription can be generalized to the non-Abelian case,

$$T(z) = \frac{1}{\beta} \sum_{a=1}^{|G|} : J^a(z) J^a(z) : \ , \tag{1.43}$$

where the normal ordering of two currents is defined by

$$\sum_a : J^a(z) J^a(z) : \equiv \lim_{z \to w} \left(\sum_a J^a(z) J^a(w) - \frac{\tilde{k}|G|}{(z-w)^2} \right) , \tag{1.44}$$

$|G| = \dim G$, and β is a constant. The currents $J^a(z)$ have to transform as the $(1,0)$ primary fields,

$$T(z) J^a(w) \sim \frac{J^a(w)}{(z-w)^2} + \frac{\partial J^a(w)}{z-w} , \tag{1.45}$$

which implies the commutation relations

$$[L_m, J^a_n] = -n J^a_{m+n} , \tag{1.46}$$

where L_n are computed from (1.24) and (1.43),

$$L_n = \frac{1}{\beta} \sum_{m=-\infty}^{+\infty} : J^a_{m+n} J^a_{-m} : \ . \tag{1.47}$$

The SS stress tensor (1.43) satisfies the correct OPE (1.27) only if

$$\beta = 2\tilde{k} + C_A , \tag{1.48}$$

where we have introduced the quadratic Casimir eigenvalue C_A of the adjoint representation, $f^{acd} f^{bcd} = C_A \delta^{ab}$. The associated central charge is

$$c = \frac{\tilde{k}|G|}{\tilde{k} + C_A/2} . \tag{1.49}$$

The SS construction thus implies that all the Virasoro generators are contained in the AKM enveloping algebra.

It is worth mentioning that both numbers, $\tilde{k}$ and $C_A/2$, introduced above do depend upon the normalization used for the structure constants f^{abc}. Given the trace

$$\text{tr}\left(t^a_{(r)} t^b_{(r)} \right) = l_r \delta^{ab} \tag{1.50}$$

in an arbitrary $\mathcal{G}$-representation (r) of dimension d_r, the diagonal sum over $a, b = 1, \ldots, |G|$ gives

$$C_r d_r = l_r |G| , \tag{1.51}$$

where the quadratic Casimir eigenvalue C_r in the representation (r) has been introduced, $C_r \delta^{ij} = \sum_a \left(t^a_{(r)} t^a_{(r)} \right)^{ij}$. If the sum were restricted to the Cartan subalgebra of $\mathcal{G}$ ($a, b = 1, \ldots, r_G$), we would get instead

$$\sum_{k=1}^{d_r} \mu_{(k)}^2 = l_r r_G \ , \tag{1.52}$$

where r_G is the rank of the group G, and μ are the weights of the representation (r) of dimension d_r. For the adjoint representation, one has $d_A = |G|$ and $C_A = l_A = r_G^{-1} \sum_{a=1}^{|G|} \alpha_{(a)}^2$, where the α's are the roots of $\mathcal{G}$. Denoting by ψ the highest root, one can introduce the normalization-*independent* quantity, $\tilde{h}_G \equiv C_A/\psi^2$, known as the dual Coxeter number,

$$\tilde{h}_G = \frac{C_A}{\psi^2} = \frac{1}{r_G}\left(n_L + \left(\frac{S}{L}\right)^2 n_S\right) \ , \tag{1.53}$$

where n_S and n_L are the number of short (S) and long (L) roots of the algebra, respectively. The $(S/L)^2$ is just the ratio of their lengths squared (the roots of simple Lie algebras come at most in two lengths). The Lie algebras associated to Dynkin diagrams with only *single* lines have roots all of the same length, and they are known as the A, D, E series of Lie algebras, i.e. $SU(n)$, $SO(2n)$ and $E_{6,7,8}$. These algebras are usually referred to as the simply laced Lie algebras. One finds from (1.53) that

$$\mathbf{SU(n)} \ (n \geq 2) \ : \ \tilde{h}_{SU(n)} = n \ , \quad l_{(n)} = \tfrac{1}{2}\psi^2 \ ;$$

$$\mathbf{SO(n)} \ (n \geq 4) \ : \ \tilde{h}_{SO(n)} = n - 2 \ , \quad l_{(n)} = \psi^2 \ ;$$

$$\mathbf{E_6} \ : \ \tilde{h}_{E_6} = 12 \ , \quad l_{(27)} = 3\psi^2 \ ; \quad \mathbf{E_7} \ : \ \tilde{h}_{E_7} = 18 \ , \quad l_{(57)} = 6\psi^2 \ ;$$

$$\mathbf{E_8} \ : \ \tilde{h}_{E_8} = 30 \ , \quad l_{(248)} = 30\psi^2 \ ;$$

$$\mathbf{Sp(2n)} \ (n \geq 1) \ : \ \tilde{h}_{Sp(2n)} = n + 1 \ , \quad l_{(2n)} = \tfrac{1}{2}\psi^2 \ ;$$

$$\mathbf{G_2} \ : \ \tilde{h}_{G_2} = 4 \ , \quad l_{(7)} = \psi^2 \ ; \quad \mathbf{F_4} \ : \ \tilde{h}_{F_4} = 9 \ , \quad l_{(26)} = 3\psi^2 \ , \tag{1.54}$$

where the index l_r has been tabulated for some representations of low dimensions. The dual Coxeter number is always an integer. The normalization-independent quantity

$$k = \frac{2\tilde{k}}{\psi^2} \tag{1.55}$$

is known as the *level* of the AKM algebra. As is shown below, k is quantized to be an integer for the unitary highest weight representations. In terms of the integers k and $\tilde{h}_G$, equation (1.49) takes the form

$$c = \frac{k|G|}{k + \tilde{h}_G} \ . \tag{1.56}$$

To derive the quantization rule for the level k, let us consider the simple case of $G = SU(2)$ first. The normalization $\psi^2 = 2$ corresponds to the $SU(2)$

structure constants $f^{ijk} = \sqrt{2}\varepsilon^{ijk}$. Because of $\sqrt{2}$ in the $SU(2)$ commutation relations, we should take

$$I^{\pm} = \frac{1}{\sqrt{2}}\left(J_0^1 \pm iJ_0^2\right) , \quad I^3 = \frac{1}{\sqrt{2}}J_0^3 , \tag{1.57}$$

to get the conventionally normalized $su(2)$ algebra $[I^+, I^-] = 2I^3$, $[I^3, I^{\pm}] = \pm I^{\pm}$, where $2I^3$ is known to have integer eigenvalues in any finite-dimensional representation. The AKM algebra (1.40) implies that the operators

$$\tilde{I}^+ = \frac{1}{\sqrt{2}}\left(J_{-1}^1 + iJ_{-1}^2\right), \quad \tilde{I}^- = \frac{1}{\sqrt{2}}\left(J_{+1}^1 - iJ_{+1}^2\right), \quad \tilde{I}^3 = \frac{1}{\sqrt{2}}J_0^3 - \frac{1}{2}k , \tag{1.58}$$

satisfy the $su(2)$ algebra as well: $[\tilde{I}^+, \tilde{I}^-] = 2\tilde{I}^3$, $[\tilde{I}^3, \tilde{I}^{\pm}] = \pm\tilde{I}^{\pm}$. Hence, $2\tilde{I}^3 = 2I^3 - k$ also has integer eigenvalues. It follows that $k \in \mathbf{Z}$ for the unitary highest weight representations.

To generalize this argument to any Lie algebra $\mathcal{G}$, it is sufficient to use the canonical $su(2)$ subalgebra (in the Chevalley basis) defined by

$$I^{\pm} = E_0^{\pm\psi} , \quad I^3 = \psi \cdot H_0/\psi^2 , \tag{1.59}$$

which is generated by the highest root ψ of $\mathcal{G}$. Equation (1.40) now implies that

$$\tilde{I}^{\pm} = E_{\mp 1}^{\pm\psi} , \quad \tilde{I}^3 = \left(\psi \cdot H_0 - \tilde{k}\right)/\psi^2 \tag{1.60}$$

also form an $su(2)$ subalgebra. This means that the level $k = 2\tilde{k}/\psi^2 = 2I^3 - 2\tilde{I}^3$ is to be quantized as an integer. Yet another proof of the quantization condition on k will be given in the next subsection.

5.1.3 WZNW Models and Topological Quantization

In the previous subsections we considered CFT with AKM symmetry, whose dynamics was determined by the SS construction. No Lagrangian local QFT was introduced. In fact, symmetry considerations prevail over Lagrangian formulations in 2d CFT, partly because there are many different 2d Lagrangians corresponding to the same CFT defined by its correlation functions.

A simple rational CFT action with AKM symmetry is given by the 2d WZNW model,

$$kI_{\text{WZNW}}(g) = \frac{k}{8\pi}\int_{S^2} d^2x \, \text{tr}\left(\partial_m g \partial^m g^{-1}\right)$$
$$+ \frac{k}{12\pi}\int_{B,\partial B=S^2} d^3y \, \varepsilon^{mnl} \, \text{tr}\left(g^{-1}\partial_m g g^{-1}\partial_n g g^{-1}\partial_l g\right) . \tag{1.61}$$

The scalar field g is valued in a (semisimple and compact) Lie group G, and it satisfies the boundary condition $g(0, \tau) = g(2\pi, \tau)$. The second WZ-term in (1.61) is given by the integral over the three-dimensional ball B whose boundary ∂B is identified with the compactified 2d Euclidean spacetime (or world-sheet) S^2. [2] The integrand of the second term is locally a total derivative, so that it represents a topological term that may be rewritten as the two-dimensional action over the sphere S^2 by using Stokes theorem. The whole WZNW term in (1.61) is properly normalized, in order to contribute an integer multiple of 2π to the action (cf. the standard definition of the instanton winding number in gauge field theories [5]). The action (1.61) is obviously (classically) scale invariant. The relative coefficient between the first (kinetic) and second (WZNW) terms in (1.61) is dictated by the extra *infinite-dimensional* symmetry associated with the AKM current algebra,

$$g(\zeta^+, \zeta^-) \to \bar{\Omega}(\zeta^-) g(\zeta^+, \zeta^-) \Omega(\zeta^+) , \qquad (1.62)$$

where $\Omega(\bar{\Omega})$ is an arbitrary group-valued function, and $\zeta^\pm = x_1 \pm x_2$. The symmetry (1.62) is the 2d analogue of the chiral symmetry well known in 4d field theories [55]. After introducing the complex variables z and $\bar{z}$ as $z = e^{ix^+}$ and $\bar{z} = e^{ix^-}$ (cf. Subsect. 5.1.1), the symmetry (1.62) takes the form

$$g(z, \bar{z}) \to \bar{\Omega}(\bar{z}) g(z, \bar{z}) \Omega(z) . \qquad (1.63)$$

Due to the boundary condition $g(0) = g(2\pi)$, the coordinates z and $\bar{z}$ are well defined (no cuts in the complex plane). The transformation law (1.63) is nothing but the AKM symmetry $\hat{G} \times \hat{\bar{G}}$ that survives in the quantized WZNW theory too. This symmetry is characterized by the currents

$$J(z) = J^a(z) t^a = k g^{-1} \partial_z g = \sum_{n=-\infty}^{+\infty} J_n z^{-n-1} ,$$

$$\bar{J}(\bar{z}) = \bar{J}^a(\bar{z}) t^a = -k(\partial_{\bar{z}} g) g^{-1} = \sum_{n=-\infty}^{+\infty} \bar{J}_n \bar{z}^{-n-1} , \qquad (1.64)$$

satisfying the conservation law

$$\bar{\partial} J = \partial \bar{J} = 0 . \qquad (1.65)$$

Equation (1.65) amounts to the equations of motion in the WZNW model. The currents J and $\bar{J}$ generate two commuting AKM algebras (1.38) with the central extension (level) k. The associated stress tensor is just given by the SS construction (1.43) of central charge (1.49) or (1.56).

The action (1.61) is the particular case of a more general 2d field theory described by the action

---

[2] Unlike the kinetic NLSM term, the WZNW term is *real* both in Minkowski spacetime *and* in Euclidean spacetime.

$$kI(g) = \frac{1}{2\lambda^2} \int \mathrm{tr}\,(g^{-1}\partial g)^2 + k\Gamma\,, \qquad (1.66)$$

where the WZ term Γ is just $\frac{1}{12\pi}\int_B \mathrm{tr}\,(g^{-1}\mathrm{d}g)^3 \equiv \int_B H$. The theory (1.66) becomes conformally invariant when the coupling constant takes its critical value, $\lambda^2 = 4\pi/k$, corresponding to the exact fixed point of the RG β-function in the scalar QFT (1.66). From the even more general point of view, the field theory (1.66) is just the very particular case of 2d NLSM with a generalized WZ term or torsion (Chap. 2).

To make clear the topological nature of the WZ term in (1.66), we now describe a *topological* quantization of the WZ coefficient k by using the notions of integer homology and DeRham cohomology [44, 268]. The kinetic term in (1.66) is defined in terms of the field g that maps the sphere S^2 into a group manifold G, while the WZ term is defined as an integral of the (2+1)-form H over the (2+1)-dimensional ball B with the boundary $\partial B = S^2$. The form H must be *closed*, but not necessarily *exact*. The closure of H guarantees the *local* existence of a 2-form b such that $H = \mathrm{d}b$, according to the Poincaré lemma, but it does *not* guarantee that the two-dimensional action $\int_{S^2} b$ will be globally well defined because of possible topological obstructions. The topological obstructions are indeed possible when the group manifold G has a non-trivial third homology class, which is implicit in the definition of the 'truly' WZ term. To avoid confusion, one should clearly distinguish between the forms defined on the group manifold G and their pull-backs to B or S^2.

To explain the relation between the integer homology and the WZ term, let us define the k-cycles [3] $\mathcal{C}$ of a given (generically d-dimensional) manifold $\mathcal{W}$ as its k-dimensional submanifolds with no boundary, $\partial\mathcal{C} = 0$. Linear combinations of cycles with integer coefficients are also cycles. The boundary of a boundary is always zero: $\partial^2 = 0$.

The *integer homology classes* $H_k(\mathcal{W}) \equiv H_k(\mathcal{W}, \mathbf{Z})$ are the equivalence classes of the cycles: two cycles $\mathcal{C}_1$ and $\mathcal{C}_2$ are said to be in the same homology class if $\mathcal{C}_1 - \mathcal{C}_2 = \partial\mathcal{B}$, where $\mathcal{B}$ is a $(k+1)$-dimensional submanifold (then $\mathcal{C}_1$ and $\mathcal{C}_2$ are just the opposite boundaries of $\mathcal{B}$). The homology classes form an Abelian group under the addition of cycles,

$$H_k(\mathcal{W}) = \underbrace{\mathbf{Z} \oplus \ldots \oplus \mathbf{Z}}_{b_k(\mathcal{W})\ \text{factors}} \oplus \mathbf{Z}_{p_1} \oplus \ldots \oplus \mathbf{Z}_{p_l}\,, \qquad (1.67)$$

where we have decomposed $H_k(\mathcal{W})$ into a sum of independent $\mathbf{Z}$-factors, each one being generated by a fundamental cycle $\mathcal{C}_i$. The simplest example is provided by the torus with two independent non-trivial cycles and $H_1(\mathcal{W}) = \mathbf{Z} \oplus \mathbf{Z}$. The number of $\mathbf{Z}$-factors in (1.67) is known as the kth Betti number (for the torus, $b_1 = 2$). In general, there may be so-called *torsion subgroups* in $H_k(\mathcal{W})$ that are also explicitly indicated in (1.67). They are generated by

---

[3] We use the same letter k for both the dimension of a cycle and the coefficient of the WZ term since they can hardly be confused.

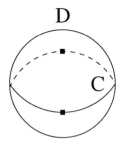

Fig. 5.1. The two-sphere with antipodal points identified (RP^3). The one-cycle C, given by the right half of the equator, belongs to a non-trivial homology class. The cycle $2C$, given by the boundary of the upper hemisphere D, belongs to the trivial homology class

some fundamental cycles $\mathcal{E}_i$ but, in contrast to the $\mathcal{C}_i$-cycles, the $p_i\mathcal{E}_i$ (no sum!) with a positive integer p_i is in the trivial class. In other words, $\mathcal{E}_i$ cannot be expressed as a boundary of a $(k+1)$-dimensional submanifold, while $p_i\mathcal{E}_i$ can (see the example in Fig. 5.1) [44].

Any k-cycle $\mathcal{C}$ can be expanded in terms of the fundamental cycles $\mathcal{C}_i$ and $\mathcal{E}_i$,

$$\mathcal{C} = \sum_{i=1}^{b_k(\mathcal{W})} n_i\mathcal{C}_i + \sum_{i=1}^{l} m_i\mathcal{E}_i + \partial\mathcal{B}_0 , \tag{1.68}$$

with integers n_i and m_i, $0 \le m_i \le p_i - 1$. Since $p_i\mathcal{E}_i$ is trivial, $p_i(m_i\mathcal{E}_i) = \partial\mathcal{B}_i$ for some $(k+1)$-dimensional submanifolds $\mathcal{B}_i$, we can rewrite (1.68) in the form

$$\mathcal{C} = \sum_{i=1}^{b_k(\mathcal{W})} n_i\mathcal{C}_i + \partial\mathcal{B}, \quad \mathcal{B} = \mathcal{B}_0 + \sum_{i=1}^{l} \mathcal{B}_i/p_i . \tag{1.69}$$

Clearly, the submanifold $\mathcal{B}$ is not unique, being defined modulo a cycle multiplied by the fraction $1/p$, where p is the least common multiplier of $\{1, p_1, \ldots, p_l\}$.

The DeRham cohomology classes $H^k(\mathcal{W}, \mathbf{R})$ are defined as the equivalence classes of closed k-forms: two k-forms Ω_1 and Ω_2 are said to be equivalent (i.e. in the same cohomology class) if $\Omega_1 - \Omega_2 = d\Lambda$ for some $(k-1)$-form Λ. The non-trivial cohomology classes are generated by forms that are closed but not exact. It is the Stokes theorem, $\int_\mathcal{C} d\Omega = \int_{\partial\mathcal{C}} \Omega$, that provides the natural duality relation between the homology and cohomology classes. The precise statement is known as the *DeRham theorem*: there exists a basis for $H^k(\mathcal{W}, \mathbf{R})$ which consists of the k-forms ω_i satisfying

$$\int_{\mathcal{C}_i} \omega_j = \delta^{ij} , \qquad i, j = 1, \ldots, b_k(\mathcal{W}) , \tag{1.70}$$

where $\mathcal{C}_i$ are the fundamental cycles of $H_k(\mathcal{W})$. In particular, the dimension of $H^k(\mathcal{W}, \mathbf{R})$ is the Betti number $b_k(\mathcal{W})$. The ω_i are called the fundamental k-forms. Using the decomposition (1.69), we find for a general k-cycle $\mathcal{C}$ that the integral of the fundamental form ω_j over $\mathcal{C}$ is always integer-valued.

We are now in a position to apply the decomposition (1.69) to the WZNW model. The image of the world-sheet $g(S^2)$ in the manifold G is a submanifold without boundary, i.e. a two-cycle. Applying (1.69) to $g(S^2)$ yields

$$g(S^2) = \sum_{i=1}^{b_2(G)} n_i \mathcal{C}_i + \partial B , \quad n_i \in \mathbf{Z} . \tag{1.71}$$

We can now represent the WZ functional $k\Gamma$ in the form

$$k\Gamma[g] = k \sum_{i=1}^{b_2(G)} n_i \Gamma[\mathcal{C}_i] + k\Gamma[\partial B] , \tag{1.72}$$

where $\Gamma[\mathcal{C}_i]$ are real parameters of the WZNW model, and $\Gamma[\partial B] \equiv \int_B \breve{H}$. Since the 3-form $\breve{H}$ is not exact, the $\Gamma[\partial B]$ is *ambiguous* because of the ambiguity in the definition of B that is only defined up to adding a cycle multiplied by $1/p$. In Euclidean QFT this ambiguity is harmless if it only affects a phase of the path integral,

$$\exp \left\{ -k \int_{(1/p)\tilde{B}} \breve{H} \right\} = 1 , \tag{1.73}$$

for any three-cycle $\tilde{B}$. It is easy to find a solution to (1.73) by using the fact that the integrals of the fundamental 3-forms $\omega_j \in H^3(G, \mathbf{R})$ over any cycle $\mathcal{C}$ are integer-valued. Therefore, we can choose $\breve{H} = 2\pi i \omega_j$ if k/p is an integer, in order to satisfy (1.73). The number of the independent WZ terms is, therefore, given by the Betti number $b_3(G)$. The quantization condition on the coefficient k of the WZ term thus precisely means that k should be an integer multiple of p.

In the case of a 2d manifold $\mathcal{W}$ with the topology of a sphere S^2, the situation is much simpler because there are no torsion subgroups ($p = 1$). The coefficient k of the WZ term can therefore be any integer. The action (1.66) is quantum-mechanically well defined in the sense that the theory it represents does not depend on the way it is parametrized. The topological approach described above is complementary to the algebraic approach used in 2d CFT, which gives rise to the same quantization condition (see the previous subsection). In the $O(n)$ NLSM (Sect. 2.1), the topological quantization of the WZNW coefficient also follows from requiring the absence of a global $O(n)$ anomaly [44].

5.2 Super-WZNW Models and Their Symmetries

Most of the results discussed in the previous section can be generalized to supersymmetric extensions of the Virasoro and AKM (current) algebras.

Their definitions and the associated representation theory can be developed along similar lines [45]. The superconformally and super-AKM invariant field theories play an important role in superstring theory (Chap. 6). In Subsect. 5.2.1 we introduce the N-extended (chiral) superconformal algebras, where $1 \leq N \leq 4$ is the number of 2d (chiral) supersymmetry charges. The $N = 2$ extended superconformal symmetry is of particular interest for superstring compactification, since it implies the $N = 1$ spacetime supersymmetry in the compactified four-dimensional superstrings [170]. The supersymmetric WZNW models are discussed in Subsect. 5.2.2.

5.2.1 Superconformal and Super-AKM Algebras

The supersymmetric extensions of the Virasoro algebra can be obtained by generalizing the conformal transformations to superconformal ones. The latter can be defined as the special superdiffeomorphisms acting in the superspace $\mathbf{z} = (z, \theta)$ and preserving the supercovariant derivative $\mathbf{D} = \partial/\partial\theta + \theta\partial_z$ [298, 299]. This approach results in the $N = 1$ Chiral Superconformal Algebra (SCA). This algebra (with a central extension) is the symmetry algebra of the fermionic (NSR) strings (Chap. 6). The general $N = 1$ SCA can be defined with an arbitrary value of central charge, while it reads [300, 301, 302]

$$T(z)T(w) \sim \frac{3\hat{c}/4}{(z-w)^4} + \frac{2T(w)}{(z-w)^2} + \frac{\partial T(w)}{z-w} \, ,$$

$$T(z)G(w) \sim \frac{\frac{3}{2}G(w)}{(z-w)^2} + \frac{\partial G(w)}{z-w} \, , \qquad (2.1)$$

$$G(z)G(w) \sim \frac{\hat{c}}{(z-w)^3} + \frac{2T(w)}{z-w} \, ,$$

where the anticommuting (Grassmann) field $G(z)$ has been introduced, and $\hat{c} \equiv \frac{2}{3}c$. The conventional normalization for a single free chiral scalar superfield $\phi(\mathbf{z}) = \phi(z) + \theta\psi(z)$ assigns the central charge $\hat{c} = 1$, which implies $c = 1 + \frac{1}{2} = \frac{3}{2}$, similarly to a single bosonic chiral scalar field $\phi(z)$ having central charge $c = 1$. The second line of (2.1) is equivalent to the statement that the *supercharge* $G(z)$ is of conformal dimension $h = 3/2$.

The OPE (2.1) are equivalent to the (anti)commutation relations

$$[L_m, L_n] = (m - n)L_{m+n} + \frac{\hat{c}}{8}(m^3 - m)\delta_{m+n,0} \, ,$$

$$[L_m, G_n] = \left(\frac{m}{2} - n\right) G_{m+n} \, ,$$

$$\{G_m, G_n\} = 2L_{m+n} + \frac{\hat{c}}{2}\left(m^2 - \frac{1}{4}\right)\delta_{m+n} \, , \qquad (2.2)$$

in terms of the moments (modes) L_n of $T(z)$ and that of $G(z)$,

$$G_n = \oint_0 \frac{dz}{2\pi i} z^{n+1/2} G(z) \ . \tag{2.3}$$

Given *integer* modding ($n \in \mathbf{Z}$) of G_n, (2.2) is called Ramond (R) algebra, whereas for *half-integer* modding, ($n \in \mathbf{Z} + 1/2$), one gets Neveu-Schwarz (NS) algebra.

In radial quantization (Subsect. 5.1.1) the dilatation operator L_0 plays the role of the Hamiltonian. It is a corollary of (2.2) that $G_0^2 = L_0 - \hat{c}/16$. Therefore, global supersymmetry is not broken when the ground states are of dimension $h = \hat{c}/16$.

The two holomorphic fields $T(z) \equiv T_B(z)$ of $h_B = 2$ and $G(z) \equiv 2T_F(z)$ of $h_F = 3/2$ can also be formally represented by one superfield $T(z, \theta) = T_F(z) + \theta T_B(z)$, with the OPE [303, 304, 305]

$$T(\mathbf{z}_1)T(\mathbf{z}_2) \sim \frac{\hat{c}}{4z_{12}^3} + \frac{3\theta_{12}}{2z_{12}^2} T(\mathbf{z}_2) + \frac{\mathbf{D}_2 T(\mathbf{z}_2)}{2z_{12}} + \frac{\theta_{12}}{z_{12}} \partial_2 T(\mathbf{z}_2) \ , \tag{2.4}$$

where the following definitions have been used:

$$z_{12} = z_1 - z_2 - \theta_1 \theta_2 \ , \quad \theta_{12} = \theta_1 - \theta_2 \ ,$$

$$\mathbf{D}_2 = \partial/\partial\theta_2 + \theta_2 \partial_2 \ . \tag{2.5}$$

The superanalogue $\phi(\mathbf{z})$ of a conformal field $\phi_B(z)$ of dimension $h + 1/2$,

$$\phi(\mathbf{z}) \equiv \phi(z, \theta) = \phi_F(z) + \theta \phi_B(z) \ , \tag{2.6}$$

has the components $\phi_F(z)$ and $\phi_B(z)$ of conformal dimensions h and $h + 1/2$, respectively, and its OPE with $T(\mathbf{z})$ takes the form [303, 304, 305]

$$T(\mathbf{z}_1)\phi(\mathbf{z}_2) \sim h \frac{\theta_{12}}{z_{12}^2} \phi(\mathbf{z}_2) + \frac{1}{2z_{12}} \mathbf{D}_2 \phi(\mathbf{z}_2) + \frac{\theta_{12}}{z_{12}} \partial_2 \phi(\mathbf{z}_2) \ . \tag{2.7}$$

The superfields $\phi(\mathbf{z})$ are called $N = 1$ superconformal fields (see [167, 306, 307, 308, 309] for more).

A Super-Conformal Field Theory (SCFT) has, in fact, more structure beyond the one given by its superfields [300], since it also contains the fields intertwining between the NS- and R-sectors. The latter are double-valued with respect to the fermionic parts of the superfields, while they are usually called *spin fields*. The superconformal fields defined above are in one-to-one correspondence to the NS ground states. The Ramond ground states $|h^{\pm}\rangle_R$, where $|h^-\rangle_R = G_0 |h^+\rangle_R$, are created from the NS vacuum by the spin fields $S^{\pm}(z)$ of dimension h: $|h^{\pm}\rangle_R = S^{\pm}(0) |0\rangle_{NS}$. The (conserved) fermion parity operator $\Gamma = (-1)^F$, anticommuting with the fermionic parts of the superfields and commuting with their bosonic parts, is called the chirality operator [300]. Since G_0 anticommutes with Γ, the paired ground states in the R-sector have opposite chirality. In a generic SCFT, spin fields of opposite chirality are non-local with respect to each other because of the appearance of fermionic fields in their OPE. A projection onto the $\Gamma = 1$ sector of SCFT gives the *local* SCFT called the spin model [300], whereas the projection itself is known as the Gliozzi-Scherk-Olive (GSO) projection [310].

Generally speaking, the N-extended (chiral) SCA is an algebra that contains (i) the Virasoro algebra with central extension (T, c), (ii) N chiral fermionic generators $G^\alpha(z)$ of dimension $3/2$ and (iii) $2T(w)\delta^{\alpha\beta}/(z - w)$ on the right-hand side of the OPE for $G^\alpha(z)G^\beta(w)$. The $N = 2$ extended SCA [311, 312, 313] contains, in particular, the $U(1)$ current $J(z)$, in addition to the anticipated $T(z)$ and $G^\alpha(z)$, $\alpha = 1, 2$. The OPE $T(z)T(w)$ and $T(w)G^\alpha(w)$ are as above, whereas the rest of the $N = 2$ SCA reads

$$G^\alpha(z)G^\beta(w) \sim \left[\frac{2\tilde{c}}{(z - w)^3} + \frac{2T(w)}{z - w}\right]\delta^{\alpha\beta} + \mathrm{i}\left[\frac{2J(w)}{(z - w)^2} + \frac{\partial J(w)}{z - w}\right]\varepsilon^{\alpha\beta} ,$$

$$T(z)J(w) \sim \frac{J(w)}{(z - w)^2} + \frac{\partial J(w)}{z - w} ,$$

$$J(z)G^\alpha(w) \sim \mathrm{i}\varepsilon^{\alpha\beta}\frac{G^\beta(w)}{z - w} ,$$

$$J(z)J(w) \sim \frac{\tilde{c}}{(z - w)^2} , \tag{2.8}$$

where $\tilde{c} \equiv c/3 = \hat{c}/2$. The normalization of central charge $\tilde{c}$ is fixed by the condition $\tilde{c} = 1$ for a free chiral scalar $N = 2$ superfield containing two scalars (each of $c = 1$) and two MW fermions (each of $c = 1/2$). It is sometimes convenient to deal with

$$G^\pm(z) = \frac{1}{\sqrt{2}}\left(G^1(z) \pm \mathrm{i}G^2(z)\right) \tag{2.9}$$

instead of $G^\alpha(z)$. Then the OPE $G^\pm(z)G^\pm(w)$ becomes non-singular, whereas the only singular OPE is $G^+(z)G^-(w)$ (or $G^-(z)G^+(w)$).

The super-AKM algebra is defined by adding to an AKM current $J^a(z)$ its fermionic superpartner $j^a(z)$ in the *adjoint* representation and of dimension $h = 1/2$. The corresponding OPE are

$$J^a(z)J^b(w) \sim \frac{k\delta^{ab}}{(z - w)^2} + \frac{\mathrm{i}f^{abc}J^c(w)}{z - w} ,$$

$$J^a(z)j^b(w) \sim \frac{\mathrm{i}f^{abc}j^c(w)}{z - w} ,$$

$$j^a(z)j^b(w) \sim \frac{k\delta^{ab}}{z - w} , \tag{2.10}$$

where we still use the 'canonical' normalization of the structure constants, with the highest root squared being equal to two, $\psi^2 = 2$.

The super-AKM currents can be unified into a single superfield as

$$J^a(\mathbf{z}) \equiv J_F^a(z) + \theta J_B^a(z) = j^a(z) + \theta J^a(z) \ . \tag{2.11}$$

The superspace form of (2.10) is given by

$$J^a(\mathbf{z}_1) J^b(\mathbf{z}_2) \sim \frac{k\delta^{ab}}{z_{12}} + \mathrm{i} \frac{\theta_{12}}{z_{12}} f^{abc} J^c(\mathbf{z}_2) \ . \tag{2.12}$$

The simplest representation of the super-AKM algebra (2.10) with the lowest possible level is given by the MW *free* fermions ψ^a transforming in the adjoint representation of a Lie group G. It is known as the quark model [314]. First, one notices that the currents

$$J_f^a(z) = \frac{\mathrm{i}}{2} f^{abc} \psi^b(z) \psi^c(z) \tag{2.13}$$

define a representation of $\hat{\mathcal{G}}$ at level $k = \tilde{h}_G = C_A/2$. The associated central charge comes either from the OPE satisfied by the free fermionic stress tensor $T_f = \frac{1}{2} : \psi^a \partial \psi^a :$, or from the SS constuction in terms of the currents (2.13). Its value is given by

$$c = \tfrac{1}{2}|G| = \frac{k|G|}{k + \tilde{h}_G} = \frac{\tilde{h}_G |G|}{\tilde{h}_G + \tilde{h}_G} \ . \tag{2.14}$$

Second, we choose the fermionic superpartner j_f of J_f in the form [4]

$$j_f^a(z) = \mathrm{i}\sqrt{k}\psi^a(z) \ . \tag{2.15}$$

This means that the bosonic currents $J_f^a(z)$ are equal to the fermionic bilinears,

$$J_f^a(z) = -\frac{\mathrm{i}}{2k} f^{abc} j^b(z) j^c(z) \ . \tag{2.16}$$

The set of $|G|$ free MW fermions can be used to realize the super-AKM algebra with the enveloping super-Virasoro algebra [315, 316, 317]. The spin-3/2 superpartner $G_f(z)$ of $T_f(z)$ is given by

$$G_f(z) = -\frac{1}{6\sqrt{C_A/2}} f^{abc} \psi^a \psi^b \psi^c \ . \tag{2.17}$$

It is now easy to complete the list of OPE by using the Wick rules and the Jacobi identity for structure constants. One finds

$$G_f(z) J_f^a(w) \sim \frac{1}{(z-w)^2} j_f^a(w) \ ,$$

---

[4] j_f^a should be proportional to ψ^a, since this is the only way to satisfy the dimension $h = 1/2$ of j_f^a. The normalization of j_f^a is determined by the algebra (2.10). Similar arguments are applied for the derivation of (2.16) and (2.17) below.

$$G_f(z)j_f^a(w) \sim \frac{1}{z-w} J_f^a(w) \ . \tag{2.18}$$

We conclude that both super-AKM and superconformal symmetries in 2d can be non-linearly realized in terms of the free fermions transforming in the adjoint representation of a semisimple Lie group G (without the $U(1)$ factors) [314, 315, 316, 317, 318].

A more general representation of the super-AKM algebra can be obtained by forming the linear combination [319]

$$J^a(z) = J_f^a(z) + \hat{J}^a(z) \ , \tag{2.19}$$

where the currents $\hat{J}^a(z)$ define a level-$\hat{k}$ representation of the AKM-algebra $\mathcal{G}$, $a = 1, \dots, |G|$. The currents $\hat{J}$ may be thought of as that of the WZNW theory, though we don't need their explicit form here. It is enough to know that $\hat{J}^a$ define the AKM representation that is independent of the fermionic fields, $\hat{J}^a(z)j_f^b(w) \sim O(1)$ and $\hat{J}^a(z)J_f^b(w) \sim O(1)$. We also assume that $\mathcal{G}$ is a simply-laced Lie algebra.

The currents J^a in (2.19) define a new AKM representation of level

$$k = \hat{k} + C_A/2 = \hat{k} + \tilde{h}_G \ , \tag{2.20}$$

and of central charge

$$c = \frac{\hat{k}|G|}{\hat{k} + \tilde{h}_G} + \tfrac{1}{2}|G| = \frac{3}{2}|G| - \frac{C_A}{2k}|G| \ . \tag{2.21}$$

The superconformal anomaly is given by

$$\hat{c} = \frac{2}{3}c = \left(1 - \frac{C_A}{3k}\right)|G| \ . \tag{2.22}$$

The AKM representation $J^a(z)$ can be extended to a representation of the super-AKM algebra by adding a superpartner current in the form

$$j^a(z) = i\sqrt{k}\psi^a \ , \tag{2.23}$$

where the subscript (f) at $j^a(z)$ has been omitted.

We can associate the $N = 1$ (super-Virasoro) SCA with this construction by defining [319]

$$T(z) = \frac{1}{2k}\left[: \hat{J}^a(z)\hat{J}^a(z) : - : j^a(z)\partial j^a(z) :\right] \ , \tag{2.24a}$$

$$G(z) = \frac{1}{k}\left[j^a(z)\hat{J}^a(z) - \frac{i}{6k}f^{abc}j^a(z)j^b(z)j^c(z)\right] \ . \tag{2.24b}$$

Equation (2.24) gives the $N = 1$ supersymmetric extension of the SS construction (Subsect. 5.1.2).

5.2.2 $N = 1$ and $N = 2$ Supersymmetric WZNW Models

The fundamental spinor representation in 2d is one-dimensional, i.e. a MW spinor. Therefore, a general 2d supersymmetry algebra may have p real left-handed spinor supersymmetry generators and q real right-handed generators, where p, q are positive integers (Chaps. 2 and 3). The most basic case is given by the $(1, 0)$ supersymmetry algebra that has only one MW supersymmetry charge [320, 321, 322] (see also Chap. 7). The N-extended (Majorana) 2d supersymmetry corresponds to the case $p = q = N$.

The standard pattern of a 2d field theory with the $N = 1$ super-AKM and superconformal symmetries is given by the supersymmetric WZNW model [167]. The corresponding action is the $(1, 1)$-supersymmetric generalization of the bosonic WZNW action of (1.61). It can be most easily constructed in a flat $(1, 1)$ superspace having coordinates $(\sigma^0, \sigma^1, \theta^\alpha)$, where the two-component M-spinor θ^α represents the anticommuting (Grassmann) superspace coordinates. [5]

The NLSM action on a group manifold G reads

$$S_{\text{sWZNW}}^{(1,1)} = -\frac{1}{4\lambda^2} \int d^2\sigma d^2\theta \, \text{tr} \left(\bar{D}^\alpha G D_\alpha G^{-1} \right)$$

$$-\frac{k}{16\pi} \int_0^1 dy \int d^2\sigma d^2\theta \, \text{tr} \left[\tilde{G}^{-1} \frac{\partial \tilde{G}}{\partial y} \bar{D}^\alpha \tilde{G} (\gamma_3)_{\alpha\beta} D^\beta \tilde{G}^{-1} \right] , \qquad (2.25)$$

where $G = \exp \left[it^a \phi^a (\sigma, \theta) \right]$ is the Lie group-valued superfield

$$G(\sigma, \theta) = g(\sigma) + i\bar{\theta}\psi(\sigma) + \tfrac{1}{2} i\bar{\theta}\theta F(\sigma) , \qquad (2.26)$$

and $\tilde{G}(\sigma, \theta; y)$ is the extension of $G(\sigma, \theta)$, restricted by the boundary conditions

$$\tilde{G} \big|_{y=1} = G , \quad \tilde{G} \big|_{y=0} = 0 . \qquad (2.27)$$

Quite similarly to its bosonic counterpart (1.61), the QFT defined by the action (2.25) has the RG fixed point at

$$\lambda^2 = 4\pi/k , \qquad (2.28)$$

where it is conformally invariant. Moreover, after rewriting the action (2.25) in terms of the light-cone variables $(\sigma^{\pm\pm}, \theta^\pm, D^\pm)$, it becomes clear that this action is also invariant under the $(1, 1)$ super-AKM transformations

$$G \to \Omega_\text{L} G \Omega_\text{R}^{-1} , \qquad (2.29)$$

---

[5] Our conventions in $(1, 1)$ superspace are: (light-cone) $\sigma^{\pm\pm} = \frac{1}{\sqrt{2}} \left(\sigma^0 \pm \sigma^1 \right)$, $\sigma^+ \equiv \sigma^{++}$, $\sigma^= \equiv \sigma^{--}$; ($\gamma$-matrices) $\gamma^0 = i\sigma_2$, $\gamma^1 = \sigma_1$, $\gamma_3 = \sigma_3$; (M-spinors) $\theta = C_2 \bar{\theta}^T$, $C_2 = \gamma^0$; (covariant derivatives) $D_\alpha = \partial/\partial\bar{\theta}^\alpha + i(\gamma \cdot \partial\theta)_\alpha = (-D^+, D^-)$, $D^\pm = \partial/\partial\theta^\mp + i\theta^\mp \partial/\partial\sigma^{\mp\mp}$; $\theta^\pm = (1 \pm \gamma_3)\theta$. We use 2d Lorentz eigenvalues (helicity) for both spinors and vectors.

where the Lie group-valued parameter superfields $\Omega_{L,R}$ are subject to the constraints

$$D^+\Omega_L = D^-\Omega_R = 0 . \tag{2.30}$$

In components, when using the algebraic equations of motion for the auxiliary field F and parametrizing the fermionic field ψ in terms of a new field χ as

$$\chi = g^{-1}\psi_+ + \psi_- g^{-1} , \tag{2.31}$$

the super-WZNW action (2.25) at $\lambda^2 = 4\pi/k$ amounts to a sum of the bosonic WZNW action (1.61) and the action of *free* fermions,

$$S_{\text{sWZNW}}^{(1,1)} = S_{\text{WZNW}} + \frac{ik}{16\pi} \int d^2\sigma \, \text{tr}\,(\bar{\chi}\gamma \cdot \partial\chi) . \tag{2.32}$$

The supersymmetric fermions χ thus decouple in the super-WZNW action at the fixed point [167].

The super-WZNW classical equations of motion have a general solution where all fields take the factorized form,

$$g = g_L(\sigma^=)g_R(\sigma^{\mp}) , \quad \chi_+ = \chi_+(\sigma^{\mp}) , \quad \chi_- = \chi_-(\sigma^=) . \tag{2.33}$$

The AKM transformations in components are given by

$$g \to \Omega_L g \Omega_R^{-1} , \quad \chi^- \to \Omega_L \chi^- \Omega_L^{-1} , \quad \chi^+ \to \Omega_R \chi^+ \Omega_R^{-1} , \tag{2.34}$$

while the accompanying super-AKM trasformations, in accordance with (2.29), read

$$\delta g = 0 , \quad \delta\chi^\pm = \beta^\pm , \tag{2.35}$$

where $\beta^\pm$ are two Grassmannian parameters, and

$$\beta^+ = \beta^+(\sigma^{\mp}) , \quad \beta^- = \beta^-(\sigma^=) . \tag{2.36}$$

There are thus *two* separate super-AKM invariances in the super-WZNW theory, one in the L-sector and another one in the R-sector. After a Wick rotation, they are related to the holomorphic and antiholomorphic parts of the theory, respectively.

It is not difficult to construct the $(1,0)$ supersymmetric WZNW action, when using the superspace techniques [320, 321, 322]. In a flat $(1,0)$ superspace parametrized by the coordinates $(\sigma^{\mp}, \sigma^=, \theta^+)$, with the flat $(1,0)$ supercovariant derivatives $(\partial_{\mp}, \partial_=, D_+)$, $D_+ = \partial/\partial\theta^+ + i\theta^+\partial_{\mp}$, the $(1,0)$ NLSM action on a group manifold is given by [323]

$$S_{\text{sWZNW}}^{(1,0)} = -\frac{i}{2\lambda^2} \int d^2\sigma (d\theta)^- \, \text{tr}\left[(G^{-1}D_+G)(G^{-1}\partial_=G)\right]$$

$$-\frac{ik}{8\pi} \int_0^1 dy \int d^2\sigma(d\theta)^- \, \text{tr} \left[\tilde{G}^{-1} \frac{\partial \tilde{G}}{\partial y} \left(D_+ \tilde{G}^{-1} \partial_- \tilde{G} - \partial_- \tilde{G}^{-1} D_+ \tilde{G} \right) \right] \, .$$

(2.37)

This action can be constructed, for example, by truncating the $(1,1)$ supersymmetric WZNW action. Here G and $\tilde{G}$ are the Lie group-valued $(1,0)$ superfields related to each other by (2.27).

The $(1,0)$ super-WZNW action is just the $(1,0)$ supersymmetric NLSM action (2.37) at the RG fixed point $\lambda^2 = 4\pi/k$. This theory has superconformal and super-AKM invariances in its L-sector, while conformal and AKM invariances are present in its R-sector. The super-AKM transformation law in superspace is given by [323]

$$G \to \Omega G \check{\Omega}^{-1} \, ,$$

(2.38)

where the Lie group-valued $(1,0)$ superfield parameters Ω and $\check{\Omega}$ are restricted by the conditions

$$D_+ \Omega = \partial_- \check{\Omega} = 0 \, .$$

(2.39)

The symmetry (2.38) is just the $(1,0)$ super-AKM invariance.

The simplest way of checking the super-AKM invariance of the super-WZNW theory at its RG fixed point is provided by the Polyakov-Wiegmann identity [49, 50]. In the case of the $(1,0)$ supersymmetric WZNW theory, this identity reads [323]

$$S_{\text{sWZNW}}^{(1,0)}(GH) = S_{\text{sWZNW}}^{(1,0)}(G) + S_{\text{sWZNW}}^{(1,0)}(H)$$

$$- \frac{i}{4\pi} \int d^2\sigma(d\theta)^- \, \text{tr} \left[(G^{-1} D_+ G)(\partial_- H \cdot H^{-1}) \right] \, .$$

(2.40)

The super-AKM invariant WZNW actions can also be constructed in the Hamiltonian approach by the method of *coadjoint orbits* [324]. The sWZNW Lagrangian can then be rewritten in the form of the integrated Kirillov-Kostant form defined on the orbits of a (super)AKM algebra [325]. To illustrate this approach, we now consider the construction of the $(2,0)$ supersymmetric WZNW theory [326, 327].

A flat $(2,0)$ superspace is parametrized by the coordinates

$$z^M = (\sigma^m, \zeta^+, \bar{\zeta}^+) \, , \quad \sigma^m = (\tau, \sigma) \, , \quad \zeta^+ = \zeta_1^+ + i\zeta_2^+ \, ,$$

(2.41)

where each Grassmannian coordinate $\zeta_{1,2}^+$ is a MW spinor. The spinor supersymmetric derivative is $D_+ = \partial_+ + i\bar{\zeta}^+ \partial_{\#}$. Together with the other derivatives, they from a (graded) $(2,0)$ supersymmetry algebra,

$$\{D_+, D_+\} = 0 \, , \quad \{D_+, \overline{D}_+\} = 2i\partial_{\#} \, , \quad [D_+, \partial_m] = 0 \, , \quad [\partial_m, \partial_n] = 0 \, ,$$

(2.42)

where $\partial_m \equiv (\partial_{\underline{+}}, \partial_=)$. The $(2,0)$ superfields are complex superfunctions $U(z)$, with their components being defined via the standard superspace expansion,

$$U(z) = \phi(\sigma) + \zeta^+ \beta_+(\sigma) + \bar{\zeta}^+ \psi_+(\sigma) + i\bar{\zeta}^+ \zeta^+ \omega_{\underline{+}}(\sigma) \ . \qquad (2.43)$$

The action of the most general $(2,0)$ supersymmetric NLSM in $(2,0)$ superspace is given by [322]

$$S = \frac{i}{2} \int d^3 z^= \left[K_{\underline{\bar{m}}}(\Phi, \overline{\Phi}) \partial_= \overline{\Phi}^{\underline{\bar{m}}} - K_{\underline{m}}(\Phi, \overline{\Phi}) \partial_= \Phi^{\underline{m}} \right] \ , \qquad (2.44)$$

where we have introduced the $(2,0)$ superspace measure, $d^3 z^= = d^2 \sigma d\bar{\zeta}^+ d\zeta^+$, and the vector potential function $K_{\underline{\bar{m}}} = (K_{\underline{m}})^*$, $m = 1, \ldots, m'$.

In order to get the standard Klein-Gordon and Dirac actions (in components) from the superspace action (2.44) in the free case of $K_m = (\Phi_m)^*$, we must impose the analyticity (or chirality) conditions on the superfields:

$$\overline{D}_+ \Phi^m = D_+ \overline{\Phi}^{\bar{m}} = 0 \ , \qquad \overline{\Phi}^{\bar{m}} = (\Phi^{\underline{m}})^* \ . \qquad (2.45)$$

This automatically implies that the complex structure on the NLSM target space is canonical. At the same time, the constraints (2.45) ensure that our action (2.44) is a 'physical' action in the sense that it reproduces the standard supersymmetric NLSM action in components or $(1,0)$ superfields [327].

The vector potential $K_M \equiv (K_m, K_{\bar{m}})$ determines the geometry of the Hermitian NLSM target manifold, i.e. the metric g_{MN} and the torsion H_{MNP}, as follows:

$$g_{\underline{m\bar{n}}} = K_{\underline{m}, \bar{n}} + K_{\underline{\bar{n}}, m} \ , \qquad H_{\underline{mn\bar{p}}} = \frac{1}{2} \left(K_{\underline{m}, n\bar{p}} - K_{\underline{n}, m\bar{p}} \right) ,$$

$$g_{\underline{mn}} = g_{\underline{\bar{m}\bar{n}}} = H_{\underline{mnp}} = H_{\underline{\bar{m}\bar{n}\bar{p}}} = 0 \ . \qquad (2.46)$$

Varying the action (2.44),

$$\delta S = \frac{i}{2} \int d^3 z^= \delta \Phi^{\underline{m}} \left[g_{\underline{m\bar{n}}} \partial_= \overline{\Phi}^{\bar{n}} + h_{\underline{mn}} \partial_= \Phi^n \right] + \text{h.c.} \ , \qquad (2.47)$$

yields the equations of motion,

$$\overline{D}_+ J_{\underline{m}}^+ \equiv \overline{D}_+ \left[g_{\underline{m\bar{n}}} \partial_= \overline{\Phi}^{\bar{n}} + h_{\underline{mn}} \partial_= \Phi^n \right] = 0 \ , \qquad (2.48)$$

where we have introduced the torsion potential $h_{\underline{mn}}$ as $h_{\underline{mn}} = K_{\underline{m}, n} - K_{\underline{n}, m}$. Equation (2.48) defines the conserved $(2,0)$ supercurrent $J_{\underline{m}}^+(z)$.

The variation (2.47) can be integrated, while this leads to the equivalent action

$$S = \frac{i}{2} \int d^3 z^= \int_0^1 dy \left[g_{\underline{m\bar{n}}} \partial_y \tilde{\Phi}^{\underline{m}} \partial_= \overline{\tilde{\Phi}}^{\bar{n}} + h_{\underline{mn}} \partial_y \tilde{\Phi}^{\underline{m}} \partial_= \tilde{\Phi}^n \right] + \text{h.c.} \ , \qquad (2.49)$$

where the interpolating superfields $\tilde{\Phi}^{\underline{m}}(z, y)$,

$$\tilde{\Phi}^{\underline{m}}(z, 0) = 0 \ , \quad \tilde{\Phi}^{\underline{m}}(z, 1) = \Phi^{\underline{m}}(z) \ , \quad \overline{D}_+ \tilde{\Phi}^{\underline{m}}(z, y) = 0 \tag{2.50}$$

have been introduced.

The torsion potential is, in general, only locally defined. To avoid this, we want to rewrite the torsion term in (2.49) in chiral form, by integrating over one complex Grassmannian coordinate. This results in the action [326]

$$S = \frac{\mathrm{i}}{2} \int_0^1 \mathrm{d}y \int \mathrm{d}^2\sigma \left\{ \int \mathrm{d}\zeta^+ \mathrm{d}\bar{\zeta}^+ \ g_{\underline{m}\tilde{n}} \partial_y \tilde{\Phi}^{\underline{m}} \partial_= \bar{\tilde{\Phi}}^{\tilde{n}} \right.$$

$$\left. + 2 \int \mathrm{d}\zeta^+ \ H_{\underline{m}n\tilde{p}} \partial_y \tilde{\Phi}^{\underline{m}} \partial_= \tilde{\Phi}^{\underline{n}} \overline{D}_+ \bar{\tilde{\Phi}}^{\tilde{p}} \right\} + \text{h.c.} \tag{2.51}$$

The integrand of the second term in (2.51) is chiral since the torsion 3-form H is closed. This WZ-type term is related by extended supersymmetry to the kinetic term. Both actions (2.49) and (2.51) cannot be considered independently of the constraints (2.46). It is the restrictions (2.46) that make the physics of the $(2+1)$-dimensional actions (2.49) and (2.51) two-dimensional. It is the special feature of the $(2,0)$ extended supersymmetry that the $(2,0)$ superfield dynamics is not entirely determined by the form of the action but, in addition, by the kinematical conditions imposed on the superfields (they are either chiral or antichiral in our case) — cf. Chap. 4.

Given a finite-dimensional semisimple Lie algebra $\mathcal{G}$ with structure constants $f_{\underline{AB}}{}^{\underline{C}}$ and generators $t_{\underline{A}}$, $\underline{A} = 1, \ldots, \dim \mathcal{G}$, the group elements (connected to the identity) are the exponentials of the algebra elements, $G(\phi) = \exp(\phi^{\underline{A}} t_{\underline{A}})$. The generators satisfy the usual relations

$$[t_{\underline{A}} \, , \, t_{\underline{B}}] = f_{\underline{AB}}{}^{\underline{C}} t_{\underline{C}} \ , \quad \mathrm{tr}\left(t_{\underline{A}} t_{\underline{B}}\right) = \gamma_{\underline{AB}} \ , \tag{2.52}$$

where the matrix $\gamma_{\underline{AB}}$ is supposed to be invertible, so that it may be used to raise and lower indices.

Let $\mathcal{T}$ be a differentiation of the Lie algebra $\mathcal{G}$, i.e. a linear map $\mathcal{T} : \mathcal{G} \to \mathcal{G}$ satisfying the conditions

$$\mathcal{T}\left([\phi, \psi]\right) = [\mathcal{T}(\phi), \psi] + [\phi, \mathcal{T}(\psi)] \ ,$$

$$\mathrm{tr}[\mathcal{T}(\phi)\psi] = -\mathrm{tr}[\phi\mathcal{T}(\psi)] \ , \tag{2.53}$$

for any two Lie algebra elements, $\phi \in \mathcal{G}$ and $\psi \in \mathcal{G}$. We can associate with this map a matrix $T_{\underline{AB}}$ via the equation

$$\mathcal{T}(\phi) = T_{\underline{AB}} \phi^{\underline{A}} t_{\underline{C}} \gamma^{\underline{BC}} \ , \tag{2.54}$$

and then rewrite (2.53) in the equivalent form

$$T_{\underline{AB}} = -T_{\underline{BA}} \ , \quad T_{\underline{A[B}} f_{\underline{CD]}}{}^{\underline{A}} = 0 \ . \tag{2.55}$$

Given the differentiation $\mathcal{T}$ defined on the algebra elements, we can always extend its definition to the universal enveloping algebra of $\mathcal{G}$, by using the rules

$$\mathcal{T}(1) = 0 \; ,$$

$$\mathcal{T}(\phi + \psi) = \mathcal{T}(\phi) + \mathcal{T}(\psi) \; ,$$

$$\mathcal{T}(\phi\psi) = \mathcal{T}(\phi)\psi + \phi\mathcal{T}(\psi) \; . \tag{2.56}$$

In the context of $(2,0)$ supersymmetry, the fields $\phi^{\underline{A}}$ above have to be replaced by the scalar $(2,0)$ superfields $\Phi^{\underline{A}}(z)$ This allows us to define the $(2,0)$ supercurrents,

$$\tilde{J}_=(z) = \tilde{G}^{-1}\partial_=\tilde{G} \; , \quad \tilde{J}_y(z) = \tilde{G}^{-1}\partial_y\tilde{G} \; . \tag{2.57}$$

The natural candidate for the super-WZNW action in the $(2,0)$ superspace is given by [328] (cf. equation (2.49) above)

$$S = \frac{k}{2\pi} \int \mathrm{d}^3 z^= \int_0^1 \mathrm{d}y \, \mathrm{tr} \left[\mathcal{T} \left(\tilde{J}_= \right) \tilde{J}_y \right] \; , \tag{2.58}$$

with a (z,y)-independent differentiation $\mathcal{T}$. The remarkable identity [327]

$$\delta \, \mathrm{tr}[\mathcal{T}(\tilde{J}_=)\tilde{J}_y] = \partial_y \mathrm{tr}[\tilde{G}^{-1}\delta\tilde{G}\mathcal{T}(\tilde{J}_=)] - \partial_= \mathrm{tr}[\tilde{G}^{-1}\delta\tilde{G}\mathcal{T}(\tilde{J}_y)] \tag{2.59}$$

implies that the integrand of (2.58) is a *closed* 2-form with respect to the integration variables $(\sigma^=, y)$. This form is known as the Kirillov-Kostant form [324].

Equation (2.59) allows us to rewrite the action (2.58) in the 2d form

$$S = \frac{k}{2\pi} \int \mathrm{d}^3 z^= W_{\underline{A}}(\Phi)\partial_=\Phi^{\underline{A}} \; , \tag{2.60}$$

where the interpolation-independent one-form $W \equiv W_{\underline{A}}\mathrm{d}\Phi^{\underline{A}}$ represents a cohomological class (modulo exact differentials).

The action (2.58) is trivially invariant under the $(2,0)$ loop group transformations

$$\tilde{G}\left(z,y\right) \rightarrow H\left(\sigma^{\mp},\zeta^+,\bar{\zeta}^+\right)\tilde{G}\left(z,y\right) \; . \tag{2.61}$$

The associated algebra is nothing but the $(2,0)$ super-AKM algebra. In addition, the action (2.58) is $(2,0)$ superconformally invariant, and it can be easily coupled to $(2,0)$ supergravity by replacing the flat superspace covariant derivatives by curved superspace covariant derivatives subject to the $(2,0)$ supergravity constraints. In the superconformal gauge, all the $(2,0)$ supergravity fields decouple, while the surviving $(2,0)$ supercoordinate transformations represent the $(2,0)$ superconformal symmetry of the action (2.58).

However, this is not yet the end of the story. The action (2.58) is still *not* a physical action, if $\Phi^{\underline{A}}(z)$ were chosen to be general or unconstrained

$(2,0)$ superfields, which is implicit in (2.61). In order to check this, it is enough to consider the Abelian case, where one gets $\partial_= \Phi^A = 0$ as the free equations of motion. To make contact with the general results about the $(2,0)$ NLSM introduced above, the $(2,0)$ superfields $\Phi^A(z)$ must be chiral or antichiral, $\Phi^A(z) = \left(\Phi^a, \overline{\Phi}^{\bar{a}}\right)$, in the canonical basis for the complex structure (see equation (2.64) below), but this assignment is *not* compatible with the transformations (2.61)! This is related to the fact that a general $(2,0)$ real scalar superfield $U(z)$ is decomposed into three irreducible pieces,

$$U(z) = \Phi(z) + \overline{\Phi}(z) + U'(z) , \tag{2.62}$$

where $\Phi(z)$ is chiral, $\overline{\Phi}(z)$ is antichiral, and $U'(z)$ is the non-chiral irreducible superfield, containing a vector $w'_+(\sigma)$ from the component decomposition of the U superfield, and satisfying the superspace constraint $i[D_+, \overline{D}_+]U'(z) = 0$. The Lie group-valued superfield $G(z)$ cannot be chosen to be chiral while maintaining the physical meaning of the $(2,0)$ WZNW action. Therefore, we conclude that the $(2,0)$ WZNW action, written in terms of the (anti)chiral $(2,0)$ superfields, is *not* invariant under the $(2,0)$ super-AKM transformations since they do *not* preserve the complex structure. Being written in a general coordinate system defined on the NLSM group manifold, the chirality condition takes the form [328, 329]

$$G^{-1}D_{2+}G = \mathcal{J}\left(G^{-1}D_{1+}G\right) , \tag{2.63}$$

where the $(2,0)$ superspace covariant derivatives D_{i+}, associated with the anticommuting coordinates ζ_i^+ and the complex structure $\mathcal{J}$, have been introduced. In the complex coordinate system $\Phi^M = \left(\Phi^m, \overline{\Phi}^{\bar{m}}\right)$, where the complex structure $\mathcal{J}$ is canonical and the Maurer-Cartan 1-form $G^{-1}dG \equiv L^A{}_M t_A d\Phi^M$ is diagonal,

$$\mathcal{J}^A{}_B = \begin{pmatrix} i\delta^a{}_b & 0 \\ 0 & -i\delta^{\bar{a}}{}_{\bar{b}} \end{pmatrix} , \quad L^A{}_M = \begin{pmatrix} L^a{}_m & 0 \\ 0 & L^{\bar{a}}{}_{\bar{m}} \end{pmatrix} , \tag{2.64}$$

equation (2.63) takes the form

$$G(z) = \exp\left(\Phi^a(z)t_a + \overline{\Phi}^{\bar{a}}(z)t_{\bar{a}}\right) , \quad \overline{D}_+\Phi^a = D_+\overline{\Phi}^{\bar{a}} = 0 . \tag{2.65}$$

Therefore, one can think of $\mathcal{J}$ as the linear (field-independent) map (or matrix) defined on the Lie algebra $\mathcal{G}$, with the properties

$$\mathcal{J}^2 = -1 , \quad \mathcal{J}^T = \mathcal{J} ,$$

$$\mathcal{N}(\phi, \psi) \equiv [\phi, \psi] - [\mathcal{J}(\phi), \mathcal{J}(\psi)] + \mathcal{J}([\mathcal{J}(\phi), \psi]) + \mathcal{J}([\phi, \mathcal{J}(\psi)]) = 0 . \tag{2.66}$$

The two-form $\mathcal{N}$ is known as the Nijenhuis tensor [329], whose vanishing is necessary for an almost complex structure to be globally well defined (Subsect. 4.1.1).

Being constrained by (2.63), the action (2.58) is no longer $(2,0)$ super-AKM invariant, but it remains $(1,0)$ super-AKM invariant and it still has the $(2,0)$ superconformal symmetry. [6] Indeed, the action (2.58) in terms of the constrained superfields (2.63) can be represented in the form (2.44), while it is straightforward to integrate over one real Grassmannian coordinate ζ_2^+ there, and thus get the $(1,0)$ superspace form of the action.

Both G and G^{-1} can be expanded in ζ_2^+ as

$$G = g + \zeta_2^+ g \mathcal{J} \left(g^{-1}D_+g\right) , \quad G^{-1} = g^{-1} - \zeta_2^+ \mathcal{J} \left(g^{-1}D_+g\right) g^{-1} , \quad (2.67)$$

where the $(1,0)$ group-valued real superfield $g = G|$ and the $(1,0)$ superspace covariant derivative $D_+ \equiv D_{1+}$ have been introduced. An integration over ζ_2^+ in the action (2.58) yields

$$S = \frac{k}{2\pi} \int \mathrm{d}^2\sigma \, (d\zeta)^- \mathrm{tr} \left[\left(g^{-1}D_+g\right) \mathcal{J}\mathcal{T} \left(g^{-1}\partial_=g\right)\right.$$

$$\left. + \int_0^1 \mathrm{d}y \, \left(\tilde{g}^{-1}D_+\tilde{g}\right) \mathcal{J}\mathcal{T} \left(\partial_=\tilde{g}^{-1}\partial_y\tilde{g} - \partial_y\tilde{g}^{-1}\partial_=\tilde{g}\right)\right] . \quad (2.68)$$

The action (2.68) differs from the standard super-WZNW action in $(1,0)$ superspace by the factor $\mathcal{J}\mathcal{T}$, while it is *not* possible to ignore this factor by identifying $\mathcal{J}$ and $\mathcal{T}$ (then $\mathcal{J}\mathcal{T}$ would be proportional to the unit matrix), since the defining conditions (2.56) and (2.66) on the $\mathcal{T}$ and $\mathcal{J}$, respectively, are *incompatible*. Nevertheless, it is still possible to prove the $(1,0)$ super-KM invariance of the action (2.68) when using, e.g., the Polyakov–Wiegmann identity.

The failure to incorporate the $(2,0)$ super-AKM symmetry within the $(2,0)$ WZNW theory does not seem to be surprising from the viewpoint of CFT: an investigation of the $N = 1$ super-AKM currents shows that they are *not* $N = 2$ superconformal fields with respect to the $N = 2$ superconformal algebra [326]. [7]

Having rewritten the Lie group-valued $(2,0)$ superfield $G(z)$ in the form (2.65), we get the $(2,0)$ supersymmetric loop group transformations in the form

$$G(z) \rightarrow \exp\left[\varepsilon^{\underline{a}}(z)t_{\underline{a}} + \bar{\varepsilon}^{\underline{a}}t_{\underline{a}}\right] G(z) , \quad (2.69)$$

with the $\sigma^=$-independent and chiral (antichiral) $(2,0)$ superfield parameters $\varepsilon^{\underline{a}}(\bar{\varepsilon}^{\underline{a}})$. This is still consistent with the given parameterization provided

---

[6] The $(1,0)$ supersymmetric NLSM model is invariant under the $(2,0)$ extended supersymmetry if its target space admits a complex structure. A complex structure can be defined for any even-dimensional Lie algebra.

[7] The $(1,0)$ super-AKM invariance is sufficient to guarantee the integrability and finiteness of the super-WZNW theory.

$$[t_{\underline{a}}, t_{\underline{b}}] = 0 \ . \tag{2.70}$$

Equation (2.70) holds when the group G is a product of two copies of an arbitrary Lie group with the naturally defined complex structure. The component structure of the associated $(2, 0)$ supercurrent then resembles the $(1, 0)$ counterpart.

5.3 Coset Construction and Gauged WZNW Models

The enveloping Virasoro algebra associated with an AKM algebra has the central charge $c = c_G$ given by (1.56). The central charge values are restricted to the interval

$$\operatorname{rank} G \leq c_G \leq |G| \tag{3.1}$$

for any Lie group G. In particular, given $G = SU(2)$, one has

$$c_{SU(2)} = \frac{3k}{k+2} \ , \tag{3.2}$$

so that $1 \leq c_{SU(2)} \leq 3$. Hence, the SS group construction alone is unable to describe rational CFT of central charge $c < 1$. We now want to break the SS-constructed stress tensor (1.43) into pieces, each with a smaller central charge. The basic idea is to generalize the group-based SS construction to the *cosets* G/H, with H being a subgroup of the Lie group G.

5.3.1 Goddard-Kent-Olive Construction

The Goddard-Kent-Olive (GKO) method [317] of constructing unitary representations of the Virasoro algebra on cosets always yields rational CFT.

To introduce the method, let $J_G^a(z)$ be the G currents, $a = 1, \ldots, |G|$, and $J_H^i(z)$ the H currents, $i = 1, \ldots, |H|$. We assume that the group indices run over the adjoint representation and the first $|H|$ currents in $\{J_G^a\}$ just represent the H currents $\{J_H^i\}$. The normalization of the structure constants is supposed to respect the highest root norm $\psi^2 = 2$, so that we have

$$
\begin{aligned}
J_G^a(z) J_G^b(w) &\sim \frac{k_G \delta^{ab}}{(z-w)^2} + \frac{i f^{abc}}{z-w} J_G^c(w) \ , \\
J_H^i(z) J_H^j(w) &\sim \frac{k_H \delta^{ij}}{(z-w)^2} + \frac{i f^{ijk}}{z-w} J_H^k(w) \ .
\end{aligned}
\tag{3.3}
$$

The level k_H is determined by the embedding of H into G. For instance, if the simple roots of H form a subset of the simple roots of G, then $k_H = k_G$. If $G = G_1 \otimes G_2$ and H is the diagonal subgroup, then $k_H = k_1 + k_2$.

We can exploit the SS construction (Subsect. 5.1.2) in order to construct the G and H stress tensors, T_G and T_H, in the form

$$T_G(z) = \frac{1/2}{k_G + \tilde{h}_G} \sum_{a=1}^{|G|} : J_G^a(z) J_G^a(z) : \ ,$$

$$T_H(z) = \frac{1/2}{k_H + \tilde{h}_H} \sum_{i=1}^{|H|} : J_H^i(z) J_H^i(z) : \ ,$$

$$(3.4)$$

with the Virasoro central charges

$$c_G = \frac{k_G |G|}{k_G + \tilde{h}_G} \ , \qquad c_H = \frac{k_H |H|}{k_H + \tilde{h}_H} \ . \tag{3.5}$$

Since the currents J_H^i of dimension $h = 1$ are all primary fields with respect to both stress tensors T_G and T_H of (3.4), we have

$$T_G(z) J_H^i(w) \sim \frac{J_H^i(w)}{(z-w)^2} + \frac{\partial J_H^i(w)}{z-w} \ ,$$

$$T_H(z) J_H^i(w) \sim \frac{J_H^i(w)}{(z-w)^2} + \frac{\partial J_H^i(w)}{z-w} \ .$$

$$(3.6)$$

Thereby, the OPE of $T_{G/H} \equiv T_G - T_H$ with J_H^i is non-singular. The OPE of $T_{G/H}$ with T_H is also non-singular since the T_H was constructed in (3.4) in terms of the H currents J_H^i alone. Hence, we get the orthogonal decomposition of the Virasoro algebra generated by T_G into two mutually commuting Virasoro subalgebras,

$$T_G = (T_G - T_H) + T_H = T_{G/H} + T_H \ , \qquad [T_{G/H}, T_H] = 0 \ . \tag{3.7}$$

In terms of modes, this means

$$\left[L_m^{G/H}, J_{H,n}^i \right] = \left[L_m^{G/H}, L_n^H \right] = \left[L_m^G - L_m^H, L_n^H \right] = 0 \ . \tag{3.8}$$

Hence, we find

$$\left[L_m^{G/H}, L_n^{G/H} \right] = \left[L_m^G - L_m^H, L_n^G - L_n^H \right] = \left[L_m^G, L_n^G \right] - \left[L_m^H, L_n^H \right]$$

$$= (m-n) L_{m+n}^G + \frac{c_G}{12} (m^3 - m) \delta_{m+n,0} - (m-n) L_{m+n}^H - \frac{c_H}{12} (m^3 - m) \delta_{m+n,0}$$

$$= (m-n) L_{m+n}^{G/H} + \frac{c_{G/H}}{12} (m^3 - m) \delta_{m+n,0} \ , \tag{3.9}$$

i.e. $L_m^{G/H}$ satisfy the Virasoro algebra of central charge [317]

$$c_{G/H} = c_G - c_H = \frac{k_G |G|}{k_G + \tilde{h}_G} - \frac{k_H |H|}{k_H + \tilde{h}_H} \ . \tag{3.10}$$

The stress tensor $T_{G/H}(z)$ defines a unitary representation of conformal symmetry since it is realized on a subspace of the unitary representation $T_G(z)$.

A simple example is provided by the coset $\widehat{SU(2)}_k/\widehat{U(1)}$. In this case, (3.2) and (3.10) give rise to the central charge

$$c_{SU(2)/U(1)} = \frac{3k}{k+2} - 1 = \frac{2(k-1)}{k+2} . \tag{3.11}$$

Choosing $k = 2$ corresponds to the Ising model of $c = 1/2$. The choice of $k = 1$ yields zero on the right-hand side of (3.11), which is the manifestation of the *quantum equivalence* $T_G = T_H$ between the two apparently different stress tensors: the SS-constructed stress tensor $T_{SU(2)}$ and the stress tensor of a single free (chiral) boson [330].

The interesting examples of the GKO construction are given by cosets with a diagonal embedding, e.g.

$$\widehat{G}/\widehat{H} = \widehat{SU(2)}_k \otimes \widehat{SU(2)}_l/\widehat{SU(2)}_{k+l} , \tag{3.12}$$

of central charge

$$\begin{aligned}
c_{G/H} &= \frac{3k}{k+2} + \frac{3l}{l+2} - \frac{3(k+l)}{k+l+2} \\
&= 1 - \frac{6l}{(k+2)(k+l+2)} + \frac{2(l-1)}{l+2} .
\end{aligned} \tag{3.13}$$

Taking $l = 1$ in (3.13) gives the discrete series of central charge values that correspond to the so-called unitary minimal models [45].

The $N = 1$ superconformal central charge series ($N = 1$ minimal models) are obtained by taking $l = 2$ in (3.13),

$$c_{G/H} = \frac{3k}{k+2} + \frac{3}{2} - \frac{3(k+2)}{(k+2)+2} = \frac{3}{2}\left(1 - \frac{8}{(k+2)(k+4)}\right) . \tag{3.14}$$

This observation can also be used to prove the unitarity of the $N = 1$ minimal models [317]. It is not difficult to verify that the GKO construction based on the cosets

$$\frac{\widehat{G}_k \otimes \widehat{SO(|G|)}_1}{G_{k+\tilde{h}_G}} , \tag{3.15}$$

where $\tilde{h}_G$ is the dual Coxeter number, gives rise to the $N = 1$ SCFT with super-AKM symmetry. Similarly, when using the cosets

$$\frac{\widehat{SO(|G|)}_1 \otimes \widehat{SO(|G|)}_1}{G_{2\tilde{h}_G}} \tag{3.16}$$

in the GKO construction, one gets the unitary minimal $N = 2$ SCFT [331, 332, 333, 334].

It is worth mentioning that GKO constructions are *not* unique. For example, there are different choices of cosets for the following $N = 1$ minimal models:

$$(k = 3, \quad c = 81/70) \quad : \quad (\widehat{E_8})_2/\widehat{SU(3)}_2 \otimes (\widehat{E_6})_2 \ ,$$

$$(k = 7, \quad c = 91/66) \quad : \quad (\widehat{E_8})_2/(\widehat{G_2})_2 \otimes (\widehat{F_4})_2 \ , \qquad (3.17)$$

$$(k = 8, \quad c = 7/5 \ ) \quad : \quad \widehat{Sp(6)}_1 \otimes \widehat{Sp(6)}_1/\widehat{Sp(6)}_2 \ .$$

5.3.2 Gauged WZNW Models

A large class of CFT based on cosets G/H can be described in terms of the *gauged* WZNW theories [335, 336, 337, 338, 339, 340]. In this subsection we describe this approach. The WZNW action (1.61) defined on a group manifold G is a good starting point for this construction,

$$kI(g) = \frac{k}{8\pi} \int d^2\sigma \, \mathrm{tr} \left(\partial_m g \partial^m g^{-1} \right) + \frac{k}{12\pi} \int \mathrm{tr} \left(dg g^{-1} \right)^3 \ . \qquad (3.18)$$

Let H be an anomaly-free vector subgroup of $G_L \times G_R$ that is the global symmetry of the action (3.18). Let $\tilde{H}$ be Lie algebra of the Lie group H. We assume that the Lie algebra $\mathcal{G}$ of G admits an $\mathrm{ad}(\tilde{H})$-invariant orthogonal decomposition having the form [8]

$$\mathcal{G} = \tilde{H} \oplus M \ , \qquad [\tilde{H}, \tilde{H}] \subset \tilde{H} \ , \qquad [\tilde{H}, M] \subset M \ , \qquad (3.19)$$

where $M \equiv \mathcal{G} \setminus \tilde{H}$; $g \in G$, $h \in H$, and $m \in G/H$. A Lie algebra $\mathcal{G}$-valued field J can always be written down as a sum, $J = J_H + J_M$, with respect to the decomposition (3.19).

We are now in a position to consider the modified WZNW action, where the subgroup H is gauged (cf. Sect. 2.1). This means that we add the minimal coupling to the H-valued gauge fields and make the action invariant under the gauge transformations (with the σ-dependent parameter λ) having the form

$$g \to \lambda g \lambda^{-1} \ , \qquad \lambda \in H \ ,$$

$$A_m \to \partial_m \lambda \cdot \lambda^{-1} + \lambda A_m \lambda^{-1} \ . \qquad (3.20)$$

In the light-cone variables $\sigma_\pm$, the gauged WZNW action takes the form [9]

$$kI(g, A) = kI(g) + \frac{k}{4\pi} \mathrm{tr} \int d^2\sigma \, \{ A_+ \partial_- g g^{-1}$$

$$- A_- g^{-1} \partial_+ g + A_+ g A_- g^{-1} - A_- A_+ \} \ , \qquad (3.21)$$

where the light-cone components of the gauge field $A_\pm$ in the adjoint representation of H have been introduced. It is useful to parametrize the gauge fields in terms of the H-group elements,

---

[8] For simplicity, the embedding index of H in G is assumed to be one.

[9] This form of the action is fixed by the gauge invariance, when requiring the vanishing of the H current in addition (see below).

$$A_- = \partial_- \tilde{h} \tilde{h}^{-1} , \quad A_+ = \partial_+ h h^{-1} . \tag{3.22}$$

The convenience of this parametrization becomes clear after exploiting the Polyakov-Wiegmann identity

$$I(gh) = I(g) + I(h) - \frac{1}{4\pi} \int d^2\sigma \, \mathrm{tr} \left[g^{-1} \partial_+ g \partial_- h h^{-1} \right] \tag{3.23}$$

which allows us to rewrite the action (3.21) in the form

$$I(g, A) = I(h^{-1} g \tilde{h}) - I(h^{-1} \tilde{h}) , \tag{3.24}$$

where the vector gauge invariance (3.20) is manifest.

The gauge fields in (3.21) are non-propagating. They play the role of Lagrange multipliers forcing the H currents to vanish in the classical gauged WZNW theory. To see how the action (3.21) describes the G/H coset structure, let us consider its equations of motion. Varying the action with respect to the gauge fields yields

$$(g^{-1} \nabla_+ g)_H = 0 , \quad (\nabla_- g \cdot g^{-1})_H = 0 , \tag{3.25}$$

where the new gauge-covariant derivatives $\nabla_m = \partial_m - [A_m, \ ]$ have been introduced. The subscript H above means projection onto the subspace H of G in the sense of (3.19). Equation (3.25) is only gauge-covariant when the commutation relations (3.19) are satisfied. The equation of motion for the g-field reads

$$\nabla_- (g^{-1} \nabla_+ g) = F , \tag{3.26}$$

while the first equation (3.25) implies $F = 0$. Hence, both connections A_+ and A_- are, in fact, on-shell trivial. Exploiting the gauge invariance (3.20) now allows us to choose the gauge $A_\pm = 0$ in the classical theory, in which the equations of motion are simplified to the form [323]

$$(g^{-1} \partial_+ g)_H = 0 , \quad (\partial_- g \cdot g^{-1})_H = 0 ,$$

$$\partial_- (g^{-1} \partial_+ g)_M = 0 , \Longleftrightarrow \partial_+ (\partial_- g \cdot g^{-1})_M = 0 . \tag{3.27}$$

The M currents are, therefore, non-vanishing on-shell, while the H currents do vanish on-shell, as they should. In quantum theory the situation is a bit more complicated since the change of variables (3.22) induces the Jacobian in the partition function of the gauged WZNW theory,

$$Z = \int [dg][dA_+][dA_-] \exp \left[-kI(g, A) \right] , \tag{3.28}$$

which reads

$$Z = \int [dg][dh][d\tilde{h}] \det D_+ \det D_- \exp \left[-kI(h^{-1} g \tilde{h}) \right] \exp \left[kI(h^{-1} \tilde{h}) \right] , \tag{3.29}$$

where
$$\det D_\pm = \det\{\partial_\pm - [A_\pm, \,]\} \,. \qquad (3.30)$$

The conventional QFT techniques can now be applied to calculate the determinants, [10] with the result [49, 50]

$$\det D_+ \det D_- = \exp\left[c_H I(h^{-1}\tilde{h})\right] \det \partial_+ \det \partial_- \,, \qquad (3.31)$$

where the quadratic Casimir operator eigenvalue c_H in the adjoint representation of H and the chiral determinants $\det \partial_\pm$ have been introduced. The latter can be conveniently represented as (Gaussian) integrals over the Faddeev-Popov (FP) ghost fields $b_\pm$ of (conformal) dimension one, and $c, \bar{c}$ of dimension zero, all in the adjoint representation of H. Changing the variables $(h^{-1}g\tilde{h}) \to g$ and fixing the gauge $\tilde{h} = 1$, (or, equivalently, $A_- = 0$) that does not produce FP-ghosts, one finally arrives at the following expression for the gauge-fixed partition function (after changing the variable h^{-1} back to h) [338, 339, 340]:

$$Z = \int [\mathrm{d}g][\mathrm{d}h][\mathrm{d}b_+][\mathrm{d}b_-][\mathrm{d}c][\mathrm{d}\bar{c}] \exp\left[-kI(g)\right] \exp\left[(k + c_H)I(h)\right]$$

$$\times \exp\left(-\mathrm{tr} \int \mathrm{d}^2\sigma\, b_+ \partial_- c\right) \exp\left(-\mathrm{tr} \int \mathrm{d}^2\sigma\, b_- \partial_+ \bar{c}\right) \,. \qquad (3.32)$$

The partition function (3.32) factorizes into three conformally invariant pieces that are actually coupled to each other by some constraints [338, 339, 340]. The constraints are easily derived by introducing an *external* gauge field $B_\pm$ and making use of the parametrization similar to that of (3.22),

$$B_+ = \partial_+ q \cdot q^{-1} \,, \quad B_- = \partial_- p \cdot p^{-1} \,, \qquad (3.33)$$

where p, q are in the fundamental representation of H. The identity

$$\int [\mathrm{d}g] \exp\left[-kI(g, B)\right] = \int [\mathrm{d}g] \exp\left[-kI(g)\right] \exp\left[kI(q^{-1}p)\right] \qquad (3.34)$$

implies that the naively B-dependent partition function is actually independent of B,

$$Z(B_+, B_-) = Z(0) \exp\{[k - (k + c_H) + c_H]I(q^{-1}p)\} = Z(0) \,, \qquad (3.35a)$$

or, equivalently,

$$\left.\frac{\delta Z(B)}{\delta B_-}\right|_{B=0} = \left\langle J_+^{\mathrm{tot}}(\sigma_+)\right\rangle = 0 \,, \qquad (3.35b)$$

---

[10] One should use a UV regulator preserving the gauge invariance, and divide the partition function by the (infinite) gauge volume.

and similarly for J_-^{tot}. [11] Being a function of the holomorphic coordinate $z = \exp(i\sigma_+)$, the current J_+^{tot} takes the form [340]

$$
\begin{aligned}
J^{i,\text{tot}}(z) &= k\,\text{tr}(t^i g^{-1}\partial_z g) - (k + c_H)\text{tr}(t^i h^{-1}\partial_z h) - if^{ijk}b_z^j c^k \\
&\equiv J^i(z) + \tilde{J}^i(z) + J_{\text{gh}}^i(z) ,
\end{aligned}
\tag{3.36}
$$

where t^i, $i = 1,\ldots,|H|$, are the generators of H in the adjoint representation. The relevant OPE read

$$
J^i(z)J^j(w) \sim if^{ijk}\frac{J^k(w)}{z - w} + \frac{k\delta^{ij}}{(z - w)^2} ,
$$

$$
\tilde{J}^i(z)\tilde{J}^j(w) \sim if^{ijk}\frac{\tilde{J}^k(w)}{z - w} - \frac{(k + c_H)\delta^{ij}}{(z - w)^2} ,
$$

$$
J_{\text{gh}}^i(z)J_{\text{gh}}^j(w) \sim if^{ijk}\frac{J_{\text{gh}}^k(w)}{z - w} + \frac{c_H d^{ij}}{(z - w)^2} .
\tag{3.37}
$$

The level and the SS-constructed central charge, associated with the current $J^{i,\text{tot}}(z)$, obviously vanish, while the current itself represents a first-class constraint in the gauged WZNW theory. Though we cannot impose it as an operator constraint (i.e. in the strong sense) in the Fock space of states, this constraint can be imposed in the weak sense, which is quite standard in QFT, namely, $J_m^{\text{tot}}|\text{phys}\rangle = 0$ for $m \geq 0$, where $J^{\text{tot}}(z) = \sum_{m=-\infty}^{\infty} J_m^{\text{tot}} z^{-m-1}$.

The holomorphic stress tensor for the theory (3.32) reads

$$
\begin{aligned}
T(z) &= \frac{1}{2k + c_G} : J^a(z)J^a(z) : - \frac{1}{2k + c_H} : \tilde{J}^a(z)\tilde{J}^a(z) : - : b_z^i \partial_z c^i : \\
&\equiv T^G + \tilde{T}^H + T^{\text{gh}} ,
\end{aligned}
\tag{3.38}
$$

where $a = 1,\ldots,|G|$. The OPE (3.37) imply the Virasoro OPE

$$
T^G(z)T^G(w) \sim \frac{c(G,k)}{2(z - w)^4} + \frac{2T^G(w)}{(z - w)^2} + \frac{\partial_w T^G(w)}{z - w} ,
$$

$$
\tilde{T}^H(z)\tilde{T}^H(w) \sim \frac{c(H, -k - c_H)}{2(z - w)^4} + \frac{2\tilde{T}^H(w)}{(z - w)^2} + \frac{\partial_w \tilde{T}^H(w)}{z - w} ,
$$

$$
T^{\text{gh}}(z)T^{\text{gh}}(w) \sim \frac{-2|H|}{2(z - w)^4} + \frac{2T^{\text{gh}}(w)}{(z - w)^2} + \frac{\partial_w T^{\text{gh}}(w)}{z - w} ,
\tag{3.39}
$$

where, in accordance with (1.56), the WZNW central charge is given by

$$
c(G,k) = \frac{2k|G|}{2k + c_G} .
\tag{3.40}
$$

---

[11] We restrict ourselves to holomorphic dependence.

Hence, the total central charge of the gauged WZNW theory is given by

$$c_{\text{tot}} = \frac{2k|G|}{2k + c_G} + \frac{2(-k - c_H)|H|}{2(-k - c_H) + c_H} - 2|H|$$

$$= \frac{2k|G|}{2k + c_G} - \frac{2k|H|}{2k + c_H} , \tag{3.41}$$

in agreement with the central charge of the GKO construction for the cosets! However, the stress tensors $T(z)$ above and $T^{\text{GKO}} \equiv T^{G/H} = T^G - T^H$ differ since

$$T = T^G - T^H + T' \equiv T^{\text{GKO}} + T' , \tag{3.42}$$

where

$$T' = \frac{: J^i J^i :}{2k + c_G} - \frac{: \tilde{J}^i \tilde{J}^i :}{2k + c_H} - : b^i_z \partial_z c^i : . \tag{3.43}$$

The stress tensor T' commutes with T^{GKO}, and its central charge is zero. Since any unitary highest weight representation of $c = 0$ is trivial [45], given a unitary physical spectrum in the gauged WZNW theory, the stress tensor T' should be trivial too. The actual proof of unitarity is far from obvious for this model, but, nevertheless, it can be done by using BRST methods, essentially because of merely the first-class constraints in the theory [45]. The physical states are given by the cohomology $\text{Ker}Q/\text{Im}Q$ of BRST charge Q, where $Q |\text{phys}\rangle = 0$ but $|\text{phys}\rangle \neq Q |*\rangle$, and $Q^2 = 0$. Since the BRST charge commutes with T^{tot}, T^{GKO} and T' individually, T' can be recognized as a BRST *exact* operator. Hence, there exists an operator $X(z)$ satisfying the relation

$$T' = \{Q, X(z)\} , \tag{3.44}$$

which makes the difference between $T(z)$ and T^{GKO} irrelevant in the matrix elements between the physical states. In particular, the states created by T' all have zero norm and they are orthogonal to any physical state. Hence, they actually decouple from the physical sector, and we find that

$$T(z) |\text{phys}\rangle = T^{\text{GKO}}(z) |\text{phys}\rangle . \tag{3.45}$$

The decoupling of the negative norm states and ghosts from the physical spectrum of the gauged WZNW theory was proved [340] in the case of an Abelian subgroup H. The Abelian unitarity provides strong evidence that the non-Abelian case should also be the same in this respect.

The (simple) supersymmetrization of the field-theoretical construction above is straightforward [323, 339]. The generalizations of the CFT coset space method are known for $N = 2$ extended supersymmetry [341], and $N = 4$ extended supersymmetry too [342, 343, 344, 345, 346]. Though coset space methods were invented to study representations of the linear extended superconformal algebras, they were also extended to non-linear $N = 4$ superconformal algebras [202].

6. NLSM and Strings

String theory [1, 170, 171, 172, 347, 348, 349] describes (quantized) relativistic one-dimensional objects (called strings) propagating in D-dimensional spacetime. The very idea of string theory originated from the efforts to describe quark confinement and Regge trajectories of hadrons. Superstring theory was later applied for a construction of quantum gravity. A string action defined in curved spacetime is described by NLSM. In this chapter we discuss the relation between the NLSM and strings.

String vacua are described by Conformal Field Theory (CFT) which plays an important role in string theory. Similarly, superconformal field theory (SCFT) is crucial in superstring theory [45]. The relation between strings and CFT (Chap. 5) is illustrated in Sect. 6.1 by presenting Zamolodchikov's c-theorem (see also Chap. 2). In Sect. 6.2 we discuss the conformally invariant 2d NLSM that may serve as a 2d laboratory for investigating black holes and T-duality. Section 6.3 is devoted to the construction of the classical actions (in components) for $N = 1$ (Neveu-Schwarz-Ramond) and $N = 2$ fermionic (spinning) strings in curved spacetime, and a calculation of their (quantum) conformal anomalies.

6.1 Bosonic String NLSM

A string propagating in D-dimensional spacetime sweeps out a 2d surface called a string *world-sheet*. Let X^μ, $\mu = 0, 1, \ldots, D - 1$, be spacetime coordinates and (σ^0, σ^1) the world-sheet coordinates. A string motion is thus described by the set of D functions $X^\mu(\sigma^0, \sigma^1)$ specifying the spacetime position of each point of the string. The coordinate σ^1 may be interpreted as a label along the string, whereas the coordinate σ^0 may be identified with the (2d) proper time, though this is not necessary.

Strings can be either closed or open. In this section we confine ourselves to the closed strings that satisfy the boundary condition $X^\mu(\sigma^0, 0) = X^\mu(\sigma^0, 2\pi)$. The physical description of strings implies the use of Minkowski signatures for both world-sheet and spacetime. Nevertheless, Wick rotations may be used to define a Euclidean formulation of string theory, e.g., in order to make contact with CFT and the theory of Riemann surfaces. The coordinate functions $X^\mu(\sigma)$, describing the embedding of a Riemann surface Σ

(representing a string world-sheet) into Euclidean spacetime $\mathbf{R}^D$, can then be equally considered as the set of D two-dimensional scalar fields defined on Σ.

The string action should be independent of the way the string world-sheet Σ is parametrized. This property is called *reparametrizational invariance*. This invariance is nothing but 2d general coordinate invariance, while the standard way of covariantization is given by a minimal coupling of the scalar fields $X^\mu(\sigma)$ to 2d 'gravity' described by a metric $g_{\alpha\beta}(\sigma)$. This results in the 2d reparametrization-invariant action known as the *Polyakov string action* [350],

$$S_\mathrm{P} = \frac{1}{4\pi\alpha'} \int_\Sigma \mathrm{d}^2\sigma \, \sqrt{g} g^{\alpha\beta} \partial_\alpha X^\mu \partial_\beta X_\mu \, , \tag{1.1}$$

where g is the determinant of the 2d metric $g_{\alpha\beta}$, and α' is the dimensional string constant (sometimes called the Regge slope parameter). The action (1.1) respects the global Poincaré symmetry in the target space (i.e. in space-time). The Polyakov string action is invariant under 2d diffeomorphisms *and* local Weyl (or scale) transformations

$$g_{\alpha\beta} \to \Lambda(\sigma) g_{\alpha\beta} \, . \tag{1.2}$$

The Weyl symmetry (1.2) is, however, anomalous in the quantized 2d theory (1.1), whereas the 2d reparametrization invariance can be safely used to bring the 2d metric to the conformally flat form called the *orthonormal* (or *conformal*) *gauge*,

$$g_{\alpha\beta} = \delta_{\alpha\beta} \mathrm{e}^{2\phi(\sigma)} \, . \tag{1.3}$$

The conformal factor $\phi(\sigma)$ disappears from the classical Polyakov string action (1.1) in the conformal gauge (1.3). The metric (1.3) obviously preserves its form under 2d conformal transformations, which take the form of *analytic* transformations in the conformal coordinates $(z, \bar{z})$, where $z = \sigma^0 + i\sigma^1$, on the Euclidean string world-sheet (Chap. 5). String theory thus possesses conformal symmetry arising as the remnant of the 2d reparametrizational symmetry in the conformal gauge. The 2d CFT gives us a useful framework for analysing general (i.e. action-independent) properties of strings.

The action (1.1) in the conformal gauge (1.3) is just a *free* action of D scalar fields, which can be easily quantized. The conformal (or Weyl) anomaly is given by the central charge in the associated Virasoro algebra. This anomaly is equal to D for the D free scalar fields (as usual, we concentrate on a holomorphic dependence of the fields). In the BRST quantized string theory, we should also take into account the bosonic FP ghost fields associated with the covariant gauge-fixing of the 2d reparametrization invariance. They are given by the (b, c) conformal ghost system [45] that contributes -26 to the (chiral) Virasoro central charge. This immediately yields the celebrated result that the quantized bosonic string is anomaly-free in the *critical dimension* of $D = 26$. The zero modes of the string scalars $X^\mu(\sigma)$ are supposed to be identified with spacetime coordinates. Physical spacetime has the

Minkowski signature that is dictated by the *no-ghost theorem* [351] requiring the same number of time-like directions in the critical target space and in the (b, c) ghost system, i.e. equal to one in both cases.

The critical bosonic string spacetime dimension of $D = 26$ is the maximal dimension where a consistent bosonic string theory can be defined. One can compose different CFT models to compensate the only relevant ghost contribution of $c_{\text{gh}} = -26$. For example, the (uncompactified) spacetime dimension can be equal to four, by using only four uncompactified free scalar fields and taking the rest of the central charge ($c_{\text{int}} = 22$) from a CFT representing the internal string degrees of freedom. This approach is known as string compactification. The central charge value of $c_{\text{int}} = 22$ is not the only condition to be imposed on the CFT describing the internal sector. Amongst the other fundamental physically motivated requirements are (i) unitarity, and (ii) a discrete bounded-from-below spectrum of the string Hamiltonian. The unitarity requirement apparently forces us to replace strings by *superstrings* since the bosonic string theory suffers from the presence of a tachyon in its (perturbative) spectrum. The critical dimension of superstings is $D = 10$, while four-dimensional superstrings imply the existence of an internal superstring sector described by a SCFT of central charge $c = 9$ [45]. The physical content of a four-dimensional superstring model crucially depends upon the choice of CFT describing the internal sector. Because of the huge variety of CFT models suitable for this purpose [352], there is an enormous uncertainty in any physical prediction to be made from superstring theory in its present form.

6.1.1 NLSM Approach to String LEEA

From the 2d viewpoint, the action (1.1) can be considered as the particular minimal coupling of conformal matter represented by free scalar fields to 2d gravity. In QFT the most general reparametrization-invariant action may have to be considered, as is normally required by renormalization. Since the 2d Euler characteristic $\chi(\Sigma)$ and a 2d 'cosmological' term are the only possible modifications, the slighly 'generalized' (in the 2d sense) Euclidean string action reads [353]

$$S_{\text{B}} = S_{\text{P}} + \lambda\chi(\Sigma) + \mu_0^2 \int_{\Sigma} \mathrm{d}^2 z \sqrt{g} \, , \qquad (1.4)$$

where λ and μ_0^2 are constants. The second term in (1.4) is a 2d topological invariant, while the third one explicitly breaks the Weyl invariance. The Weyl invariance can however be restored in the critical dimension $D = 26$. Away from the critical dimension, the Weyl anomaly should thus be represented by the *Liouville theory*, which is the effective field theory of quantum 2d gravity. The Liouville potential is given by an exponential of the conformal factor ϕ, which appears due to the last term in (1.4) to be rewritten in the conformal gauge (1.3).

Since the Polyakov string action also has a spacetime interpretation, there is another way of geometrical generalization in the form of NLSM [354]. Having required NLSM renormalizability, one can also write down the most general reparametrization-invariant local NLSM action in the form [355]

$$
\begin{aligned}
S_{\text{NLSM}} = & \frac{1}{4\pi\alpha'} \int_\Sigma \mathrm{d}^2 z \, \sqrt{g} g^{\alpha\beta} \partial_\alpha X^\mu \partial_\beta X^\nu G_{\mu\nu}(X) \\
& + \frac{1}{4\pi\alpha'} \int_\Sigma \mathrm{d}^2 z \, \varepsilon^{\alpha\beta} \partial_\alpha X^\mu \partial_\beta X^\nu B_{\mu\nu}(X) \\
& + \frac{1}{4\pi} \int_\Sigma \mathrm{d}^2 z \, \sqrt{g} R^{(2)} \Phi(X) - \frac{1}{\pi\alpha'} \int_\Sigma \mathrm{d}^2 z \, \sqrt{g} \, T(X) \, ,
\end{aligned}
\tag{1.5}
$$

where a symmetric *spacetime* metric $G_{\mu\nu}(X)$, an antisymmetric Kalb-Ramond field $B_{\mu\nu}(X)$, a dilaton field $\Phi(X)$, and a tachyon field $T(X)$ have been introduced (and, in fact, defined this way). The action (1.5) describes string propagation in the spacetime background of the fields that are in one-to-one correspondence with the lowest string excitation modes. The dilaton contribution in (1.5) is known as the Fradkin-Tseytlin (FT) term [355]. The FT term breaks the *classical* Weyl invariance that is supposed to be restored after taking into account quantum corrections. The 2d Weyl invariance implies 2d conformal invariance in any local field theory (Chap. 5). We ignore the tachyon ($T = 0$) in this section (see, however, Sect. 6.2).

The quantized NLSM (1.5) is conformally invariant provided $\langle T_\alpha{}^\alpha \rangle = 0$, where $T_\alpha{}^\beta$ is the 2d stress tensor. For dimensional and symmetry reasons, the trace of the 2d stess tensor can be represented in terms of finite NLSM composite operators as

$$
2\pi T_\alpha{}^\alpha = \bar{\beta}^G_{\mu\nu} \sqrt{g} g^{\alpha\beta} \partial_\alpha X^\mu \partial_\beta X^\nu + \bar{\beta}^B_{\mu\nu} \varepsilon^{\alpha\beta} \partial_\alpha X^\mu \partial_\beta X^\nu + \bar{\beta}^\Phi \sqrt{g} R^{(2)} \, ,
\tag{1.6}
$$

where the *Weyl anomaly coefficients* $\bar{\beta}^G_{\mu\nu}$, $\bar{\beta}^B_{\mu\nu}$ and $\bar{\beta}^\Phi$ are some local functions of the fields $G_{\mu\nu}, B_{\mu\nu}$ and Φ. They 'almost' coincide with the NLSM renormalization group (RG) β-functions $\beta^G_{\mu\nu}$, $\beta^B_{\mu\nu}$ and β^Φ, respectively, up to the world-sheet total derivative terms representing the differences between integrated and local expressions [356, 357, 358, 43]. The NLSM RG β-functions are subject to the ambiguities caused by field reparametrizations (Chap. 2), whereas the Weyl anomaly coefficients are invariant.

The NLSM RG β-functions can be explicitly calculated in the lowest orders of α', by using conventional quantum perturbation theory for the 2d NLSM (see Chapters 2 and 3). For example, as regards the leading terms in the purely bosonic case, one finds [91, 355]

$$
\beta^G_{\mu\nu} = R_{\mu\nu} - H^2_{\mu\nu} + 2\nabla_\mu \nabla_\nu \Phi + O(\alpha') \, ,
$$

$$
\beta^B_{\mu\nu} = 2\nabla_\rho H^\rho_{\mu\nu} - 4(\nabla_\rho \Phi) H^\rho_{\mu\nu} + O(\alpha') \, ,
$$

$$\frac{\beta^\Phi}{\alpha'} = \frac{1}{\alpha'}\frac{(D-26)}{48\pi^2} + \frac{1}{16\pi^2}\left\{4(\nabla\Phi)^2 - 4\nabla^2\Phi - R + \frac{1}{3}H^2\right\} + O(\alpha') , \quad (1.7)$$

where the Ricci tensor $R_{\mu\nu}$ and the scalar curvature R defined in terms of the D-dimensional metric G, and the totally antisymmetric field strength $H_{\mu\nu\rho} \equiv \frac{3}{2}\partial_{[\mu}B_{\nu\rho]}$, have been introduced. We use the book-keeping definitions

$$H^2_{\mu\nu} \equiv H_{\mu\lambda\rho}H^{\lambda\rho}_\nu \quad \text{and} \quad H^2 \equiv H_{\mu\nu\rho}H^{\mu\nu\rho} . \quad (1.8)$$

The vanishing of all the RG β-functions of NLSM leads to the effective equations of motion for the background fields, which generalize the Einstein equations of general relativity. In their exact (to all orders in α') form, they are just the conditions of the conformal invariance of the NLSM (1.5).

The central charge of the NLSM-realized 2d CFT is of special interest. To relate it to the NLSM Weyl anomaly coefficients, let us first note that the constancy of the dilaton β-function follows from the vanishing of other RG β-functions, in the lowest order of α' [91]. This consistency condition follows from the fact that the conformal anomaly is essentially (up to a constant normalization factor) given by the dilaton $\bar{\beta}$-function. For example, one finds from (1.7), in the lowest order in α',

$$\begin{aligned} 0 &= \nabla^\mu\left[R_{\mu\nu} - H^2_{\mu\nu} + 2\nabla_\mu\nabla_\nu\Phi\right] \\ &= \nabla_\mu\left[-2(\nabla\Phi)^2 + 2\nabla^2\Phi + \frac{1}{2}R - \frac{1}{6}H^2\right] , \end{aligned} \quad (1.9)$$

where the Riemannian Bianchi identity has been used. To identify the (appropriately normalized) 'on-shell' dilaton $\bar{\beta}$-function with the central charge of the corresponding CFT, a similar consistency condition should be valid in general. The rigorous statement is known as the Curci-Paffuti relation [359],

$$\partial_\mu\bar{\beta}^\Phi = \mathcal{O}^G_\mu\bar{\beta}^G + \mathcal{O}^H_\mu\bar{\beta}^H , \quad (1.10)$$

where $\mathcal{O}^{G,H}_\mu$ are some field-dependent first-order *linear* differential operators, whose explicit form can be calculated in perturbation theory. The original proof of (1.10) in [359] is based on the detailed structure of NLSM renormalization, but it is rather complicated. To understand the origin of the result, a simple argument can be presented by considering the first two variations of the NLSM partition function Z with respect to the conformal factor ϕ in the conformal gauge (1.3) [349],

$$\frac{\delta\ln Z}{\delta\phi(z)} \propto \langle T_\alpha{}^\alpha(z)\rangle \propto \left\langle \bar{\beta}^\Phi(z)R^{(2)}(z)\right\rangle , \quad (1.11)$$

and, hence,

$$\frac{\delta^2\ln Z}{\delta\phi(z)\delta\phi(z')} \propto \left\langle \bar{\beta}^\Phi(z)\Box_z\delta(z - z')\right\rangle . \quad (1.12)$$

Because of the symmetry of the left-hand side of (1.12) with respect to the exchange $z \leftrightarrow z'$, one finds that

$$\left\langle \left[\bar{\beta}^{\Phi}(z) - \bar{\beta}^{\Phi}(z') \right] \Box_z \delta(z - z') \right\rangle = 0 \ . \tag{1.13}$$

Equation (1.13) seems to be enough to conclude that $\bar{\beta}^{\Phi}(z) = \bar{\beta}^{\Phi}(z')$ (see Subsect. 6.1.2 for yet another argument).

One gets additional insights into general features of the NLSM-based CFT from the partial results about the NLSM RG β-functions in (1.7), by questioning the existence of an action from which the NLSM β-functions follow as the equations of motion. This action is known in string theory as the *low-energy string effective action* (cf. Chaps. 2 and 3). Let us rewrite the on-shell equations (1.7) in the more conventional form used in general relativity [91],

$$0 = \beta^G_{\mu\nu} + 8\pi^2 g_{\mu\nu} \frac{\beta^{\Phi}}{\alpha'} = \left(R_{\mu\nu} - \tfrac{1}{2} G_{\mu\nu} R \right) - T^{\text{matter}}_{\mu\nu} \ ,$$

$$0 = \beta^B_{\mu\nu} = 2\nabla_{\rho} H^{\rho}_{\mu\nu} - 2(\nabla_{\rho}\Phi) H^{\rho}_{\mu\nu} \ ,$$

$$0 = 8\pi^2 \frac{\beta^{\Phi}}{\alpha'} + \tfrac{1}{2} G^{\mu\nu} \beta^G_{\mu\nu} = 2(\nabla\Phi)^2 - \nabla^2\Phi - \frac{1}{3} H^2 \ , \tag{1.14}$$

where the spacetime matter stress-energy tensor $T^{\text{matter}}_{\mu\nu}$ has been introduced:

$$T^{\text{matter}}_{\mu\nu} \equiv H^2_{\mu\nu} - \frac{1}{6} G_{\mu\nu} H^2 - 2\nabla_{\mu}\nabla_{\nu}\Phi + 2G_{\mu\nu}\nabla^2\Phi - 2G_{\mu\nu}(\nabla\Phi)^2 \ . \tag{1.15}$$

It is not difficult to verify that (1.14) do appear as the (Euler-Lagrange) equations of motion associated with the spacetime action [91]

$$I^{\text{eff}} \propto \int \mathrm{d}^D X \sqrt{-G} e^{-2\Phi} \left[-R + 4(\nabla\Phi)^2 + \frac{1}{3} H^2 \right] \ . \tag{1.16}$$

The exponential factor in (1.16) is a direct consequence of (1.5) and the definition of the quantum string effective action, since the zero-mode of the dilaton field is multiplied by the Euler characteristic $\chi(\Sigma)$ in (1.5), while $\chi = 2$ for the Riemann sphere. In general, contributions to the string effective action from the string world-sheets of genus h should by weighted by the factor $\exp[-2(1-h)\Phi]$. The action (1.16) takes the conventional form, after the Weyl transformation of the spacetime metric,

$$G_{\mu\nu} \to \exp\left[\frac{4\Phi}{D-2} \right] G_{\mu\nu} \ . \tag{1.17}$$

The equivalent action results in

$$I^{\text{eff}} \propto \int \mathrm{d}^D X \sqrt{-G} \left\{ -R + \frac{4}{D-2}(\nabla\Phi)^2 + \frac{1}{3} \exp\left[-\frac{8\Phi}{D-2} \right] H^2 \right\} \ , \tag{1.18}$$

while it yields the bosonic terms in the gravitational sector of the *Chapline-Manton action* describing 10-dimensional $N = 1$ supergravity interacting with $N = 1$ supersymmetric Yang-Mills matter in $D = 10$ [360]. This result is also relevant to 10-dimensional critical superstrings since 2d NLSM fermions do not contribute to the one-loop RG β-functions of the 2d supersymmetric NLSM.

6.1.2 Zamolodchikov's c-Theorem

The results of the preceeding subsection raise the question of whether the Weyl invariance conditions in a generic two-dimensional NLSM or, perhaps, in any 2d QFT, may always be put into the form of Euler-Lagrange equations associated with some action? The existence proof is usually referred to as *Zamolodchikov's c-theorem* [87]. The c-theorem is valid under the following basic assumptions: (i) reparametrizational invariance, (ii) renormalizability, and (iii) locality of the 2d QFT under consideration. The first condition implies the conservation law for the QFT stress-energy tensor, $\partial_\alpha(\sqrt{g}T^{\alpha\beta}) = 0$, while it also allows us to choose the conformal gauge. The second condition tells us that the stress-energy tensor is finite. The third condition implies the positivity of energy in QFT, while it is going to be represented by the positivity of the function $E(G)$ to be defined below.

The proof of the c-theorem goes as follows [87, 361]. Consider a 2d [1] reparametrization-invariant and renormalizable QFT described by an action $S = \int d^2z\,\mathcal{L}$, and let $\{G^i\}$ be the set of all its (renormalized) couplings, with β^i being the corresponding RG β-functions,

$$\sqrt{g}T_{\alpha\beta} \equiv \frac{\delta S}{\delta g^{\alpha\beta}} \ , \quad \beta^i \equiv \frac{dG^i}{dt} \ , \tag{1.19}$$

where t is the RG parameter. It should be emphasized that we are dealing with the 2d renormalizable QFT *away* from its critical (RG) fixed point. The QFT under consideration becomes a CFT at the RG fixed point G_0^i, where $\beta^i(G_0) = 0$. Using the conformal coordinates $(z, \bar{z})$ and the notation $T(z, \bar{z}) = \sqrt{g}\,T_{zz}$, $\Theta(z, \bar{z}) = \sqrt{g}\,T_{z\bar{z}}$, let us define the following real functions of the renormalized couplings $\{G\}$:

$$C(G) = 2z^4\,\langle T(z, \bar{z})T(0,0)\rangle|_{z\bar{z}=1} \ ,$$

$$D(G) = z^3\bar{z}\,\langle T(z, \bar{z})\Theta(0,0)\rangle|_{z\bar{z}=1} \ ,$$

$$E(G) = z^2\bar{z}^2\,\langle \Theta(z, \bar{z})\Theta(0,0)\rangle|_{z\bar{z}=1} \ , \tag{1.20}$$

where $z\bar{z} = 1$ is chosen to be the normalization 'point'. Because of the conservation and 'no-renormalization' of the stress-energy tensor ($\partial_i \equiv \partial/\partial G^i$),

---

[1] It is not known whether the c-theorem may be extended to dimensions higher than two [362].

$$\frac{\mathrm{d}T_{\alpha\beta}}{\mathrm{d}t} = \frac{\partial T_{\alpha\beta}}{\partial t} + \beta^i \partial_i T_{\alpha\beta} = 0 , \tag{1.21}$$

we have the relations

$$\tfrac{1}{2}\beta^i \partial_i C = 3D - \beta^i \partial_i D , \quad \beta^i \partial_i D - D = 2E - \beta^i \partial_i E . \tag{1.22}$$

Hence, the function $I(G)$ defined by

$$I(G) \equiv C(G) - 4D(G) - 6E(G) \tag{1.23}$$

satisfies the equation

$$\frac{\mathrm{d}I(G)}{\mathrm{d}t} = \beta^i \partial_i I(G) = -12E(G) . \tag{1.24}$$

Equation (1.24) implies that the function $I(G)$ monotonically decreases under the RG flow, since the function $E(G)$ is positive definite, according to its definition in (1.20). The analogue of (1.6) near the critical point takes the form

$$\Theta = \beta^i \Lambda_i , \tag{1.25}$$

and it is valid on-shell. The normally ordered composite operators Λ_i are defined from the QFT action as $\Lambda_i = \partial_i \mathcal{L}$. It follows that

$$E(G) = \beta^i \beta^j \mathcal{G}_{ij} , \tag{1.26}$$

in accordance with (1.20), where the symmetric and positive definite 'metric' $\mathcal{G}_{ij}$ in the space of couplings,

$$\mathcal{G}_{ij} = z^2 \, \bar{z}^2 \, \langle \Lambda_i(z,\bar{z})\Lambda_j(0,0)\rangle\big|_{z\bar{z}=1} , \tag{1.27}$$

has been introduced. Equation (1.24) now yields

$$\beta^i \left(\partial_i I + 12\mathcal{G}_{ij}\beta^j\right) = 0 . \tag{1.28}$$

Because of the positivity of the metric $\mathcal{G}_{ij}$, the stationarity of the Zamolodchikov action $I(G)$, $\partial_i I(G) = 0$, implies criticality, $\beta^i = 0$, according to (1.28). One thus gets the existence proof for the action itself. The critical value of the action $I(G)$ yields the central charge of the corresponding CFT. The criticality condition $\beta^i = 0$ does not, however, imply $\partial_i I = 0$. This would be true only under the existence of the *off-shell* relation

$$\partial_i I = K_{ij}\beta^j \tag{1.29}$$

with an invertible matrix K_{ij}. As was suggested in [87], the K-matrix may exist, and it may even be the same as the 'metric' $\mathcal{G}_{ij}$. However, the explicit perturbative calculations of Chap. 2 (see also [363, 364, 365]) result in the presence of field derivatives in the K-*operator*, which questions the very existence of the off-shell relation (1.29) beyond the one-loop NLSM level.

After taking into account the total derivative terms making a difference between the β's and $\bar{\beta}$'s in NLSM, a relation similar to (1.28) can be established in terms of the Weyl anomaly coefficients, by making appropriate changes in the meaning of the Λ_i-operators in (1.25): they have to be normally ordered finite operator products coming from the renormalization of the NLSM composites, $\Lambda_i \to [\Lambda_i]$. The Curci-Paffuti relation can also be justified on the basis of Zamolodchikov's c-theorem [361]. Let us introduce a new function,

$$H_i(G) = z^3 \bar{z} \, \langle T(z,\bar{z})[\Lambda_i]\rangle|_{z\bar{z}=1} \; , \tag{1.30}$$

so that $\bar{\beta}^i H_i = D$. The RG equation generically reads

$$\frac{\mathrm{d}[\Lambda_i]}{\mathrm{d}t} = (P_i{}^j - \partial_i\beta^j)[\Lambda_j] \equiv \Gamma_i{}^j[\Lambda_j] \; , \tag{1.31}$$

where the 'matrix' $P_i{}^j$ arises due to 2d total derivative terms. It follows that

$$\beta^k \partial_k H_i - \Gamma_i{}^k H_k - H_i = 2\mathcal{G}_{ik}\bar{\beta}^k + \Gamma_i{}^k \mathcal{G}_{kn}\bar{\beta}^n - \beta^j \partial_j \mathcal{G}_{in}\bar{\beta}^n - \mathcal{G}_{ij}\beta^n \partial_n \bar{\beta}^j \; , \tag{1.32}$$

and, hence,

$$\begin{aligned}
\beta^i \partial_i D + \beta^i \partial_i (H_j \beta^j) = {} & D + 2E - \beta^i \partial_i E \\
& + \left\{ \beta^i \partial_i \bar{\beta}^j - \bar{\beta}^i \partial_i \beta^j - \bar{\beta}^i P_i{}^j \right\} \left(H_j + \mathcal{G}_{jn}\bar{\beta}^n \right) \; .
\end{aligned} \tag{1.33}$$

Comparing (1.33) with (1.22) yields that the last term in (1.33) vanishes. This means that the coefficient in the curly brackets is zero. The latter just amounts to the Curci-Paffuti relation.

6.2 Conformally Invariant NLSM and Black Holes

The NLSM approach to string theory (Subsect. 6.1.1) poses conditions on the background spacetime geometry, which generalize the Einstein equations of gravity, while they reduce to them in the limit $\alpha' \to 0$. Unfortunately, the effective string equations are only available in the form of the α'-expansion, and merely a few leading terms of this expansion are explicitly known (Chaps. 2 and 3). Similarly to general relativity, where the black hole geometry arises as an *exact* solution to Einstein's equations, the *exact* solutions to the full effective string equations of motion are needed to discuss physical consequences of strings on top of general relativity. A perturbative solution to the string equations of motion, which has the causal structure of a (Schwarzschild) black hole, was found by Witten [366]. Though being two-dimensional, the Witten solution yields important insights into some 'stringy' effects in general relativity and (string) cosmology. Since strings are extended objects, test strings feel background geometry in a way that is different from the one felt by point particles. As a result, different (dual) point particle geometries may appear to be equivalent from the string viewpoint via dual symmetry relating equivalent CFT. This is known as *T-duality* of strings (Subsect. 6.2.3).

Witten [366] used the gauged 2d NLSM with the target space $G = SL(2, \mathbf{R})$ whose $H = U(1)$ subgroup is gauged (not to be confused with the G/H coset construction in Sect. 5.3). Any 2d WZNW model on a group manifold yields a solvable 2d CFT with an AKM symmetry (Sect. 5.1). To get a non-compactified time in the target NLSM space representing space-time, one needs a *non-compact* group G indeed, with $SO(1, 2)$ or, equivalently, $SL(2, \mathbf{R})$ being the simplest non-trivial choice. However, the minus sign in the bilinear product of the $sl(2, \mathbf{R})$ algebra leads to an indefinite Fock space, which is inconsistent with unitarity. To get a physical theory, it is natural to gauge a $U(1)$ symmetry subgroup of the WZNW model over $SL(2, \mathbf{R})$ [366]. The effective space $SL(2, \mathbf{R})/U(1)$ is of dimension two, while it can be identified with 2d spacetime having one time-like and one space-like direction. Since only the spacetime transverse modes of string excitations are physical, the only propagating mode in Witten's spacetime is given by a tachyon. Remarkably, the target space geometry of Witten's gauged NLSM has all the characteristic features of a black hole (Subsect. 6.2.1). It can be verified up to fourth order in α' [367] that the NLSM RG β-functions vanish for the $SL(2, \mathbf{R})/U(1)$ geometry. The exact (non-perturbative) description of Witten's black hole (Subsect. 6.2.2) was found in [368].

6.2.1 Witten's Black Hole

Since $SL(2, \mathbf{R})$ is a non-compact group, the unitarity conditions for the gauged model $SL(2, \mathbf{R})/U(1)$ are different from those of compact groups [45]. The current algebra obtained from the compact group symmetry of the WZNW action (Chap. 5) should, therefore, be modified.

Let us consider the $SO(2, 1)$ current (AKM) algebra whose generators $(J_m^{\pm}, J_m^3)$ obey the commutation relations

$$[J_n^+, J_m^-] = kn\delta_{n+m,0} - 2J_{n+m}^3 , \tag{2.1}$$

$$[J_n^3, J_m^{\pm}] = \pm J_{n+m}^{\pm} , \quad [J_n^3, J_m^3] = -\tfrac{1}{2}kn\delta_{n+m,0} .$$

The associated Sugawara-Sommerfeld (SS) stress tensor (Subsect. 5.1.2) reads

$$T(z) = \frac{1}{k-2}G_{ab} : J^a(z)J^b(z) : , \tag{2.2}$$

where $G_{ab} = \text{diag}(+, +, -)$ is the $SO(2, 1)$ metric, while the factor (-2) in the denominator stands for the $SO(2, 1)$ Casimir eigenvalue, $\varepsilon^{acd}\varepsilon_{cd}^{\ b} = -2G^{ab}$. The central charge of the Virasoro algebra associated with the SS stress tensor (2.2) is [369]

$$c = \frac{3k}{k-2} , \tag{2.3}$$

while it amounts to changing the sign in the expression (5.3.2) for the central charge of the $SU(2)$ current algebra of level k. In the non-compact case, the coefficient k at the WZ term in the WZNW action is not quantized, since the integration in the WZ term goes over the non-compact volume which does not imply the boundary conditions needed for topological quantization (cf. Subsect. 5.1.3).

Though the non-compact $SO(2,1)$ current algebra does not have any unitary representations, they exist for the gauged case, $SO(2,1)/U(1)$, at any real value of k satisfying $k > 2$ [369]. The negative norm states of $SO(2,1)$ are removed by gauging the $U(1)$ part of the $SO(2,1)$ current algebra. The central charge of the $SL(2,\mathbf{R})/U(1)$ gauged model is (cf. (5.3.11))

$$c = \frac{3k}{k-2} - 1 = 2 + \frac{6}{k-2} \; . \tag{2.4}$$

The critical value of $c = 26$ implies $k = 9/4 > 2$ in (2.4).

The ungauged $SL(2,\mathbf{R})$ WZNW action,

$$S[G] = \frac{k}{8\pi} \int_\Sigma \sqrt{g} g^{\alpha\beta} \mathrm{tr}\left(G^{-1}\partial_\alpha G G^{-1}\partial_\beta G\right) + ik\Gamma \; , \tag{2.5}$$

with the WZ term

$$\Gamma[G] = \frac{1}{12\pi} \int_{B,\,\partial B=\Sigma} \mathrm{tr}\left(G^{-1}\mathrm{d}G \wedge G^{-1}\mathrm{d}G \wedge G^{-1}\mathrm{d}G\right) \; , \tag{2.6}$$

has the rigid $SL(2,\mathbf{R}) \times SL(2,\mathbf{R})$ symmetry by construction (Sect. 5.1). The anomaly-free $U(1)$ subgroup generated by

$$\delta G = -\varepsilon(\Lambda G + G\Lambda) \; , \quad \Lambda = \begin{pmatrix} 0 & 1 \\ -1 & 0 \end{pmatrix} \in SL(2,\mathbf{R}) \; , \tag{2.7}$$

is gauged by introducing an Abelian gauge field A_α subject to the Abelian gauge transformation $\delta A_\alpha = \partial_\alpha \varepsilon$ (Subsect. 5.3.2). The gauged WZNW action (5.3.21) in complex coordinates $(z, \bar{z})$ can be rewritten in the form

$$S[G,A] = S[G] + \frac{k}{2\pi} \int \mathrm{d}^2 z \left\{ A_{\bar{z}}\mathrm{tr}(\Lambda G^{-1}\partial G) + A_z\mathrm{tr}(\Lambda\bar{\partial}GG^{-1}) \right.$$
$$\left. + A_z A_{\bar{z}} \left[-2 + \mathrm{tr}(\Lambda G\Lambda G^{-1}) \right] \right\} \; . \tag{2.8}$$

The gauge fixing can be conveniently done by gauging away one component of G, with the remaining one being of the form

$$G = \cosh r \begin{pmatrix} 1 & 0 \\ 0 & 1 \end{pmatrix} + \sinh r \begin{pmatrix} \cos\theta & \sin\theta \\ \sin\theta & -\cos\theta \end{pmatrix} \; , \tag{2.9}$$

in terms of the independent coordinates (r, θ). Substituting (2.9) into (2.8) and performing Gaussian integration over A-fields results in the NLSM action [366]

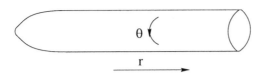

Fig. 6.1. Witten's black hole geometry in the coordinates (r, θ)

$$\tilde{S}(r,\theta) = \frac{k}{4\pi} \int d^2x \sqrt{g} g^{\alpha\beta} \left(\partial_\alpha r \partial_\beta r + \tanh^2 r \partial_\alpha \theta \partial_\beta \theta\right) , \qquad (2.10)$$

with the target space (NLSM) metric

$$ds^2 = dr^2 + \tanh^2 r d\theta^2 . \qquad (2.11)$$

For large r (or $\tanh r \to 1$), the NLSM manifold described by the metric (2.11) has the shape of a semi-infinite cigar (Fig. 6.1) that asymptotically approaches $R \times S^1$ for $r \to \infty$, the θ coordinate being periodic with period 2π. This is similar to a Euclidean black hole in D dimensions, which has the asymptotic topology $R^{D-1} \times S^1$ when $r \to \infty$.

Since the metric (2.11) is not Ricci-flat, the dilaton dynamics is important to achieve conformal invariance — see (1.7) and (1.14) — since there is no Kalb-Ramond field in the case under consideration. In accordance with (1.5), the action (2.10) should therefore be modified by the Fradkin-Tseytlin term,

$$S(r,\theta) = \tilde{S}(r,\theta) - \frac{1}{4\pi} \int d^2x \sqrt{g} R^{(2)} \Phi(r,\theta) . \qquad (2.12)$$

The solution to the dilaton equation of motion in the one-loop NLSM approximation — see the third line of (1.14) with the metric (2.11) — is given by [366]

$$\Phi(r,\theta) = \log \cosh r + \text{const.} \qquad (2.13)$$

Substituting the results (2.11) and (2.13) into the third line of (1.7) yields the central charge $c = 2 + 6/k$ which agrees with the exact formula (2.4) to leading order of $1/k$, as it should have been expected from the one-loop perturbation theory.

After the Wick rotation $\theta = it$, back to the Lorentzian signature, one gets from (2.11) the 2d spacetime metric

$$ds^2 = dr^2 - \tanh^2 r dt^2 . \qquad (2.14)$$

The coordinate transformation

$$v = + \tfrac{1}{2} \exp \left[r + \log(1 - e^{-2r}) + t \right] ,$$
$$u = - \tfrac{1}{2} \exp \left[r + \log(1 - e^{-2r}) - t \right] , \qquad (2.15)$$

yields [366]

$$ds^2 = -\frac{dudv}{1-uv} , \qquad \Phi = \log(1 - uv) . \qquad (2.16)$$

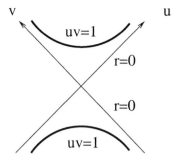

Fig. 6.2. Witten's black hole geometry in the coordinates (u, v). The curvature singularity occurs at $uv = 1$, while the event horizont is located at $uv = 0$

The 1+1 dimensional spacetime described by (2.16) has all the characteristic features of a (Schwarzschild) black hole, such as the curvature singularity at $uv = 1$ and the existence of a horizon at $uv = 0$ (Fig. 6.2).

There are also some important differences between the 2d black holes of Schwarzschild and Witten [370]. For example, Witten's black hole is in thermal equilibrium with matter, whereas the Schwarzschild black hole is not.

6.2.2 Exact String Black Hole Geometry

The exact form (to all orders in $1/k$) of the NLSM target space metric of the gauged WZNW model $SL(2, \mathbf{R})/U(1)$ can be obtained from the zero mode generator L_0 of the corresponding Virasoro algebra, after identifying L_0 with the Laplacian of the background metric coupled to the dilaton [368].

The WZNW action (2.5) can be parametrized as

$$G = e^{i\theta_L \sigma_2/2} e^{r\sigma_1/2} e^{i\theta_R \sigma_2/2} \in SL(2, \mathbf{R}) , \qquad (2.17)$$

in terms of the real coordinates (r, θ_L, θ_R), where $0 \leq r \leq \infty$, $0 \leq \theta_L < 2\pi$, $-2\pi \leq \theta_R < 2\pi$, and σ_i are Pauli matrices. The σ_2 generator can be identified with the generator of the gauge $U(1)$ subgroup, while the local gauge transformations are given by

$$\theta_L \to \theta_L + \phi_L , \quad \theta_R \to \theta_R + \phi_R , \quad A \to A + \partial\phi_L , \quad \bar{A} \to \bar{A} + \bar{\partial}\phi_R , \quad (2.18)$$

where $(\phi_L)^* = \phi_R$. The complete (BRST-invariant) gauge-fixed action associated with the gauged WZNW model (2.8), in the Lorentz gauge $\partial_\alpha A^\alpha = 0$, is given by [368]

$$S = \frac{k}{4\pi} \int d^2z \left(\bar{\partial}r\partial r - \bar{\partial}\theta_L\partial\theta_L - \bar{\partial}\theta_R\partial\theta_R - 2\cosh r \, \bar{\partial}\theta_L\partial\theta_R \right)$$

$$- \frac{k}{4\pi} \int d^2z \, \partial\phi\bar{\partial}\phi + \int d^2z \, (b\bar{\partial}c + \bar{b}\partial\bar{c}) , \qquad (2.19)$$

where the gauge slice has been parametrized by $A^\alpha = \varepsilon^{\alpha\beta}\partial_\beta\phi$ with $\phi = \phi_L - \phi_R$, and the chiral spin-one ghost system (b, c) [45] has been introduced.

It is straightforward (Sect. 5.1) to calculate the holomorphic $SL(2, \mathbf{R})$ currents out of the action (2.19). One finds [368]

$$J^3(z) = k(\partial\theta_L + \cosh r \partial\theta_R) , \quad J^\pm(z) = k e^{\pm i\theta_L}(\partial r \pm i \sinh r \partial\theta_R) , \quad (2.20)$$

which satisfy the $SL(2, \mathbf{R})$ current (AKM) algebra (2.1). The BRST-invariant stress tensor

$$T(z) = \frac{1}{k-2}G_{ab}J^a J^b + \frac{k}{4}(\partial\phi)^2 + b\partial c \qquad (2.21)$$

obeys the Virasoro algebra (5.1.27) with the central charge (2.4).

The BRST-invariant vertex operator $T_{mn}^l(r, \theta_L, \theta_R)$, creating the state $|l, w\rangle$ with $SL(2, \mathbf{R})$ 'isospin' l and the J^3-eigenvalue w, was calculated in [368],

$$T_{mn}^l(r, \theta_L, \theta_R) = \rho_{w_L w_R}^l(\cosh r) \exp(i w_L \theta_L + i w_R \theta_R) , \qquad (2.22)$$

where $\rho_{ww'}^l(x)$ are the Jacobi functions whose integral representation is given by

$$\rho_{mn}^l(\cosh r) = \oint \frac{dz}{2\pi i z} z^{m-l}\left(\cosh\frac{r}{2} + z\sinh\frac{r}{2}\right)^{l+n}\left(z\cosh\frac{r}{2} + \sinh\frac{r}{2}\right)^{l-n} .$$
$$(2.23)$$

The quantum numbers w_L and w_R are just the eigenvaules of J_0^3 and $\bar{J}_0^3$, respectively, while they are subject to the constraints $w_L + w_R = nk$ and $w_L - w_R = m$, where (m, n) are integers to be interpreted as the discrete momentum of the string in the θ direction and the winding number, respectively.

The zero modes of the $SL(2, \mathbf{R})$ current operators in the representation (2.22) read

$$J^3 = i\frac{\partial}{\partial\theta_L} , \quad J^\pm = e^{\pm i\theta_L}\left[\frac{\partial}{\partial r} \pm \frac{i}{\sinh r}\left(\frac{\partial}{\partial\theta_R} - \cosh r \frac{\partial}{\partial\theta_L}\right)\right] , \qquad (2.24)$$

whereas the zero (left-moving) mode L_0, defined according to (5.1.24) with the SS stress tensor (2.21) in the representation (2.22), is given by

$$L_0 = -\frac{2}{k-2}\left[\frac{\partial^2}{\partial r^2} + \coth r \frac{\partial}{\partial r}\right.$$
$$\left. + \frac{1}{\sinh^2 r}\left(\frac{\partial^2}{\partial\theta_L^2} - 2\cosh r\frac{\partial^2}{\partial\theta_L\partial\theta_R} + \frac{\partial^2}{\partial\theta_R^2}\right)\right] - \frac{1}{k}\frac{\partial^2}{\partial\theta_L^2} , \qquad (2.25)$$

and similarly for the right-moving zero mode $\bar{L}_0$ after interchanging θ_L and θ_R.

As was already mentioned in the beginning of this section, the only propagating string mode in the 2d black hole background is given by a tachyon.

The tachyonic state $|T(r, \theta_L, \theta_R)\rangle$ satisfies the standard (closed) string on-shell condition [170, 349]

$$(\bar{L}_0 - L_0)|T(r, \theta_L, \theta_R)\rangle = \left(\frac{\partial^2}{\partial\theta_R^2} - \frac{\partial^2}{\partial\theta_L^2}\right)|T(r, \theta_L, \theta_R)\rangle = 0 , \qquad (2.26)$$

which implies $T(r, \theta_L, \theta_R) = T(r, \theta) + \tilde{T}(r, \tilde{\theta})$ with $\theta = \frac{1}{2}(\theta_L + \theta_R)$ and $\tilde{\theta} = \frac{1}{2}(\theta_L - \theta_R)$. The tachyonic state $|T(r, \theta)\rangle$ representing the momentum mode is obviously the eigenstate of $(J^3 + \bar{J}^3)$ with zero eigenvalue, whereas the tachyonic state $\big|\tilde{T}(r, \tilde{\theta})\big\rangle$ representing the winding mode is the eigenstate of $(J^3 - \bar{J}^3)$ with zero eigenvalue. Both states are related by T-duality (Subsect. 6.2.3).

On the one hand, in the NLSM approach to bosonic string theory (Subsect. 6.1.1), the (2d) tachyonic effective action to be added to (1.16) reads

$$S_T = \int d^2x\, e^\Phi \sqrt{G}\left(G^{\mu\nu}\partial_\mu T\partial_\nu T - 2T^2\right) . \qquad (2.27)$$

On the other hand, the tachyonic action is fully determined by the zero-mode operator L_0 as

$$S_T = \int d^2x\, e^\Phi \sqrt{G}\left(TL_0T - 2T^2\right) . \qquad (2.28)$$

The consistency of (2.27) and (2.2.8) implies that L_0 of (2.25) should be identified with the NLSM Laplacian associated with (2.27), i.e.

$$L_0 = -\frac{1}{2e^\Phi\sqrt{G}}\partial_\mu e^\Phi \sqrt{G}G^{\mu\nu}\partial_\nu . \qquad (2.29)$$

This amounts to the statement that the NLSM of the form (1.5) is 2d conformally invariant, i.e. a CFT.

Being applied to the tachyonic field $T(r, \theta)$, (2.29) also determines the metric and the dilaton [368],

$$ds^2 = \frac{1}{2}(k - 2)\left[dr^2 + \beta^2(r)d\theta^2\right] , \qquad (2.30)$$

and

$$\Phi(r, \theta) = \log\left[\sinh\frac{r}{\beta(r)}\right] , \qquad (2.31)$$

respectively, where

$$\beta(r) = \frac{2}{\sqrt{\coth^2\frac{r}{2} - \frac{2}{k}}} . \qquad (2.32)$$

It is now straightforward to verify that the results of the preceeding subsection are reproduced to leading order (in $1/k$). Equations (2.30), (2.31) and 2.32) give the *exact* (2d black hole) solution to the background string equations of motion [368].

The *same* exact results are obtained from analysing the quantum LEEA of the gauged WZNW NLSM with $G/H = SL(2, \mathbf{R})/U(1)$ [371, 372]. As was demonstrated in Sect. 5.1, quantizing the 2d (ungauged) WZNW action (5.1.61) on a group manifold G essentially results in the shift of the WZ coefficient (level) k to $k + c_G$. This means that the LEEA of the WZNW model is proportional to the classical WZNW action with $k \to (k + c_G)$. As was also demonstrated in Subsect. 5.3.2, the gauged WZNW model on G/H is formally equivalent to the sum of two (ungauged) WZNW models, the one on the group G with level k and another one on the group H with level $-(k + c_H)$, modulo a ghost contribution — see (5.3.24) and (5.3.36). Accordingly, the LEEA of the gauged WZNW model on $SL(2, \mathbf{R})/U(1)$, in the parametrization (2.17) supplemented by $A = \partial\rho_L$ and $\bar{A} = \bar{\partial}\rho_R$, is given by [371, 372]

$$S(r, \theta, \tilde{\theta}, \rho, \tilde{\rho}) = \frac{(k-2)}{2\pi} \int d^2z \left[\tfrac{1}{2}\partial r \bar{\partial} r - a(r)\partial\kappa\bar{\partial}\kappa - b(r)\partial\tilde{k}\bar{\partial}\tilde{k} \right.$$

$$\left. + a(r)(\partial\kappa\bar{\partial}\tilde{\kappa} - \bar{\partial}\kappa\partial\tilde{\kappa}) + \frac{2k}{k-2}(\partial\kappa\bar{\partial}\kappa - 2\partial\kappa\bar{\partial}\theta + \partial\theta\bar{\partial}\theta) \right] ,$$

$$(2.33)$$

where we have used the notation $\theta = \tfrac{1}{2}(\theta_L - \theta_R)$, $\tilde{\theta} = \tfrac{1}{2}(\theta_L + \theta_R)$, and similarly for ρ and $\tilde{\rho}$ in terms of (ρ_L, ρ_R), and we have introduced the functions $\kappa = \theta + \rho$, $\tilde{\kappa} = \tilde{\theta} - \tilde{\rho}$, and

$$a(r) = \cosh r - 1 , \quad b(r) = \cosh r + 1 . \qquad (2.34)$$

The Gaussian integration over κ and $\tilde{\kappa}$ in (2.33), and an expansion of the result up to second order in the derivatives ∂ and $\bar{\partial}$, give rise to the exact generalization of the semiclassical result (2.10) in the form [371, 372]

$$S(r, \theta) = \frac{(k-2)}{4\pi} \int d^2z \left[\partial r \bar{\partial} r + f(r)\partial\theta\bar{\partial}\theta \right] , \qquad (2.35)$$

with the function

$$f(r) = \frac{a(r)\tanh^2\frac{r}{2}}{1 - \frac{2}{k}\tanh^2\frac{r}{2}} . \qquad (2.36)$$

The dilaton contribution of (2.12) in the effective action approach originates from the determinant coming from an integration over the gauge fields A and $\bar{A}$, while it has essentially the same form as in (2.13) [371, 372]. This confirms the results of [366, 368] from the different (Lagrangian) point of view.

6.2.3 T-duality of Strings and NLSM

Being extended objects, strings feel background spacetime geometry differently from point particles. In particular, strings naturally possess T-duality [373] having no analogue for point particles.

Consider a free closed string, whose motion is described by coordinates $X^\mu(\sigma, \tau)$ with the boundary condition $X^\mu(\sigma + 2\pi, \tau) = X^\mu(\sigma, \tau)$. Since $X^\mu(\sigma, \tau)$ satisfies a free 2d wave equation, it can be decomposed into left-moving and right-moving parts,

$$X^\mu(\sigma, \tau) = X_L^\mu(\sigma + \tau) + X_R^\mu(\sigma - \tau) \ , \tag{2.37}$$

whose mode expansions are given by

$$X_L^\mu(\sigma + \tau) = x_L^\mu + \sqrt{\frac{\alpha'}{2}} p_L^\mu(\sigma + \tau) + i\sqrt{\frac{\alpha'}{2}} \sum_{n \neq 0} \frac{1}{n} \alpha_n^\mu e^{-in(\sigma + \tau)} \ ,$$

$$X_R^\mu(\sigma - \tau) = x_R^\mu - \sqrt{\frac{\alpha'}{2}} p_R^\mu(\sigma - \tau) + i\sqrt{\frac{\alpha'}{2}} \sum_{n \neq 0} \frac{1}{n} \tilde{\alpha}_n^\mu e^{in(\sigma - \tau)} \ , \tag{2.38}$$

respectively. Accordingly, their canonically conjugated momenta are

$$P_\tau^\mu = \frac{1}{2\pi\sqrt{2\alpha'}} \left\{ p_L^\mu + p_R^\mu + \sum_{n \neq 0} \left(\tilde{\alpha}_n^\mu e^{in(\sigma - \tau)} + \alpha_n^\mu e^{-in(\sigma + \tau)} \right) \right\} \ ,$$

$$P_\sigma^\mu = \frac{1}{2\pi\sqrt{2\alpha'}} \left\{ p_L^\mu - p_R^\mu + \sum_{n \neq 0} \left(\alpha_n^\mu e^{-in(\sigma + \tau)} - \tilde{\alpha}_n^\mu e^{in(\sigma - \tau)} \right) \right\} \ . \tag{2.39}$$

Let us now assume that one of the string coordinates, say, $X(\sigma, \tau)$, is compactified on a circle, $X \approx X + 2\pi R$, with compactification radius R. Given a well-defined (scalar) field $\phi(X)$ on the circle, it should satisfy the condition $\phi(X) = \phi(X + 2\pi R)$. Having expanded the field $\phi(X)$ in Fourier series,

$$\phi(X) = \sum_{n \in Z} \phi_n e^{iPX} \ , \tag{2.40}$$

the single-valuedness of the field $\phi(X)$ leads to a quantization of the momentum P,

$$P = \frac{n}{R} \ , \quad n \in \mathbf{Z} \ . \tag{2.41}$$

The first line of (2.39) yields the total momentum

$$P = \int_0^{2\pi} d\sigma \, P_\tau = \frac{1}{\sqrt{2\alpha'}} (p_L + p_R) \ , \tag{2.42}$$

so that (2.41) and (2.42) together imply the quantization rule

$$p_L + p_R = \sqrt{2\alpha'} \frac{n}{R} \ . \tag{2.43}$$

More information can be obtained from the string one-loop vacuum amplitude, which amounts to considering a closed string of length $2\pi R$, propagating for a time lapse 2π. Compactifying the time coordinate, $\tau \approx \tau + 2\pi$, attributes a torus topology to the Euclidean string world-sheet. String theory should be invariant under global reparametrizations of the torus — this important symmetry is called *modular invariance*. Modular invariance implies an additional quantization rule on (p_L, p_R), which can be seen as follows [45, 373]. The translation operator on the Euclidean world-sheet is given by $\exp(2\pi i R P)$, whereas the similar (Euclidean) time translation operator is $\exp(-\theta H)$ where H is the Hamiltonian. Their product, $V(R, \theta) \sim \exp(2\pi i R P) \exp(-\theta H)$, is the locally defined vertex operator on the torus. Modular invariance means 'democracy' between any time directions on the torus. This implies that a string of length R, propagating for a time lapse $2\pi R$, should be indistinguishable from the string just considered above. As regards the vertex operator V, this amounts to the symmetry $R \leftrightarrow \alpha'/R$. When being applied to (2.39)–(2.43), it gives rise to the relation

$$p_L - p_R = \sqrt{\frac{2}{\alpha'}} m R , \quad m \in \mathbf{Z} . \tag{2.44}$$

Combining (2.43) and (2.44) results in

$$
\begin{aligned}
p_L &= \frac{1}{\sqrt{2}} \left(\frac{\sqrt{\alpha'}}{R} n + \frac{R}{\sqrt{\alpha'}} m \right) , \\
p_R &= \frac{1}{\sqrt{2}} \left(\frac{\sqrt{\alpha'}}{R} n - \frac{R}{\sqrt{\alpha'}} m \right) .
\end{aligned}
\tag{2.45}
$$

It is easy to verify that the zero modes, L_0 and $\bar{L}_0$, of the Virasoro algebra are invariant under the T-duality:

$$\frac{R}{\sqrt{\alpha'}} \leftrightarrow \frac{\sqrt{\alpha'}}{R} \quad \text{and} \quad m \leftrightarrow n . \tag{2.46}$$

In fact, the whole underlying CFT is invariant under the transformations (2.46). For example, let us consider the one-loop partition function of the compactified bosonic string,

$$Z = \int_\Gamma d^2\tau \, \hat{Z}(\tau, \bar{\tau}) \sum_{p_L, p_R} \text{Tr} \exp \left[i\pi\tau \bar{L}_0 - i\pi\bar{\tau} L_0 \right] , \tag{2.47}$$

where τ is the torus modular parameter, Γ is the fundamental region in the complex τ-plane, the trace (Tr) is taken over the space spanned by all string excitations $(\alpha, \tilde{\alpha})$, and $\hat{Z}(\tau, \bar{\tau})$ stands for the contribution of all string coordinates except of the compactified one. The modular transformations are given by the torus mapping class group $SL(2, \mathbf{Z})$, while this symmetry is manifest in (2.47).

T-duality essentially means an exchange of the string momentum modes with the string winding modes. Being extended object, a closed bosonic string can wrap about a circle, which makes T-duality possible. The T-dulaity transfromations (2.46) also imply the existence of a fundamental scale in string theory.

The simple considerations above about T-duality can be generalized to the strings described by NLSM [373]. In fact, the formal part of this story was already discussed in Sect. 2.5 where general NLSM (or string theory) actions with isometries were considered. The T-duality transformations of the spacetime metric and torsion potential are given by (2.5.4), whereas the dilaton transformation law is given by the shift described by (2.5.6). We conclude that the apparently different (T-dual) spacetime backgrounds related by (2.5.4) and (2.5.6) are equivalent in string theory [373]. This property of string theory clearly distinguishes it from general relativity. It also implies that, being non-invariant under T-duality, the formalism of Riemannian geometry is not fully adequate for string theory. In the context of Witten's black hole, where merely the 2d metric and dilaton are involved, the T-duality with respect to the exchange of momentum and winding string modes is manifest (Subsects. 6.2.1 and 6.2.2). In particular, the apparent singularity of the momentum string mode in the cigar-type geometry at $r = 0$ (Fig. 6.1) is absent in the dual trumpet-type geometry for the winding string mode [368].

6.3 Non-Chiral Fermionic Strings in Curved Spacetime

In this section we construct non-chiral $N = 1$ and $N = 2$ fermionic (spinning) string actions in a curved spacetime background (with torsion) [374]. This amounts to the construction of the *locally* $N = 1$ and $N = 2$ supersymmetric NLSM in 2d. Their perturbative anomaly structure is derived (Subsect. 6.3.3). The chiral (1,0) supersymmetric (heterotic) string models are considered in Chap. 7. Since $N = 4$ supersymmetry on the string worldsheet implies negative (or zero) critical dimension (in a flat background) and non-unitary amplitudes, it is not considered at all (see, however, [45]).

The standard component action of the $N = 1$ fermionic string, known as the Neveu-Schwarz-Ramond (NSR) model in *flat* spacetime [375, 376], was constructed in [377, 378]. A similar action for $N = 2$ strings was found in [379]. From the technical viewpoint, all those actions describe minimal coupling of 2d, N-extended supergravity to 2d, N-extended scalar matter. Their generalization to the case of matter self-interaction given by NLSM is presented below, in components. We use the Noether procedure in the $N = 1$ case (Subsect. 6.3.1), and $N = 2$ superconformal tensor calculus [380, 381] in the $N = 2$ case. See [382, 304] for the superspace description of fermionic strings.

6.3.1 $N = 1$ Neveu-Schwarz-Ramond NLSM

Our starting point is the 2d, $N = 1$ rigidly supersymmetric NLSM with torsion (Sect. 3.1). Let us rewrite its component action (3.1.18b) in the form [2]

$$
\begin{aligned}
I_0[\phi, \lambda] = \tfrac{1}{2} \int \mathrm{d}^2 x \Big[& G_{ij} \partial_a \phi^i \partial^a \phi^j + H_{ij} \varepsilon^{ab} \partial_a \phi^i \partial_b \phi^j + \mathrm{i} G_{ij} \bar{\lambda}^i \gamma^a (D_a \lambda)^j \\
& + \mathrm{i} H_{ijk} \bar{\lambda}^i \gamma_3 \gamma^a (\partial_a \phi^j) \lambda^k + \tfrac{1}{6} R_{ijkl} (\bar{\lambda}^i \lambda^k)(\bar{\lambda}^j \lambda^l) \\
& - \tfrac{1}{4} D_l H_{ijk} (\bar{\lambda}^i \gamma_3 \lambda^j)(\bar{\lambda}^k \lambda^l) - \tfrac{1}{4} H_{ijm} H^m{}_{kl} (\bar{\lambda}^i \gamma_3 \lambda^j)(\bar{\lambda}^k \gamma_3 \lambda^l) \Big] ,
\end{aligned}
$$
$$(3.1)$$

where the curvature R_{ijkl} and the covariant derivatives D are constructed in terms of the metric $G_{ij}(\phi)$, whereas $H_{ij}(\phi)$ is the potential of the torsion H_{ijk}. By construction, the action I_0 is invariant under 2d rigid supersymmetry transformations,

$$
\delta \phi^i = \bar{\varepsilon} \lambda^i , \quad \delta \lambda^i = -\mathrm{i}(\partial\!\!\!/ \phi^i)\varepsilon + \tfrac{1}{2}\varepsilon(\Gamma^i_{jk} \bar{\lambda}^j \lambda^k - H^i{}_{jk} \bar{\lambda}^j \gamma_3 \lambda^k) . \tag{3.2}
$$

We use the notation

$$
\eta^{ab} = \mathrm{diag}(+, -) , \quad \varepsilon^{01} = \varepsilon_{10} = 1 ,
$$

$$
\{\gamma^a, \gamma^b\} = 2\eta^{ab} , \quad \gamma^a \gamma^b = \eta^{ab} + \varepsilon^{ab} \gamma_3 , \quad \gamma_3 = \gamma^0 \gamma^1 . \tag{3.3}
$$

Latin lower-case letters denote flat 2d vector indices, Greek lower-case letters denote curved 2d vector indices, while they are related via the 'zweibein' $e^a_\mu(x)$,

$$
e^a_\mu e^{b\mu} = \eta^{ab} , \quad e^a_\mu e_{a\nu} = g_{\mu\nu}(x) . \tag{3.4}
$$

The matrices $\gamma^\mu \equiv e^\mu_a \gamma^a$ are x-dependent, while they satisfy the relation

$$
\gamma^\mu \gamma^\nu = g^{\mu\nu} + E^{\mu\nu} \gamma_3 , \tag{3.5}
$$

where $E^{\mu\nu} = e^{-1} \varepsilon^{\mu\nu}$ and $e = \det(e^a_\mu)$. Here are some useful identities:

$$
\varepsilon^{ab} \gamma_b = \gamma_3 \gamma^a , \quad \gamma^a \gamma^b \gamma^c = \gamma^c \gamma^b \gamma^a , \quad \gamma_a \gamma^b \gamma^a = 0 . \tag{3.6}
$$

Let us also rewrite the NLSM action (3.1) in the equivalent form, in terms of the 2d MW spinors $\lambda^i_\pm$,

$$
\begin{aligned}
I_0[\phi, \lambda_\pm] = \tfrac{1}{2} \int \mathrm{d}^2 x \Big[& G_{ij} \partial_a \phi^i \partial^a \phi^j + H_{ij} \varepsilon^{ab} \partial_a \phi^i \partial_b \phi^j + \mathrm{i} G_{ij} \bar{\lambda}^i_+ (\partial\!\!\!/^+ \lambda_+)^j \\
& + \mathrm{i} G_{ij} \bar{\lambda}^i_- (\partial\!\!\!/^- \lambda_-)^j + \tfrac{1}{4} R^+_{ijkl} (\bar{\lambda}^i_+ \gamma_a \lambda^j_+)(\bar{\lambda}^k_- \gamma^a \lambda^l_-) \Big] ,
\end{aligned}
$$
$$(3.7)$$

---

[2] We find it convenient to introduce in this section a new notation, which is different from the one used in Sect. 3.1, in order to accommodate the extra 2d supergravity structure.

with the supersymmetry transformation laws

$$\delta\phi^i = \delta_+\phi^i + \delta_-\phi^i = \bar{\varepsilon}_+\lambda^i_- + \bar{\varepsilon}_-\lambda^i_+ \, ,$$

$$\delta\lambda^i_\pm = -\mathrm{i}(\partial\!\!\!/\phi^i)\varepsilon_\mp - \Gamma_\pm{}^i{}_{jk}\lambda^j_\pm(\delta_\pm\phi^k) \, , \tag{3.8}$$

where we have introduced the notation

$$\Gamma_\pm{}^i{}_{jk} = \Gamma^i_{jk} \pm H^i{}_{jk} \, , \quad R^{\pm i}{}_{jkl} = R^i{}_{jkl}(\Gamma_\pm) \, , \tag{3.9a}$$

and

$$D^\pm_a\lambda^i_\pm = \partial_a\lambda^i_\pm + \Gamma_\pm{}^i{}_{jk}\lambda^j_\pm\partial_a\phi^k \, , \quad \lambda^i_\pm = \tfrac{1}{2}(1\pm\gamma_3)\lambda^i \, . \tag{3.9b}$$

Let us consider now a variation of the action, δI_0, with an x-dependent (local) infinitesimal parameter $\varepsilon(x)$, and then expand δI_0 in power series of λ. The fifth-order term $\delta I_0^{(5)}$ vanishes, like in the rigid case, since it does not contain $\partial_a\varepsilon$. Calculation of $\delta I_0^{(1)}$ yields

$$\delta I^{(1)} = \tfrac{1}{2}\int \mathrm{d}^2x \left[2(\partial_a\phi^i)\bar{\lambda}^j\gamma^b\gamma^a(\partial_b\varepsilon)G_{ij} + \partial_a\left\{(\partial^a\phi^i)(\bar{\varepsilon}\lambda^j)G_{ij}\right.\right.$$

$$\left.\left.+(\varepsilon^{ab}\partial_b\phi^i)(\bar{\varepsilon}\gamma_3\lambda^j)G_{ij} - 2(\varepsilon^{ab}\partial_b\phi^i)(\bar{\varepsilon}\lambda^j)H_{ij}\right\}\right] \, . \tag{3.10}$$

The calculation of $\delta I_0^{(3)}$ is more involved [374]. The following Fierz identities are useful here:

$$(\bar{\lambda}^i\lambda^k)(\bar{\varepsilon}\gamma_a\lambda^j) + (\bar{\varepsilon}\lambda^k)(\bar{\lambda}^i\gamma_a\lambda^j) + (\bar{\varepsilon}\lambda^i)(\bar{\lambda}^k\gamma_a\lambda^j) = 0 \, ,$$

$$(\bar{\lambda}^i\gamma_3\gamma^a\varepsilon)(\bar{\lambda}^k\lambda^l) - (\bar{\varepsilon}\lambda^k)(\bar{\lambda}^l\gamma_3\gamma^a\lambda^i) - (\bar{\varepsilon}\lambda^l)(\bar{\lambda}^k\gamma_3\gamma^a\lambda^i) = 0 \, ,$$

$$(\bar{\lambda}^i\gamma_3\lambda^j)(\bar{\lambda}^l\gamma^a\varepsilon) - (\bar{\lambda}^i\gamma^a\lambda^j)(\bar{\lambda}^l\gamma_3\varepsilon) + (\bar{\varepsilon}\lambda^i)(\bar{\lambda}^l\gamma^a\gamma_3\lambda^j)$$

$$-(\bar{\varepsilon}\lambda^j)(\bar{\lambda}^l\gamma^a\gamma_3\lambda^i) - (\bar{\varepsilon}\lambda^l)(\bar{\lambda}^i\gamma_3\gamma^a\lambda^j) = 0 \, . \tag{3.11}$$

By the use of (3.11) we find

$$\delta I_0^{(3)} = \tfrac{1}{2}\int \mathrm{d}^2x \left[-\tfrac{\mathrm{i}}{3}H_{ijk}(\bar{\lambda}^i\gamma_3\lambda^j)(\bar{\lambda}^k\partial\!\!\!/\varepsilon) - \tfrac{2\mathrm{i}}{3}H_{ijk}(\bar{\lambda}^i\gamma_3\gamma^a\lambda^j)(\bar{\lambda}^k\partial_a\varepsilon)\right.$$

$$\left.-\partial_a\left\{\tfrac{\mathrm{i}}{6}H_{ijk}(\bar{\lambda}^i\gamma_3\lambda^j)(\bar{\lambda}^k\gamma^a\varepsilon) + \tfrac{\mathrm{i}}{3}H_{ijk}(\bar{\lambda}^i\gamma_3\gamma^a\lambda^j)(\bar{\varepsilon}\lambda^k)\right\}\right] \, . \tag{3.12}$$

Equation (3.12) determines the Noether supercurrent, while the total derivative term may be ignored. This implies that, in local supersymmetry, the NLSM (3.1) has to be modified by the Noether term,

$$I_N = \tfrac{1}{2}\int \mathrm{d}^2x \left[2(\partial_a\phi^i)(\bar{\lambda}^j\gamma^b\gamma^a\psi_b)G_{ij} - \tfrac{\mathrm{i}}{3}H_{ijk}(\bar{\lambda}^i\gamma_3\gamma^a\lambda^j)(\bar{\lambda}^k\psi_a)\right.$$

$$\left.-\tfrac{\mathrm{i}}{3}H_{ijk}(\bar{\lambda}^i\gamma^a\lambda^j)(\bar{\lambda}^k\gamma_3\psi_a)\right] \, , \tag{3.13}$$

where the 2d Majorana gravitino field ψ_a has been introduced, whose 2d supersymmetry transformation law reads

$$\delta\psi_a = -\partial_a\varepsilon + \dots \; , \tag{3.14}$$

where the dots stand for higher-order corrections. In deriving (3.13) we have also used the identity

$$2H_{ijk}(\bar{\lambda}^i\gamma_3\gamma^a\lambda^j)(\bar{\lambda}^k\partial_a\varepsilon) + H_{ijk}(\bar{\lambda}^i\gamma_3\lambda^j)(\bar{\lambda}^k\slashed{\partial}\varepsilon)$$

$$= H_{ijk}(\bar{\lambda}^i\gamma_3\gamma^a\lambda^j)(\bar{\lambda}^k\partial_a\varepsilon) + H_{ijk}(\bar{\lambda}^i\gamma^a\lambda^j)(\bar{\lambda}^k\gamma_3\partial_a\varepsilon) \; . \tag{3.15}$$

The next step in the Noether procedure is a variation of the Noether term (3.13). This gives rise to the 2d energy-momentum tensor of the NLSM, and thus requires 2d gravity as yet another compensating field, along with the gravitino, in local 2d supersymmetry. Adding miminal interaction with 2d gravity into the sum of (3.7) and (3.13), together with the minimal covariantization of the supersymetry transformation laws with respect to the 2d supergravity fields, e^a_μ and ψ_a , turn out to be enough to establish the local invariance of the action, without adding any other (non-minimal) terms [374]. The final, locally supersymmetric action is given by [374]

$$I = \tfrac{1}{2}\int \mathrm{d}^2x \left[eg^{\mu\nu}G_{ij}\partial_\mu\phi^i\partial_\nu\phi^j + H_{ij}\varepsilon^{\mu\nu}\partial_\mu\phi^i\partial_\nu\phi^j + \mathrm{i}eG_{ij}\bar{\lambda}^i\gamma^\mu(D_\mu\lambda)^j \right.$$

$$+ \mathrm{i}eH_{ijk}\bar{\lambda}^i\gamma_3\gamma^\mu(\partial_\mu\phi^j)\lambda^k + \tfrac{e}{6}R_{ijkl}(\bar{\lambda}^i\lambda^k)(\bar{\lambda}^j\lambda^l)$$

$$- \tfrac{e}{4}D_lH_{ijk}(\bar{\lambda}^i\gamma_3\lambda^j)(\bar{\lambda}^k\lambda^l) - \tfrac{e}{4}H_{ijm}H^m{}_{kl}(\bar{\lambda}^i\gamma_3\lambda^j)(\bar{\lambda}^k\gamma_3\lambda^l)$$

$$+ 2e\left(\partial_\mu\phi^i + \tfrac{1}{2}\bar{\psi}_\mu\lambda^i\right)(\bar{\lambda}^j\gamma^\nu\gamma^\mu\psi_\nu)G_{ij}$$

$$\left. - \tfrac{\mathrm{i}e}{3}H_{ijk}(\bar{\lambda}^i\gamma_3\gamma^\mu\lambda^j)(\bar{\lambda}^k\psi_\mu) - \tfrac{\mathrm{i}e}{3}H_{ijk}(\bar{\lambda}^i\gamma^\mu\lambda^j)(\bar{\lambda}^k\gamma_3\psi_\mu)\right] \; . \tag{3.16}$$

The 2d local $N = 1$ supersymmetry transformation laws read

$$\delta\phi^i = \bar{\varepsilon}\lambda^i \; ,$$

$$\delta\lambda^i = -\mathrm{i}\left[\slashed{\partial}\phi^i + (\bar{\lambda}^i\psi_\nu)\gamma^\nu\right]\varepsilon + \tfrac{1}{2}\varepsilon\left[\Gamma^i_{jk}\bar{\lambda}^j\lambda^k - H^i{}_{jk}\bar{\lambda}^j\gamma_3\lambda^k\right] \; ,$$

$$\delta e^a_\mu = 2\mathrm{i}\bar{\varepsilon}\gamma^a\psi_\mu \; , \tag{3.17}$$

$$\delta\psi_\mu = -D_\mu\varepsilon \equiv -\left(\partial_\mu + \tfrac{1}{2}\omega_\mu\gamma_3\right)\varepsilon \; ,$$

where the 2d (Lorentzian) spin-connection ω_μ contains the gravitino-induced 2d torsion [377, 378],

$$\omega_\mu = \omega^{(0)}_\mu + K_\mu \; , \quad K_\mu = 2\mathrm{i}e\varepsilon_{\mu\nu}\bar{\psi}^\nu\gamma^\rho\psi_\rho \; ,$$

$$\omega_\mu^{(0)} = \tfrac{1}{2}\varepsilon^{ab}e_a^\nu \left[\partial_\mu e_{b\nu} - \partial_\nu e_{b\mu} + e_b^\sigma e_\mu^c \partial_\sigma e_{c\nu} \right] . \tag{3.18}$$

It is worth mentioning that the spin connection-dependent terms in the covariant derivative, which are present in the kinetic terms for the spinors λ in (3.16), drop out from the action due to the antisymmetry property of the Majorana spinor bilinears, $\bar{\lambda}^i\gamma^\mu\lambda^j = -\bar{\lambda}^j\gamma^\mu\lambda^i$.

The reason for the absence of non-miminal terms in the action (3.16), i.e. beyond the Noether term and minimal coupling, is related to the independence of the supersymmetry transformation laws (3.17) upon the (NLSM) matter fields. It is not difficult to verify that the action (1.16) and the transformation laws (3.17) reduce to the standard ones of the NSR model [377, 378] in the case of

$$G_{ij} = \delta_{ij} \quad \text{and} \quad H_{ijk} = 0 . \tag{3.19}$$

The action (1.16) is also automatically invariant under 2d general coordinate transformations with parameters $\xi_\mu(x)$,

$$\delta e_\mu^a = \xi^\lambda \partial_\lambda e_\mu^a , \quad \delta\psi_\mu = \xi^\lambda \partial_\lambda \psi_\mu + \psi_\lambda \partial_\mu \xi^\lambda ,$$

$$\delta\phi^i = \xi^\lambda \partial_\lambda \phi^i , \quad \delta\lambda^i = \xi^\lambda \partial_\lambda \lambda^i , \tag{3.20}$$

2d local Lorentz rotations with parameter $l(x)$,

$$\delta e_\mu^a = l\varepsilon_b^a e_\mu^b , \quad \delta\psi_\mu = -\tfrac{1}{2}l\gamma_3\psi_\mu , \quad \delta\lambda^i = -\tfrac{1}{2}l\gamma_3\lambda^i , \tag{3.21}$$

and local Weyl (scale) transformations with parameter $\Omega(x)$,

$$\delta e_\mu^a = \Omega e_\mu^a , \quad \delta\psi_\mu = \tfrac{1}{2}\Omega\psi_\mu , \quad \delta\lambda^i = -\tfrac{1}{2}\Omega\lambda^i , \tag{3.22}$$

as should have been expected. In fact, the action (3.16) also has 'accidental' superconformal invariance,

$$\delta\psi_\mu = \gamma_\mu\pi , \tag{3.23}$$

with spinor parameter $\pi(x)$, which can be verified by using Fierz identities in the terms containing the 2d torsion.

By the use of local symmetries, one can (locally) gauge away the whole 2d supergravity, i.e. set up a gauge

$$e_\mu^a = \delta_\mu^a , \quad \psi_\mu = 0 , \tag{3.24}$$

thus reproducing the original NLSM action (3.1) out of (3.16). In quantum theory, both the conformal symmetry (3.22) and the superconformal symmetry (3.23) are anomalous (Subsect. 6.3.3).

In terms of the MW spinors $\lambda_\pm$, the action (3.16) reads as follows:

$$I = \tfrac{1}{2} \int \mathrm{d}^2 x \, e \left[g^{\mu\nu} G_{ij} \partial_\mu \phi^i \partial_\nu \phi^j + e^{-1} H_{ij} \varepsilon^{\mu\nu} \partial_\mu \phi^i \partial_\nu \phi^j + i G_{ij} \bar{\lambda}^i_+ (\not{D}^+ \lambda_+)^j \right.$$

$$+ i G_{ij} \bar{\lambda}^i_- (\not{D}^- \lambda_-)^j + \tfrac{1}{4} R^+_{ijkl} (\bar{\lambda}^i_+ \gamma_\mu \lambda^j_+)(\bar{\lambda}^k_- \gamma^\mu \lambda^l_-)$$

$$+ 2 \left(\partial_\mu \phi^i + \tfrac{1}{2} \bar{\psi}_\mu \lambda^i_+ + \tfrac{1}{2} \bar{\psi}_\mu \lambda^i_- \right)$$

$$\times \left(\bar{\lambda}^j_+ \left[g^{\mu\nu} + e^{-1} \varepsilon^{\mu\nu} \right] \psi_\nu + \bar{\lambda}^j_- \left[g^{\mu\nu} - e^{-1} \varepsilon^{\mu\nu} \right] \psi_\nu \right) G_{ij}$$

$$+ \tfrac{2i}{3} H_{ijk} (\bar{\lambda}^i_+ \gamma^\mu \lambda^j_+)(\bar{\lambda}^k_+ \psi_\mu) - \tfrac{2i}{3} H_{ijk} (\bar{\lambda}^i_- \gamma^\mu \lambda^j_-)(\bar{\lambda}^k_- \psi_\mu) \right] ,$$

(3.25)

while the supersymmetry transformation laws (3.17) are given by

$$\delta e^a_\mu = 2i\bar{\varepsilon}\gamma^a \psi_\mu , \quad \delta \psi_\mu = -D_\mu \varepsilon ,$$

$$\delta \phi^i = \delta_+ \phi^i + \delta_- \phi^i = \bar{\varepsilon}_+ \lambda^i_- + \bar{\varepsilon}_- \lambda^i_+ ,$$

(3.26)

$$\delta \lambda^i_\pm = -i \left[\not{\partial} \phi^i + \left(\bar{\lambda}^i_+ \psi_\nu + \bar{\lambda}^i_- \psi_\nu \right) \gamma^\nu \right] \varepsilon_\mp - \Gamma_\pm{}^i{}_{jk} \lambda^j_\pm (\delta_\pm \phi^k) .$$

6.3.2 $N = 2$ Fermionic String NLSM

In this subsection we illustrate the (component) methods of *superconformal tensor calculus* in the case of off-shell non-chiral 2d, $N = 2$ supersymmetry [380, 381] by constructing the invariant NLSM action for an $N = 2$ fermionic string [374]. This amounts to a derivation of miminal coupling of 2d, $N = 2$ supergravity to the 2d, $N = 2$ NLSM (with torsion).

The $N = 2$ superconformal tensor calculus in 2d is based on the gauge theory for the superconformal algebra of $OSp(2|2) \times OSp(2|2)$. Some of its gauge field strengths (curvatures) are to be constrained further, in order to eliminate as much of the independent gauge fields as possible, without going on-shell. The superconformal algebra comprises eight bosonic (P_a, K_a, M, D, A, G) and eight fermionic (Q^i_α, S^i_α) generators, where A and G are chiral (internal symmetry) generators, M is a 2d Lorentz generator, D generates dilatations, P_a generates translations, K_a generates special conformal transformations, Q^i_α are $N = 2$ supersymmetry (Majorana) generators and S^i_α are $N = 2$ superconformal (Majorana) generators, $a = 0, 1$, $\alpha = 1, 2$, $i = 1, 2$. The proper constraints [380, 381] allow one to express the gauge fields ($f^a_\mu, b_\mu, \phi^i_{\alpha\mu}$) associated with the ($K_a, D, S^i_\alpha$) generators, respectively, in terms of the remaining gauge fields. See e.g., [380] for more details.

The relevant 2d, $N = 2$ off-shell matter supermultiplets are given by (i) a chiral (untwisted) $N = 2$ multiplet $\Sigma_\lambda = \{\phi, \chi, \mathcal{F}\}$, comprising the complex scalar ϕ of (Weyl) weight λ, the two-component complex spinor χ and the complex auxiliary field $\mathcal{F}$, (ii) a real vector supermultiplet $V_\lambda = \{\hat{C}, \zeta, \mathcal{H}, B_2, B_3, \boldsymbol{B}_m, \tilde{\lambda}, \hat{D}\}$ comprising the real scalars ($\hat{C}, B_2, B_3, \hat{D}$), the complex scalar $\mathcal{H}$, the two-component complex spinors ($\zeta, \tilde{\lambda}$) and the real

2d vector V_m. As regards $N = 2$ susy in 2d, there exists yet another super-multiplet, $\Sigma_\lambda^{\mathrm{T}} = \{\pi, \beta, G\}$, known as the *twisted* chiral multiplet [203]. It has the same components as the untwisted multiplet, while their Q- and S- transformation laws are obtained from the untwisted ones after the substitution

$$\varepsilon_Q \to P_-\varepsilon_Q + P_+\varepsilon_Q^c = \varepsilon_{Q-} + \varepsilon_{Q+}^c \ ,$$

$$\varepsilon_S \to P_+\varepsilon_S + P_-\varepsilon_S^c = \varepsilon_{S+} + \varepsilon_{S-}^c \ , \tag{3.27}$$

where ε_Q and ε_S are the parameters of local Q and S supersymmetry, respectively, $P_\pm = \frac{1}{2}(1 \pm \gamma_3)$ are chiral 2d projectors, and ε^c is the Majorana conjugated spinor. The gauge transformations of the twisted chiral multiplet components are given by [374]

$$\delta\pi = \varepsilon^a \partial_a \pi + \tfrac{i}{2}\lambda\pi\alpha + \lambda\pi\lambda_D + i\overline{(\varepsilon_{Q-} + \varepsilon_{Q+}^c)}\beta \ ,$$

$$\delta\beta = \varepsilon^a \partial_a \beta + \lambda_M \sigma_{01}\beta + \tfrac{i}{4}\gamma_3\beta\lambda_A + \tfrac{i}{2}(\lambda - 1/2)\beta\alpha + (\lambda + 1/2)\beta\lambda_D$$

$$+ \lambda\pi(\varepsilon_{S+} + \varepsilon_{S-}^c) + \tfrac{1}{2}\slashed{\partial}\pi(\varepsilon_{Q-} + \varepsilon_{Q+}^c) + \tfrac{1}{2}(\varepsilon_{Q-}^c + \varepsilon_{Q+})G \ , \tag{3.28}$$

$$\delta G = \varepsilon^a \partial_a G + \tfrac{i}{2}(\lambda - 1)G\alpha + (\lambda + 1)G\lambda_D$$

$$+ i\overline{(\varepsilon_{Q+} + \varepsilon_{Q-}^c)}\slashed{\partial}\beta - 2i\lambda\overline{(\varepsilon_{S-} + \varepsilon_{S+}^c)}\beta \ ,$$

where $\varepsilon^a, \lambda_M, \lambda_D, \lambda_A, \alpha, \varepsilon_G, \varepsilon_S$ are the parameters of general coordinate, Lorentz, Weyl, chiral, A-internal, Q-susy, and S-susy, respectively. The full form of these transformations is obtained after replacing the flat derivatives in (3.28) by the covariant ones. An off-shell closure of the transformation algebra on the twisted chiral multiplet is a consequence of its known closure [380] on the untwisted chiral multiplet. In other words, a twisted chiral multiplet is equivalent to the untwisted one. However, if both types of chiral multiplets are present in the NLSM action where they interact with each other, it results in some new geometrical features of the NLSM, e.g., a non-vanishing torsion (see below).

The $N = 2$ superconformal tensor calculus [380, 381] gives us the invariant 'action' for a vector multiplet of Weyl weight λ, whose Lagrangian 'density' reads [3]

$$\mathcal{L} = e\left[-\hat{D} + \Box\hat{C} + i\psi \cdot \gamma(\tilde{\lambda} - \slashed{D}\zeta) + i\mathcal{H}(\bar{\psi}_\mu \sigma^{\mu\nu}\psi_\nu^c) + \text{h.c.}\right] \ , \tag{3.29}$$

where the superconformal generalization $\Box$ of the standard 2d wave operator $\Box$ can be obtained from inspecting the transformation properties of $eg^{\mu\nu}D_\nu\hat{C}$ [380]. We find [374]

$$\Box\hat{C} = e^{-1}\partial_\mu(eg^{\mu\nu}D_\nu\hat{C}) - \lambda b^\mu(D_\mu\hat{C}) + \tfrac{1}{2}\lambda\hat{C}R + i\lambda\hat{C}\left(\bar{\psi}_\mu\sigma^{\mu\nu}\phi_\nu + \right.$$

$$+ \bar{\phi}_\mu\sigma^{\mu\nu}\psi_\nu) - \tfrac{1}{2}\left[i(\bar{\phi}\cdot\gamma\zeta) - i(\bar{\psi}_\mu D^\mu\zeta) - i(\bar{\psi}_\mu\sigma^{\mu\nu}\slashed{D}\hat{C}\psi_\nu) + \text{h.c.}\right] \tag{3.30}$$

---

[3] This local generalization of the D-term is obtained by the Noether procedure.

Let us consider some $N = 2$ untwisted chiral multiplets, $\Sigma_i = \{\phi_i, \chi_i, \mathcal{F}_i\}_{\lambda=0}$, where $i = 1, 2, \ldots, n$, as well as some $N = 2$ twisted chiral multiplets, $\Sigma_i^{\mathrm{T}} = \{\pi_p, \beta_p, G_p\}_{\lambda=0}$, where $p = 1, 2, \ldots, m$. Let $K(\phi, \phi^*, \pi, \pi^*)$ be a differentiable function that can be considered as the 2d, $N = 2$ counterpart to the 4d, $N = 1$ Kähler potential (Subsect. 3.4.2). To construct the locally $N = 2$ supersymmetric NLSM action, we define a vector multiplet V whose first component is given by the function K, i.e. $\hat{C} = K(\phi, \phi^*, \pi, \pi^*)$. The other components of the vector multiplet V can be found by using the known (and model-independent) Q-supersymmetry transformation laws of the vector multiplet components, which are provided by superconformal tensor calculus [380, 381]. The density (3.29), in terms of the components of $V = K$, yields the NLSM Lagrangian that we are looking for.

For example, we have

$$\delta K = \frac{\partial K}{\partial \phi_i} \mathrm{i}(\bar{\varepsilon}\chi_i) + \frac{\partial K}{\partial \pi_p} \mathrm{i}(\bar{\varepsilon}P_+\beta_P) + \frac{\partial K}{\partial \pi_q^*} \mathrm{i}(\bar{\varepsilon}P_-\beta_q^c) + \ldots = -\frac{\mathrm{i}}{2}\bar{\varepsilon}\zeta + \ldots , \quad (3.31)$$

and, hence,

$$\zeta = -2\frac{\partial K}{\partial \phi_i}\chi_i - 2\frac{\partial K}{\partial \pi_p}P_+\beta_P - 2\frac{\partial K}{\partial \pi_q^*}P_-\beta_q^c . \quad (3.32)$$

It is straightforward (though lengthy) to calculate the remaining components $(B_2, B_3, \hat{B}_m, \mathcal{H}, \tilde{\lambda}, \hat{C})$ of $V = K$. To write down the final action in compact form, we introduce the notation

$$\begin{aligned}
\phi_+^A &\equiv (\phi_i, \pi_p) , \quad \phi_-^A \equiv (\phi_i, \pi_p^*) , \\
\bar{\phi}_+^A &\equiv (\phi_i^*, \pi_q^*) , \quad \bar{\phi}_-^A \equiv (\phi_i^*, \pi_p) , \\
\lambda_+^A &\equiv \{\chi_{i+}, \beta_{p+}\} , \quad \lambda_-^A \equiv \{\chi_{i-}, \beta_{p-}^c\} ; \\
\bar{\lambda}_+^A &\equiv \{\bar{\chi}_{i+}, \bar{\beta}_{p+}\} , \quad \bar{\lambda}_-^A \equiv \{\bar{\chi}_{i-}, \bar{\beta}_{p-}^c\} ,
\end{aligned} \quad (3.33)$$

where, by definition,

$$\bar{\phi}^A \equiv \phi^{\bar{A}} , \quad \text{and} \quad \bar{\lambda}^A \equiv \lambda^{\bar{A}} . \quad (3.34)$$

Note that this notation doubles the number of independent scalar fields.

After eliminating the auxiliary fields $\mathcal{F}$ and G, the invariant $N = 2$ string NLSM action takes the form [374]

$$\begin{aligned}
I = \int \mathrm{d}^2 x \, e \Big\{ &g_{\bar{A}B}\partial_\mu \phi_+^{\bar{A}}\partial^\mu \phi_+^B + h_{\bar{A}B}E^{\mu\nu}\partial_\mu \phi_+^{\bar{A}}\partial_\nu \phi_+^B + \Big[\mathrm{i}g_{\bar{A}B}\bar{\lambda}_+^{\bar{A}}(\hat{\slashed{D}}^{\mathrm{cov}}\lambda_+)^B \\
&+ \mathrm{i}g_{\bar{A}B}\bar{\lambda}_-^{\bar{A}}(\hat{\slashed{D}}^{\mathrm{cov}}\lambda_-)^B - \frac{\mathrm{i}}{2}g_{\bar{A}B}\partial_\nu \phi_+^{\bar{A}}(\bar{\psi}_\mu \gamma^\nu \gamma^\mu \lambda_+^B) \\
&- \frac{\mathrm{i}}{2}g_{\bar{A}B}\partial_\nu \phi_-^{\bar{A}}(\bar{\psi}_\mu \gamma^\nu \gamma^\mu \lambda_-^B) + \mathrm{h.c.}\Big] + 2\hat{R}_{\bar{A}B\bar{C}D}(\lambda_+^{\bar{A}}\gamma^\mu \lambda_+^B)(\lambda_-^{\bar{C}}\gamma_\mu \lambda_-^D) \\
&+ \Big[(\lambda_+^{\bar{A}}\gamma_\nu \lambda_+^B)(\bar{\psi}_\mu \gamma^\nu \gamma^\mu \lambda_+^C)H_{\bar{A},BC}^+ + (\lambda_-^{\bar{A}}\gamma_\nu \lambda_-^B)(\bar{\psi}_\mu \gamma^\nu \gamma^\mu \lambda_-^C)H_{\bar{A},BC}^- + \mathrm{h.c.}\Big] \Big\}
\end{aligned} \quad (3.35)$$

where we have introduced the metric

$$g_{\bar{A}B} = \begin{pmatrix} \frac{\partial^2 K}{\partial \phi^* \partial \phi} & 0 \\ 0 & -\frac{\partial^2 K}{\partial \pi^* \partial \pi} \end{pmatrix} \; , \tag{1.36a}$$

and the torsion potentials

$$h_{\bar{A}B} = \begin{pmatrix} 0 & -\frac{\partial^2 K}{\partial \phi^* \partial \pi} \\ \frac{\partial^2 K}{\partial \pi^* \partial \phi} & 0 \end{pmatrix} \; , \quad \text{or} \quad \tilde{h}_{\bar{A}B} = \begin{pmatrix} 0 & \frac{\partial^2 K}{\partial \phi^* \partial \pi^*} \\ -\frac{\partial^2 K}{\partial \pi \partial \phi} & 0 \end{pmatrix} \; . \tag{3.36b}$$

In deriving these equations we have also used the identities

$$\begin{aligned} g_{\bar{A}B} \partial_\mu \phi_+^{\bar{A}} \partial^\mu \phi_+^B &= g_{\bar{A}B} \partial_\mu \phi_-^{\bar{A}} \partial^\mu \phi_-^B \; , \\ h_{\bar{A}B} \varepsilon^{\mu\nu} \partial_\mu \phi_+^{\bar{A}} \partial_\nu \phi_+^B &\approx \tilde{h}_{\bar{A}B} \varepsilon^{\mu\nu} \partial_\mu \phi_-^{\bar{A}} \partial_\nu \phi_-^B \; . \end{aligned} \tag{3.37}$$

with the latter being valid up to a total derivative.

The curvature and torsion in (3.35) are (cf. [196])

$$H^+_{\bar{A},BC} = \frac{1}{2}\left(\frac{\partial h_{\bar{A}C}}{\partial \phi_+^B} - \frac{\partial h_{\bar{A}B}}{\partial \phi_+^C} \right) \; , \quad H^-_{\bar{A},BC} = \frac{1}{2}\left(\frac{\partial h_{\bar{A}C}}{\partial \phi_-^B} - \frac{\partial h_{\bar{A}B}}{\partial \phi_-^C} \right) \; ,$$

$$\begin{aligned} \hat{R}_{\bar{A}B\bar{C}D} =& \frac{\partial^4 K}{\partial \phi_+^{\bar{A}} \partial \phi_+^B \partial \phi_-^{\bar{C}} \partial \phi_-^D} - g^{j^*i} \frac{\partial^3 K}{\partial \phi_j^* \partial \phi_+^B \partial \phi_-^D} \cdot \frac{\partial^3 K}{\partial \phi_i \partial \phi_+^{\bar{A}} \partial \phi_-^{\bar{C}}} \\ &+ g^{pq^*} \frac{\partial^3 K}{\partial \pi_q^* \partial \phi_-^{\bar{C}} \partial \phi_+^B} \cdot \frac{\partial^3 K}{\partial \pi_p \partial \phi_+^{\bar{A}} \partial \phi_-^{\bar{D}}} \; . \end{aligned} \tag{3.38}$$

The covariant derivatives read [374]

$$(\hat{D}_\mu{}^{\text{cov}} \lambda_+)^B = D_\mu \lambda_+^B - \tfrac{1}{2} P_+ \!\!\not{D} \phi_+^B \psi_\mu + \Gamma^B_{CD(+)} \partial_\mu \phi_-^C \lambda_+^D \; , \tag{3.39}$$

where D_μ is the standard (spinor) covariant derivative [139], while the connection (with torsion) is given by

$$\Gamma^B_{CD(+)} = g^{B\bar{A}} \Gamma_{CD(+),\bar{A}} \; , \quad \Gamma_{CD(+),\bar{A}} = \frac{\partial^3 K}{\partial \phi^C \partial \phi_+^D \partial \phi_+^{\bar{A}}} - \frac{\partial h_{\bar{A}D}}{\partial \phi_-^C} \; . \tag{3.40}$$

Similarly, we find

$$(\hat{D}_\mu{}^{\text{cov}} \lambda_-)^B = D_\mu \lambda_-^B - \tfrac{1}{2} P_- \!\!\not{D} \phi_-^B \psi_\mu + \Gamma^B_{CD(-)} \partial_\mu \phi_-^C \lambda_-^D \; , \tag{3.41}$$

where

$$\Gamma^B_{CD(-)} = g^{B\bar{A}} \Gamma_{CD(-),\bar{A}} \; , \quad \Gamma_{CD(-),\bar{A}} = \frac{\partial^3 K}{\partial \phi_-^{\bar{A}} \partial \phi_+^D \partial \phi^C} + \frac{\partial \tilde{h}_{\bar{A}D}}{\partial \phi_+^C} \; . \tag{3.42}$$

The NLSM action (3.35) is invariant under *all* local 2d, $N = 2$ superconformal symmetries by construction. Therefore, there exists a classical gauge where all the $N = 2$ supergravity gauge fields are absent. In this (superconformal) gauge, the $N = 2$ string NLSM action (3.35) reduces to the rigidly $N = 2$ supersymmetric 2d NLSM (with torsion, or a WZ-type term) of [203]. In quantum theory, some of the superconformal symmetries are anomalous (Subsect. 6.3.3).

The gauge-fixed 2d, $N = 2$ NLSM [203] has the *generalized* Kähler gauge invariance under the local transformations

$$\delta K = \Lambda(\phi, \pi) + \Omega(\phi, \pi^*) + \text{h.c.} \tag{3.43}$$

of the generalized Kähler potential $K(\phi, \phi^*, \pi, \pi^*)$, with arbitrary functions Λ and Ω of their arguments. As is well known in four dimensions [383], given a Kähler gauge *dependence* in the 4d, $N = 1$ NLSM minimally coupled to 4d, $N = 1$ supergavity, there exists a mechanism of compensation in supergravity that gives rise to a topological quantization of the gravitational (Newton's) coupling constant. For instance, the Kähler gauge dependence may be compensated by local chiral transformations in 4d, $N = 1$ supergravity [383]. At first sight, the Kähler gauge dependence is also present in our 2d, locally $N = 2$ supersymmetric NLSM (3.35) where it originates from two sources, the Q-supercovariant derivative $\hat{D}_\mu$ and the contorsion tensor (bilinear in the gravitino field) in the gravitational spin-connection,

$$\omega_\mu(e_\lambda^a, b_\nu, \psi_\rho) = \omega_\mu(e_\lambda^a, b_\nu) + \tfrac{i}{2}\varepsilon_\mu^\nu \left[(\bar\psi_\nu \gamma \cdot \psi) - (\bar\psi \cdot \gamma\psi_\nu) \right] . \tag{3.44}$$

In fact, all those Kähler-dependent terms in the action (3.35) turn out to be proportional to

$$\tilde{h}_{\bar A B}(\lambda_-^{\bar A}\psi_\mu)(\bar\psi_\nu\lambda_-^B)E^{\mu\nu} + h_{\bar A B}(\lambda_+^{\bar A}\psi_\mu)(\bar\psi_\nu\lambda_+^B)E^{\mu\nu}$$

$$= \tfrac{1}{2}E^{\mu\nu}(\bar\psi_\mu\gamma^\rho\psi_\nu)\left[\frac{\partial^2 K}{\partial\phi_i\partial\pi_q^*}(\bar\beta_{q+}\gamma_\rho\chi_{i+}) - \frac{\partial^2 K}{\partial\phi_i\partial\pi_p}(\bar\beta_{p-}^c\gamma_\rho\chi_{i-}) - \text{h.c.} \right] , \tag{3.45}$$

while they mutually cancel each other [374]. This means the Kähler gauge independence of the $N = 2$ fermionic string NLSM, as well as the absence of topological quantization of 2d gravity coupling.

6.3.3 Anomalies of Bosonic and Fermionic Strings

The N-extended fermionic string actions in curved spacetime background, which were constructed in Subsects. 6.3.1 and 6.3.2, can be interpreted as 2d, N-extended supersymmetric NLSMs (with torsion) coupled to 2d, N-extended supergravity. As was demonstrated in the preceeding subsections, these classical NLSM actions are, in fact, invariant under the full 2d, N-extended superconformal symmetry. After quantization, the superconformal symmetry is generically broken, while its breaking implies that some of the 2d supergravity fields become propagating. If this truly happens, it clearly destroys the consistency of the string actions whose fundamental (reparametrizational and superconformal) invariances are to be preserved.

The off-shell (closed bosonic) string (Euclidean) effective action of the massless string modes reads [355]

$$\Gamma[G, H, \ldots] = \sum_\chi e^{\bar\sigma \chi} \int [dg_{\mu\nu}][d\phi_i] e^{-I} , \qquad (3.46a)$$

where I is the bosonic string NLSM action,

$$I = \frac{1}{2\pi\alpha'} \int d^2x \left\{ \tfrac{1}{2}\sqrt{g} g^{\mu\nu} \partial_\mu \phi^i \partial_\nu \phi^j G_{ij}(\phi) + \tfrac{1}{2}\varepsilon^{\mu\nu} \partial_\mu \phi^i \partial_\nu \phi^j H_{ij}(\phi) + \ldots \right\} ,$$

$$(3.46b)$$

while the dots stand for all higher-derivative terms consistent with the fundamental symmetries (we ignore the dilaton contribution for simplicity). The sum over the Euler characteristics χ of the string world-sheet is explicitly written in (3.46a), whereas we restrict ourselves to the tree level calculations ($\chi = 2$) below.

The derivation of Γ can be naturally divided into two tasks: first, one computes the LEEA of the 2d NLSM (with torsion or a generalized WZNW term) on a curved string world-sheet of given topology χ and 2d metric $g_{\mu\nu}$, and, second, one 'averages' the LEEA over all 2d topologies and 2d metrics. The first problem reduces to the computation of the LEEA in the NLSM background-field approach (Chap. 2),

$$\exp\{-W[G, H.g]\} = \int [D\xi] \exp\{-I[\phi + \eta(\xi), g]\} , \qquad (3.47)$$

where ξ is the NLSM quantum field (Sect. 2.2). The general structure of W is dictated by dimensional reasons and symmetries,

$$W = -\frac{1}{\varepsilon}\beta \int d^2x \sqrt{g} R(x) + \gamma \int d^2x d^2x' (R\sqrt{g})_x \Box_{xx'}^{-1} (R\sqrt{g})_{x'} , \qquad (3.48)$$

where we have used dimensional regularization, $\varepsilon = 2 - d$, and we have introduced the Green function of the covariant Laplacian, $\Box$, and $R = R_{\lambda\mu}{}^{\mu\lambda}$. The dimensionless (anomaly) coefficients β and γ are actually proportional to a single Silly coefficient [355] so that $\beta = 4\gamma$, while both can also be calculated in NLSM quantum perturbation theory (in α' series). The second term of (3.48) is the conformal (or Weyl) anomaly. In this subsection, we calculate the leading contributions to the conformal anomaly for $N = (0,0)$, $N = (1,1)$ and $N = (2,2)$ fermionic strings [374].

When using the conformally flat 2d metric, $g_{\mu\nu} = e^{2\sigma}\delta_{\mu\nu}$, we have

$$g^{\mu\nu} = e^{-2\sigma}\delta_{\mu\nu} , \quad \sqrt{g} = e^{2\sigma} , \quad R = -2e^{-2\sigma}\Box\sigma . \qquad (3.49)$$

In d dimensions, with $g_{\mu\nu} = \delta_{\mu\nu} + h_{\mu\nu}$, we have

$$\sqrt{g}g^{\mu\nu} = \left(1 + \tfrac{1}{2}h + \tfrac{1}{8}h^2 - \tfrac{1}{4}h_{\rho\lambda}^2\right) - h_{\mu\nu}\left(1 + \tfrac{1}{2}h\right) + h_{\mu\lambda}h_{\lambda\nu} \qquad (3.50)$$

up to second order in the h fields. Hence, given $h_{\mu\nu} = 2\sigma\delta_{\mu\nu}$ and $h \equiv h_{\mu\mu} = 2\sigma d$, we find (in the same order)

$$\sqrt{g}g^{\mu\nu} = \left\{1 + \varepsilon\sigma(x) + \frac{1}{2}\varepsilon(d-4)\sigma^2(x)\right\}\delta_{\mu\nu} . \qquad (3.51)$$

Similarly, we wind for the 'zweibein' that

$$e^{1/2}e_a^\mu = \left(1 + \tfrac{1}{2}\varepsilon\sigma\right)\delta_{a\mu} , \quad e^{-1/2}e_{a\mu} = \left(1 - \tfrac{1}{2}\varepsilon\sigma\right)\delta_{a\mu} . \qquad (3.52)$$

Let us now rescale the NLSM quantum field, $\xi \to \sqrt{2\pi\alpha'}\xi$, and use the conformally flat metric on the string world-sheet (the 2d Lorentz invariance is known to be non-anomalous, so that it can be used to bring the world-sheet metric into the conformally flat form). The part of the background-quantum splitted NLSM action, relevant for two-loop perturbative calculations, reads

$$I_{\text{int}} = \int d^2x \left[\frac{2\pi\alpha'}{6}R_{abcd}(\phi)\partial_\mu\xi^a\partial_\mu\xi^d\xi^b\xi^c + \tfrac{1}{3}(2\pi\alpha')^{1/2}\varepsilon^{\mu\nu}H_{abc}(\phi)\partial_\mu\xi^a\partial_\nu\xi^b\xi^c \right.$$

$$+ \varepsilon\sigma(x)\tfrac{1}{2}\partial_\mu\xi^a\partial_\mu\xi^a + \varepsilon\sigma(x)\frac{2\pi\alpha'}{6}R_{abcd}(\phi)\partial_\mu\xi^a\partial_\mu\xi^d\xi^b\xi^c$$

$$\left. + \frac{2\pi\alpha'}{4}D_aH_{bcd}(\phi)\varepsilon^{\mu\nu}\partial_\mu\xi^b\partial_\nu\xi^c\xi^d\xi^a \right] ,$$

$$(3.53)$$

where the values of $\{\phi^i\}$ are the coordinates of the embedding (background) spacetime where our test string propagates. Accordingly, the conformal anomaly associated with the second term in (3.48) reads

$$4\gamma \int d^2x\, \sigma(x)\Box\sigma(x) . \qquad (3.54)$$

The first term in (3.48) can be absorbed into the world-sheet topology-counting coupling constant $\tilde{\sigma}$ in (3.46a), so that it may be ignored. The relevant one- and two-loop graphs contributing to (3.54) are depicted in Fig. 6.3. Unlike Chap. 2, tadpoles may also contribute here, because of the compactness of the (Euclidean) string world-sheet and the related absence of IR divergences. The last term in (3.53) can actually be ignored in the two-loop approximation, due to the total antisymmetry of the torsion tensor H_{abc}.

The one-loop calculation of the graph in Fig. 6.3(a) yields [172]

$$\beta^{(1)} = \frac{D}{24\pi} , \quad \gamma^{(1)} = \frac{D}{96\pi} , \qquad (3.55)$$

The two-loop diagrams of Fig. 6.3 yield [374]

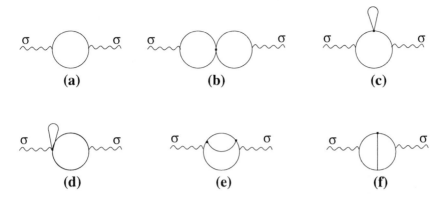

Fig. 6.3. The one- and two-loop graphs contributing to the conformal anomaly of the closed bosonic string NLSM with torsion

$$\textbf{(b)} \ = \ -\frac{\pi\alpha'\varepsilon^2 R}{3(2\pi)^6} \int d^2k\,\sigma(k)\sigma(-k) \int d^dp d^dq$$

$$\times \frac{(p\cdot k - p^2)^2(q\cdot k - q^2) + (p\cdot k - p^2)(p\cdot q)(k\cdot q - q^2)}{p^2(k-p)^2 q^2(k-q)^2} \ ,$$

$$\textbf{(c)} \ = \ \frac{\pi\alpha'\varepsilon^2 R}{3(2\pi)^4}G(0) \int d^2k\,\sigma(k)\sigma(-k) \int d^dp \frac{(p\cdot k - p^2)^2}{(k-p)^2 p^2} \ ,$$

$$\textbf{(d)} \ = \ -\frac{\pi\alpha'\varepsilon^2 R}{3(2\pi)^4}G(0) \int d^2k\,\sigma(k)\sigma(-k) \int d^dp \frac{(p\cdot k - p^2)^2}{(k-p)^2 p^2} \ ,$$

$$\textbf{(e)} + \textbf{(f)} \ = \ \frac{2\varepsilon^2\pi\alpha' H^2_{abc}}{(2\pi)^6}\varepsilon^{\mu\nu}\varepsilon^{\rho\lambda} \int d^2k\,\sigma(k)\sigma(-k) \int d^dp d^dq$$

$$\times \frac{1}{p^2(k-p)^2 q^2(k-p-q)^2}$$

$$\times \left\{ \frac{(k\cdot p - p^2)(k\cdot q - q^2)p_\mu(k-q)_\nu(k-p)_\rho q_\lambda}{(k-q)^2} \right.$$ $$\left. + \frac{(k\cdot p - p^2)q_\mu(k-p)_\nu q_\rho(k-p)_\lambda}{(k-p)^2} \right\} \ ,$$

(3.56)

where $G(x-y)$ is the scalar propagator (2.3.2), and

$$\sigma(x) = \frac{1}{(2\pi)^2} \int d^2p\,\sigma(p)\exp(-ip\cdot x) \ . \tag{3.57}$$

The quantum ambiguities related to the uncertainty in the definition of $\varepsilon^{\mu\nu}\varepsilon^{\rho\lambda}$ in d dimensions (Sect. 2.4) do not play any role here, since only the well-defined residues at the $1/\varepsilon^2$ poles contribute in (3.56). We are, therefore, allowed to use the 2d identity

$$\varepsilon^{\mu\nu}\varepsilon^{\rho\lambda} = \delta^{\mu\lambda}\delta^{\nu\rho} - \delta^{\mu\rho}\delta^{\nu\lambda} \ . \tag{3.58}$$

As is clear from (3.56), though being separately non-vanishing, the tadpole contributions cancel each other. The rest of (3.56) gives rise to the two-loop conformal anomaly contribution

$$\gamma^{(2)} = \frac{\alpha'}{64\pi}\left(-R + \tfrac{1}{3}H^2\right) = \frac{\alpha'}{64\pi}\left(-\hat{R} - \tfrac{2}{3}H^2\right) \ , \tag{3.59}$$

where we have used the notation (2.4.37) and introduced the scalar curvature $\hat{R}$ with the torsion H — see (2.2.23). Note that (3.59) agrees with (1.16).

Having obtained W in the leading approximation in α' for the closed bosonic string, it is not difficult to calculate the effective action Γ of the massless string modes (i.e. of the metric and Kalb-Ramond field) in the string tree-level approximation ($\chi = 2$), by using the identity [355]

$$\int \mathrm{d}^2x\mathrm{d}^2x'(R\sqrt{g})_x \Box_{xx'}^{-1}(R\sqrt{g})_{x'} = 16\pi \tag{3.60}$$

in the case of a two-sphere. We find

$$\Gamma[G, H] \approx \int \frac{\mathrm{d}^D\phi}{(2\pi\alpha')^{D/2}}\sqrt{G(\phi)}\left\{1 + \tfrac{\alpha'}{4}\left[-R + \tfrac{1}{3}H^2\right]\right\} \ , \tag{3.61}$$

in agreement with Subsects. 2.4.1 and 6.1.1. The presence of the cosmological term in (3.61) is due to a tachyon in the spectrum of the closed bosonic string, while both the tachyon and the cosmological term are known to be absent for superstrings. After taking into account the bosonic ghost contributions to the conformal anomaly [45, 172], we find

$$\begin{aligned}
\beta &= \frac{D-8}{24\pi} + \frac{\alpha'}{16\pi}\left(-\hat{R} - \tfrac{2}{3}H^2\right) + \dots \ , \\
\gamma &= \frac{D-26}{96\pi} + \frac{\alpha'}{64\pi}\left(-\hat{R} - \tfrac{2}{3}H^2\right) + \dots \ .
\end{aligned} \tag{3.62}$$

The critical dimension of the bosonic string NLSM is therefore given by

$$D_{\mathrm{cr}} = 26 + \alpha' H^2 + O\left([\alpha']^2\right) \ , \tag{3.63}$$

where we have used the fact that the vanishing NLSM β-function conditions imply $\hat{R} = 0$. Note that D_{cr} *differs* from its flat value of 26.

The two-loop anomaly calculation in the $N = 1$ and $N = 2$ fermionic string NLSM goes in a similar way. We expand the 2d supersymmetric NLSM action $I_S[e_\mu^a, \phi + \eta(\xi), \psi]$ up to fourth-order terms in the quantized fields $\xi^i(x)$ and $\psi_\alpha^i(x)$, and rescale them as $\xi \to (2\pi\alpha')^{1/2}\xi$ and $\psi \to e^{-1/4}(2\pi\alpha')^{1/2}\psi$. The relevant fermionic-dependent terms in the background-field expansion of the fermionic string NLSM are given by

$$I_{\text{F, int}} = \int d^2x \left[\tfrac{i}{4}\varepsilon\sigma\bar\psi^a\gamma_\mu\partial_\mu\psi^a - \tfrac{1}{2}(2\pi\alpha')^{1/2}\left(1 - \tfrac{1}{2}\varepsilon\sigma\right)H_{abc} \right.$$

$$\times\, i\bar\psi^a\gamma_\mu\varepsilon^{\mu\nu}\psi^b\partial_\nu\xi^c + \tfrac{1}{4}(2\pi\alpha')\left(1 + \tfrac{1}{2}\varepsilon\sigma\right)R_{acdf}i\bar\psi^a\gamma_\mu\psi^c\xi^d\partial_\mu\xi^f$$

$$\left. +\tfrac{2\pi\alpha'}{16}\hat R_{abcd}\bar\psi^a(1+\gamma_3)\psi^c\bar\psi^b(1+\gamma_3)\psi^d \right] .$$

$$(3.64)$$

These terms are supposed to be added to those in (3.53).

The superconformal gauge is defined by

$$N = 1: \qquad e^a_\mu = e^\sigma\delta_{a\mu} , \quad \psi_\mu = \tfrac{1}{2}\gamma_\mu\lambda , \tag{3.65a}$$

with Majorana spinor λ, or

$$N = 2: \qquad e^a_\mu = e^\sigma\delta_{a\mu} , \quad \psi_\mu = \tfrac{1}{2}\gamma_\mu\lambda , \quad A^\mu = \tfrac{1}{2}\varepsilon^{\mu\nu}\partial_\nu\rho , \tag{3.65b}$$

with Dirac spinor λ, respectively.

By supersymmetry, the anomalous Lagrangian in (3.54) should be extended as

$$N = 1: \qquad \tfrac{1}{2}(\partial_a\sigma)^2 \rightarrow \tfrac{1}{2}(\partial_a\sigma)^2 + \tfrac{i}{2}\bar\lambda\slashed\partial\lambda , \tag{3.66a}$$

or

$$N = 2: \qquad \tfrac{1}{2}(\partial_a\sigma)^2 \rightarrow \tfrac{1}{2}(\partial_a\sigma)^2 - \tfrac{1}{2}(\partial_a\rho)^2 + i\bar\lambda\slashed\partial\lambda , \tag{3.66b}$$

respectively. The one- and two-loop fermionic graphs are depicted in Fig. 6.4, where the vertices with dots contain the factor $\sqrt{\alpha'}H_{abc}$.

The one-loop results are given by [384, 3]

$$N = 1: \quad \beta^{(1)} = \tfrac{D}{16\pi} , \quad \gamma^{(1)} = \tfrac{D}{64\pi} ,$$
$$N = 2: \quad \beta^{(1)} = \tfrac{D}{8\pi} , \quad \gamma^{(1)} = \tfrac{D}{32\pi} . \tag{3.67}$$

Including the (conformal and superconformal) ghost contributions [45, 172] yields

$$N = 1: \quad \beta^{(1)}_{\text{total}} = \tfrac{D-2}{16\pi} , \quad \gamma^{(1)}_{\text{total}} = \tfrac{D-10}{64\pi} ,$$
$$N = 2: \quad \beta^{(1)}_{\text{total}} = \tfrac{D}{8\pi} , \quad \gamma^{(1)}_{\text{total}} = \tfrac{D-2}{32\pi} . \tag{3.68}$$

The two-loop contributions, corresponding to the graphs (b) and (c) in Fig. 6.4, vanish. The vertices appearing in the rest of the diagrams in Fig. 6.4 can be read off from the effective Lagrangian

$$L_{\text{eff}} = \tfrac{\varepsilon\sigma}{2}\partial_\mu\xi^a\partial_\mu\xi^a + \tfrac{i\varepsilon\sigma}{4}\bar\psi^a\slashed\partial\psi^a - \tfrac{1}{2}(1 - \tfrac{1}{2}\varepsilon\sigma)(2\pi\alpha')^{1/2}H_{abc}i\bar\psi^a\gamma^\nu\gamma_3\psi^b\partial_\nu\xi^c. \tag{3.69}$$

In fact, *all* the two-loop graphs of Fig. 6.4 do *not* contribute to the anomaly, either in the $N = 1$ or $N = 2$ case [374]. For example, the graphs (d) and (g) are proportional to

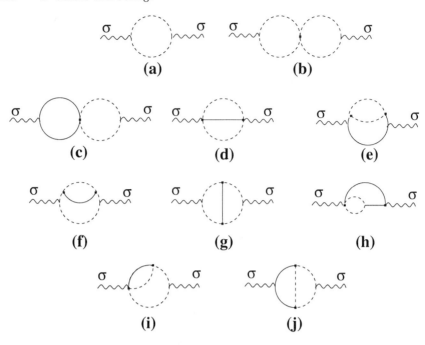

Fig. 6.4. The one- and two-loop fermionic graphs contributing to the superconformal anomaly of the fermionic string NLSM

$$\lim_{\varepsilon \to +0} \varepsilon^2 \int d^d p d^d q \frac{(k \cdot q - p \cdot q - q^2)(k \cdot p - p^2 - p \cdot q)}{(k - p - q)^2 p^2 q^2} = 0 . \tag{3.70}$$

Similarly, we find

$$(e) \sim \lim_{\varepsilon \to +0} \varepsilon^2 \int d^d p d^d q \frac{(p \cdot k - p^2)^2 \left[2(p \cdot q)(p^2 - p \cdot q) - p^2(q \cdot p - q^2) \right]}{p^4 (k - p)^2 q^2 (p - q)^2} = 0 ,$$

$$(f, i) \sim \lim_{\varepsilon \to +0} \varepsilon^2 \int d^d p d^d q \frac{(p \cdot q - p^2)(k \cdot q + q^2 - k \cdot p - p \cdot q)}{p^2 (k + q)^2 (q - p)^2} = 0 ,$$

$$(h) \sim \lim_{\varepsilon \to +0} \varepsilon^2 \int d^d p d^d q (k \cdot p - p^2) \frac{p_\nu (k - p)_\rho \operatorname{tr} \left[\slashed{q} \gamma^\nu (\slashed{p} + \slashed{q}) \gamma^\rho \right]}{p^2 (k - p)^2 q^2 (p + q)^2} = 0 ,$$

$$(j) \sim \lim_{\varepsilon \to +0} \varepsilon^2 \int d^d p d^d q \frac{(k \cdot p - p^2) p_\mu (k - p)_\nu}{p^2 (k - p)^2 q^2}$$

$$\times \operatorname{tr} \left\{ \gamma^\mu \slashed{q} \gamma^\nu \left[\frac{(\slashed{k} + \slashed{q} - \slashed{p})}{(k + q - p)^2} + \frac{(\slashed{q} - \slashed{p})}{(q - p)^2} \right] \right\} = 0 . \tag{3.71}$$

We thus get the following perturbative results:

$$N = 1: \quad \beta = \frac{D-2}{16\pi} - \frac{\alpha' H^2}{6\cdot 4\pi} + \cdots , \qquad \gamma = \frac{D-10}{64\pi} - \frac{\alpha' H^2}{6\cdot 16\pi} + \cdots ,$$

$$N = 2: \quad \beta = \frac{D}{8\pi} - \frac{\alpha' H^2}{6\cdot 4\pi} + \cdots , \qquad \gamma = \frac{D-2}{32\pi} - \frac{\alpha' H^2}{6\cdot 16\pi} + \cdots ,$$

(3.72)

which imply, in particular, the real critical dimensions

$$N = 1: \quad D_{\mathrm{cr}} = 10 + \frac{2\alpha' H^2}{3} + \cdots ,$$

$$N = 2: \quad D_{\mathrm{cr}} = 4 + \frac{2\alpha' H^2}{3} + \cdots ,$$

(3.73)

to leading order in α', where we have used the fact that, in the case of $N = (2, 2)$ strings, the parameter D in the second line of (3.72) denotes the *complex* dimension of the background space (see Sect. 6.4).

As an example, let us consider the background spacetime given by the product $SU(2) \times M_d$ of the three-dimensional sphere of radius r and d-dimensional flat space M_d. The vanishing generalized scalar curvature, $\hat{R} = 0$, implies in this case

$$\frac{6r^2}{\alpha'} = |n| , \quad H^2 = -R = \frac{1}{r^2} , \tag{3.74}$$

where the WZ coefficient n is an integer. Hence, the critical dimension (3.63) in the bosonic string case is given by

$$N = 0: \quad D_{\mathrm{cr}} = d_{\mathrm{cr}} + 3 = 26 + \frac{6}{|n|} . \tag{3.75}$$

This is to be compared with the exact results of [385] for the the critical dimension of the closed bosonic string propagating on the $SU(2) \times M_d$ background, which were obtained by using the CFT techniques (Chap. 5), i.e. by calculating the corresponding central charge. One finds [385]

$$D_{\mathrm{cr}} = 26 + \frac{6}{|n|} \frac{1}{(1 + 2/|n|)} , \tag{3.76}$$

in agreement with the perturbative result (3.75).

6.4 $N = (2, 2)$ String Amplitudes and Integrability

The interacting *closed* $(2, 2)$ world-sheet supersymmetric string theory can be interpreted as a theory of Self-Dual Gravity (SDG) [386]. Similarly, the *open* $(2, 2)$ string theory can be interpreted as a Self-Dual Yang-Mills (SDYM) theory [387]. Since open strings can 'create' closed strings which, in their turn, can interact with open ones, there are quantum corrections to the effective field equations of the open $(2,2)$ string theory. Because of the 'topological' nature of the (2,2) string theories, *only* 3-point tree string amplitudes are non-vanishing and local. As a result, quantum perturbative corrections in the *mixed* theory of open *and* closed (2,2) strings are still under control. In particular, the SDYM equations receive corrections from diagrams with internal gravitons, so that they become the YM self-duality equations on a Kähler background [387]. Therefore, they should respect integrability [479].

Contrary to the SDYM equations and naive expectations, the effective gravitational equations of motion in the mixed (open/closed) (2,2) string theory get modified in such a way that the resulting 'spacetime' is no longer self-dual [387]. Accordingly, the integrability property seems to be lost in the mixed (2,2) string theory. In this section we discuss the exact effective field equations of motion in the *mixed* theory, and show that the integrability is maintained in the case of an Abelian gauge group [388]. The effective field theory is *not* the Einstein-Maxwell system describing the standard interaction of the non-linear graviton with a photon. Instead, it is of the Born-Infeld type [389], being non-linear with respect to both gravitational *and* 'electromagnetic' fields. The NLSM interpretation of the Born-Infeld action, in connection with partial susy breaking, is discussed in Sect. 9.2.

In Subsect. 6.4.1 we summarize basic facts about closed and open (2,2) strings. The mixed theory is discussed in Subsect. 6.4.2.

6.4.1 Closed and Open $(2, 2)$ Strings

The $(2, 2)$ strings are $N = 2$ fermionic strings with two world-sheet super-symmetries, both for the left-moving and right-moving degrees of freedom (Subsect. 6.3.2). [4] The critical open and closed $(2, 2)$ strings live in four real dimensions, with the signature $2 + 2$. The $N = 2$ strings cannot be defined in $1 + 3$ dimensions since the latter are incompatible with a complex structure (Subsect. 6.3.2). The $N = 2$ string physical spectrum consists of a single massless particle that can be assigned in the adjoint of a gauge group G in the open string case.

The only non-vanishing $(2, 2)$ string tree scattering amplitudes are 3-point trees, while all higher n-point functions vanish due to kinematical reasons in $2 + 2$ dimensions. Tree-level calculations of string amplitudes do not require the (heavy) general techniques of the BRST quantization [393] or topological methods [394]. The vertex operator for a $(2, 2)$ closed string particle of momentum k reads in $(2, 2)$ world-sheet superspace as

$$V_{\rm c} = \frac{\kappa}{\pi} \exp \left\{ {\rm i} \left(k \cdot \bar{Z} + \bar{k} \cdot Z \right) \right\} , \qquad (4.1)$$

where κ is the $(2,2)$ closed string coupling constant, and $Z^i(x, \bar{x}, \theta, \bar{\theta})$ are complex $(2, 2)$ chiral superfields. We use here complex coordinates $(x, \bar{x})$ on

---

[4] The $(2, 1)$ and $(2, 0)$ heterotic strings can also be defined [390, 391, 392] either in $2 + 1$ or $1 + 1$ dimensional spacetime. We do not consider them here.

the string world-sheet, while $(z^i, \bar{z}^{\bar{i}})$ denote complex coordinates of the $(2,2)$ string target space, $i = 1, 2$.

When using the $(2,2)$ super-Möbius invariance of the $(2,2)$ super-Riemann sphere, it is not difficult to calculate the correlation function of three V_c. One finds [386]

$$A_{ccc} = \kappa c_{23}^2 \ , \quad \text{where} \quad c_{23} \equiv \left(k_2 \cdot \bar{k}_3 - \bar{k}_2 \cdot k_3 \right) \ . \tag{4.2}$$

One can check that the A_{ccc} is totally symmetric on-shell, while it is merely invariant under the subgroup $U(1,1) \cong SL(2, \mathbf{R}) \otimes U(1)$ of the full 'Lorentz' group $SO(2,2) \cong SL(2, \mathbf{R}) \otimes SL(2, \mathbf{R})'$ in $2+2$ dimensions.

Since all higher correlators vanish [386], the local 3-point function (4.2) alone determines the *exact* effective action having the form [386]

$$S_{\mathrm{P}} = \int \mathrm{d}^{2+2} z \ \left(\frac{1}{2} \eta^{i\bar{j}} \partial_i \phi \bar{\partial}_{\bar{j}} \phi + \frac{2\kappa}{3} \phi \partial \bar{\partial} \phi \wedge \partial \bar{\partial} \phi \right) \ , \tag{4.3}$$

which is known as the *Plebański* action for Self-Dual Gravity (SDG). Hence, the massless 'scalar' of the closed string theory can be identified with a deformation of the Kähler potential K of self-dual (= Kähler + Ricci-flat) gravity [386], where

$$K = \eta_{i\bar{j}} z^i \bar{z}^{\bar{j}} + 4\kappa\phi \ , \quad \eta_{i\bar{j}} = \eta^{i\bar{j}} = \begin{pmatrix} 1 & 0 \\ 0 & -1 \end{pmatrix} \ . \tag{4.4}$$

The $(2,2)$ closed string target space metric is therefore given by

$$g_{i\bar{j}} = \partial_i \bar{\partial}_{\bar{j}} K = \eta_{i\bar{j}} + 4\kappa \partial_i \bar{\partial}_{\bar{j}} \phi \ . \tag{4.5}$$

Similarly, in the open $(2,2)$ string case, when using the $(2,2)$ superspace vertex

$$V_o = g \exp \left\{ \mathrm{i} \left(k \cdot \bar{Z} + \bar{k} \cdot Z \right) \right\} \ , \tag{4.6}$$

assigned to the boundary of the $(2,2)$ supersymmetric upper half-plane (or $(2,2)$ superdisk) with proper boundary conditions, one finds that the open string 3-point function [387]

$$A_{ooo} = -\mathrm{i} g c_{23} f^{abc} \ , \tag{4.7}$$

is essentially the 'square root' of A_{ccc}, as it should (the f^{abc} are the structure constants of G). The A_{ooo} can be obtained from the effective action [387]

$$S_{\mathrm{DNS}} = \int \mathrm{d}^{2+2} z \ \eta^{i\bar{j}} \left(\frac{1}{2} \partial_i \varphi^a \bar{\partial}_{\bar{j}} \varphi^a - \mathrm{i} \frac{g}{3} f^{abc} \varphi^a \partial_i \varphi^b \bar{\partial}_{\bar{j}} \varphi^c \right) + \cdots \ . \tag{4.8}$$

Requiring all the higher-point amplitudes to vanish in the *field* theory (2.8) determines the additional local n-point interactions, $n > 3$, denoted by dots in (4.8). The full action S_{DNS} is known as the Donaldson-Nair-Schiff (DNS) action [395, 396]. The DNS equation of motion yields the *Yang* equation [397] of the SDYM,

$$\eta^{i\bar{j}}\bar{\partial}_{\bar{j}}\left(e^{-2ig\varphi}\partial_i e^{2ig\varphi}\right) = 0 , \tag{4.9}$$

where the matrix φ is Lie algebra-valued, $\varphi = \varphi^a t^a$, and the Lie algebra generators t^a of G are taken to be anti-Hermitian. The DNS action is known to be dual (in the field theory sense) to the Leznov-Parkes (LP) action [398, 399], which has only *cubic* interaction, and whose equation of motion also describes the SDYM. The LP effective action arises from the open (2,2) string theory after taking into account the world-sheet instanton corrections [400]. The higher n-point functions, $n > 3$, vanish in the LP field theory [399].

6.4.2 Mixed $(2,2)$ String Amplitudes

When open strings join together, they form closed strings. In turn, the closed strings can interact with the open strings. Therefore, the open string theory has *mixed* (open/closed) amplitudes too. In particular, the coupling constants of the closed and open strings are related,

$$\kappa \sim \sqrt{\hbar}\, g^2 . \tag{4.10}$$

The only non-vanishing 3-point mixed amplitude is given by [387]

$$A_{ooc} = \frac{\kappa}{\pi}\delta^{ab}c_{23}^2\int_{-\infty}^{+\infty}\mathrm{d}x\,\frac{1}{x^2+1} = \kappa\delta^{ab}c_{23}^2 , \tag{4.11}$$

where the integration over the position x of one of the open string vertices goes along the border of the upper half-plane (= real line). All higher n-point mixed amplitudes, $n \geq 4$, are believed to be zero, like the purely open or closed string ones. The additional (mixed) term in the (2,2) open string effective field theory action, which is supposed to reproduce the A_{ooc}, reads as follows [387]:

$$S_{\mathrm{M}} = \int \mathrm{d}^{2+2}z \left(2\kappa\phi\partial\bar{\partial}\varphi^a \wedge \partial\bar{\partial}\varphi^a\right) . \tag{4.12}$$

The complete non-Abelian effective action can be determined by demanding all higher-point amplitudes to vanish in the *field* theory describing the mixed $(2,2)$ strings, order by order in n. Rescaling ϕ by a factor of 4κ, and φ by a factor of g, one finds [387]

$$
\begin{aligned}
S_{\mathrm{tot}} =& \frac{1}{16\kappa^2}\int \mathrm{d}^{2+2}z\left[\frac{1}{2}\eta^{i\bar{j}}\partial_i\phi\bar{\partial}_{\bar{j}}\phi + \frac{1}{6}\phi\partial\bar{\partial}\phi\wedge\partial\bar{\partial}\phi\right] + \frac{1}{g^2}\int \mathrm{d}^{2+2}z\,\eta^{i\bar{j}} \\
&\times\left[-\frac{1}{2}\mathrm{Tr}\left(\partial_i\varphi\bar{\partial}_{\bar{j}}\varphi\right) - \frac{2i}{3!}\mathrm{Tr}\left(\bar{\partial}_{\bar{j}}\varphi[\partial_i\varphi,\varphi]\right) + \frac{2^2}{4!}\mathrm{Tr}\left(\bar{\partial}_{\bar{j}}\varphi[[\partial_i\varphi,\varphi],\varphi]\right)\right] \\
&+ \frac{1}{g^2}\int \mathrm{d}^{2+2}z\left[\frac{1}{2}\partial\bar{\partial}\phi\wedge\mathrm{Tr}\left(\varphi\partial\bar{\partial}\varphi\right) + \frac{2i}{3!}\partial\bar{\partial}\phi\wedge\mathrm{Tr}\left(\varphi[\partial\varphi,\bar{\partial}\varphi]\right)\right] + \dots
\end{aligned}
\tag{4.13}
$$

Despite the complicated form of the action (4.13), its equations of motion can be written down in a simple geometrical form [387], namely,

$$g^{i\bar{j}}(\phi)\bar{\partial}_{\bar{j}}\left(e^{-2i\varphi}\partial_i e^{2i\varphi}\right) = 0 , \qquad (4.14)$$

and

$$-\det g_{i\bar{j}} = +1 + \frac{2\kappa^2}{g^2}\mathrm{Tr}\left(F_{i\bar{j}}F^{i\bar{j}}\right) , \qquad (4.15)$$

where $F_{i\bar{j}}$ is the YM field strength of the YM gauge fields

$$A \equiv e^{-i\varphi}\partial e^{i\varphi} , \qquad \bar{A} \equiv e^{i\varphi}\bar{\partial}e^{-i\varphi} , \qquad (4.16)$$

$g_{i\bar{j}} = \eta_{i\bar{j}} + \partial_i\bar{\partial}_{\bar{j}}\phi$ is a Kähler metric, $g^{i\bar{j}}$ is its inverse, and the indices $(i, \bar{j})$ are raised and lowered by using the totally antisymmetric Levi-Civita symbols ε^{ij}, $\varepsilon^{\bar{i}\bar{j}}$, and ε_{ij}, $\varepsilon_{\bar{i}\bar{j}}$ ($\varepsilon_{12} = \varepsilon^{12} = 1$).

Equation (4.14) is just the Yang equation (of motion) of the DNS action describing the SDYM on a curved Kähler background, as should have been expected. It is the gravitational equation (3.6) that is of interest. Associated with the Kähler metric

$$ds^2 = 2g_{i\bar{j}}dz^i d\bar{z}^{\bar{j}} \equiv 2K_{,i\bar{j}}dz^i d\bar{z}^{\bar{j}} , \qquad (4.17)$$

is the fundamental (Kähler) closed two-form

$$\Omega = g_{i\bar{j}}dz^i \wedge d\bar{z}^{\bar{j}} \equiv K_{,i\bar{j}}dz^i \wedge d\bar{z}^{\bar{j}} , \qquad (4.18)$$

where K is the (locally defined) Kähler potential, and all subscripts after a comma denote partial differentiations. We regard the complex coordinates $(z^i, \bar{z}^{\bar{i}})$ as independent variables, so that our complexified 'spacetime' $\mathcal{M}$ is locally a direct product of two two-dimensional complex manifolds $\mathcal{M} \cong M_2 \otimes \bar{M}_2$, where both M_2 and $\bar{M}_2$ are endowed with complex structures, i.e. they possess closed non-degenerate two-forms ω and $\bar{\omega}$, respectively The normalization of the holomorphic two-forms ω and $\bar{\omega}$ is fixed by the flat 'spacetime' limit where $\omega = dz^1 \wedge dz^2$ and $\bar{\omega} = d\bar{z}^{\bar{1}} \wedge d\bar{z}^{\bar{2}}$. Hence, the effective equations of motion (4.14) and (4.15) in the mixed (2,2) string theory can be rewritten in truly geometrical form as [387]

$$\Omega \wedge F = 0 , \qquad (4.19)$$

and

$$\Omega \wedge \Omega + \frac{4\kappa^2}{g^2}\mathrm{Tr}(F \wedge F) = 2\omega \wedge \bar{\omega} , \qquad (4.20)$$

where F is the YM Lie algebra-valued field strength two-form satisfying

$$\omega \wedge F = \bar{\omega} \wedge F = 0 . \qquad (4.21)$$

Equations (4.19) and (4.21) are just the self-dual Yang-Mills equations in Kähler 'spacetime'. They are, therefore, integrable and their solutions describe Yang-Mills instantons [401]. In particular, one can always locally change the flat SDYM equations of motion into the SDYM equations on a curved Kähler background by a diffeomorphism transformation compatible with the Kähler structure.

The integrability condition for the gravitational equations of motion in the complexified 'spacetime' is known to be precisely equivalent to the (anti)self-duality of the *Weyl* curvature tensor [206]. The Penrose twistor construction [206] transforms the problem of solving the non-linear partial differential equations of conformally self-dual gravity into the standard Riemann-Hilbert problem of patching together certain holomorphic data.

In the case under investigation, the self-duality of the Weyl tensor in Kähler space is equivalent to vanishing scalar curvature [402, 403]. The Ricci tensor is related to the Kähler metric as

$$R_{i\bar{j}} = \partial_i \bar{\partial}_{\bar{j}} \log \det(g_{k\bar{k}}) \ . \tag{4.22}$$

Equation (4.15) or (4.20) now yields

$$R_{i\bar{j}} = \partial_i \bar{\partial}_{\bar{j}} \log \left[1 + \frac{2\kappa^2}{g^2} \mathrm{Tr}(F_{i\bar{j}} F^{i\bar{j}}) \right] \ , \tag{4.23}$$

and, hence,

$$R = g^{m\bar{n}} \partial_m \bar{\partial}_{\bar{n}} \log \left[1 + \frac{2\kappa^2}{g^2} \mathrm{Tr}(F_{ij} F^{i\bar{j}}) \right] \ , \tag{4.24}$$

while both do *not* vanish on-shell. It is also obvious that the 'matter' stress-energy tensor, which is equal to the Einstein tensor in accordance with (4.23) and (4.24), does *not* vanish either. This is to be compared to the standard gravitational equations of motion in the case of the Einstein-Yang-Mills coupled system that is described by the standard action given by the sum of the Einstein-Hilbert and Yang-Mills terms. The standard YM stress-energy tensor is quadratic with respect to the YM field strength, while it vanishes under the SDYM condition. In our case, the YM stress-energy tensor is not even polynomial in the YM field strength, so that it has to correspond to the non-polynomial (in F) effective action.

To address integrability properties of (4.20), let us rewrite first (4.15) in the form

$$\det(g_{i\bar{j}}) + \frac{2\kappa^2}{g^2} \mathrm{Tr} \det(F_{i\bar{j}}) = -1 \ , \tag{4.25}$$

where both determinants are two-dimensional. Given an *Abelian* field strength F satisfying the self-duality condition (4.19) or, equivalently, $g_{1\bar{1}} F_{2\bar{2}} + g_{2\bar{2}} F_{1\bar{1}} - g_{1\bar{2}} F_{2\bar{1}} - g_{2\bar{1}} F_{1\bar{2}} = 0$, there is the remarkable identity

$$\det(g) + \frac{2\kappa^2}{g^2} \det(F) = \det \left(g + \frac{\kappa\sqrt{2}}{g} F \right) \ . \tag{4.26}$$

In addition, (4.16) in the Abelian case implies $A = i\partial\varphi$, $\bar{A} = -i\bar{\partial}\varphi$, and, hence,

$$F = 2i\partial\bar{\partial}\varphi \; . \tag{4.27}$$

Taken together, they allow us to represent (4.25) as the Plebański heavenly equation

$$\det\left(\partial\bar{\partial}\mathcal{K}\right) = -1 \; , \tag{4.28}$$

with a *complex* potential

$$\mathcal{K} \equiv K + i\frac{2\sqrt{2}\kappa}{g}\varphi \; , \tag{4.29}$$

whose imaginary part is a harmonic function (because of the self-duality of F), and of order $\hbar^{1/2}g$ because of (4.10). Equation (4.28) can be recognized as the consistency condition for the linear system

$$L_{\bar{1}}\psi \equiv \left[\bar{\partial}_{\bar{1}} + i\lambda\bar{B}_{\bar{1}}\right]\psi \equiv \left[\bar{\partial}_{\bar{1}} + i\lambda\left(\mathcal{K}_{,2\bar{1}}\partial_1 - \mathcal{K}_{,1\bar{1}}\partial_2\right)\right]\psi = 0 \; ,$$
$$L_{\bar{2}}\psi \equiv \left[\bar{\partial}_{\bar{2}} + i\lambda\bar{B}_{\bar{2}}\right]\psi \equiv \left[\bar{\partial}_{\bar{2}} + i\lambda\left(\mathcal{K}_{,2\bar{2}}\partial_1 - \mathcal{K}_{,1\bar{2}}\partial_2\right)\right]\psi = 0 \; , \tag{4.30}$$

where λ is a complex spectral parameter. The linear equations (4.30) describe fibring of the associated twistor space in the sense of Penrose [206].

Hence, the Frobenius integrability is still kept in the mixed $N = 2$ string theory, similarly to the usual case of the Plebański heavenly equation with a real Kähler potential. The generalized 'metric' defined with respect to the complex Kähler potential is not real. Its only use is making the integrability apparent, while the true metric in (4.17) is real, of course.

When trying to generalize these results to the non-Abelian situation, one arrives at an obstruction, since the crucial relation (4.26) is no longer valid. If, nevertheless, one wants to impose that relation, one arrives at

$$F^a \wedge F^b = 0 \; , \quad \text{when} \quad a \neq b \; . \tag{4.31}$$

Equation (4.31) implies that different directions in the YM group space do not 'see' each other. Hence, insisting on integrability sends us back to the Abelian case.

It should be noticed that the solutions to the gravitational equations of motion (4.28) are all stationary with respect to the Born-Infeld type effective action

$$S = \int d^{2+2}z \sqrt{-\det\left(g_{i\bar{j}} + \frac{\kappa\sqrt{2}}{g}F_{i\bar{j}}\right)} \; . \tag{4.32}$$

The action S is *not* the standard Born-Infeld action [389] since the determinant in (4.32) is two-dimensional, not four-dimensional.

It is worth mentioning that the infinite hierarchy of conservation laws and the infinite number of symmetries [404] exist as a consequences of Penrose's twistor construction, when it is formally applied to our 'almost self-dual' gravity with a complex Kähler potential. The underlying symmetry is known to be the loop group $S^1 \to \text{SDiff}(2)$ of the area-preserving (holomorphic) diffeomorphisms (of a 2-plane), which can be considered as a 'large N limit' (W_∞) of the W_N symmetries in 2d CFT [405, 406]. The area-preserving holomorphic diffeomorphisms,

$$\partial_i \bar{\partial}_{\bar{j}} \mathcal{K}(z, \bar{z}) \to \partial_i \xi^k(z) \partial_k \bar{\partial}_{\bar{k}} \mathcal{K}(\xi, \bar{\xi}) \bar{\partial}_{\bar{j}} \bar{\xi}^{\bar{k}}(\bar{z}) \ , \tag{4.33}$$

leave both (4.28) and (4.32) invariant since

$$\left| \det(\partial_i \xi^k) \right| = 1 \tag{4.34}$$

by definition.

7. Chiral (1,0) NLSM and Heterotic Strings

In the NLSM approach to the calculation of the bosonic string LEEA (Sect. 6.1) one considers propagation of a single (test) string in the background of its massless modes (described by a 2d NLSM). Quantum consistency of the NLSM implies its 2d conformal invariance. In the case of the NSR string, the non-chiral (1,1) supersymmetric 2d NLSM should be considered along similar lines. As regards the so-called *heterotic* strings [407, 408], one should deal with a chiral, (1,0) supersymmetric 2d NLSM [321, 409, 410, 411, 412, 413]. We call the (1,0) supersymmetric 2d NLSM *heterotic* too.

In this chapter we systematically describe the (1,0) superspace techniques in application to quantum perturbation theory for the heterotic NLSM and related heterotic string theory. The results include a derivation of multi-loop superconformal anomalies in the heterotic NLSM in the critical dimension $D = 10$ [414, 415, 416, 417], and the Lagrangian chiral coset construction of consistent heterotic string models in lower dimensions [323].

7.1 Anomalies of Heterotic NLSM in (1, 0) Superspace

In Subsect. 7.1.1 we introduce a (1,0) superspace in 2d [320, 322, 418, 419]. In Subsect. 7.1.2 we consider the background-field method for the heterotic NLSM. Subsection 7.1.3 is devoted to the (perturbative) supergravitational and superconformal anomalies relevant to the heterotic string.

7.1.1 (1, 0) Superspace

(1,0) superspace is parametrized by two bosonic and one fermionic coordinates,

$$z^A = \left\{ \sigma^{\pm}, \sigma^{=}, \theta^{+} \right\} , \qquad (1.1)$$

where $(+, -)$ denote the helicity $\pm\frac{1}{2}$, while the helicity notation $(\mp, =)$ is used for the (contravariant) vector indices $(++, --)$, respectively. One has $\lambda^{+} = \lambda_{-}$ and $\lambda_{+} = -\lambda^{-}$ for 2d (MW) spinors λ.

The $(1,0)$ flat superspace covariant derivatives are given by

$$D_A \equiv \{D_+, \partial_{\mp}, \partial_=\} , \quad D_+ = \frac{\partial}{\partial\theta^+} + i\theta^+ \partial_{\mp} . \tag{1.2}$$

They satisfy the algebra

$$\{D_+, D_+\} = 2i\partial_{\mp} , \quad \partial_{\mp}\partial_= = \Box ,$$
$$[\partial_a, D_+] = [\partial_a, \partial_b] = 0 . \tag{1.3}$$

The $(1,0)$ superspace integration is defined by

$$\int d^3z^- = \int d^2\sigma d\theta_+ = \int d^2\sigma \frac{\partial}{\partial\theta^+} , \tag{1.4}$$

while we have (up to a surface term)

$$\int d^2\sigma d\theta_+ L = \int d^2\sigma \left(D_+ L \right)\Big|_{\theta=0} . \tag{1.5}$$

The $(1,0)$ superspace delta-function is given by

$$\delta^3_-(z, z') = \delta^2(\sigma, \sigma')(\theta^+ - \theta'^+) \equiv \delta^2(\sigma, \sigma')\delta(\theta - \theta') , \tag{1.6}$$

while it has the following properties:

$$\delta(\theta - \theta')D_+\delta(\theta - \theta') = \delta(\theta - \theta') ,$$

$$\delta(\theta - \theta')\delta(\theta - \theta') = 0 , \quad \delta_{12}\overleftarrow{D}_2 = D_1\delta_{12} . \tag{1.7}$$

The two basic, bosonic and fermionic, $(1,0)$ superfields are

$$\phi(\sigma, \theta) = A(\sigma) + \theta^+\lambda_+(\sigma) ,$$
$$\Psi_-(\sigma, \theta) = \eta_-(\sigma) + \theta^+ F(\sigma) . \tag{1.8}$$

Since the full $(1,0)$ superspace action is a scalar, the corresponding superfield Lagrangian is a spinor of (lower) charge $(-)$.

A two-dimensional $(1,0)$ supergravity in a curved $(1,0)$ superspace is described by the Lorentz-covariant and supercovariant derivatives

$$\nabla_A \equiv (\nabla_+, \nabla_=, \nabla_{\mp}) = E_A{}^M D_M + \omega_A \hat{M} \equiv E_A + \Omega_A , \tag{1.9}$$

where we have introduced the supervielbein $E_A{}^M(z)$ and the superconnection $\omega_A(z)$ in $(1,0)$ superspace. The covariant derivatives (1.9) satisfy the off-shell $(1,0)$ supergravity constraints [322]

$$\{\nabla_+, \nabla_+\} = 2i\nabla_{\mp} , \quad [\nabla_+, \nabla_=] = -2i\Sigma^+\hat{M} ,$$

$$[\nabla_+, \nabla_{\mp}] = 0 , \quad [\nabla_{\mp}, \nabla_=] = -\Sigma^+\nabla_+ - R\hat{M} , \tag{1.10}$$

where we have introduced the 2d Lorentz generator $\hat{M}$,

$$[\hat{M}, \psi_{\pm}] = \pm \tfrac{1}{2} \psi_{\pm} , \tag{1.11}$$

and the $(1,0)$ superfield strength Σ^+ of the 2d, $(1,0)$ supergravity. The superfield R in (1.10) is defined by

$$R = 2\nabla_+ \Sigma^+ . \tag{1.12}$$

The second line of (1.10) is, in fact, a consequence of the first line via the use of Bianchi identities [322].

Because of the constraints (1.10), the only independent superfield components of the supervielbein $E_A{}^M$ and the superconnection ω_A are $(E_+{}^=, E_-{}^+)$, $(E_+{}^\doubleplus, E_+{}^+)$ and $E_-{}^=$. The covariance of the derivatives (1.9) under $(1,0)$ diffeomorphisms and local Lorentz rotations implies the transformation law

$$\nabla_A' = e^K \nabla_A e^{-K} , \quad K = K^N D_N + \Lambda_l \hat{M} . \tag{1.13}$$

It yields

$$\delta E_A{}^N = -\nabla_A K^N + K^L D_L E_A{}^N + E_A{}^L K^P \{D_P, D_L\}^N + \Lambda_l [\hat{M}, E_A{}^N] , \tag{1.14}$$

and, in particular,

$$\delta E_+{}^\doubleplus = K^N D_N E_+{}^\doubleplus + 2\mathrm{i} E_+{}^+ K^+ - E_+{}^N D_N K^\doubleplus + \tfrac{1}{2} \Lambda_l E_+{}^\doubleplus . \tag{1.15}$$

As is clear from (1.15), the parameter K^+ can be used to gauge away $E_+{}^\doubleplus$, i.e. to set $E_+{}^\doubleplus = 0$. In this manifestly $(1,0)$ supersymmetric gauge the superfield parameter K^+ is obviously no longer independent, while the $K^\doubleplus$ transformation has to be accompanied by the compensating transformation with [418]

$$K^+ = -\tfrac{\mathrm{i}}{2} (E_+{}^+)^{-1} \nabla_+ K^\doubleplus . \tag{1.16}$$

It is convenient to introduce the Lorentz-invariant scalar S and the *Lorentz compensator* L as [418]

$$E_+{}^+ (E_-{}^=)^{1/2} \equiv e^{-S} , \quad E_+{}^+ (E_-{}^=)^{-1/2} \equiv e^L . \tag{1.17}$$

Under Lorentz transformations, we have

$$L \to L + \Lambda_l , \tag{1.18}$$

so that it is always possible to choose a gauge $L = 0$. In this (manifestly $(1,0)$ supersymmetric) gauge, a supersymmetry transformation has to be accompanied by the compensating Lorentz transformation with the parameter

$$\begin{aligned}
\Lambda_l &= (E_+{}^+)^{-1} \nabla_+ K^+ - \tfrac{1}{2} (E_-{}^=)^{-1} \nabla_- K^= \\
&= \tfrac{1}{2} \left(\nabla_\doubleplus K^\doubleplus - \nabla_- K^= \right) + \dots .
\end{aligned} \tag{1.19}$$

A solution to the constraints (1.10) in the (1,0) supersymmetric gauge $E_+{}^\pm = L = 0$ reads [322]

$$E_+ = \mathrm{e}^{-S/2}\hat{E}_+ , \quad E_= = \mathrm{e}^{-S}\left[\hat{E}_= + \tfrac{1}{2}\hat{C}_{+=}{}^\pm\hat{E}_+\right] , \tag{1.20}$$

where we have introduced the *Beltrami superdifferentials* [322]

$$\hat{E}_+ = D_+ + H_+{}^= \partial_= , \quad \hat{E}_= = \partial_= + H_={}^\pm \partial_\pm , \quad \hat{E}_\pm = -\tfrac{i}{2}\{\hat{E}_+, \hat{E}_+\} , \tag{1.21}$$

and the anholonomic coefficients $\hat{C}_{AB}{}^C$ defined by the algebra

$$\{\hat{E}_A, \hat{E}_B\} = \hat{C}_{AB}{}^C \hat{E}_C . \tag{1.22}$$

For example, in the linearized approximation, we have

$$
\begin{aligned}
\nabla_+ &= (1 - \tfrac{1}{2}S)D_+ + H_+{}^=\partial_= - \left(D_+ S + \partial_= H_+{}^=\right)\hat{M} , \\
\nabla_\pm &= (1 - S)\partial_\pm + i\left[\tfrac{1}{2}(\partial_= H_+{}^=) + (D_+ S)\right]D_+ - i(D_+ H_+{}^=)\partial_= \\
&\quad - \left(\partial_\pm S - iD_+\partial_= H_+{}^=\right)\hat{M} , \\
\nabla_= &= (1 - S)\partial_= - \tfrac{i}{2}\left(D_+ H_={}^\pm\right)D_+ + H_={}^\pm\partial_\pm \\
&\quad + \left(\partial_= S + \partial_\pm H_={}^\pm\right)\hat{M} .
\end{aligned}
\tag{1.23}
$$

The unconstrained superfields $(H_+{}^=, H_={}^\pm)$ and S are just the pre-potentials of the (1,0) supergravity in (1,0) superspace. Amongst their components,

$$H_+{}^= = \rho_+{}^= + \theta^+ h_\pm{}^= , \quad H_={}^\pm = h_={}^\pm + \theta^+ \Psi_={}^+ ,$$

$$S = h + \theta^+ \Psi_\pm{}^+ , \tag{1.24}$$

are a 2d graviton field (h), a 2d gravitino field (Ψ), and a gauge field (ρ). The covariant superfield strength of 2d, (1,0) supergravity has the expansion

$$\Sigma^+ = \tfrac{i}{2}\left[D_+\left(\partial_\pm H_={}^\pm + 2\partial_= S\right) + \partial_=^2 H_+{}^=\right] + \dots . \tag{1.25}$$

The curved (1,0) superspace density is given by

$$E = [\mathrm{Ber}(E_A{}^M)]^{-1} = \mathrm{e}^{3S/2}\left[1 + iH_={}^\pm\left(D_+ H_+{}^= + H_+{}^= \partial_= H_+{}^=\right)\right]^{-1} . \tag{1.26}$$

The (1,0) superspace general coordinate transformations in the linearized approximation (after taking into account the compensating transformations in the gauge $E_+{}^\pm = L = 0$) read

$$\delta H_+{}^= = -D_+ K^= + \dots , \quad \delta H_={}^\pm = -\partial_= K^\pm + \dots ,$$

$$\delta S = \tfrac{1}{2} \left(\partial_{\underline{+}} K^{\mp} + \partial_{-} K^{=} \right) + \cdots .$$ (1.27)

The local scale transformations with parameter Λ_s in $(1,0)$ superspace are defined by [411, 412]

$$E_+{}^{M'} = e^{-\Lambda_s/2} E_+{}^M , \quad E_-{}^{M'} = e^{-\Lambda_s} E_-{}^M ,$$
$$E_{\underline{+}}{}^{M'} = e^{-\Lambda_s} E_{\underline{+}}{}^M - i(D_+ e^{-\Lambda_s}) E_+{}^M .$$ (1.28)

It follows from (1.28) that the gauge $E_+{}^{\underline{+}} = L = 0$ is invariant under the scale transformations, whereas the superfield S is shifted as

$$\delta S = \Lambda_s .$$ (1.29)

The superfield S is called the *scale compensator*.

The linearized scale transformations take the form

$$\delta S = \Lambda_s , \quad \delta H_+{}^= = 0 , \quad \delta H_-{}^{\underline{+}} = 0 .$$ (1.30)

It is worth mentioning that the $(1,0)$ supergravity strength Σ^+ is dependent upon S, so that Σ^+ is not a scale-invariant superfield. The condition $S = 0$ can be interpreted as the $(1,0)$ superconformal gauge.

The rigid $(1,0)$ supersymmetric 2d actions of the $(1,0)$ superfields ϕ and Ψ of (1.8) are given by [320]

$$I[\phi] = \tfrac{i}{2} \int d^3 z^- \, D_+ \phi \partial_- \phi , \quad I[\Psi] = -\tfrac{1}{2} \int d^3 z^- \, \Psi_- D_+ \Psi_- ,$$ (1.31)

respectively. Equation (1.31) allows us to evaluate the corresponding super-propagators in $(1,0)$ superspace,

$$\langle \phi_1 \phi_2 \rangle = \frac{D_+}{\Box} \delta_{12}^{(3)} , \quad \langle \Psi_1 \Psi_2 \rangle = \frac{i\partial_- D_+}{\Box} \delta_{12}^{(3)} .$$ (1.32)

The covariantization of the free actions (1.31) with respect to $(1,0)$ supergravity in $(1,0)$ superspace amounts to replacing the flat superspace covariant derivatives by the curved superspace covariant derivatives, and changing the integration measure as

$$d^3 z^- \rightarrow d^3 z^- E .$$ (1.33)

In what follows we find it more appropriate to associate the Lorentz minus charge with the whole $(1,0)$ curved superspace measure, $d^3 z^- E \rightarrow d^3 z E^-$.

7.1.2 Background-Field Method in the Heterotic NLSM

The action of the heterotic string propagating in the background of its massless bosonic modes (a metric $g_{\mu\nu}$, an antisymmetric tensor $b_{\mu\nu}$, a dilaton Φ, and the gauge bosons A_μ^{IJ} transforming in a linear representation of some subgroup G' of the non-anomalous gauge group G in D spacetime dimensions) reads

$$
I_{\text{het.}}[X, \Psi] = -\frac{1}{4\pi\alpha'} \int d^3 z E^- \left\{ i\nabla_+ X^\mu \nabla_- X^\nu [g_{\mu\nu}(X) + b_{\mu\nu}(X)] \right.
$$
$$
\left. + \Psi_-^I [\delta^{IJ}\nabla_+ + A_+^{IJ}(X)]\Psi_-^J + \alpha' \Phi(X)\Sigma^+ \right\} ,
$$
(1.34)

where $\mu = 0, 1, \ldots, D - 1$. In (1.34) we have introduced heterotic fermions Ψ_-^I with $I = 1, 2, \ldots, N$, and heterotic gauge fields

$$
A_+^{IJ}(X) = A_\mu^{IJ}(X)\nabla_+ X^\mu .
$$
(1.35)

The background-quantum splitting of the $(1,0)$ supergravity superfields goes along standard lines: the constraints (1.10) for the full (background-quantum) covariant derivatives are solved in terms of the background-covariant (classical) derivatives and the quantum superfields $(H_+{}^=, H_-{}^{\ddagger}, S)$. The quantum superfields L and $E_+{}^{\ddagger}$ are algebraically gauged away, without introducing propagating ghosts. The quantum scale transformations allow the extra gauge $S = 0$ that also does not imply propagating ghosts. After the background-quantum splitting, the quantum gauge invariance can be fixed by requiring *all* quantum $(1,0)$ supergravity superfields to vanish. [1] Hence, the $(1,0)$ supergravity superfields can be considered as the background only.

Fixing the gauge symmetry by the vanishing quantum $(1,0)$ supergravity superfields induces the Faddeev-Popov (FP) ghost action in $(1,0)$ superspace [418],

$$
I_{\text{FP}}[b, c] = \int d^3 z^- E \left\{ b_-{}^{\ddagger}\nabla_+ c^= + b_+{}^=\nabla_- c^{\ddagger} \right\} ,
$$
(1.36)

where we have merely written down the propagating ghosts originating from the gauge-fixing of $(1,0)$ superdiffeomorphisms. The complete $(1,0)$ ghost system [420] also includes the non-propagating ghosts associated with the gauge-fixing of the $(1,0)$ Weyl and Lorentz gauge symmetries, and the gauge $E_+{}^{\ddagger} = 0$ used in the preceeding subsection. Our goal in this section is the calculation of the anomalous part of the renormalized quantum effective action of the heterotic NLSM in the background of the $(1,0)$ supergravity (Subsect. 7.1.3). In this subsection we develop some technical tools for this purpose.

To integrate over the quantum fields in the action (1.34), in the covariant (in D dimensions) way, we use the covariant background-field method for NLSM (Sects. 2.2 and 3.1), which implies the use of geodesics satisfying the equation

$$
\frac{d^2 X^\mu}{ds^2} + \Gamma_{\nu\rho}^\mu \frac{dX^\nu}{ds}\frac{dX^\rho}{ds} = 0 ,
$$
(1.37)

---

[1] We ignore here possible topological obstructions, and assume the absence of the supergravitational and superconformal anomalies – see Subsect. 7.1.3.

with the boundary conditions $X^\mu|_{s=0} = X^\mu$ and $X^\mu|_{s=1} = X^\mu + \pi^\mu$, where we have introduced the NLSM background superfields (X) and the quantum fluctuations (π). The covariant quantum NLSM superfields ξ^μ are the normal coordinates in the NLSM target space or, equivalently,

$$\xi^\mu = \left.\frac{dX^\mu}{ds}\right|_{s=0} = \xi^\mu(s)|_{s=0} \ . \tag{1.38}$$

The doubly covariant (with respect to the worldsheet $(1,0)$ superdiffeomorphisms and the NLSM target space ($=$spacetime) reparametrizations) derivatives in $(1,0)$ superspace are

$$\mathcal{D}_A \equiv (\mathcal{D}_+, \mathcal{D}_-) = \nabla_A + \Gamma_A \ , \quad \Gamma_{A\mu}{}^\nu \equiv \Gamma_{\sigma\mu}{}^\nu \nabla_A X^\sigma \ . \tag{1.39}$$

The NLSM 'geodesic' derivative with a covariant completion (in the NLSM-sense) satisifes the standard relations (Sect. 2.2)

$$D(s)T_{\mu\cdots} = \mathcal{D}_\nu T_{\mu\cdots}\xi^\nu \ , \quad D(s)\xi^\mu(s) = 0 \ . \tag{1.40}$$

We also use the standard notation for the Cristoffel symbols $\Gamma_{\mu\nu}{}^\lambda$ and the Riemann-Cristoffel curvature $R_{\nu\sigma\mu\tau}$,

$$\Gamma_{\mu\tau,\nu} = \tfrac{1}{2}(\partial_\mu g_{\tau\nu} + \partial_\tau g_{\mu\nu} - \partial_\nu g_{\mu\tau}) \ ,$$

$$R_{\nu\sigma,\mu\tau} = \partial_\nu \Gamma_{\tau\sigma,\mu} - \partial_\sigma \Gamma_{\nu\tau,\mu} + 2\Gamma^\lambda_{\tau[\sigma}\Gamma_{\nu]\lambda,\mu} \ . \tag{1.41}$$

We define the derivative $\mathcal{D}_{+\!\!\!+}$ as

$$2i\mathcal{D}_{+\!\!\!+} = \{\mathcal{D}_+, \mathcal{D}_+\} \ , \quad (\mathcal{D}_{+\!\!\!+} \neq \nabla^{\text{cov.}}_{+\!\!\!+} \ !) \ , \tag{1.42}$$

where $\nabla^{\text{cov.}}_{+\!\!\!+}$ is defined as the minimally covariantized $\nabla_{+\!\!\!+}$-derivative with respect to the Γ-connection, $\nabla^{\text{cov.}}_{+\!\!\!+} = \nabla_{+\!\!\!+} + \Gamma_{+\!\!\!+}$.

It is not difficult to verify the following consequences of the definitions above:

$$\begin{aligned}
\mathcal{D}_+\xi^\mu &= \nabla_+\xi^\mu + \Gamma^\mu_{\tau\sigma}\nabla_+X^\sigma\xi^\tau \ , \\
\mathcal{D}_-\xi^\mu &= \nabla_-\xi^\mu + \Gamma^\mu_{\tau\sigma}\nabla_-X^\sigma\xi^\tau \ , \\
\mathcal{D}_{+\!\!\!+}\xi^\mu &= \nabla^{\text{cov}}_{+\!\!\!+}\xi^\mu - \tfrac{i}{2}R_{\sigma\nu,}{}^\mu{}_\tau\nabla_+X^\sigma\nabla_+X^\nu\xi^\tau \ , \\
D(s)\mathcal{D}_A\xi^\mu(s) &= \xi^\nu(s)\xi^\tau(s)R_{\nu\sigma,}{}^\mu{}_\tau\nabla_A X^\sigma \ ,
\end{aligned} \tag{1.43}$$

and

$$\nabla^{\text{cov.}}_{+\!\!\!+}\xi^\mu = \nabla_{+\!\!\!+}\xi^\mu + \Gamma_{\nu\tau}{}^\mu\nabla_{+\!\!\!+}X^\nu\xi^\tau \ . \tag{1.44}$$

Using $\nabla^{\text{cov.}}_{+\!\!\!+}$ as the definition of $\mathcal{D}_{+\!\!\!+}$ would give rise the anticommutator

$$\{\mathcal{D}_+, \mathcal{D}_+\} = 2i\nabla^{\text{cov.}}_{+\!\!\!+} + R \cdot \nabla_+X\nabla_+X - \text{term} \ , \tag{1.45}$$

instead of (1.42). Though being equivalent, this notation is rather inconvenient.

The manifestly covariant formalism of the background-quantum field expansion in the sector of the heterotic fermions can be developed by using the prescription

$$\Psi^I(s): \quad \Psi^I(0) = \Psi^I \ , \quad \Psi^I(1) = \Psi^I + \Delta^I \ , \quad D^2(s)\Psi^I_-(s) = 0 \ , \quad (1.46)$$

where Δ^I describe the fermionic quantum fluctuations over the fermionic background Ψ^I, whereas $\Psi^I(s)$ interpolate between Ψ^I and $\Psi^I + \Delta^I$ in a gauge-covariant way provided that the derivative $D(s)$ is defined with a gauge-covariant completion, i.e.

$$D(s)\Psi^I_- = \left[\delta^{IJ}\frac{d}{ds} + A^{IJ}_\mu \frac{dX^\mu}{ds}\right]\Psi^J_- = \frac{d\Psi^I_-}{ds} + A^{IJ}_\mu \xi^\mu \Psi^J_- \ . \quad (1.47)$$

The spinors $\chi^I_- \equiv D(s)\Psi^I_-(s)\big|_{s=0}$, satisfying the equation

$$D(s)\chi^I_-(s) = 0 \ , \quad (1.48)$$

can be used as supercovariant and gauge-covariant quantum fermionic superfields. This way of background-quantum splitting in the gauge sector of the heterotic string is equivalent to the use of the (generalized) *Schwinger* gauge $A_\mu \xi^\mu = 0$.

The gauge-covariantized spinor derivatives in a curved $(1,0)$ superspace read

$$(\mathcal{D}_+\Psi_-)^I \equiv \left(\delta^{IJ}\nabla_+ + A^{IJ}_\mu \nabla_+X^\mu\right)\Psi^J_- \ . \quad (1.49)$$

Here are some useful identities:

$$D(s)\mathcal{D}_+\Psi^I_- = \mathcal{D}_+\chi^I_- + F^{IJ}_{\mu\nu}\xi^\mu\nabla_+X^\nu\Psi^J_- \ ,$$

$$D(s)F^{IJ}_{\mu\nu} = \xi^\rho\mathcal{D}_\rho F^{IJ}_{\mu\nu} \ , \quad (1.50)$$

where $\mathcal{D}_\rho$ is the standard covariant and gauge-covariant derivative, whereas $F^{IJ}_{\mu\nu}$ is the corresponding gauge field strength,

$$F^{IJ}_{\mu\nu} = \partial_\mu A^{IJ}_\nu - \partial_\nu A^{IJ}_\mu + A^{IK}_\mu A^{KJ}_\nu - A^{IK}_\nu A^{KJ}_\mu \ . \quad (1.51)$$

The action (1.34) consists of the four different terms, so that we describe the background-quantum expansion of each term separately,

$$I_{\text{het.}} = I_{\text{grav.}} + I_{\text{WZ}} + I_{\text{H}} + I_{\text{FT}} \ . \quad (1.52)$$

The background-quantum field expansion of any term in (1.52) has the generic structure (cf. Chaps. 2 and 3)

$$I[X+\pi(\xi), \Psi+\Delta(\chi)] = I[X,\Psi] + \sum_{n=1}^{\infty} I^{(n)} \ , \quad I^{(n)} = \frac{1}{n!}\frac{d^n I(s)}{ds^n}\bigg|_{s=0} \ . \quad (1.53)$$

(i) In the gravitational sector we have

$$I_{\text{grav.}} = -\frac{1}{4\pi\alpha'} \int d^3z^- E i \nabla_+ X^\mu \nabla_- X^\nu g_{\mu\nu}(X) \ . \tag{1.54}$$

It is straightforward to calculate $I_{\text{grav.}}^{(n)}$. As regards the first six leading terms, we find

$$I_{\text{grav.}}^{(1)} = -\frac{1}{2\pi\alpha'} \int d^3z^- E i \mathcal{D}_+ \xi^\mu \nabla_- X^\nu g_{\mu\nu} \ ,$$

$$I_{\text{grav.}}^{(2)} = -\frac{1}{4\pi\alpha'} \int d^3z^- E i \left\{ \mathcal{D}_+ \xi^\mu \mathcal{D}_- \xi^\nu g_{\mu\nu} + R_{\lambda\sigma,\nu\tau} \nabla_+ X^\sigma \nabla_- X^\nu \xi^\lambda \xi^\tau \right\} \ ,$$

$$I_{\text{grav.}}^{(3)} = -\frac{1}{4\pi\alpha'} \int d^3z^- E i \left\{ \tfrac{2}{3} \xi^\lambda \xi^\tau R_{\lambda\sigma,\nu\tau} \nabla_+ X^\sigma \mathcal{D}_- \xi^\nu \right.$$
$$\left. + \tfrac{2}{3} \xi^\lambda \xi^\tau R_{\lambda\sigma,\nu\tau} \nabla_+ X^\sigma \mathcal{D}_+ \xi^\nu + \tfrac{1}{3} \mathcal{D}_\mu R_{\lambda\sigma,\nu\tau} \nabla_+ X^\sigma \nabla_- X^\nu \xi^\mu \xi^\lambda \xi^\tau \right\} \ ,$$

$$I_{\text{grav.}}^{(4)} = -\frac{1}{4\pi\alpha'} \int d^3z^- E i \left\{ \tfrac{1}{4} \mathcal{D}_\mu R_{\lambda\sigma,\nu\tau} \nabla_+ X^\sigma \mathcal{D}_- \xi^\nu \xi^\mu \xi^\lambda \xi^\tau \right.$$
$$+ \tfrac{1}{4} \mathcal{D}_\mu R_{\lambda\sigma,\nu\tau} \nabla_- X^\sigma \mathcal{D}_+ \xi^\nu \xi^\mu \xi^\lambda \xi^\tau + \tfrac{1}{3} R_{\lambda\sigma,\nu\tau} \xi^\lambda \xi^\tau \mathcal{D}_+ \xi^\sigma \mathcal{D}_- \xi^\nu$$
$$\left. + \left[\tfrac{1}{3} R_{\lambda\sigma,\nu\tau} R_{\delta\varepsilon,}{}^\nu{}_\gamma + \tfrac{1}{12} \mathcal{D}_\delta \mathcal{D}_\gamma R_{\lambda\sigma,\varepsilon\tau} \right] \nabla_+ X^\sigma \nabla_- X^\varepsilon \xi^\lambda \xi^\tau \xi^\delta \xi^\gamma \right\}$$

$$I_{\text{grav.}}^{(5)} = -\frac{1}{4\pi\alpha'} \int d^3z^- E i \left\{ \tfrac{1}{6} \mathcal{D}_\mu R_{\lambda\sigma,\nu\tau} \mathcal{D}_+ \xi^\sigma \mathcal{D}_- \xi^\nu \xi^\mu \xi^\lambda \xi^\tau + \dots \right\} \ ,$$

$$I_{\text{grav.}}^{(6)} = -\frac{1}{4\pi\alpha'} \int d^3z^- E i \left\{ \left(\tfrac{1}{20} \mathcal{D}_\alpha \mathcal{D}_\mu R_{\lambda\sigma,\nu\tau} + \tfrac{2}{45} R_{\lambda\gamma,\nu\tau} R_{\alpha\sigma,}{}^\gamma{}_\mu \right) \mathcal{D}_+ \xi^\sigma \right.$$
$$\left. \times \mathcal{D}_- \xi^\nu \xi^\alpha \xi^\mu \xi^\lambda \xi^\tau + \dots \right\} \ ,$$
$$\tag{1.55}$$

where the dots stand for extra terms (with the factors ∇X) that are not relevant for a calculation of anomalies (Subsect. 7.1.3).

(ii) As regards the WZ-sector,

$$I_{\text{WZ}} = -\frac{1}{4\pi\alpha'} \int d^3z^- E i \nabla_+ X^\mu \nabla_- X^\nu b_{\mu\nu}(X) \ , \tag{1.56}$$

the background-dependent vertices can only depend upon the field strength H of the antisymmetric tensor b,

$$H_{\tau\mu\nu} = \tfrac{1}{2} \left(\partial_\tau b_{\mu\nu} + \partial_\mu b_{\nu\tau} + \partial_\nu b_{\tau\mu} \right) = \tfrac{3}{2} \partial_{[\tau} b_{\mu\nu]} \ . \tag{1.57}$$

We find

$$I_{WZ}^{(1)} = -\frac{1}{2\pi\alpha'} \int d^3z^- \, Ei\xi^\tau \nabla_+ X^\mu \nabla_- X^\nu H_{\tau\mu\nu}(X) \ ,$$

$$I_{WZ}^{(2)} = -\frac{1}{4\pi\alpha'} \int d^3z^- \, Ei \left\{ \xi^\tau \mathcal{D}_+ \xi^\mu \nabla_- X^\nu H_{\tau\mu\nu} \right.$$

$$\left. + \xi^\tau \nabla_+ X^\mu \mathcal{D}_- \xi^\nu H_{\tau\mu\nu} + \xi^\lambda \xi^\tau \nabla_+ X^\mu \nabla_- X^\nu \mathcal{D}_\lambda H_{\tau\mu\nu} \right\} \ ,$$

$$I_{WZ}^{(3)} = -\frac{1}{4\pi\alpha'} \int d^3z^- \, Ei \left\{ \tfrac{2}{3} R^\mu{}_{\gamma,\rho[\delta} H_{\nu]\tau\mu} \xi^\tau \xi^\rho \xi^\gamma \nabla_+ X^\delta \nabla_- X^\nu \right.$$

$$+ \tfrac{2}{3} H_{\tau\mu\nu} \xi^\tau \mathcal{D}_+ \xi^\mu \mathcal{D}_- \xi^\nu + \tfrac{2}{3} \xi^\tau \xi^\lambda \left(\mathcal{D}_+ \xi^\mu \nabla_- X^\nu + \mathcal{D}_- \xi^\nu \nabla_+ X^\mu \right)$$

$$\left. \times \mathcal{D}_\lambda H_{\tau\mu\nu} + \tfrac{1}{3} \xi^\tau \xi^\lambda \xi^\sigma \nabla_+ X^\mu \nabla_- X^\nu \mathcal{D}_\sigma \mathcal{D}_\lambda H_{\tau\mu\nu} \right\} \ ,$$

$$I_{WZ}^{(4)} = -\frac{1}{4\pi\alpha'} \int d^3z^- \, Ei \left\{ \tfrac{1}{2} R^\mu{}_{\gamma,\rho[\delta} H_{\nu]\tau\mu} \xi^\tau \xi^\rho \xi^\gamma (\nabla_+ X^\delta \mathcal{D}_- \xi^\nu \right.$$

$$+ \mathcal{D}_+ \xi^\nu \nabla_- X^\delta) + \tfrac{1}{6} \xi^\lambda \xi^\tau \xi^\rho \xi^\gamma \nabla_+ X^\delta \nabla_- X^\nu \mathcal{D}_\lambda (R^\mu{}_{\gamma,\rho[\delta} H_{\nu]\tau\mu})$$

$$+ \tfrac{1}{2} \xi^\lambda \xi^\tau \mathcal{D}_+ \xi^\mu \mathcal{D}_- \xi^\nu \mathcal{D}_\lambda H_{\tau\mu\nu} + \xi^\alpha \xi^\beta \xi^\tau \xi^\lambda \nabla_+ X^\nu \nabla_- X^\rho$$

$$\times \left(\tfrac{1}{3} \mathcal{D}_\lambda H_{\tau\mu[\rho} R_{\nu]\alpha,}{}^\mu{}_\beta - \tfrac{1}{12} \mathcal{D}_\alpha \mathcal{D}_\beta \mathcal{D}_\lambda H_{\tau\nu\rho} \right)$$

$$\left. + \tfrac{1}{4} \xi^\tau \xi^\lambda \xi^\gamma \left(\mathcal{D}_+ \xi^\mu \nabla_- X^\nu + \mathcal{D}_- \xi^\nu \nabla_+ X^\mu \right) \mathcal{D}_\gamma \mathcal{D}_\lambda H_{\tau\mu\nu} \right\} \ ,$$

$$I_{WZ}^{(5)} = -\frac{1}{4\pi\alpha'} \int d^3z^- \, Ei\xi^\lambda \xi^\tau \xi^\gamma \mathcal{D}_+ \xi^\mu \mathcal{D}_- \xi^\nu$$

$$\times \left[\tfrac{1}{5} \mathcal{D}_\gamma \mathcal{D}_\lambda H_{\tau\mu\nu} + \tfrac{2}{15} R^\rho{}_{\gamma,\lambda[\mu} H_{\nu]\tau\rho} \right] + \ldots \ ,$$

$$I_{WZ}^{(6)} = -\frac{1}{4\pi\alpha'} \int d^3z^- \, Ei\xi^\alpha \xi^\beta \xi^\lambda \xi^\tau \mathcal{D}_+ \xi^\rho \mathcal{D}_- \xi^\nu \left\{ \tfrac{1}{18} \mathcal{D}_\alpha \mathcal{D}_\beta \right.$$

$$\left. \times \mathcal{D}_\gamma H_{\tau\rho\nu} + \tfrac{1}{18} \mathcal{D}_\lambda R^\mu{}_{\beta,\alpha[\rho} H_{\nu]\tau\mu} + \tfrac{1}{9} H_{\tau\mu[\rho} R_{\nu]\alpha,}{}^\mu{}_\beta \right\} + \ldots \ ,$$

$$\text{(1.58)}$$

where we have explicitly written down only those terms that are going to be relevant for our calculation of perturbative anomalies in Subsect. 7.1.3. All (anti)symmetrizations above are defined with unit weight.

It is sometimes convenient to introduce the connection and the curvature with torsion (Sect. 2.2),

$$\Gamma_{\sigma\tau}^{(\pm)\mu} = \Gamma_{\sigma\tau}^\mu \pm H_{\sigma\tau}^\mu \ , \qquad R_{\mu\nu,\rho\sigma}^{(\pm)} = R_{\mu\nu,\rho\sigma}(\Gamma^{(\pm)}) \ , \qquad \text{(1.59)}$$

as well as the associated covariant derivatives,

$$\mathcal{D}_A^{(\pm)} \xi^\mu \equiv \mathcal{D}_A \xi^\mu \pm H_{\tau\sigma}^\mu \nabla_A X^\sigma \xi^\tau \ . \qquad \text{(1.60)}$$

With the definitions (1.59) and (1.60), the leading terms in the expansions (1.55) and (1.58) can be rewritten as follows:

$$I^{(1)}_{\text{grav.+WZ}} = -\frac{1}{2\pi\alpha'} \int d^3z^- \, Ei\mathcal{D}^{(+)}_+ \xi^\mu \nabla_= X^\nu g_{\mu\nu} \ ,$$

$$I^{(2)}_{\text{grav.+WZ}} = -\frac{1}{4\pi\alpha'} \int d^3z^- \, \Big\{ \mathcal{D}^{(+)}_+ \xi^\mu \mathcal{D}^{(-)}_= \xi^\nu g_{\mu\nu} \tag{1.61}$$

$$+ R^{(+)}_{\nu\sigma,\tau\mu} \nabla_+ X^\sigma \nabla_= X^\tau \xi^\nu \xi^\mu \Big\} \ ,$$

where we have used yet another identity

$$D(s)\mathcal{D}^{(+)}_+ \xi^\mu = R^{(+)\mu}_{\nu\sigma,\ \tau} \nabla_+ X^\sigma \xi^\nu \xi^\tau \ . \tag{1.62}$$

The torsion H does not explicitly appear in (1.61). However, it is not the case in the higher orders in ξ. For example,

$$I^{(3)}_{\text{grav.+WZ}} = -\frac{1}{4\pi\alpha'} \int d^3z^- \, Ei \Big\{ \tfrac{1}{3} \xi^\lambda \xi^\tau R^{(+)}_{\lambda\sigma,\nu\tau} \nabla_+ X^\sigma \mathcal{D}^{(-)}_= \xi^\nu$$

$$+ \tfrac{2}{3} \mathcal{D}^{(+)}_+ \xi^\lambda \mathcal{D}^{(-)}_= \xi^\nu \xi^\tau H_{\nu\tau\lambda} + \tfrac{1}{3} \mathcal{D}^{(+)}_+ \xi^\mu \xi^\lambda \xi^\tau R^{(-)}_{\lambda\sigma,\mu\tau} \nabla_= X^\sigma$$

$$+ \tfrac{1}{3} \mathcal{D}^{(+)}_+ \xi^\sigma R^{(+)}_{\nu\sigma,\tau\mu} \nabla_= X^\tau \xi^\nu \xi^\mu + \tfrac{1}{3} R^{(+)}_{\nu\sigma,\tau\mu} \nabla_+ X^\sigma \mathcal{D}^{(+)}_= \xi^\tau \xi^\nu \xi^\mu$$

$$+ \tfrac{1}{3} \mathcal{D}^{(+)}_\lambda R^{(+)}_{\nu\sigma,\tau\mu} \xi^\lambda \xi^\nu \xi^\mu \nabla_+ X^\sigma \nabla_= X^\tau \Big\} \ . \tag{1.63}$$

(iii) The Fradkin-Tseytlin term in (1.34) has the following background-quantum field expansion:

$$I_{\text{FT}} = -\frac{1}{4\pi\alpha'} \int d^3z^- \, E\alpha' \Sigma^+ \Phi \ ,$$

$$I^{(1)}_{\text{FT}} = -\frac{1}{4\pi\alpha'} \int d^3z^- \, E\alpha' \Sigma^+ \xi^\lambda \partial_\lambda \Phi \ ,$$

$$I^{(2)}_{\text{FT}} = -\frac{1}{4\pi\alpha'} \int d^3z^- \, E\alpha' \Sigma^+ \tfrac{1}{2!} \xi^\gamma \xi^\lambda D_\lambda D_\gamma \Phi \ , \tag{1.64}$$

$$I^{(n)}_{\text{FT}} = -\frac{1}{4\pi\alpha'} \int d^3z^- \, E\alpha' \Sigma^+ \tfrac{1}{n!} \xi^{\lambda_1} \cdots \xi^{\lambda_n} D_{\lambda_1} \cdots D_{\lambda_n} \Phi \ .$$

(iv) Finally, the 'heterotic' term

$$I_{\text{H}}[\Psi, X] = -\frac{1}{4\pi\alpha'} \int d^3z^- \, E\Psi^I_- (D_+ \Psi_-)^I \tag{1.65}$$

in (1.34), after the substitutions $\Psi \to \Psi + \Delta(\chi)$, $X \to X + \pi(\xi)$, yields

$$I_{\rm H}^{(1)} = -\frac{1}{4\pi\alpha'}\int {\rm d}^3 z^- \, E\Psi_-^I F_{\mu\nu}^{IJ}\xi^\mu\nabla_+ X^\nu\Psi_-^J \,,$$

$$I_{\rm H}^{(2)} = -\frac{1}{4\pi\alpha'}\int {\rm d}^3 z^- \, E\left\{\chi_-^I \mathcal{D}_+\chi_-^I + 2\chi_-^I F_{\mu\nu}^{IJ}\xi^\mu\nabla_+ X^\nu\Psi_-^J \right.$$
$$\left. -\tfrac{1}{2}\Psi_-^I\Psi_-^J \mathcal{D}_\lambda F_{\mu\nu}^{IJ}\nabla_+ X^\nu\xi^\lambda\xi^\mu - \tfrac{1}{2}\Psi_-^I\Psi_-^J F_{\mu\nu}^{IJ}\xi^\mu \mathcal{D}_+\xi^\nu\right\}\,,$$

$$I_{\rm H}^{(3)} = -\frac{1}{4\pi\alpha'}\int {\rm d}^3 z^- \, E\left\{\chi_-^I F_{\mu\nu}^{IJ}\xi^\mu\nabla_+ X^\nu\chi_-^J + \chi_-^I\xi^\lambda\xi^\mu \mathcal{D}_\lambda F_{\mu\nu}^{IJ} \right.$$
$$\times \nabla_+ X^\nu\Psi_-^J + \chi_-^I F_{\mu\nu}^{IJ}\xi^\mu \mathcal{D}_+\xi^\nu\Psi_-^J - \tfrac{1}{3}\Psi_-^I\Psi_-^J \mathcal{D}_\lambda F_{\mu\nu}^{IJ}\mathcal{D}_+\xi^\mu\xi^\lambda\xi^\nu$$
$$\left. -\tfrac{1}{6}\Psi_-^I\Psi_-^J\xi^\mu\xi^\tau\xi^\lambda\nabla_+ X^\nu\left(\mathcal{D}_\tau \mathcal{D}_\lambda F_{\mu\nu}^{IJ} - R_{\lambda\nu,}{}^\gamma{}_\tau F_{\mu\gamma}^{IJ}\right)\right\}\,,$$

$$I_{\rm H}^{(4)} = -\frac{1}{4\pi\alpha'}\int {\rm d}^3 z^- \, E\left\{\tfrac{1}{2}\chi_-^I\chi_-^J\xi^\nu \mathcal{D}_+\xi^\mu F_{\mu\nu}^{IJ} - \tfrac{1}{2}\chi_-^I\chi_-^J\xi^\lambda\xi^\mu \right.$$
$$\times \mathcal{D}_\lambda F_{\mu\nu}^{IJ}\mathcal{D}_+\xi^\nu + \tfrac{2}{3}\chi_-^I\xi^\lambda\xi^\mu \mathcal{D}_\lambda F_{\mu\nu}^{IJ}\mathcal{D}_+\xi^\nu\Psi_-^J + \tfrac{1}{3}\chi_-^I\xi^\gamma\xi^\lambda\xi^\mu$$
$$\times \left(\mathcal{D}_\gamma \mathcal{D}_\lambda F_{\mu\nu}^{IJ} + F_{\mu\rho}^{IJ}R_{\gamma\nu,}{}^\rho{}_\lambda\right)\nabla_+ X^\nu\Psi_-^J - \tfrac{1}{24}\Psi_-^I\Psi_-^J\xi^\mu\xi^\lambda\xi^\gamma$$
$$\times \left(3\mathcal{D}_\mu \mathcal{D}_\lambda F_{\gamma\nu}^{IJ} + F_{\mu\rho}^{IJ}R_{\lambda\nu,}{}^\rho{}_\gamma\right)\mathcal{D}_+\xi^\nu - \tfrac{1}{24}\Psi_-^I\Psi_-^J\xi^\gamma\xi^\tau\xi^\lambda\xi^\mu$$
$$\left. \times \left(\mathcal{D}_\gamma \mathcal{D}_\tau \mathcal{D}_\lambda F_{\mu\nu}^{IJ} + F_{\tau\rho}^{IJ}\mathcal{D}_\gamma R_{\lambda\nu,}{}^\rho{}_\mu + 3\mathcal{D}_\gamma F_{\tau\rho}^{IJ}R_{\lambda\nu,}{}^\rho{}_\mu\right)\nabla_+ X^\nu\right\}\,,$$

$$I_{\rm H}^{(5)} = -\frac{1}{4\pi\alpha'}\int {\rm d}^3 z^- \, E\left\{-\tfrac{1}{3}\chi_-^I\chi_-^J\xi^\mu\xi^\lambda \mathcal{D}_\lambda F_{\mu\nu}^{IJ}\mathcal{D}_+\xi^\nu + \ldots\right\}\,,$$

$$I_{\rm H}^{(6)} = -\frac{1}{4\pi\alpha'}\int {\rm d}^3 z^- \, E\left\{-\tfrac{1}{24}\chi_-^I\chi_-^J\xi^\gamma\xi^\lambda\xi^\mu\left(3\mathcal{D}_\gamma \mathcal{D}_\lambda F_{\mu\nu}^{IJ}\right.\right.$$
$$\left.\left. +F_{\gamma\rho}^{IJ}R_{\lambda\nu,}{}^\rho{}_\mu\right)\mathcal{D}_+\xi^\nu + \ldots\right\}\,.$$

$$(1.66)$$

The free actions of the quantum $(1,0)$ superfields ξ^μ and χ_-^I in the expansions (1.55) and (1.66) determine their superpropagators in $(1,0)$ superspace. We find $(2\pi\alpha' = 1)$

$$\langle\xi^\mu(z)\xi^\nu(z')\rangle = g^{\mu\nu}\frac{D_+}{\Box}\delta_-^{(3)}(z,z')$$
$$= g^{\mu\nu}\int\frac{{\rm d}^2 p}{(2\pi)^2}\frac{1}{-p^2}D_+ {\rm e}^{ip\cdot(x-x')}\delta_-(\theta-\theta')\,,$$

$$\langle\chi_-^I(z)\chi_-^J(z')\rangle = \delta^{IJ}\frac{i\partial_= D_+}{\Box}\delta_-^{(3)}(z,z')$$
$$= \delta^{IJ}\int\frac{{\rm d}^2 p}{(2\pi)^2}\frac{p_=}{p^2}D_+ {\rm e}^{ip\cdot(x-x')}\delta_-(\theta-\theta')\,.$$

$$(1.67)$$

Fig. 7.1. The self-energy type function that determines the supergravitational and superconformal anomalies of the heterotic string in the background of its massless bosonic modes

The background-field expansions (i)–(iv) and the superpropagators (1.67) define the covariant Feynman rules in $(1,0)$ superspace for the heterotic NLSM. They can be used for the calculation of the heterotic NLSM supergravitational and superconformal anomalies that are of interest in heterotic string theory (see Subsect. 7.1.3)

7.1.3 Supergravitational and Superconformal Anomalies

We assume that the reader is familiar with anomalies in QFT. The discovery of anomalies is often referred to the original papers [421, 422] (see e.g., [423, 424, 425] for a review or an introduction). The conformal and gravitational anomalies in various dimensions were investigated in [426, 427, 428, 429].

The supergravitational and superconformal anomalies of the heterotic NLSM in a curved superspace of $(1,0)$ supergravity can be extracted from the self-energy graphs (Fig. 7.1).

The results of a calculation of the graphs depicted in Fig. 7.1 are going to be dependent upon the $(1,0)$ superfield pre-potentials of the background $(1,0)$ supergravity, $H_+{}^=$ and $H_-{}^{\dagger}$, in a (gauge) non-covariant way. The gauge invariance can, nevertheless, be restored by adding some *local* counterterms to be dependent upon $H_+{}^=$, $H_-{}^{\dagger}$ and S. The final form of the anomaly depends upon the scale compensator S in a gauge-invariant way. In other words, a supergravitational anomaly can be 'traded' for a conformal anomaly, in agreement with the standard (Adler-Rosenberg) procedure for computing Lorentz-invariant anomalies [424, 425]. The explicit Feynman rules formulated in Subsect. 7.1.2 are enough up to five loops.

The relevant part of the heterotic NLSM action (without ghosts, and with $2\pi\alpha' = 1$) can be rewritten in the form

$$I = I_0 + I_{\text{int.}} , \quad I_{\text{int.}} = I_0' + I_1 , \tag{1.68}$$

with

$$I_0 = -\frac{1}{2} \int d^3z^- \left[iD_+\xi^\mu \partial_= \xi^\nu g_{\mu\nu} + \chi^I_- D_+\chi^I_- \right] , \tag{1.69}$$

$$
\begin{aligned}
I_0' = \ -\frac{1}{2} \int d^3z^- \Big\{ &\big(\xi^\rho A_{\rho\mu\nu} + \xi^\lambda\xi^\rho B_{\lambda\rho\mu\nu} + \xi^\gamma\xi^\lambda\xi^\rho \mathcal{D}_{\gamma\lambda\rho\mu\nu} \\
&+ \xi^\tau\xi^\gamma\xi^\lambda\xi^\rho M_{\gamma\lambda\rho\tau\mu\nu} \big) iD_+\xi^\mu \partial_= \xi^\nu \\
&+ \big(\xi^\mu C^{IJ}_{\mu\nu} + \xi^\lambda\xi^\mu E^{IJ}_{\lambda\mu\nu} + \xi^\gamma\xi^\lambda\xi^\mu K^{IJ}_{\gamma\lambda\mu\nu} \big) \chi^I_-\chi^J_- D_+\xi^\nu \Big\} ,
\end{aligned}
$$

$$I_1 = -\frac{1}{2} \int d^3 z^- \left\{ \left(g_{\mu\nu} + \xi^\rho A_{\rho\mu\nu} + \xi^\lambda \xi^\rho B_{\lambda\rho\mu\nu} + \xi^\gamma \xi^\lambda \xi^\rho \mathcal{D}_{\gamma\lambda\rho\mu\nu} \right. \right.$$

$$\left. + \xi^\tau \xi^\lambda \xi^\gamma \xi^\rho M_{\gamma\lambda\rho\tau\mu\nu} \right) \left[iH_+^= \partial_- \xi^\mu \partial_- \xi^\nu + \frac{1}{2} D_+ \xi^\mu (D_+ H_-^{\,\ddagger}) D_+ \xi^\nu \right.$$

$$\left. + i D_+ \xi^\mu H_-^{\,\ddagger} \partial_{\ddagger} \xi^\nu \right] + \chi_-^I H_+^= \partial_- \chi_-^I + \left(\xi^\mu C_{\mu\nu}^{IJ} + \xi^\lambda \xi^\mu E_{\lambda\mu\nu}^{IJ} \right.$$

$$\left. + \xi^\gamma \xi^\lambda \xi^\mu K_{\gamma\lambda\mu\nu}^{IJ} \right) \chi_-^I \chi_-^J H_+^= \partial_- \xi^\nu + \frac{1}{2\pi} \sum_{n=2}^{6} \frac{1}{n!} \mathcal{D}_{\mu_1} \cdots \mathcal{D}_{\mu_n} \Phi$$

$$\left. \times \xi^{\mu_1} \cdots \xi^{\mu_n} \left(\frac{i}{2} D_+ \partial_{\ddagger} H_-^{\,\ddagger} + \frac{i}{2} \partial_-^2 H_+^= \right) \right\} ,$$

(1.70)

where we have introduced the notation

$$A_{\rho\mu\nu} = \frac{2}{3} H_{\rho\mu\nu} , \quad B_{\lambda\rho\mu\nu} = \frac{1}{3} R_{\lambda\mu\nu\rho} + \frac{1}{2} \mathcal{D}_\lambda H_{\rho\mu\nu} ,$$

$$\mathcal{D}_{\gamma\lambda\rho\mu\nu} = \frac{1}{6} \mathcal{D}_\gamma R_{\lambda\mu\nu\rho} + \frac{1}{5} \mathcal{D}_\gamma \mathcal{D}_\lambda H_{\rho\mu\nu} + \frac{2}{15} R^\tau{}_{\lambda,\gamma[\mu} H_{\gamma]\rho\tau} ,$$

$$M_{\gamma\lambda\rho\tau\mu\nu} = \frac{1}{20} \mathcal{D}_\gamma \mathcal{D}_\lambda R_{\rho\mu\nu\tau} + + \frac{2}{45} R_{\rho\delta,\nu\tau} R_{\gamma\mu,}{}^\delta{}_\lambda + \frac{1}{18} \mathcal{D}_\gamma \mathcal{D}_\lambda \mathcal{D}_\rho H_{\tau\mu\nu}$$

$$+ \frac{1}{18} \mathcal{D}_\gamma R^\delta{}_{\rho,\lambda[\mu} H_{\nu]\tau\delta} + \frac{1}{9} \mathcal{D}_\rho H_{\tau\delta[\mu} R_{\nu]\gamma,}{}^\delta{}_\lambda ,$$

(1.71)

$$C_{\mu\nu}^{IJ} = -\frac{1}{2} F_{\mu\nu}^{IJ} , \quad E_{\lambda\mu\nu}^{IJ} = -\frac{1}{3} \mathcal{D}_\lambda F_{\mu\nu}^{IJ} ,$$

$$K_{\gamma\lambda\mu\nu}^{IJ} = -\frac{1}{24} \left(3 \mathcal{D}_\gamma \mathcal{D}_\lambda F_{\mu\nu}^{IJ} + F_{\gamma\rho}^{IJ} R_{\lambda\nu,}{}^\rho{}_\mu \right) .$$

Note that the (1,0) superfield $H_+^=$ is fermionic, whereas the (1,0) superfield $H_-^{\,\ddagger}$ is bosonic.

The supergravitational anomaly in (1,0) superspace takes the form

$$\frac{1}{32\pi} \int d^3 z^- \left\{ \gamma_1 D_+ H_+^= \frac{\partial_=^4}{\Box} H_+^= - i\gamma_2 D_+ H_-^{\,\ddagger} \frac{\partial_{\ddagger}^3}{\Box} H_-^{\,\ddagger} \right\} , \quad (1.72)$$

where γ_1 and γ_2 are background-dependent coefficients to be determined. The form of (1.72) is dictated by (1,0) supersymmetry and the Wess-Zumino consistency condition for the anomaly. Of course, this form also appears in explicit calculations. The coefficients $\gamma_{1,2}$ can be calculated in quantum perturbation theory, order-by-order in α'. The (1,0) supergraph techniques are essentially based on the D-algebra of the (1,0) superspace covariant derivatives, which allows one to reduce the multi-loop (1,0) superspace integrals to standard multi-loop integrals in 2d momentum space [418]. The 2d momentum integrals can be computed by using the standard Feynman parametrization.

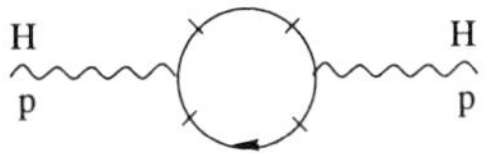

Fig. 7.2. The one-loop $(1,0)$ supergraph contributing to the supergravitational anomaly. Dashes stand for the $\partial_=$-derivatives or the corresponding momenta

The One-Loop Anomaly of the Heterotic String. The one-loop supergravitational anomalies (with ghost contributions) were calculated in [418]. For example, the one-loop supergraph depicted in Fig. 7.2 gives rise to the following momentum integral:

$$
\frac{1}{(2\pi)^2} \int d^2k \, \frac{k_=^2 (k_= + p_=)^2}{k^2 (k+p)^2}
$$

$$
= \frac{1}{(2\pi)^2} \int_0^1 d\alpha \int d^2k \, \frac{(k_= - \alpha p_=)^2 (k_= + (1-\alpha)p_=)^2}{(k + \alpha(1-\alpha)p^2)^2}
$$

$$
= p_=^4 \int_0^1 d\alpha \, \alpha(1-\alpha) \int \frac{d^2k}{(2\pi)^2} \frac{1}{(k^2 + p^2)^2} = -\frac{i}{24\pi} \frac{p_=^4}{p^2} \ .
$$

(1.73)

It can be shown (for dimensional grounds) that all graphs of the type depicted in Fig. 7.3 merely imply *local* contributions to the anomaly, so that they can be disregarded.

The one-loop $(1,0)$ ghost contributions for the heterotic string are well known [407, 408], so that we merely summarize the results. The one-loop anomalous effective action is given by [418]

$$
W_{\text{one-loop}} = \frac{1}{96\pi} \int d^3 z^- \left\{ (D - 26 + \tfrac{1}{2}N)D_+ H_+^= \frac{\partial_=^4}{\Box} H_+^= \right.
$$

$$
\left. - \tfrac{3}{2}i(D - 10)D_+ H_-^\pm \frac{\partial_\pm^3}{\Box} H_-^\pm \right\} \ .
$$

(1.74)

The action (1.74) is not invariant under the linearized background gauge transformations, while

$$
\gamma_1^{\text{one-loop}} = \tfrac{1}{3}(D - 26 + \tfrac{1}{2}N) \neq \tfrac{1}{2}(D - 10) = \gamma_2^{\text{one-loop}} \ .
$$

(1.75)

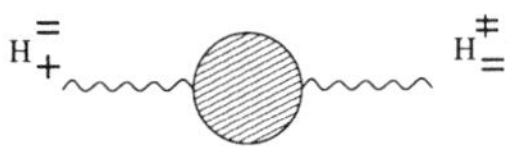

$$
H_+^= \sim\!\!\!\sim\!\!\!\sim \bigotimes \sim\!\!\!\sim\!\!\!\sim H_-^\pm
$$

Fig. 7.3. The $(1,0)$ supergraphs that are irrelevant for the supergravitational anomaly

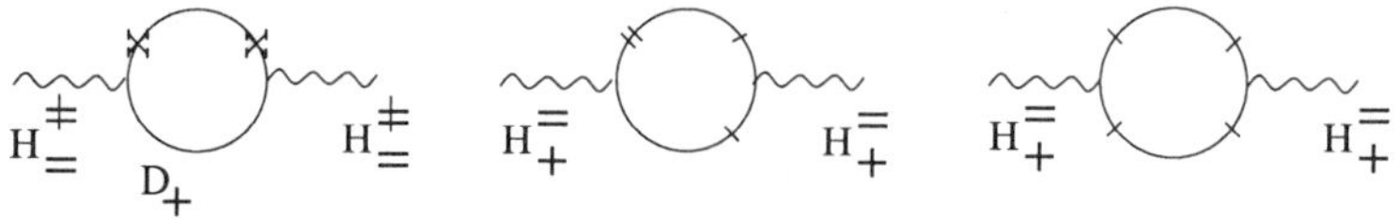

Fig. 7.4. The one-loop $(1,0)$ supergraphs describing the dilaton contribution to the anomaly of the heterotic NLSM. Dashes and crosses stand for the ∂_- and $\partial_{\pm}$ derivatives, respectively

However, after adding some local finite counterterms of the type

$$H_+ {}^= \Box H_- {}^\pm , \quad S\partial_-^2 H_+{}^= , \quad S\partial_{\pm} D_+ H_-{}^\pm , \quad S\partial_- D_+ S , \tag{1.76}$$

one can rewrite the action (1.74) in the manifestly gauge-invariant form, in terms of the (1,0) supergravity field strength Σ^+ [418],

$$W_{\text{one-loop, modified}} \;=\; \frac{1}{16\pi}(D-10)\int d^3 z^- \, \Sigma^+ \left(\frac{D_+}{\Box}\right) \Sigma^+ , \tag{1.77}$$

provided that $N = D + 22$, i.e. $\gamma_1 = \gamma_2 \equiv \gamma$.

The action (1.77) represents the conformal anomaly of the heterotic string in a flat background (Σ^+ is S-dependent!). The anomaly vanishes provided that

$$D = 10 , \quad \text{and} \quad N = 32 , \tag{1.78}$$

as it should [407, 408].

The one-loop dilaton contribution to the heterotic NLSM anomaly is dictated by the (1,0) graphs depicted in Fig. 7.4. We find

$$W_\Phi^{1-\text{loop}} \;=\; \frac{\mathcal{D}^2 \Phi}{8(4\pi)^2} \int d^3 z^- \left\{ iD_+ H_- {}^\pm \frac{\partial_{\pm}^3}{\Box} H_- {}^\pm - D_+ H_+ {}^= \frac{\partial_{\pm}^4}{\Box} H_+ {}^= \right\} ,$$

which implies

$$\gamma_\Phi^{1-\text{loop}} \;=\; -\frac{1}{4\pi}\mathcal{D}^2 \Phi . \tag{1.79}$$

Two-Loop NLSM Corrections to the Anomaly. The corrections to the supergravitational anomaly of the heterotic NLSM, which are dependent of the B-tensor (see (1.71) for the definition of B), are determined by the diagrams schematically depicted in Fig. 7.5.

The corresponding momentum integrals (after performing the D-algebra) are given by

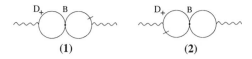

Fig. 7.5. The two-loop B-dependent $(1,0)$ supergraphs contributing to the anomaly of the heterotic NLSM

$$
I_{(1)}(p) = \int \frac{d^2k d^2q}{(2\pi)^4} \frac{q_=^2 (q_= + p_=)(k_= + p_=)}{(q^2 - m^2)[(q + p)^2 - m^2]}
$$

$$
\times \frac{k^2}{(k^2 - m^2)[(k + p)^2 - m^2]} = \frac{1}{64\pi^2} \frac{p_=^4}{p^2} + O(m^2) ,
$$

$$
I_{(2)}(p) = \int \frac{d^2k d^2q}{(2\pi)^4} \frac{q_=(q_= + p_=)}{(q^2 - m^2)[(q + p)^2 - m^2]}
$$

$$
\times \frac{k^2 (k_= + p_=)^2}{(k^2 - m^2)[(k + p)^2 - m^2]} = -\frac{1}{32\pi^2} \frac{p_=^4}{p^2} + O(m^2) ,
$$

(1.80)

where we have introduced the mass parameter m as the IR regulator.

Similarly, the A^2-tensor dependent (see (1.71) for the definition of A) corrections originate from the graphs depicted in Fig. 7.6. The other A^2-dependent graphs depicted in Fig. 7.7 do not contribute to the anomaly.

It is easy to verify that the momentum integral $I(p) \equiv I_{(f)} = I_{(g)} = I_{(h)}$ corresponding to the supergraphs (f,g,h) in Fig. 7.6 is given by

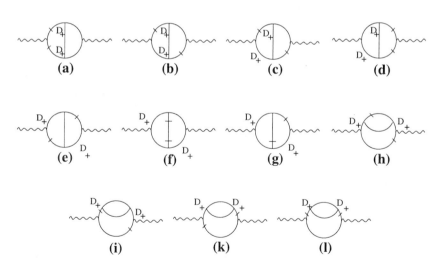

Fig. 7.6. The two-loop A^2-dependent $(1,0)$ supergraphs contributing to the anomaly of the heterotic NLSM

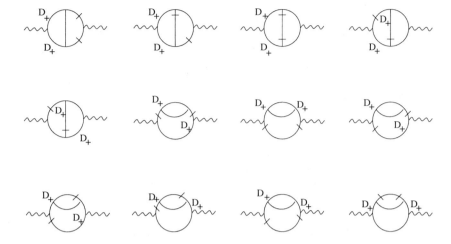

Fig. 7.7. The two-loop A^2-dependent $(1,0)$ supergraphs that do not contribute to the anomaly of the heterotic NLSM

$$I(p) = \int \frac{d^2k \, d^2q}{(2\pi)^4} \frac{k_= q_= (k_= + q_= + p_=)^2}{k^2 q^2 (k + q + p)^2} = \frac{1}{96\pi^2} \frac{p_=^4}{p^2} \ . \tag{1.81}$$

The finite anomalous parts of the diagrams (a,b,c,d,e,i,k,l) in Fig. 7.6 also turn out to be proportional to $I(p)$, though they require more integrations by parts. It is straightforward to get altogether [415]

$$\gamma^{\text{two}-\text{loop}} = \frac{1}{16\pi} \left(-R + \tfrac{1}{3}H^2 \right) \ . \tag{1.82}$$

We find no dilaton contributions at the given order of perturbation theory.

The conformal anomaly coefficient γ is related to the central charge of the Virasoro algebra in the *on-shell* heterotic string theory, i.e. on its effective spacetime equations of motion [77, 43] (cf. Subsects. 6.1.1 and 6.3.3). The effective Lagrangian $L_{\text{eff.}}$ of the bosonic massless modes of the heterotic string is proportional to the so-called 'averaged' conformal anomaly $\tilde{\beta}^{(\Phi)}$ (in the critical dimension $D = 10$) [356, 357, 358],

$$2\gamma + \alpha'(\mathcal{D}_\mu \Phi)^2 = \tilde{\beta}^{(\Phi)} = \beta^{(\Phi)} - \tfrac{1}{4}\beta^{(g)}_{\mu\nu} g^{\mu\nu} = \frac{D-10}{2} + \frac{\alpha'}{2} L_{\text{eff.}} \ , \tag{1.83}$$

where $\tilde{\beta}^{(\Phi)}$ is defined by the vacuum expectation value of the $(1,0)$ supertrace T_- of the 2d stress-energy tensor of the heterotic string,

$$\langle T \rangle_- = \frac{1}{4\pi} \tilde{\beta}^{(\Phi)} \Sigma^+ + \dots \ . \tag{1.84}$$

Here $\beta^{(g)}_{\mu\nu}$ is the metric RG β-function, $\beta^{(\Phi)}$ is the dilaton RG β-function, while the dots stand for the terms proportional to the heterotic string effective equations of motion. Let us recall that the RG β-functions of the heterotic NLSM are defined modulo spacetime field redefinitions [430].

To first order in α', the NLSM metric RG β-function is given by (Subsects. 2.4.1 and 3.2.1)

$$\beta_{\mu\nu}^{(g)} = \alpha'\left(R_{\mu\nu} - H_{\mu\nu}^2\right) \ , \quad H_{\mu\nu}^2 \equiv H_{\mu\lambda\rho}H_\nu{}^{\lambda\rho} \ . \tag{1.85}$$

As regards the heterotic string effective action, we thus find

$$S_{\text{eff.}} = \int \mathrm{d}^{10}X\sqrt{g}L_{\text{eff.}} = \frac{1}{2}\int \mathrm{d}^{10}X\sqrt{g}\left\{-R - 4D^2\Phi + 4(D_\mu\Phi)^2 + \tfrac{1}{3}H^2\right\} \ . \tag{1.86}$$

This agrees with the bosonic part of the Chapline-Manton action [360] of 10d supergravity, as well as the results of [431]. The kinetic terms of the Yang-Mills bosons appear in the next order of perturbation theory (see below).

The difference between 2γ and $\tilde{\beta}^{(\Phi)}$ in (1.83) is given by the (tree-level) Fradkin-Tseytlin term because of its non-minimal coupling to $(1,0)$ supergravity in (1.34) [432]. Indeed, let us consider the variation of the NLSM effective action,

$$\Gamma_{\text{eff.}} = I_{\text{het.}} + W_{\text{anomalous}} \ , \tag{1.87}$$

with respect to the superconformal factor S. The variation of the first term in (1.87) does not vanish since

$$\delta I_{\text{FT}} = -\frac{1}{4\pi\alpha'}\int \mathrm{d}^3z^-E\alpha'(\delta S)\mathrm{i}\nabla_+\nabla_-\Phi(X)$$
$$= -\frac{1}{4\pi}\int \mathrm{d}^3z^-E(\delta S)\mathrm{i}(\nabla_+X^\mu\nabla_-X^\nu D_\mu D_\nu\Phi + \nabla_+\nabla_-X^\mu D_\mu\Phi) \ . \tag{1.88}$$

By using the classical equation of motion for X^μ,

$$\mathrm{i}\nabla_+\nabla_-X^\mu = \tfrac{1}{2}\alpha'D^\mu\Phi\Sigma^+ \ , \tag{1.89}$$

we can rewrite (1.88) in the form

$$\delta I_{\text{FT}} = -\frac{1}{4\pi}\int \mathrm{d}^3z^-E\delta S\left(\mathrm{i}\nabla_+X^\mu\nabla_-X^\nu D_\mu D_\nu\Phi \right. $$
$$\left. +\tfrac{1}{2}\alpha'D^\mu\Phi D_\mu\Phi\Sigma^+\right) \ . \tag{1.90}$$

By definition, the averaged supertrace of the (1,0) stress-energy tensor is given by

$$\langle T\rangle_- = -2\frac{\delta\Gamma_{\text{eff.}}}{\delta S} \ . \tag{1.91}$$

The first relation in (1.83) now follows from (1.90) and (1.91).

The additional non-covariant terms in the heterotic string LEEA are given by the Lorentz and gauge Chern-Simons type contributions [321, 409, 410,

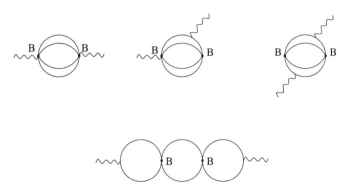

Fig. 7.8. The three-loop B^2-dependent corrections to the anomaly of the heterotic NLSM

433]. It is worth mentioning here that the non-covariant contributions to the anomaly can be absorbed into the torsion [433, 434]. This implies that the *form* of the superconformal anomaly does not change to all orders, in agreement with the Adler-Bardeen theorem of QFT [435]. The consistency between the covariant (superconformal) and canonical (super-Virasoro) anomalies was verified in [436].

Three-Loop NLSM Corrections to the Anomaly. The B^2-dependent contributions to the anomaly are generated by the diagrams schematically depicted in Fig. 7.8, modulo a distribution of the 2d derivatives on the internal lines.

The BA^2-dependent contributions appear from the diagrams of Fig. 7.9. As in Fig. 7.8, we only display the 'generating' graphs – in fact, each of them represents a series of similar graphs with various distributions of the 2d derivatives on the internal lines.

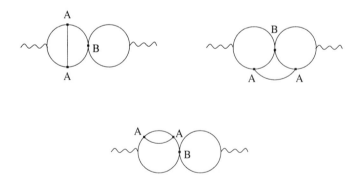

Fig. 7.9. The three-loop BA^2-dependent corrections to the anomaly of the heterotic NLSM

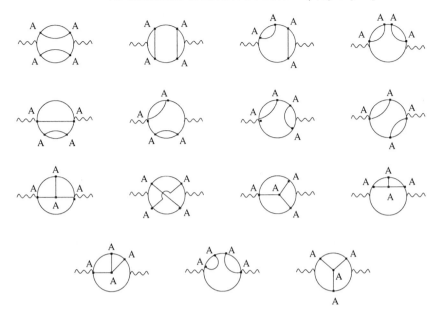

Fig. 7.10. The three-loop A^4-dependent corrections to the anomaly of the heterotic NLSM

The generating graphs for the A^4-type corrections are depicted in Fig. 7.10. The C^2-type corrections originate from the graphs in Fig. 7.11.

As an example, let us consider the first graph in Fig. 7.8. It generates the three Feynman diagrams depicted in Fig. 7.12, after taking into account all possible (inequivalent) distributions of the derivatives ∂_- on the internal lines. The background dependence of all diagrams in Fig. 7.12 is the same. We explicitly indicate the background B^2-dependence in the first diagram of Fig. 7.12 only.

It is straightforward to perform the D-algebra in $(1,0)$ superspace for each of the $(1,0)$ supergraphs depicted in Fig. 7.12. This gives rise to three different multi-loop integrals in the conventional (two-dimensional) momentum space,

Fig. 7.11. The three-loop C^2-dependent corrections to the anomaly of the heterotic NLSM

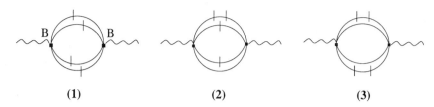

Fig. 7.12. The three-loop diagrams generated by the first graph in Fig. 7.8 with the ∂_- derivatives on the internal lines

$$J_1(p) = \int \frac{d^2k d^2q d^2l}{(2\pi)^6} \frac{k_-(k_- + q_- + l_- + p_-)q_-l_-}{(k^2 - m^2)\left[(k+q+l+p)^2 - m^2\right](q^2 - m^2)(l^2 - m^2)}$$

$$J_2(p) = \int \frac{d^2k d^2q d^2l}{(2\pi)^6} \frac{k_-^2 q_- l_-}{(k^2 - m^2)\left[(k+q+l+p)^2 - m^2\right](q^2 - m^2)(l^2 - m^2)}$$

$$J_3(p) = \int \frac{d^2k d^2q d^2l}{(2\pi)^6} \frac{k_-^2 q_-^2}{(k^2 - m^2)\left[(k+q+l+p)^2 - m^2\right](q^2 - m^2)(l^2 - m^2)}$$

$$(1.92)$$

By the use of the equations

$$m^2 \int \frac{d^2k}{(2\pi)^2} \frac{1}{(k^2 - m^2)[(k+p)^2 - m^2]} = \frac{1}{4\pi} + O(p^2/m^2) ,$$

$$p_+ J_- = p_- J_+ , \quad J_{-(+)} \equiv \int \frac{d^2k}{(2\pi)^2} \frac{k_{-(+)}}{(k^2 - m^2)[(k+p)^2 - m^2]} , \quad (1.93)$$

it is not difficult to verify that the divergent contributions from the second and third diagrams in Fig. 7.12 cancel against the divergences of the second and third diagrams in Fig. 7.8.

As regards the first graph in Fig. 7.12, we have

$$J_1(p) = -\frac{1}{6\pi(4\pi)^2} \frac{p_-^4}{p^2} . \qquad (1.94)$$

The last graph in Fig. 7.8 has a background dependence proportional to the Ricci tensor squared. However, its contribution to the anomaly vanishes.

After taking into account the Bianchi identities for the curvature tensor and integrating by parts, the total three-loop contribution to the anomaly takes the form [415]

$$\gamma^{\text{three-loop}} = -\frac{1}{16(4\pi)^2} \Big[R_{\mu\nu\lambda\rho}^2 + (F_{\mu\nu}^{IJ})^2 - 2R_{\mu\nu\lambda\rho}H_\sigma{}^{\mu\nu}H^{\lambda\rho\sigma}$$

$$+ \tfrac{2}{3}H^4 - 2(H_{\mu\nu}^2)^2 \Big] , \qquad (1.95)$$

where we have used the notation $R_{\mu\nu\lambda\rho}^2 = R_{\mu\nu\lambda\rho}R^{\mu\nu\lambda\rho}$ and

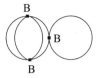

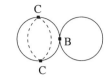

Fig. 7.13. Four-loop contributions to the anomaly of the heterotic NLSM

$$H^4 = H_{\mu\nu\rho}H^{\lambda\nu\sigma}H_\sigma{}^{\mu\gamma}H^\rho{}_{\lambda\gamma} \ , \quad (H^2_{\mu\nu})^2 = H^2_{\mu\nu}(H^2)^{\mu\nu} \ . \tag{1.96}$$

We do not display the dilaton-dependent graphs since their contributions cancel altogether at the given order in α'. This conclusion agrees with the calculations of [437, 438].

Equation (1.95) gives rise to the following correction to the low-energy effective action of the heterotic NLSM:

$$L^{(1)}_{\text{eff.}} = -\frac{\alpha'}{16}\left[R^2_{\mu\nu\lambda\rho} + (F^{IJ}_{\mu\nu})^2 - 2R_{\mu\nu\lambda\rho}H_\sigma{}^{\mu\nu}H^{\lambda\rho\sigma} + \tfrac{2}{3}H^4 - 2(H^2_{\mu\nu})^2\right] \ , \tag{1.97}$$

in agreement with the results of Subsect. 2.4.2 about the *bosonic* NLSM β-functions. The correction (1.97) also agrees with the low-energy effective action derived from the heterotic string tree scattering amplitudes [439].

Four- and Five-Loop Corrections to the Heterotic NLSM Anomaly. In what follows we set the torsion $H_{\mu\nu\lambda} = 0$ for simplicity.

The four-loop contributions to the heterotic NLSM anomaly γ and, hence, to the heterotic string low-energy effective action, come from the diagrams schematically depicted in Fig. 7.13. Their background dependence is given by

$$\gamma_{\text{four-loop}} = \text{const.}(\alpha')^3 R^{\mu\nu}\left[R_{\mu\lambda\rho\gamma}R_\nu{}^{\lambda\rho\gamma} + F_{\mu\lambda}{}^{IJ}F_\nu{}^{IJ\lambda}\right] \ . \tag{1.98}$$

Equation (1.98) agrees with the known fact that the anomaly vanishes provided that the R and F tensors coincide (in the proper basis) [440].

As regards the five-loop corrections to γ in the gravitational sector, there are only two 'basic' generating graphs shown in Fig. 7.14. The external legs are supposed to be attached to them in all possible ways, as well as the derivatives on the internal lines.

Actual calculations of the Feynman diagrams corresponding to the first graph **(a)** in Fig. 7.14 can be significantly simplified by ignoring all contributions whose background dependence is proportional to $\beta^{(g),1}_{\mu\nu} \sim R_{\mu\nu}$ or

(a)   **(b)**

Fig. 7.14. The five-loop 'basic' generating graphs (without external legs) contributing to the anomaly of the heterotic NLSM in the gravitational sector

$\beta_{\mu\nu}^{(g),2} \sim R_{\mu\alpha\beta\gamma} R_{\nu}{}^{\alpha\beta\gamma}$, because those contributions can be eliminated by re-definitions of the NLSM metric. The final result in the gravitational sector of the heterotic NLSM reads [415]

$$\gamma_{\text{five−loop}} = \frac{\text{const.}}{(4\pi)^4} \left\{ R_\alpha{}^{\nu\gamma}{}_\beta R_{\mu\nu\lambda\rho} R^{\alpha\delta\gamma\mu} R^\beta{}_{\gamma\delta}{}^\rho + \tfrac{1}{2} R_{\alpha\beta\nu\lambda} R_{\mu\rho}{}^{\nu\lambda} R^{\alpha\delta\gamma\mu} R^\beta{}_{\gamma\delta}{}^\rho \right\}$$
(1.99)

A calculation of the constant in (1.99) amounts to the five-loop momentum integral, which is rather involved. Fortunately, this constant can be fixed indirectly, by using (1.83) and the results of [441, 442] for the heterotic string low-energy effective action:

$$
\begin{aligned}
L_{\text{eff.}}^{(3),g(a)} = \tfrac{1}{16} (\alpha')^3 \zeta(3) & \left\{ R_\alpha{}^{\nu\lambda}{}_\beta R_{\mu\nu\lambda\rho} R^{\alpha\delta\gamma\mu} R^\beta{}_{\gamma\delta}{}^\rho \right. \\
& \left. + \tfrac{1}{2} R_{\alpha\beta\nu\lambda} R_{\mu\rho}{}^{\nu\lambda} R^{\alpha\delta\gamma\mu} R^\beta{}_{\gamma\delta}{}^\rho \right\} ,
\end{aligned}
$$
(1.100)

which imply that const. $= \tfrac{1}{4}\zeta(3)$ in (1.99), where we have used our convention $2\pi\alpha' = 1$.

The Lagrangian (1.100) coincides (up to a total derivative and Ricci tensor dependent terms) with the standard contribution to the heterotic string LEEA obtained from the heterotic string tree amplitudes [175, 439],

$$L_{\text{eff.}}^{(3),g(a)} = \frac{\zeta(3)}{3 \cdot 2^{12}} (\alpha')^3 \underbrace{t^{\mu\nu\lambda\rho\theta\eta\xi\omega} \cdot t^{\alpha\beta\gamma\delta\varepsilon\kappa\tau\zeta}}_{g-\text{product}} R_{\mu\nu\alpha\beta} R_{\lambda\rho\gamma\delta} R_{\theta\eta\varepsilon\kappa} R_{\xi\omega\tau\zeta} ,$$
(1.101)

where the t-tensor is well known in string theory [170]:

$$
\begin{aligned}
t^{\mu_1\nu_1\mu_2\nu_2\mu_3\nu_3\mu_4\nu_4} = & \left[-\tfrac{1}{2} \left(g^{\mu_1\mu_2} g^{\nu_1\nu_2} - g^{\mu_1\nu_2} g^{\nu_1\mu_2} \right) \left(g^{\mu_3\mu_1} g^{\nu_3\nu_4} - g^{\mu_3\nu_4} g^{\nu_3\mu_4} \right) \right. \\
& \left. + \tfrac{1}{2} \left(g^{\nu_1\mu_2} g^{\nu_2\mu_3} g^{\nu_3\mu_4} g^{\nu_4\mu_1} + \text{antisymm. of } [\mu_i\nu_i] \right) \right] \\
& + [(1324) + (1342) \text{ permutations}] .
\end{aligned}
$$
(1.102)

Similarly, the contribution to the anomaly γ due to the second graph in Fig. 7.14 gives rise to another R^4-term in the heterotic string LEEA (cf. [439]),

$$L_{\text{eff.}}^{(3),g(b)} = -\frac{3}{2^{11}} (\alpha')^3 t^{\alpha\theta\gamma\delta\mu\nu\lambda\rho} R_{\alpha\beta\varepsilon\tau} R_{\gamma\delta}{}^{\tau\varepsilon} R_{\mu\nu\eta\omega} R_{\lambda\rho}{}^{\omega\eta} .$$
(1.103)

In the Yang-Mills sector of the heterotic NLSM, the five-loop contributions to the anomaly arise from the basic graphs depictured in Fig. 7.15. Their total contribution to γ has the form

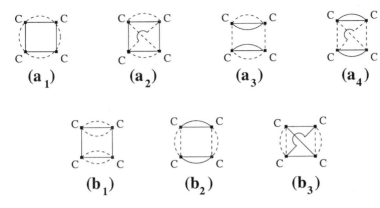

Fig. 7.15. The five-loop 'basic' generating graphs (without external legs) contributing to the anomaly of the heterotic NLSM in the Yang-Mills sector

$$\gamma^{\text{five-loop, YM}} = (\alpha')^4 \left\{ a_1 \text{tr}(F_{\mu\nu}F^{\nu\lambda}F_{\lambda\rho}F^{\rho\mu}) + a_2 \text{tr}(F_{\mu\nu}F^{\nu\lambda}F^{\rho\mu}F^{\lambda\rho}) \right.$$
$$+ a_3 \text{tr}(F_{\mu\nu}F^{\mu\nu}F_{\rho\lambda}F^{\rho\lambda}) + a_4 \text{tr}(F_{\mu\nu}F_{\rho\lambda}F^{\mu\nu}F^{\rho\lambda})$$
$$+ b_1 \text{tr}(F_{\mu\nu}F^{\nu\rho})\text{tr}(F_{\rho\lambda}F^{\lambda\mu}) + b_2 \text{tr}(F_{\mu\nu}F_{\rho\lambda})\text{tr}(F^{\mu\nu}F^{\rho\lambda})$$
$$\left. + b_3 \text{tr}(F_{\mu\nu}F_{\rho\lambda})\text{tr}(F^{\mu\rho}F^{\nu\lambda}) \right\} ,$$

$$(1.104)$$

whose numerical coefficients $(a_1, a_2, a_3, a_4, b_1, b_2, b_3)$ are determined by certain finite five-loop momentum integrals. Remarkably, all the a-coefficients in (1.104) actually vanish [415].

Equation (1.103) is to be compared to the corresponding F^4-dependent contribution to the heterotic string LEEA [433, 439],

$$L_{\text{eff.}}^{(3), \, YM} = -\frac{3}{2^{11}}(\alpha')^3 t^{\mu\nu\rho\lambda\alpha\beta\gamma\delta} \text{tr}(F_{\mu\nu}F_{\rho\lambda})\text{tr}(F_{\alpha\beta}F_{\gamma\lambda}) . \qquad (1.105)$$

By using (1.102) we can rewrite (1.105) in the form

$$L_{\text{eff.}}^{(3), \, YM} = -\frac{3}{2^{10}}(\alpha')^3 \left[8 \, \text{tr}(F_{\mu\nu}F^{\nu\rho})\text{tr}(F_{\rho\lambda}F^{\lambda\mu}) \right.$$
$$- 2 \, \text{tr}(F_{\mu\nu}F_{\rho\lambda})\text{tr}(F^{\mu\nu}F^{\rho\lambda}) + 4 \, \text{tr}(F_{\mu\nu}F_{\rho\lambda})\text{tr}(F^{\mu\rho}F^{\nu\lambda})$$
$$\left. - \text{tr}(F_{\mu\nu}F^{\mu\nu})\text{tr}(F_{\rho\lambda}F^{\rho\lambda}) \right] .$$

$$(1.106)$$

The anomaly (1.103) can also be rewritten in the form

$$\gamma^{\text{five-loop, YM}} = (\alpha')^4 \left[b_1 \text{tr}(F_{\mu\nu}F^{\nu\rho})\text{tr}(F_{\rho\lambda}F^{\lambda\mu}) + b_2 \text{tr}(F_{\mu\nu}F_{\rho\lambda})\text{tr}(F^{\mu\nu}F^{\rho\lambda}) \right.$$
$$\left. + b_3 \text{tr}(F_{\mu\nu}F_{\rho\lambda})\text{tr}(F^{\mu\rho}F^{\nu\lambda}) \right] .$$

$$(1.107)$$

The structure of (1.107) agrees with (1.106) since the last term in (1.106) is proportional to the lower-order contribution whose coefficient is dependent upon parametrization and therefore it can be set to zero.

The mixed R^2F^2-contributions to the anomaly γ are supposed to complete the terms already present in (1.103) and (1.106) to the $(R^2 + F^2)^2$ combination,

$$
\begin{aligned}
L_{\text{eff.}}^{(3),\ YM+g(b)} &= -\frac{3}{2^{11}}(\alpha')^3 \left[\text{tr}(F_{\mu\nu}F_{\rho\lambda}) - \text{tr}(R_{\mu\nu}R_{\rho\lambda}) \right] \\
&\times t^{\mu\nu\rho\lambda\alpha\beta\gamma\delta} \left[\text{tr}(F_{\alpha\beta}F_{\gamma\delta}) - \text{tr}(R_{\alpha\beta}R_{\gamma\delta}) \right] ,
\end{aligned}
\tag{1.108}
$$

in order to achieve the correspondence with the results of [439]. We did not attempt to verify it in the NLSM approach.

The methods presented in this section can be applied, in principle, up to any given order in α'. As we demonstrated above, the results of the NLSM approach in curved superspace of $(1,0)$ supergravity agree with the known results about the heterotic string LEEA obtained from tree string amplitides. From the technical viewpoint, the calculation of the anomaly requires one more order in α', when being compared to the RG β-function approach (Chaps. 2 and 3). However, the scalar background dependence of the NLSM anomaly is easier to calculate than the NLSM β-functions.

Because of (1.72), cancellation of all supergravitational anomalies in the heterotic NLSM implies the two conditions on the background fields of the heterotic string, $\gamma_1 = \gamma_2 = 0$. Our calculations in this section were performed about flat $(1,0)$ superspace, i.e. in the *non-covariant* way with respect to $(1,0)$ superdiffeomorphisms. Though it was possible to rewrite the results in covariant form by the use of some local counterterms, the covariance of the final results implies the existence of manifestly covariant methods of quantum calculations in a curved $(1,0)$ superspace, which may be useful for a derivation of general statements about the structure of the anomalies in the heterotic NLSM. The covariant techniques, based on the proper-time methods in $(1,0)$ superspace, were developed in [443, 416].

7.2 Chiral Coset Construction of Heterotic Strings

The geometrical covariant description of the four-dimensional ($D = 4$) heterotic string models and the manifest covariant realization of their internal and world-sheet symmetries can be achieved by the use of two-dimensional (2d) Lagrangian chiral field theory in a curved $(1,0)$ superspace [323]. The important constituents of this construction are $(1,0)$ supersymmetric chiral bosons (called *leftons* and *rightons*) and chiral non-movers (called *notons*). The $D = 4$ heterotic strings have two standard descriptions depending upon whether their internal degrees of freedom are represented by chiral bosons or chiral fermions [444]. A $D = 4$ heterotic string in the free fermion approach [445, 446] only exists in the superconformal gauge, while it cannot be considered as the gauge-fixed form of some covariant action in a curved $(1, 0)$ superspace. This happens because the OPE defining the super-AKM and super-Virasoro algebras in the fermionic picture only exist in quantum theory, not in the classical one. A resolution of this problem provides the motivation for introducing chiral bosons.

The covariant action of a 2d chiral (Abelian) boson (lefton) was suggested in [447]. To enforce the chirality constraint, the new auxiliary field (Lagrange multiplier) and the new diffeomorphism type symmetry (known as *Siegel's symmetry*) were introduced in [447]. The Siegel symmetry is anomalous in quantum theory. This equally applies to the 2d (Abelian) righton action [448].

To cancel the Siegel anomaly, it was suggested in [449] that the consistent action of chiral bosons should consist of two parts, namely, the original Siegel action and another term depending upon classically non-moving fields (notons). Being propagating at the quantum level, the notons are used to cancel the anomaly. This way of Siegel's anomaly cancellation is known as the *Hull mechanism*.

The Lagrangian description of $D = 4$, toroidally compactified, heterotic strings by the use of the $(1, 0)$ supersymmetric (Abelian) rightons or leftons was initiated in [450] and further developed in [451]. The coupling of the $(1, 0)$ supersymmetric (Abelian) leftons and rightons to an external Abelian gauge superfield was found in [452].

To construct 2d covariant Lagrangians for some 'realistic' $D = 4$ heterotic string models, the $(1, 0)$ supersymmetric *non-Abelian* lefton and righton actions are needed. Some important steps towards the non-Abelian construction were made in [453, 454], where the relevant chiral actions on *group* manifolds and their coupling to the external non-Abelian vector supermultiplets were constructed. The non-Abelian construction of the heterotic strings on the *coset* manifolds is given in this section.

The standard description of superstring compactification is based on the existence of (extended) superconformal structure (Chaps. 5 and 6), whereas much less attention is usually paid to the underlying covariant world-sheet actions. The minimal discrete series in 2d (super)conformal field theory are most easily constructed in the GKO approach which, in turn, can be realized by using the gauged WZNW models defined on the appropriate cosets G/H (Sect. 5.3). It should therefore be possible to construct the chiral 2d covariant actions describing the non-Abelian (or 'realistic') $D = 4$ heterotic strings generalizing the toroidally compactified (Abelian) ones.

In this section, we present the invariant actions describing $(1, 0)$ supersymmetric non-Abelian leftons and rightons on coset manifolds. Our presentation covers both chiral and non-chiral $(1, 0)$ supersymmetric WZNW actions. The supergravitational, chiral and Siegel anomalies of the $D \leq 10$ heterotic string models are calculated by using the Lagrangian realization of the GKO coset construction. The general conditions for the anomaly cancellation are found, and various anomaly-free solutions are discussed. Amongst them are some 'realistic' solutions, such as Gepner's models [295] and the $D = 4$ heterotic strings compactified on orbifolds. The coupling of the heterotic string coset models to the bosonic massless modes of the heterotic string and the Siegel anomalies of the associated NLSM are also discussed.

Subsection 7.2.1 is devoted to the covariant description of the $(1,0)$ supersymmetric non-Abelian leftons and rightons. Subsection 7.2.2 deals with t he non-chiral $(1,0)$ supersymmetric WZNW models on the cosets G/H. In Subsect. 7.2.3 we introduce the $(1,0)$ supersymmetric chiral coset actions. In Subsect. 7.2.4 we discuss supergravitational, Siegel and chiral anomalies. The general invariant action in a curved $(1,0)$ superspace, suitable for compactified heterotic strings, in terms of the coset-valued, $(1,0)$ supersymmetric rightons and leftons, is also presented in Subsect. 7.2.4. Some anomaly-free solutions are given in Subsect. 7.2.5. A propagation of the $D = 4$ heterotic string in the background of its massless bosonic modes and the Siegel anomaly of the associated heterotic NLSM are investigated in Subsect. 7.2.6.

7.2.1 Non-Abelian (1, 0) Supersymmetric Chiral Bosons

Our starting point is the level-k WZNW model (Chap. 5) defined on the $(1,0)$ locally supersymmetric world-sheet, in terms of the matrix $(1,0)$ scalar superfields g taking their values in a compact Lie group $\mathbf{G}$,

$$
kS(g) = -\frac{ik}{2\pi} \int d^3z \, E^- \mathrm{Tr} \left\{ \left(g^{-1} \nabla_+ g \right) \left(g^{-1} \nabla_= g \right) \right.
$$

$$
\left. + \int_0^1 dy \left(\tilde{g}^{-1} \frac{\partial \tilde{g}}{\partial y} \right) \left[\nabla_+ \tilde{g}^{-1} \nabla_= \tilde{g} - \nabla_= \tilde{g}^{-1} \nabla_+ \tilde{g} \right] \right\} , \tag{2.1}
$$

where we have introduced the notation

$$
g(z) = \exp \left[i\varphi^I(z) t^I \right] , \quad I = 1, \ldots, \dim \mathbf{G} ,
$$

$$
[t^I, t^J] = i f^{IJK} t^K , \quad f^{IJK} f^{IJL} = 2c_{\mathbf{G}} \delta^{KL} , \quad \mathrm{Tr} \left(t^I t^J \right) = 2\delta^{IJ} . \tag{2.2}
$$

In (3.2) t^I denote the generators of the Lie algebra $\mathbf{g}$. The extension $\tilde{\varphi}$ of the field φ obeys the boundary conditions

$$
\tilde{\varphi}^I(z, y) \big|_{y=1} = \varphi^I(z) , \quad \tilde{\varphi}^I(z, y) \big|_{y=0} = 0 . \tag{2.3}
$$

Being applied to the action (2.1), the Polyakov-Wiegmann identity [49, 50] yields

$$
S(gh) = S(g) + S(h) - \frac{i}{\pi} \int d^3z \, E^- \mathrm{Tr} \left\{ \left(g^{-1} \nabla_+ g \right) \left(\nabla_= h \cdot h^{-1} \right) \right\} . \tag{2.4}
$$

The identity (2.4) is useful in proving the invariance of the action $S(g)$ under the transformations

$$g \to \Omega g \hat{\Omega}^{-1} \,, \tag{2.5}$$

whose **G** group-valued $(1,0)$ superfield parameters, Ω and $\hat{\Omega}$, are restricted by the conditions

$$\nabla_+ \Omega = \nabla_= \hat{\Omega} = 0 \,. \tag{2.6}$$

Equation (2.5) implies the $(1,0)$ super-AKM invariance of the heterotic string theory on group manifolds [295].

The action (2.1) is non-chiral in the sense that the scalar component of $g(z)$ has independent left- and right-propagating modes. The non-Abelian $(1,0)$ supersymmetric chiral boson model was constructed in [453, 454]. We adopt here the most straightforward approach that guarantees the Siegel gauge invariance. To describe it, we begin with the lefton case [323].

The action of $(1,0)$ non-Abelian leftons $g_L(z)$ on the group manifold $\mathbf{G}_L$ can be obtained from (2.1) by introducing a new set of covariant derivatives D_A instead of the old ones ∇_A, [2]

$$D_+ = \nabla_+ + \Lambda_+{}^= \nabla_= - \left(\nabla_= \Lambda_+{}^= \right) \hat{M} \,, \quad D_\pm = -\frac{\mathrm{i}}{2} \{ D_+, D_+ \} \,,$$

$$D_= = \nabla_= \,, \quad [D_=, D_+] = 2\mathrm{i}\, \Sigma_L^+ \hat{M} \,,$$

$$\Sigma_L^+ = \Sigma^+ + \frac{\mathrm{i}}{2} \nabla_= \nabla_= \Lambda_+{}^= \,, \tag{2.7}$$

where $\hat{M}$ is the generator of 2d Lorentz transformations, Σ^+ is the superfield strength of $(1,0)$ supergravity, and $\Lambda_+{}^=$ is the superfield Lagrange multiplier whose equation of motion forces the scalars g_L to be left-moving.

The non-Abelian $(1,0)$ lefton action takes the form

$$k_L S_L(g_L) = -\frac{\mathrm{i} k_L}{2\pi} \int \mathrm{d}^3 z \, E^- \mathrm{Tr} \left\{ L_+ L_= + \Lambda_+{}^= L_= L_= \right.$$

$$\left. + \int_0^1 \mathrm{d}y \left(\tilde{g}_L^{-1} \frac{\partial \tilde{g}_L}{\partial y} \right) \left[\nabla_+ \tilde{L}_= - \nabla_= \tilde{L}_+ \right] \right\} \,, \tag{2.8}$$

where we have used the notation

$$L_A = g_L^{-1} \nabla_A g_L \,. \tag{2.9}$$

The action (2.8) possesses the Siegel symmetry,

$$g_L^{-1} \delta g_L = \epsilon_+ L_= \,, \quad \delta \Lambda_+{}^= = -D_+ \epsilon_\pm \,. \tag{2.10}$$

---

[2] The curved $(1,0)$ superspace covariant derivatives D_A should not be confused with the flat $(1,0)$ superspace covariant derivatives introduced in Subsect. 7.1.1.

This invariance becomes obvious when one notices that the action of the Λ-transformation on the covariant derivatives (2.7) is equivalent to the combined action of the general coordinate and Lorentz transformations,

$$\delta D_A = [\epsilon_+ D_= + L\hat{M}, D_A] \ , \quad L = -\frac{1}{2} \nabla_= \epsilon_+ \ , \tag{2.11a}$$

and the super-Weyl transformations,

$$\delta \nabla_+ = \tfrac{1}{2} \tilde{\Omega} \nabla_+ + (\nabla_+ \tilde{\Omega})\hat{M} \ ,$$

$$\delta \nabla_= = \tilde{\Omega} \nabla_= - (\nabla_= \tilde{\Omega})\hat{M} \ , \tag{2.11b}$$

$$\delta \Sigma^+ = \tfrac{3}{2} \tilde{\Omega} \Sigma^+ - i\nabla_= \nabla_+ \tilde{\Omega} \ ,$$

with the parameter $\tilde{\Omega} = \tfrac{1}{2} \nabla_= \epsilon_+$.

The superfield $\Lambda_+{}^=$ can be gauged away by the use of the symmetry (2.10). However, this symmetry is anomalous in quantum theory. The only consistent and manifestly covariant way to cancel this Siegel anomaly is the Hull mechanism [449] which amounts to adding to the action (2.8) an extra term,

$$k_{\mathrm{L}} S_{\mathrm{L}}(g_{\mathrm{L}}) \to k_{\mathrm{L}} S_{\mathrm{L}}(g_{\mathrm{L}}) + S_{\mathrm{L-N}}(\Psi_-) \ , \tag{2.12}$$

depending upon the classically non-propagating superfields (notons) Ψ_- contributing to the Siegel anomaly. The noton action $S_{\mathrm{L-N}}$ reads [449, 455]

$$S_{\mathrm{L-N}} = -\int \mathrm{d}^3 z \, E^- \Psi_-^i \left(\nabla_+ + \Lambda_+{}^= \nabla_= \right) \Psi_-^i \ , \quad i = 1, \ldots, N_\Psi \ , \tag{2.13}$$

while it also has Siegel's symmetry. The action on the right-hand side of (2.12) can be thought of as the truly $(1,0)$ supersymmetric lefton action.

The non-Abelian $(1,0)$ leftons can be coupled to a Siegel gauge-invariant vector $(1,0)$ supermultiplet $A_-(z) = i A_-^I (z) t_{\mathrm{L}}^I$. The corresponding action is given by [453, 454]

$$k_{\mathrm{L}} S_{\mathrm{L}} \left(g_{\mathrm{L}}; A_= \right) = -\frac{i k_{\mathrm{L}}}{2\pi} \int \mathrm{d}^3 z E^- \mathrm{Tr} \left\{ L_+ \left(L_= + 2A_= \right) + \Lambda_+^= \left(L_= + A_= \right)^2 \right.$$

$$\left. + \int_0^1 \mathrm{d}y \left(\tilde{g}_{\mathrm{L}}^{-1} \frac{\partial \tilde{g}_{\mathrm{L}}}{\partial y} \right) \left[\nabla_+ \tilde{L}_= - \nabla_= \tilde{L}_+ \right] \right\} \ . \tag{2.14}$$

Equation (2.14) is invariant under the following Siegel transformations [453]:

$$g_{\mathrm{L}}^{-1} \delta g_{\mathrm{L}} = \epsilon_+ \left(L_= + A_= \right) \ , \quad \delta A_= = 0 \ , \tag{2.15}$$

where $\Lambda_+{}^=$ transforms according to (2.10). This symmetry can be most easily understood after parametrizing the superfields $A_=$ as follows:

$$A_= = \nabla_= \Omega_{\rm L} \cdot \Omega_{\rm L}^{-1} , \tag{2.16}$$

in terms of the **G**-valued matrix $(1,0)$ superfields $\Omega_{\rm L}(z)$, and rewriting the action (2.14) in the equivalent form

$$S_{\rm L}\left(g_{\rm L}, A_=\right) = S_{\rm L}\left(g_{\rm L}\Omega_\Lambda\right) - S\left(\Omega_{\rm L}\right) , \tag{2.17}$$

where we have used the identity (2.4). It is worth emphasizing here that the subscript 'L' is absent in the last term of (2.17).

The action (2.14) may be useful for an investigation of the low-energy dynamics of four-dimensional heterotic strings. Indeed, since the superfields $A_=(z)$ are inert under the transformations (2.15), they can be identified with a condensate of the heterotic string massless modes.

To covariantly describe the $(1,0)$ supersymmetric non-Abelian rightons $g_{\rm R}(z)$ on a group manifold $\mathbf{G}_{\rm R}$ [453, 454], we again have to appropriately modify the covariant derivatives, $\nabla_A \to \mathcal{D}_A$, this time as follows:

$$D_{--} \equiv \nabla_= - \frac{i}{2}\nabla_+\left(\Lambda_=^{\ddagger}\right)\nabla_+ + \Lambda_=^{\ddagger}\nabla_{\ddagger} + \left(\nabla_{\ddagger}\Lambda_=^{\ddagger}\right)\hat{M} ,$$

$$D_{++} \equiv \nabla_+ , \quad [D_{--}, D_{++}] = 2i\Sigma_R^+ \hat{M} , \tag{2.18}$$

$$\Sigma_R^+ \equiv \Sigma^+ + \frac{i}{2}\nabla_+\nabla_{\ddagger}\Lambda_=^{\ddagger} ,$$

where $\Lambda_=^{\ddagger}$ is the Lagrange $(1,0)$ superfield multiplier forcing the superfields $g_{\rm R}(z)$ to be right-moving on the equations of motion. The non-Abelian righton action is given by

$$k_{\rm R} S_{\rm R}\left(g_{\rm R}\right) = -\frac{ik_{\rm R}}{2\pi}\int d^3z\, E^- {\rm Tr}\left\{ R_+ R_= - i\Lambda_=^{\ddagger} R_+ \nabla_+ R_+ \right.$$

$$\left. + \frac{4}{3}i\Lambda_=^{\ddagger} R_+ R_+ R_+ - \int_0^1 dy\left(\frac{\partial \tilde{g}_{\rm R}}{\partial y}\tilde{g}_{\rm R}^{-1}\right)\left[\nabla_+\tilde{R}_= - \nabla_=\tilde{R}_+\right]\right\} , \tag{2.19}$$

where $R_A = \nabla_A g_{\rm R} \cdot g_{\rm R}^{-1}$. Equation (2.19) is invariant under the Siegel transformations

$$\delta g_{\rm R} \cdot g_{\rm R}^{-1} = i\epsilon_=\left(\nabla_+ R_+ - R_+ R_+\right) + \frac{i}{2}\left(\nabla_+\epsilon_=\right)R_+ ,$$

$$\delta\Lambda_=^{\ddagger} = D_{--}\epsilon_= . \tag{2.20}$$

The model (2.19) can be generalized further, by introducing the coupling to a non-Abelian chiral vector supermultiplet $A_+(z) = iA_+^I(z)t_{\rm R}^I$ [454],

$$k_R S_R (g_R; A_+) = -\frac{ik_R}{2\pi} \int d^3 z \, E^- \text{Tr}\Big\{ R_- (R_+ + 2A_+)$$

$$- i\Lambda_-{}^\mp (R_+ + A_+) \nabla_+ (R_+ + A_+)$$

$$+ \frac{4i}{3} \Lambda_-{}^\mp (R_+ + A_+)^2 \left(R_+ - \frac{1}{2} A_+ \right) \tag{2.21}$$

$$- \int_0^1 dy \left(\frac{\partial \tilde{g}_R}{\partial y} \tilde{g}_R^{-1} \right) \left[\nabla_+ \tilde{R}_- - \nabla_- \tilde{R}_+ \right] \Big\} \quad .$$

The Polyakov-Wiegmann identity (2.4) yields in this case

$$S_R (g_R; A_+) = S_R (\Omega_R g_R) - S (\Omega_R) \ ,$$

$$A_+ \equiv \Omega_R^{-1} \nabla_+ \Omega_R \ . \tag{2.22}$$

Since A_+ (and, hence, Ω_R) are supposed to be inert under the Siegel transformations, the Siegel symmetry of the action (2.21) can be easily checked,

$$\delta g_R \cdot g_R^{-1} = i\epsilon_- \nabla_+ (A_+ + R_+)$$

$$+ i\epsilon_- \left(A_+^2 - R_+^2 \right) + \tfrac{i}{2} \left(\nabla_+ \epsilon_- \right) (A_+ + R_+) \ , \tag{2.23}$$

where we have used (2.20) and (2.22). The transformation law of the superfield $\Lambda_-{}^\mp$ is given by (2.20).

Similarly to the left chiral symmetry (2.10), the right chiral symmetry (2.20) is also anomalous in quantum theory. However, in contrast to the lefton case, it is *not* possible to construct a (1,0) supersymmetric noton-type action (coupled to the Lagrange multiplier $\Lambda_-{}^\mp$) that would cancel the anomaly of the Siegel symmetry (2.20). The required action, nevertheless, exists if *both* leftons *and* rightons are involved [456]. After being coupled to both Lagrange multipliers, $\Lambda_+{}^=$ and $\Lambda_-{}^\mp$, the action in question describes some classically non-propagating scalar superfields, and it has a complicated Siegel symmetry. The explicit form of this action is [456]

$$S_{\text{LR-N}} = i \int d^3 z \, E^- \frac{1}{1 - i\Lambda^2} \left[D_{--} W^\alpha + \frac{i}{2} \left(\nabla_+ \Lambda_-{}^\mp \right) \nabla_+ W^\alpha \right]$$

$$\times \left[D_+ W^\alpha - i\nabla_+ \left(\Lambda_-{}^\mp \Lambda_+{}^= \nabla_+ W^\alpha \right) \right] \ , \tag{2.24}$$

where

$$\Lambda^2 = \Lambda_-{}^\mp \nabla_+ \Lambda_+{}^= \ , \quad \alpha = 1, \ldots, N_W \ . \tag{2.25}$$

Equation (2.24) is the (1,0) supersymmetric generalization of Hull's noton action [449]. Therefore, the total action of leftons and rightons is given by

$$S_{\text{tot.}} = k_L S_L (g_L) + k_R S_R (g_R) + S_{\text{L-N}} + S_{\text{LR-N}} \ . \tag{2.26}$$

Is is worth mentioning the existence of yet another, classically equivalent form of the $(1,0)$ supersymmetric Hull noton action (2.24), known as the $(1,0)$ supersymmetric bosonized lefton-righton Thirring model [456] with particular scalar coupling,

$$S'_{\text{LR}-\text{N}} = i \int d^3z \, E^- \left\{ \nabla_+ W^\alpha \left(\nabla_- W^\alpha + 2P^\alpha_{\underline{=}} \right) \right.$$

$$+ \Lambda_+^{\underline{=}} \left(\nabla_- W^\alpha + P^\alpha_{\underline{=}} \right)^2 + \nabla_- Z^\alpha \left(\nabla_+ Z^\alpha + 2Q^\alpha_+ \right) \qquad (2.27)$$

$$\left. - i\Lambda_-^{\pm} \left(\nabla_+ Z^\alpha + Q^\alpha_+ \right) \nabla_+ \left(\nabla_+ Z^\alpha + Q^\alpha_+ \right) + Q^\alpha_+ P^\alpha_{\underline{=}} \right\} \,,$$

where Z^α, $P^\alpha_{\underline{=}}$ and Q^α_+ play the role of auxiliary variables. The action (2.27) is invariant under the following Siegel transformations:

$$\delta W^\alpha = \epsilon_{\pm} \left(\nabla_- W^\alpha + P^\alpha_{\underline{=}} \right) \,, \qquad \delta P^\alpha_{\underline{=}} = \delta Q^\alpha_+ = 0 \,,$$

$$\delta Z^\alpha = i\epsilon_{\pm} \nabla_+ \left(\nabla_+ Z^\alpha + Q^\alpha_+ \right) + \tfrac{i}{2} \left(\nabla_+ \epsilon_- \right) \left(\nabla_+ Z^\alpha + Q^\alpha_+ \right) \,, \qquad (2.28)$$

while it also possesses the local invariance,

$$\delta W^\alpha = +K^\alpha \,, \qquad \delta Z^\alpha = +K^\alpha \,,$$

$$\delta Q^\alpha_+ = -\nabla_+ K^\alpha \,, \qquad \delta P^\alpha_{\underline{=}} = -\nabla_- K^\alpha \,. \qquad (2.29)$$

Let us now describe the quantization of the 'improved' Hull noton model (2.27). First, we introduce the new variables, $\tilde{W}^\alpha(z)$ and $\tilde{Z}^\alpha(z)$, as

$$\tilde{W}^\alpha(z) = W^\alpha(z) - 2i \int d^3z' \, E^-(z') G(z,z') \nabla'_+ P^\alpha_{\underline{=}}(z') \,,$$

$$\qquad (2.30)$$

$$\tilde{Z}^\alpha(z) = Z^\alpha(z) - 2i \int d^3z' \, E^-(z') G(z,z') \nabla'_- Q^\alpha_+(z') \,,$$

where the scalar Green function $G(z,z')$ has been defined by the equations

$$2i\nabla_+ \nabla_- G(z,z') = \delta_-(z,z') \,,$$

$$\qquad (2.31)$$

$$\delta_-(z,z') = (E^-)^{-1} \delta^{(2)}(\sigma,\sigma')(\theta - \theta') \,.$$

The definitions (2.30) imply [416]

$$\nabla_- \tilde{W}^\alpha = \nabla_- W^\alpha + P^\alpha_{\underline{=}} \,, \qquad \nabla_+ \tilde{Z}^\alpha = \nabla_+ Z^\alpha + Q^\alpha_+ \,. \qquad (2.32)$$

The action (2.27) can be rewritten in terms of the new variables (2.30) as follows:

$$S'_{\text{LR-N}} = i \int d^3z\, E^- \Big\{ (D_+ \tilde{W}^\alpha)(D_- \tilde{W}^\alpha) + (D_{++} \tilde{Z}^\alpha)(D_{--} \tilde{Z}^\alpha)$$

$$- 2i \int d^3z\, E^-(z) \Big[\int d^3z'\, E^-(z')[(\nabla_+ P_=^\alpha(z)) G(z,z')(\nabla'_+ P_=^\alpha(z'))$$

$$+ (\nabla_- Q_+^\alpha(z)) G(z,z')(\nabla'_- Q_+^\alpha(z'))] + Q_+^\alpha P_=^\alpha \Big] .$$

$$(2.33)$$

The superfields $\tilde{W}^\alpha$ and $\tilde{Z}^\alpha$ minimally transform under the Siegel transformations,

$$\delta \tilde{W}^\alpha = \epsilon_{\mp} \nabla_- \tilde{W}^\alpha \ , \quad \delta \tilde{Z}^\alpha = -\epsilon_= \nabla_+ \tilde{Z}^\alpha + \frac{i}{2}(\nabla_+ \epsilon_=) \nabla_+ \tilde{Z}^\alpha \ , \quad (2.34)$$

while they are inert under the transformations (2.29).

The gauge invariance (2.29) can be partially fixed by imposing the gauge $P_=^\alpha = 0$. In quantum theory this gauge fixing leads to the FP factor $[\det(\nabla_-)]^\alpha$. Changing the variables, $Q_+^\alpha = \nabla_+ \rho^\alpha$, gives rise to yet another FP determinant, $[\det(\nabla_+)]^\alpha$, in the quantum measure. Putting these all together yields the vacuum energy in the theory (3.27). The final result is conveniently written down in terms of the ghost-type superfields, $\Phi = \left\{ b_+^\alpha,\ b_=^\alpha,\ c^\alpha,\ \bar{c}^\alpha \right\}$ in the form $\exp(iW_{\text{gh}}) = \int [d\ldots] \exp(iS_{\text{gh}})$, where

$$S_{\text{gh}} = i \int d^3z\, E^- \Big[(D_+ \tilde{W}^\alpha)(D_- \tilde{W}^\alpha) + (D_{++} \tilde{Z}^\alpha)(D_{--} \tilde{Z}^\alpha)$$

$$+ (\nabla_+ \rho^\alpha)(\nabla_- \rho^\alpha) + b_=^\alpha \nabla_+ \bar{c}^\alpha + b_+^\alpha \nabla_= c^\alpha \Big] .$$

$$(2.35)$$

Only the first two terms in (2.35) transform under the Siegel tranformations.

To demonstrate the equivalence of our model (2.27) to the original Hull theory (2.24), we use the equations of motion for the superfields $P_=^\alpha$ and Q_+^α and rewrite the theory in terms of the variables $\tilde{W}^\alpha$, $\tilde{Z}$, $\Lambda_+^=$ and $\Lambda_-^\mp$. Substituting the solutions, $P_=^\alpha = P_=^\alpha(W, Z, \Lambda)$ and $Q_+^\alpha = Q_+^\alpha(W, Z, \Lambda)$, into (2.27) yields (2.24), with the field W^α being replaced by $W^\alpha - Z^\alpha$.

The singular term $(1 - i\Lambda^2)^{-1}$ also appears in the action (2.24), while it makes the theory ill defined because of the apparent singularity at $i\Lambda^2 = 1$. To describe one of the possible ways to avoid this singularity, we partially fix the gauge invariance (2.29) by imposing the gauge condition

$$P_=^\alpha = 0 . \tag{2.36}$$

Note that this gauge does *not* spoil the Siegel invariance (2.28).

In the gauge (2.36) the equations of motion take the Λ-independent form,

$$\nabla_- W^\alpha = 0 \ , \quad \nabla_- Z^\alpha = 0 \ ,$$

$$\nabla_+ W^\alpha + Q_+^\alpha = 0 \ , \quad \nabla_+ Z^\alpha + Q_+^\alpha = 0 \ , \tag{2.37}$$

while no restrictions on $\Lambda_+^=$ and $\Lambda_-^\mp$ arise.

The gauge condition (2.36) does not completely fix the invariance (2.29) since there is a residual invariance with the parameter K^α satisfying the constraint $\nabla_- K^\alpha = 0$. Therefore, by using the invariance (2.29) with the parameter $K^\alpha = -W^\alpha$ (this invariance survives in the gauge (2.36) due to (2.37)), we arrive at the vanishing of W^α. Equation (2.37) now yields $Q_+^\alpha = 0$ and $Z^\alpha = \text{const}$.

7.2.2 Non-Chiral $(1,0)$ Supersymmetric Coset Models

From the viewpoint of string theory (Chap. 6), the WZNW models describe string propagation on group manifolds. The GKO method (Subsect. 5.3.1) generalizes the SS group construction (Subsect. 5.1.2) to the cosets $\mathbf{G/H}$, and it allows one to obtain a broader class of 2d CFT describing (super)string compactification. The bosonic coset models are realized in the 2d Lagrangian approach as gauged WZNW models (Subsect. 5.3.2). In this subsection we describe non-chiral $(1,0)$ supersymmetric gauged WZNW theories as a prerequisite for introducing the chiral ones.

Let $\mathbf{H}$ be a subgroup of $\mathbf{G}$, and $\mathbf{h}$ be the corresponding Lie algebra. The ad($\mathbf{h}$)-invariant orthogonal decomposition of the Lie algebra $\mathbf{g}$ is given by

$$\mathbf{g} = \mathbf{h} \oplus \mathbf{m}, \quad [\mathbf{h}, \mathbf{h}] \subset \mathbf{h}, \quad [\mathbf{h}, \mathbf{m}] \subset \mathbf{m}. \tag{2.38}$$

Gauging the diagonal subgroup $\mathbf{H}$ of the global $\mathbf{G}_L \times \mathbf{G}_R$ symmetry of the action $S(g)$ gives rise to the action

$$kS\left(g, \Gamma_+, \Gamma_=\right) = kS(g) + \frac{ik}{\pi} \int d^3z \, E^- \text{Tr}\left\{\nabla_= g \cdot g^{-1}\Gamma_+ \right.$$
$$\left. - \Gamma_= g^{-1}\nabla_+ g + g^{-1}\Gamma_+ g\Gamma_= - \Gamma_+\Gamma_=\right\}, \tag{2.39}$$

where the gauge superfields Γ_+ and $\Gamma_=$ are valued in the adjoint representation of $\mathbf{H}$. The action (2.39) is invariant under the transformations

$$g \to \Lambda g \Lambda^{-1}, \quad \Lambda \in \mathbf{H},$$

$$\Gamma_A \to \nabla_A \Lambda \cdot \Lambda^{-1} + \Lambda\Gamma_A\Lambda^{-1}. \tag{2.40}$$

To make the invariance manifest, the gauge superfields should be parametrized in terms of the $\mathbf{H}$-group valued scalar superfields $\tilde{h}$ and h as follows:

$$\Gamma_= = \nabla_= \tilde{h} \cdot \tilde{h}^{-1}, \quad \Gamma_+ = \nabla_+ h \cdot h^{-1}. \tag{2.41}$$

This results in the action

$$S\left(g, \Gamma_+, \Gamma_=\right) = S\left(h^{-1}g\tilde{h}\right) - S\left(h^{-1}\tilde{h}\right), \tag{2.42}$$

where the identity (2.4) has been used. The symmetry (2.40) is now obvious.

To understand how the action (2.39) describes the **G/H** coset structure, let us consider the equations of motion. Varying (2.39) with respect to Γ_+ and Γ_- yields

$$\left(g^{-1}\hat{\nabla}_+ g\right)_\mathbf{h} = 0 \ , \quad \left(\hat{\nabla}_= g \cdot g^{-1}\right)_\mathbf{h} = 0 \ , \tag{2.43}$$

where we have introduced the new derivatives $\hat{\nabla}_A$ as

$$\hat{\nabla}_A \equiv \nabla_A - [\Gamma_A, \dots] \ , \quad \hat{\nabla}_{\mmlToken{+}} \equiv \frac{\mathrm{i}}{2}\left\{\hat{\nabla}_+, \hat{\nabla}_+\right\} \ ,$$

$$[\hat{\nabla}_=, \hat{\nabla}_+] = 2\mathrm{i}\Sigma^+ \hat{M} - W_- \ , \quad W_- \equiv \nabla_= \Gamma_+ - \nabla_+ \Gamma_= + [\Gamma_+, \Gamma_=] \ . \tag{2.44}$$

In (2.43) $(\dots)_\mathbf{h}$ denotes the projection onto the subspace $\mathbf{h}$ of $\mathbf{g}$ in the sense of (2.38). It is worth noticing that (2.43) is gauge covariant if and only if the commutation relations (2.37) are fulfilled! The equation of motion for $g(z)$ reads

$$\hat{\nabla}_=\left(g^{-1}\hat{\nabla}_+ g\right) = W_- \ . \tag{2.45}$$

Equations (2.43) and (2.45) imply $W_- = 0$. Hence, both connections Γ_+ and Γ_- are, in fact, trivial on-shell. Exploiting the gauge invariance (2.40) allows us to choose the gauge $\Gamma_+ = \Gamma_= = 0$. In this gauge the equations of motion (2.43) and (2.45) are simplified to

$$\left(g^{-1}\nabla_+ g\right)_\mathbf{h} = \left(\nabla_= g \cdot g^{-1}\right)_\mathbf{h} = 0 \ ,$$

$$\nabla_=\left(g^{-1}\nabla_+ g\right)_\mathbf{m} = 0 \quad \Longleftrightarrow \quad \nabla_+\left(\nabla_= g \cdot g^{-1}\right)_\mathbf{m} = 0 \ . \tag{2.46}$$

The **m**-currents are thus non-zero on-shell, whereas the **h**-currents vanish on-shell, as they should.

At the end of this subsection we derive the component form of the (1,0) superfield action (2.39). In the Wess-Zumino gauge for the symmetry (2.40),

$$\Gamma_+ | = 0 \ , \tag{2.47}$$

where $|$ means taking the first component of a superfield or an operator, the field components of the relevant (1,0) superfields are

$$\nabla_+ \Gamma_+ | \equiv \mathrm{i}V_{\mmlToken{+}} \ , \quad \Gamma_= | \equiv V_= \ , \quad \nabla_+ \Gamma_= | \equiv \lambda_- \ ,$$

$$g | = U \ , \quad \left(g^{-1}\nabla_+ g\right)_\mathbf{m} | = \mathrm{i}\psi_+ \ , \quad \left(g^{-1}\nabla_+ g\right)_\mathbf{h} | = \mathrm{i}\beta_+ \ . \tag{2.48}$$

Calculation of the component form of the action (2.39) is now straightforward. The component action in the superconformal gauge reads

$$S\left(g, \Gamma_+, \Gamma_=\right) = S_{\text{WZNW}}\left(U, V_=, V_+\right) - \frac{i}{\pi} \int d^2\sigma \, \text{Tr}\left[\psi_+ \partial_= \psi_+\right]$$

$$- \frac{i}{\pi} \int d^2\sigma \, \text{Tr}\left[\beta_+ \left(\partial_= - 2V_=\right)\beta_+ + i\lambda_- \beta_+\right] , \tag{2.49}$$

where the S_{WZNW} denotes the gauged (level-k) bosonic WZNW action. The last term in (2.49) ensures the *non-dynamical* nature of the Majorana-Weyl 2d spinors β_+ and λ_- in the adjoint representation of **H** (i.e. these spinors vanish on-shell).

7.2.3 Chiral $(1,0)$ Supersymmetric Coset Models

We are now in a position to discuss the $(1,0)$ supersymmetric *chiral* coset models. In this subsection we construct their invariant actions and introduce their coupling to the background gauge superfields in a way that is consistent with the Siegel symmetry.

G/H Lefton Models. The simplest way to obtain the $(1,0)$ supersymmetric lefton coset action is to replace the derivatives ∇_A by the derivatives D_A in the action (2.39). The derivatives D_A have been defined in (2.7). The new action is given by

$$k_{\text{L}} S_{\text{L}}\left(g_{\text{L}}, \Gamma_+, \Gamma_=, \Lambda_+^=\right) = k_{\text{L}} S_{\text{L}}(g_{\text{L}}) + \frac{ik_{\text{L}}}{\pi} \int d^3z \, E^- \text{Tr}\left\{\nabla_= g_{\text{L}} \cdot g_{\text{L}}^{-1} \Gamma_+\right.$$

$$\left. - \Gamma_= g_{\text{L}}^{-1}\left(\nabla_+ g_{\text{L}} + \Lambda_+^= \nabla_= g_{\text{L}}\right) + g_{\text{L}}^{-1} \Gamma_+ g_{\text{L}} \Gamma_= - \Gamma_+ \Gamma_=\right\} , \tag{2.50}$$

where the $S_{\text{L}}(g_{\text{L}})$ has been defined in Subsect. 7.3.1.

To display the symmetries of the action (2.50), it is convenient to introduce the following parametrization of the connection superfields $\Gamma_=$ and Γ_+:

$$\Gamma_= = \nabla_= \tilde{h} \cdot \tilde{h}^{-1} , \quad \Gamma_+ = D_+ h \cdot h^{-1} , \quad \text{where} \quad h, \tilde{h} \in \mathbf{H} \tag{2.51}$$

which is different from the similar parametrization used in Subsect. 7.2.2. The action (2.50) can be rewritten in the form

$$S_{\text{L}}\left(g_{\text{L}}, \Gamma_+, \Gamma_=, \Lambda_+^=\right) = S_{\text{L}}\left(h^{-1} g_{\text{L}} \tilde{h}\right) - S_{\text{L}}\left(h^{-1} \tilde{h}\right) , \tag{2.52}$$

whose gauge invariances are manifest, namely,
(i) the **H**-gauge invariance:

$$g_{\text{L}} \to \Lambda g_{\text{L}} \Lambda^{-1} , \quad \Lambda \in \mathbf{H} , \quad \Gamma_A \to D_A \Lambda \cdot \Lambda^{-1} + \Lambda \Gamma_A \Lambda^{-1} , \tag{2.53}$$

(ii) the Siegel invariance:

$$\delta g_{\text{L}} = \epsilon_+ \nabla_= g_{\text{L}} , \quad \delta \Lambda_+^= = -D_+ \epsilon_+ ,$$

$$\delta\Gamma_= = \nabla_= \left(\epsilon_{\mp}\Gamma_=\right) , \qquad \delta\Gamma_+ = \epsilon_{\mp}\nabla_=\Gamma_+ . \tag{2.54}$$

The equations of motion in the theory (2.50) are

$$\left(\hat{\nabla}_{=g_L} \cdot g_L^{-1}\right)_h = \left(g_L^{-1}\hat{D}_+g_L\right)_h = 0 ,$$

$$\hat{D}_+\left(\hat{\nabla}_{=g_L} \cdot g_L^{-1}\right) = g_L W_=^L g_L^{-1} ,$$

$$\mathrm{Tr}\left\{\left(\hat{\nabla}_{=g_L} \cdot g_L^{-1}\right)^2\right\} = 0 , \tag{2.55}$$

where we have introduced the gauge $(1,0)$ superfield strength $W_=^L$ as

$$W_=^L = D_=\Gamma_+ - D_+\Gamma_= + [\Gamma_+,\Gamma_=] , \tag{2.56}$$

and the gauge-covariant derivatives $\hat{D}_A$ according to the pattern (2.44) for the derivatives $\hat{\nabla}_A$. Varying the action with respect to the Lagrange multiplier $\Lambda_+^=$ results in the third line of (2.55), where we have also used the first equation of (2.55).

Equation (2.55) tells us that the $(1,0)$ connection Γ_A is trivial on-shell, $\Gamma_A = D_A h \cdot h^{-1}$. In the gauge $\Gamma_A = 0$, $\Lambda_+^= = 0$, we are left with the only non-zero current $\left(g_L^{-1}\nabla_{+g_L}\right)_m$ on-shell, as it should.

In Subsect. 7.2.1 we learned that the non-Abelian leftons can be coupled to the Siegel-gauge invariant Lie algebra-valued background $(1,0)$ superfields via the action (2.14). The natural question now arises whether it is possible to generalize this coupling to the $(1,0)$ leftons living in a coset space. This indeed turns out to be possible [323].

First, we note that the superfield g_L is inert under the transformation (2.53) if and only if g_L takes its values in the centralizer of $\mathbf{H}$ in $\mathbf{G}$, which will be denoted by $\mathbf{H}^C$. Hence, we can introduce extra gauging with respect to $\mathbf{H}^C$. Let $A_=(z)$ be the background vector superfield belonging to the Lie algebra $\mathbf{h}^C$ of the Lie group $\mathbf{H}^C$. By definition of $\mathbf{H}^C$ we have

$$[\mathbf{h}^C,\mathbf{h}] = 0 , \qquad \mathbf{h}^C \subset \mathbf{m} . \tag{2.57}$$

The appropriate generalization of the action (2.50) reads

$$k_L S_L \left(g_L,\Gamma_+,\Gamma_=,\Lambda_+^=;A_=\right) = k_L S_L \left(g_L;A_=\right)$$

$$+ \frac{ik_L}{\pi} \int d^3 z E^- \mathrm{Tr}\left\{\nabla_{=g_L} \cdot g_L^{-1}\Gamma_+ - \Gamma_=g_L^{-1}\left(\nabla_{+g_L} + \Lambda_+^=\nabla_{=g_L}\right)\right.$$

$$\left. + g_L^{-1}\Gamma_{+g_L}\left(\Gamma_= + A_=\right) - \Gamma_+\Gamma_=\right\} . \tag{2.58}$$

Here $S_L\left(g_L; A_=\right)$ is the action (2.14), with $A_=$ being restricted to the subalgebra $\mathbf{h}^C$ of $\mathbf{g}$. It is not difficult to check the invariance of the action (2.58) under the gauge transformations (2.53) and the Siegel transformations (2.15) supplemented by the variations (2.54) of the gauge superfields Γ_-, Γ_+ and $\Lambda_+=$. The superfield $A_=$ is inert under both transformations.

G/H Righton Models. An action for the $(1,0)$ supersymmetric righton coset models can be constructed in full analogy with the lefton case considered above. Replacing the derivatives ∇_A by the new ones $\mathcal{D}_A$ in accordance with (2.18), and using the latter in the action (2.39) yield the new action

$$k_R S_R\left(g_R, \Xi_+, \Xi_=, \Lambda_=^{\ddagger}\right) = k_R S_R(g_R) + \frac{ik_R}{\pi} \int d^3 z\, E^- \operatorname{Tr}\{\Xi_+$$

$$\times \left(\nabla_= g_R - i\left(\nabla_+ \Lambda_=^{\ddagger}\right)\nabla_+ g_R + \Lambda_=^{\ddagger}\nabla_{\pm} g_R\right)g_R^{-1} - \Xi_= g_R^{-1}\nabla_+ g_R$$

$$+ g_R^{-1}\Xi_+ g_R \Xi_= - \Xi_+ \Xi_=\} \ . \tag{2.59}$$

This action is invariant under the gauge transformations

$$g_R \to \Lambda g_R \Lambda^{-1}\ , \quad \Lambda \in \mathbf{H}\ ,$$

$$\Xi_A \to \mathcal{D}_A\Lambda \cdot \Lambda^{-1} + \Lambda\Xi_A\Lambda^{-1}\ , \tag{2.60}$$

and the Siegel transformations (2.20) supplemented by the following variations of the vector $(1,0)$ superfields Ξ_+ and $\Xi_=$:

$$\delta\Xi_+ = -\frac{i}{2}\nabla_+\left[\nabla_+\left(\epsilon_=\Xi_+\right) + \epsilon_=\nabla_+\Xi_+\right]\ ,$$

$$\delta\Xi_= = \epsilon_=\nabla_{\pm}\Xi_= - \frac{i}{2}\left(\nabla_+\epsilon_=\right)\nabla_+\Xi_=\ . \tag{2.61}$$

As before, the connection Ξ_A is trivial on-shell, $\Xi_A = \mathcal{D}_A h \cdot h^{-1}$, where $h \in \mathbf{H}$. In the gauge $\Xi_A = \Lambda_=^{\ddagger} = 0$, the equations of motion are given by

$$\left(g_R^{-1}\nabla_+ g_R\right) = 0\ , \quad \left(\nabla_= g_R \cdot g_R^{-1}\right)_{\mathbf{h}} = 0\ . \tag{2.62}$$

Therefore, the only current that does not vanish on-shell is the **m-current** $(\nabla_= g_R \cdot g_R^{-1})_{\mathbf{m}}$, as it should.

The coupling of the **G/H** rightons to the background gauge $(1,0)$ superfields $A_+(z)$ valued in the centralizer $\mathbf{h}^C$ of $\mathbf{h}$ in $\mathbf{g}$ is described by the action

$$k_R S_R\left(g_R, \Xi_+, \Xi_=, \Lambda_=^{\ddagger}; A_+\right) = k_R S_R\left(g_R; A_+\right)$$

$$+ \frac{ik_R}{\pi}\int d^3 z\, E^-\operatorname{Tr}\left\{\Xi_+\left(\nabla_= g_R - i\left(\nabla_+\Lambda_=^{\ddagger}\right)\nabla_+ g_R + \Lambda_=^{\ddagger}\nabla_{\pm} g_R\right)g_R^{-1}\right.$$

$$-\Xi_= g_{\mathrm{R}}^{-1}\nabla_+ g_{\mathrm{R}} + g_{\mathrm{R}}^{-1}\left(\Xi_+ - A_+\right)g_{\mathrm{R}}\Xi_= - \Xi_+\Xi_=\Big\} \ . \tag{2.63}$$

Here $S_{\mathrm{R}}\left(g_{\mathrm{R}}; A_+\right)$ is the action (2.21), with A_+ being restricted to the sub-algebra $\mathbf{h}^C$ of $\mathbf{g}$. The action (2.63) is invariant under the transformations (2.23), (2.60) and (2.61), while the superfield A_+ is inert under both of them.

The chiral gauged (1,0) supersymmetric WZNW actions constructed above differ from those of [454]. The Siegel transformations proposed in [454] are not standard (i.e. non-minimal), whereas in our approach they take the standard (in Siegel's sense) or minimal form.

7.2.4 Anomalies of (1,0) Supersymmetric Coset Models

The consistency of QFT requires the absence of anomalies for all local symmetries. Anomalies of the coset models can be understood by using the algebraic results of 2d CFT about the WZNW models defined on group manifolds (Chap. 5). To make this point as clear as possible, we propose here the alternative derivation of the anomalies, which is based on the field-theoretical description. The novel feature of this approach is the treatment of the anomalies of Siegel symmetries [323].

Supergravitational Anomalies of the Non-chiral Coset Models. In classical theory, the (1,0) supersymmetric coset models (2.42) are invariant under the general coordinate and local Lorentz transformations in a curved (1,0) superspace with the parameters $K = K^M\partial_M + \Lambda\hat{M}$, $K^M = \left(K^{\ddagger}, K^{=}, K^{+}\right)$, and the super-Weyl transformations (2.11b). In quantum theory, all these symmetries become potentially anomalous, while the anomalies exhibit themselves in a non-vanishing variation of the quantum effective action. In fact, not all of the symmetries are dangerous since some anomalies can be absorbed by adding local counterterms to the quantum effective action. As was demonstrated in [418, 443], only the general coordinate transformations with the two independent parameters $K^{\ddagger}$ and $K^{=}$ are to be taken into account. The variation of the quantum effective action W with respect to these transformations is given by

$$\delta_{\mathrm{g.-c.}} W = \frac{1}{12\pi}\int \mathrm{d}^3 z\, E^-\left(\nu_{\mathrm{L}}{}^s K^{=}\nabla_= + \nu_{\mathrm{R}}\, K^{\ddagger}\nabla_{\ddagger}\right)\Sigma^+ \ , \tag{2.64}$$

where we have introduced the left $(\nu_{\mathrm{L}}{}^s)$ and right (ν_{R}) central charges. Their values for the WZNW models defined on the group $(\mathbf{G})$ manifolds (Chap. 5) are [167, 385, 457]

$$\nu_{\mathrm{L}}{}^s[\mathbf{G}] = \frac{2k\,\dim \mathbf{G}}{2k + c_2} + \frac{1}{2}\dim \mathbf{G} \ , \quad \nu_{\mathrm{R}}[\mathbf{G}] = \frac{2k\,\dim \mathbf{G}}{2k + c_2} \ , \tag{2.65}$$

where we have introduced the quadratic Casimir operator eigenvalue, c_2, in the adjoint representation of $\mathbf{G}$.

We now show in the functional approach that the corresponding quantities for the non-chiral $\mathbf{G/H}$ coset $(1,0)$ supersymmetric models (2.39) are given by

$$\nu_{\mathrm{L}}{}^{s}[\mathbf{G/H}] = \nu_{\mathrm{L}}{}^{s}[\mathbf{G}] - \nu_{\mathrm{L}}{}^{s}[\mathbf{H}] , \quad \nu_{\mathrm{R}}[\mathbf{G/H}] = \nu_{\mathrm{R}}[\mathbf{G}] - \nu_{\mathrm{R}}[\mathbf{H}] , \quad (2.66)$$

in exact agreement with the result of the coset GKO construction (Sect. 5.3). To prove (2.66), we use the results of Subsect. 5.3.2 and the results of [339, 340] for the bosonic cosets.

Our starting point is the component form (2.49) of the $(1,0)$ supersymmetric coset action. The first term in (2.49) is the gauged (level-k) bosonic WZNW model, so that its contribution to the central charges is given by [339, 340]

$$\nu_{\mathrm{R}}[\mathbf{G/H}] = \nu_{\mathrm{L}}[\mathbf{G/H}] = \frac{2k \dim \mathbf{G}}{2k + c_2} - (\mathbf{G} \leftrightarrow \mathbf{H}) . \quad (2.67)$$

The second term in (2.49) is the free action of $(\dim \mathbf{G} - \dim \mathbf{H})$ lower plus spinors. These fields contribute to the left central charge only, while they complete it with the supersymmetric expression

$$\nu_{\mathrm{L}}{}^{s}[\mathbf{G/H}] = \nu_{\mathrm{L}}[\mathbf{G/H}] + \frac{1}{2}(\dim \mathbf{G} - \dim \mathbf{H}) . \quad (2.68)$$

As far as the last term in the component action (2.49) is concerned, at first sight, it describes a coupling of the chiral fermions to a gauge chiral superfield $V_{=}$. Therefore, after quantization, there is room for *chiral* anomalies in addition to the anticipated supergravitational anomalies. Fortunately, the anomalies due to the last term in (2.49) actually do not arise. To explain this, we note that at the classical level the last term in the component action (2.49), in fact, describes non-moving fermions. In quantum theory the contribution of the non-moving fermions to the quantum effective action is also trivial,

$$\int [\mathrm{d}\beta_+][\mathrm{d}\lambda_-] \exp\left\{\frac{k}{\pi}\int \mathrm{d}^2\sigma \operatorname{Tr}\left[\beta_+\left(\partial_= - 2V_=\right)\beta_+ + \mathrm{i}\lambda_-\beta_+\right]\right\} = 1 , \quad (2.69)$$

where we have used a shift $\lambda_- \to \lambda_- - \mathrm{i}\left(\partial_=\beta_+ + [\beta_+, V_=]\right)$.

We use the standard quantum effective action for the theory (2.39) in $(1,0)$ superspace. In functional form it reads

$$\exp\left(\mathrm{i}W_{\mathbf{G/H}}\right) = \int [\mathrm{d}g][\mathrm{d}\Gamma_=][\mathrm{d}\Gamma_+]\mu(\Gamma) \exp\left\{\mathrm{i}kS\left(g, \Gamma_+, \Gamma_=\right)\right\} , \quad (2.70)$$

where we have introduced the FP measure

$$\mu(\Gamma) = \delta\left[f(\Gamma)\right] \Delta_{\mathrm{FP}}(\Gamma) . \quad (2.71)$$

This measure results from the gauge fixing of the local gauge invariance (2.40) by the use of a gauge $f(\Gamma) = 0$. The explicit form of the gauge is specified below.

The superfield redefinition (2.41) in the functional integral (2.70) yields the measure

$$[d\Gamma_-][d\Gamma_+] = [dh][d\tilde{h}] \det[\hat{\nabla}_+] \det[\hat{\nabla}_-] , \qquad (2.72)$$

where the gauge-covariant superderivatives $\hat{\nabla}_+$ and $\hat{\nabla}_-$ have been defined in (2.44).

The product of the chiral determinants in (2.72) can be rewritten by introducing the new ghost-type superfields $\Phi = \{b_+, b_-, c, \bar{c}\}$ valued in the adjoint representation of $\mathbf{H}$, as follows:

$$
\begin{aligned}
\det\left[\hat{\nabla}_+\right] \det\left[\hat{\nabla}_-\right] &= \int [d\Phi] \exp\left\{ i \int d^3 z\, E^- \mathrm{Tr}\left[b_+ \hat{\nabla}_- c + b_- \hat{\nabla}_+ \bar{c}\right]\right\} \\
&= \exp\left\{-i c_2 S\left(h \cdot \tilde{h}^{-1}\right)\right\} \int [d\Phi] \exp\left\{ i \int d^3 z\, E^- \right. \\
&\quad \left. \times \mathrm{Tr}\left[b_+ \nabla_- c + b_- \nabla_+ \bar{c}\right]\right\} ,
\end{aligned}
$$
$$(2.73)$$

where the techniques similar to those of Subsect. 5.3.2 or in [339, 340] have been used to derive the last identity.

A convenient choice of the gauge fixing function $f(\Gamma)$ in (2.71) is given by $\tilde{h} = \tilde{h}(\Gamma_-) = \mathrm{const.}$, because the FP determinant $\Delta_{\mathrm{FP}}(\Gamma)$ does not contribute in this gauge. After a shift $g \to hg$, (2.70) yields

$$
\begin{aligned}
\exp(i\, W_{\mathbf{G}/\mathbf{H}}) &= \int [dg][dh] \exp\left\{i[kS(g) - (k + c_2)S(h)]\right\} \\
&\quad \times \int [d\Phi] \exp\left\{ i \int d^3 z\, E^- \mathrm{Tr}\left[b_+ \nabla_- c + b_- \nabla_+ \bar{c}\right]\right\} .
\end{aligned}
$$
$$(2.74)$$

In (2.74) the ghost superfields are only coupled to the (1,0) supergravity pre-potentials. Hence, the standard values of their central charges are valid, $\nu_{\mathrm{L}} = -3 \dim \mathbf{H}$ and $\nu_{\mathrm{R}} = -2 \dim \mathbf{H}$. By using (2.65) and (2.74), and taking into account the relation $c_2(\mathbf{G}) = c_2(\mathbf{H})$ that is the simple corollary of the decomposition (2.38), we get the total anomaly coefficients of (2.66), namely,

$$
\frac{2k \dim \mathbf{G}}{2k + c_2} + \frac{2(k + c_2) \dim \mathbf{H}}{2(k + c_2) - c_2} - 2 \dim \mathbf{H} = \frac{2k \dim \mathbf{G}}{2k + c_2} - \frac{2k \dim \mathbf{H}}{2k + c_2} , \quad \text{etc.}
$$

Siegel Anomalies of Chiral Coset Models. Let us consider again the general action (2.26), with the rightons and leftons being defined on a group manifold. We also replace the Hull notion action (2.24) by the classically equivalent action (2.27),

$$S_{\text{tot.}}[\mathbf{G}] = k_{\text{L}} S_{\text{L}}(g_{\text{L}}) + k_{\text{R}} S_{\text{R}}(g_{\text{R}}) + S_{\text{L-N}} + S'_{\text{LR-N}} . \tag{2.75}$$

The action (2.75) possesses left and right local Siegel symmetries. In the background-field quantization method, the action (2.75) should, therefore, be accompanied by the corresponding ghost action,

$$S_{\text{L}}^{\text{gh.}} + S_{\text{R}}^{\text{gh.}} = \int \mathrm{d}^3 z \, E^- \left\{ \beta_- {}^{\ddagger}D_+ \kappa_+ + \beta_+ {}^{=}D_{--} \kappa_- \right\} . \tag{2.76}$$

The first three terms in (2.75) contain the derivatives, D_A or $\mathcal{D}_A$, with the minimal coupling to the Siegel multipliers, whereas the last one, $S'_{\text{LR-N}}$, does not. As was demonstrated in Subsect. 7.2.1, the path integral quantization of the theory $S'_{\text{LR-N}}$ leads to the 'effective' action (2.33), with every term being minimally coupled to the certain derivative in the list $\{D_A, \mathcal{D}_A, \nabla_A\}$. This is important for justifying separate discussions of the left and right Siegel symmetries.

An anomalous variation of the quantum effective action under the Siegel ϵ_+-transformation is given by

$$\delta_{\text{L}} W_{\mathbf{G}} = \frac{\nu_{\text{L}}}{12\pi} \int \mathrm{d}^3 z \, E^- \epsilon_+ \nabla_= \Sigma_{\text{L}}^+ , \tag{2.77}$$

where the $(1,0)$ superfield strength Σ_{L}^+ was introduced in (2.7). The values of ν_{L} associated with the actions $k_{\text{L}} S_{\text{L}}$, $S_{\text{L-N}}$, $S_{\text{L}}^{\text{gh.}}$ and $S_{\widetilde{W}}$, respectively, [3] are given by

$$\nu_{\text{L}}\left(k_{\text{L}} S_{\text{L}}\right) = \frac{2k_{\text{L}} \dim \mathbf{G}_{\text{L}}}{2k_{\text{L}} + c_{2,\text{L}}} ,$$

$$\nu_{\text{L}}\left(S_{\text{L-N}}\right) = \frac{1}{2} N_\Psi ,$$

$$\nu_{\text{L}}\left(S_{\text{L}}^{\text{gh.}}\right) = -26 , \tag{2.78}$$

$$\nu_{\text{L}}\left(S_{\widetilde{W}}\right) = N_W .$$

The first line of (2.78) is a simple consequence of the fact [454] that the component spinor fields of $g_{\text{L}}(z)$ do *not* couple to the component multiplier $\Lambda_+{}^=$ of $\Lambda_+{}^=(z)$. We thus see that our theory is going to be free from the anomaly of the Siegel ϵ_+-symmetry provided that the condition

$$\frac{2k_{\text{L}} \dim \mathbf{G}_{\text{L}}}{2k_{\text{L}} + c_{2,\text{L}}} + \frac{1}{2} N_\Psi + N_W = 26 \tag{2.79}$$

is satisfied.

---

[3] The action $S_{\widetilde{W}}$ is defined by the first term in (2.33) that contributes to the left Siegel anomaly.

An anomalous variation of the quantum effective action under the other (Siegel) ϵ_- -transformation is

$$\delta_R W_G = \frac{\nu_R}{12\pi} \int d^3 z \, E^- \nabla_= \nabla_{\not=} \Sigma_R^+ \, , \tag{2.80}$$

where the $(1,0)$ superfield strength Σ_R^+ has been introduced in (2.18). Only the actions $k_R S_R$, $S_R^{gh.}$ and $S_{\tilde{Z}}$ to be defined as the second term in (2.33) contribute to this anomaly. We find

$$\nu_R \left(k_R S_R \right) = \frac{2k_R \dim \mathbf{G_R}}{2k_R + c_{2,R}} + \frac{1}{2} \dim \mathbf{G_R} \, ,$$

$$\nu_R \left(S_R^{gh.} \right) = -15 \, , \tag{2.81}$$

$$\nu_R \left(S_{\tilde{Z},R} \right) = N_W \, ,$$

respectively. The first line of (2.81) can be justified by noticing that the component spinor fields of $g_R(z)$ are notons, while they couple to the first component of the Siegel multiplier $\Lambda_-^+(z)$. Requiring anomaly freedom with respect to the ϵ_- -symmetry gives rise to the second condition

$$\frac{2k_R \dim \mathbf{G_R}}{2k_R + c_{2,R}} + \frac{1}{2} \dim \mathbf{G_R} + \frac{3}{2} N_W = 26 - 11 = 15 \, . \tag{2.82}$$

To obtain similar conditions for the $(1,0)$ supersymmetric chiral $\mathbf{G/H}$ coset models with rightons and leftons, we replace the group actions $k_L S_L (g_L)$ and $k_R S_R (g_R)$ in (2.75) by their coset generalizations (2.50) and (2.59), respectively,

$$S_{\text{tot.}} [\mathbf{G/H}] = k_L S_L \left(g_L, \Gamma_+, \Gamma_=, \Lambda_+^= \right) + k_R S_R \left(g_L, \Xi_+, \Xi_=, \Lambda_-^+ \right) \tag{2.83}$$
$$+ S_{L-N} + S'_{RL-N} \, ,$$

and use the first two lines of (2.78). As a result, the conditions of the Siegel anomaly freedom take the form

$$\left[\frac{2k_L \dim \mathbf{G_L}}{2k_L + c_{2,L}} - (\mathbf{G_L} \leftrightarrow \mathbf{H_L}) \right] + \frac{1}{2} N_\Psi + N_W = 26 \, , \tag{2.84a}$$

$$\left[\frac{2k_R \dim \mathbf{G_R}}{2k_R + c_{2,R}} + \frac{1}{2} \dim \mathbf{G_R} - (\mathbf{G_R} \leftrightarrow \mathbf{H_R}) \right] + \frac{3}{2} N_W = 15 \, . \tag{2.84b}$$

Heterotic String Actions in $D < 10$. We are now in a position to describe the 2d Lagrangian realization of the $D < 10$ heterotic strings, with their internal chiral degrees of freedom being valued on homogeneous spaces. A generic heterotic string action (NLSM) consists of three parts,

$$S_{\text{HS}} = S_{\text{string}} + S_{\text{YM}} + S_{\text{tot.}} [\mathbf{G/H}] \; . \tag{2.85}$$

The first term is the usual string action,

$$S_{\text{string}} = \int d^3 z \, E^- [i\eta_{\underline{ab}}(\nabla_+ X^{\underline{a}})(\nabla_= X^{\underline{b}})] \; , \tag{2.86}$$

where $\eta_{\underline{ab}}$ denotes flat (Minkowskian) metric of D-dimensional spacetime, and $\underline{a}, \underline{b} = 0, 1, \dots D - 1$. The second term in the action (2.85) describes the $(1,0)$ supersymmetric heterotic fermions,

$$S_{\text{YM}} = - \int d^3 z \, E^- \eta_-^{\hat{I}} \nabla_+ \eta_-^{\hat{I}} \; , \quad \text{where} \quad \hat{I} = 1, \dots, N_F \; . \tag{2.87}$$

The third term has already been discussed above.

After quantization the theory (2.85) suffers from the supergravitational and Siegel anomalies. The anomaly-freedom in the Siegel symmetry occurs when the conditions (2.84) are satisfied. In this case one can choose the background gauge $\Lambda_+^{=} = \Lambda_-^{\ddagger} = 0$ in the functional integral defining the quantum effective action of our theory. Then our results above can be directly applied to a calculation of the left and right central charges of the $(1,0)$ supersymmetric $\mathbf{G/H}$ coset models. After taking into account the standard contributions to the central charges from the 'spacetime' action (2.86) and the 'Yang-Mills' action (2.87), putting all the contributions to the supergravitational anomaly coefficients together yields the following anomaly-freedom conditions:

$$\left[\frac{2k_L \dim \mathbf{G}_L}{2k_L + c_{2,L}} + \frac{1}{2} \dim \mathbf{G}_L - (\mathbf{G}_L \leftrightarrow \mathbf{H}_L) \right] + \frac{3}{2} N_W = \frac{3}{2}(10 - D) \; ,$$

$$\left[\frac{2k_R \dim \mathbf{G}_R}{2k_R + c_{2,R}} - (\mathbf{G}_R \leftrightarrow \mathbf{H}_R) \right] + \frac{1}{2} N_F = 26 - D \; . \tag{2.88}$$

After being restricted to $D = 4$ heterotic strings, they become

$$\left[\frac{2k_L \dim \mathbf{G}_L}{2k_L + c_{2,L}} + \frac{1}{2} \dim \mathbf{G}_L - (\mathbf{G}_L \leftrightarrow \mathbf{H}_L) \right] + \frac{3}{2} N_W = 9 \; ,$$

$$\left[\frac{2k_R \dim \mathbf{G}_R}{2k_R + c_{2,R}} - (\mathbf{G}_R \leftrightarrow \mathbf{H}_R) \right] + \frac{1}{2} N_F = 22 \; . \tag{2.89}$$

In fact, the level k' of $\hat{\mathbf{H}}$ is determined by the embedding of $\mathbf{H}$ into $\mathbf{G}$. For instance, when the simple roots of $\mathbf{H}$ are a subset of the simple roots of $\mathbf{G}$ with the level number k, then $k' = k$. If $\mathbf{G}$ is the direct product of $\mathbf{G}_{k_1}$ and $\mathbf{G}_{k_2}$ and $\mathbf{H}$ is the diagonal subgroup, then $k' = k_1 + k_2$ [45].

Equations (2.84) and (2.88) can be further generalized by taking more than one righton series (each series with its own Lagrange multiplier enforcing the chirality condition). The Siegel anomaly cancellation condition in (2.84b) should be satisfied by each righton series separately. Hence, each righton series must be supplemented by the appropriate notons. As regards the central charge, the righton series *additively* contribute to the second lines of (2.88) and (2.89). Those features may be imporant for phenomenological applications of the heterotic strings.

The righton series can be used to bosonize *all* the heterotic fermions, in accordance with the second line of (2.89). This is impossible when using only one righton set. Similar considerations apply to multiple lefton series.

7.2.5 Anomaly-Free Solutions

Note that the standard heterotic string in $D = 10$ with $N_F = 32$ heterotic fermions, and without rightons and leftons (and notons) satisfies the general equations (2.88). By further taking into account modular invariance, this leads to the well-known gauge groups Spin(32)/$\mathbf{Z}_2$ and $E_8 \otimes E_8'$ [407, 408].

If $D < 10$, (3.84) determines the number N_Ψ of spinor notons and the number N_W of non-moving scalar W-notons. Simultaneously, it also puts restrictions on the choice of left and right cosets, $\mathbf{G_L/H_L}$ and $\mathbf{G_R/H_R}$, respectively. When the number N_W increases, the constraints (2.84) become rather severe.

Let us consider, for example, the Abelian case with $N_L \neq 0$ and $N_R = 0$. This corresponds to a toroidal compactification of the heterotic string [444]. Equations (2.84a) and (2.88) in D spacetime dimensions with N_L Abelian leftons yield

$$D + N_L = 10 , \quad D + \tfrac{1}{2}N_F = 26 , \quad \tfrac{1}{2}N_\Psi + N_L = 26 . \tag{2.90}$$

In $D = 4$ dimensions we thus find

$$N_L = 6 , \quad N_F = 44 , \quad N_\Psi = 40 . \tag{2.91}$$

This gives rise to $N = 4$ extended spacetime supersymmetry and the gauge group of rank 22, e.g. $SO(44)$ or $E_8 \otimes E_8' \otimes E_6$ for the four-dimensional heterotic strings. The lefton group is $\mathbf{G_L} = [U(1)]^6$.

The general non-Abelian case corresponds to $N_L \neq 0$ and $N_R \neq 0$. At the present level of understanding, it is highly desirable to have $N = 1$ spacetime supersymmetry in the $D = 4$ heterotic string theory [170]. This implies the presence of $N = 2$ superconformal symmetry on the string world-sheet [458]. It is, therefore, natural to use the $N = 2$ minimal models to saturate the central charge condition by leftons. The $N = 2$ minimal models are exactly solvable in 2d CFT (Chap. 5), while they can be easily realized in the GKO construction by the use of the coset $\dfrac{\widehat{SU(2)_k}}{U(1)}$, or the appropriate gauge WZNW chiral Lagrangian of the (1,0) supersymmetric leftons.

Tensoring the $N = 2$ minimal models in heterotic string theory is subject to the restriction [458]

$$\sum_{i=1}^{i_0} \frac{3k_i}{k_i + 2} = 9 \ , \tag{2.92}$$

whose 168 different solutions can be found, e.g. in [459, 460]. The modular invariance of the string partition function can be implemented by using the standard trick of 'embedding the spin connection into the gauge connection' [170]. [4] The Gepner construction (2.92) uses the sum of the $N = 2$ minimal models for rightons. The $(2,0)$ world-sheet supersymmetry can then be extended to $(2,2)$ world-sheet supersymmetry [462].

$N = 1$ spacetime supersymmetry in the $D = 4$ heterotic string theory can also be achieved via KS construction [341, 45] which associates an $N = 2$ superconformal model with any Hermitian symmetric space (Sect. 5.3). [5] There are several irreducible cases that give rise to the special value of the central charge, $c = 9$, e.g.,

$$\frac{\widehat{SU(10)}_1}{\widehat{SU(9)}_1 \otimes \widehat{U(1)}} \ . \tag{2.93}$$

In both interesting cases (Gepner construction and Kazama-Suzuki models) an embedding of $\mathbf{G_L/H_L}$ into $\mathbf{G_R/H_R}$ is necessary,

$$(\mathbf{G_R/H_R}) = (\mathbf{G_L/H_L}) \otimes (\mathbf{G_R/H_R})' \ , \tag{2.94}$$

while it gives rise (at $k = 1$) to the equations

$$\begin{aligned} \text{rank}(\mathbf{G_R/H_R})' + \tfrac{1}{2}\dim(\mathbf{G_R/H_R})' &= 6 \ , \\ \tfrac{1}{2}\dim(\mathbf{G_R/H_R}) + 14 &= N_F \ . \end{aligned} \tag{2.95}$$

Amongst the simple groups we find the solution

$$(\hat{\mathbf{G}}_\mathbf{R}/\hat{\mathbf{H}}_\mathbf{R})' = \widehat{SU(3)}_1 \ , \quad N_F = 22 + 2i_0 \ . \tag{2.96}$$

For instance, in the case of the $\mathbf{3}^5$ Gepner model [458] we have $N_F = 32$. Amongst the Hermitian irreducible globally symmetric spaces we find only two solutions for $(\hat{\mathbf{G}}_\mathbf{R}/\hat{\mathbf{H}}_\mathbf{R})'$,

$$\frac{\widehat{SO(8)}_1}{\widehat{SO(6)}_1 \otimes \widehat{SO(2)}} \quad \text{and} \quad \frac{\widehat{SU(5)}_1}{\widehat{SU(3)}_1 \otimes \widehat{SU(2)}_1 \otimes \widehat{U(1)}} \ . \tag{2.97}$$

Taking Q sets of non-Abelian rightons living on $SU(3)$ of rank 2 (and level $k = 1$) in the $D = 4$ heterotic string theory leads to the anomaly-free restriction

---

[4] This is considered e.g., from the viewpoint of the world-sheet string action, in [461].

[5] The $(1,0)$ supersymmetric world-sheet description of KS models is available in [326].

$$\tfrac{1}{2}N_F + 2Q = 22 \ . \tag{2.98}$$

In fact, there exist many possibilites for the number of heterotic fermions not exceeding 44. The same is true for the gauge group of rank ≤ 22, e.g. $SO(44)$, $[SU(3)]^{11}$, etc.

7.2.6 No Siegel Anomaly in the $D = 4$ Heterotic NLSM

In the previous subsections we described the construction of the 2d covariant *free* heterotic string actions for $D < 10$. Now we would like to briefly discuss the coupling to the background fields corresponding to the bosonic massless modes of the heterotic string, i.e. the heterotic NLSM.

The action of the NLSM can be divided into three parts, $S_\sigma = S_{\mathrm{I}} + S_{\mathrm{II}} + S_{\mathrm{III}}$. The first part is standard (Sect. 7.1),

$$S_{\mathrm{I}} = \mathrm{i} \int \mathrm{d}^3 z \, E^- \left\{ [\, g_{\underline{mn}}(X) + b_{\underline{mn}}(X) \,](\nabla_+ X^{\underline{m}})(\nabla_- X^{\underline{n}}) \right.$$
$$\left. -\mathrm{i}\phi(X)\Sigma^+ \right\} \ , \tag{2.99}$$

where we have introduced the NLSM metric $g_{\underline{mn}}(X)$, the torsion two-form $b = b_{\underline{mn}}(X) \, \mathrm{d}X^{\underline{m}} \wedge \mathrm{d}X^{\underline{n}}$, and the dilaton $\phi(X)$. The second part of the action describes the coupling to the $SO(N_F)$ gauge fields $B_{\underline{m}}^{\hat{I}\hat{J}}(X)$,

$$S_{\mathrm{II}} = -\int \mathrm{d}^3 z \, E^- \eta_-^{\hat{I}} [\, \delta^{\hat{I}\hat{J}}\nabla_+ \, + \, (\nabla_+ X^{\underline{m}})B_{\underline{m}}^{\hat{I}\hat{J}}(X)]\eta_-^{\hat{J}} \ . \tag{2.100}$$

The third part is more complicated: it describes the coupling to some additional spacetime vectors, $A_{\underline{m}}(X)$ and $C_{\underline{m}}(X)$, and scalars $\zeta^{\hat{I}\hat{J}}(X)$,

$$S_{\mathrm{III}} = k_{\mathrm{L}} S_{\mathrm{L}}(g_{\mathrm{L}}, \Gamma_+, \Gamma_-; A_-) + k_{\mathrm{R}} S_{\mathrm{R}}(g_{\mathrm{R}}, \Xi_+, \Xi_-; A_+) + S_{\mathrm{L-N}} + S_{\mathrm{LR-N}} \ , \tag{2.101}$$

where we have introduced the notation

$$A_- = (\nabla_- X^{\underline{m}})A_{\underline{m}}(X) + \mathrm{i}\eta_-^{\hat{I}}\eta_-^{\hat{J}}\zeta^{\hat{I}\hat{J}}(X) \ ,$$
$$A_+ = (\nabla_+ X^{\underline{m}})C_{\underline{m}}(X) \ . \tag{2.102}$$

For the theory to be consistent, the spacetime fields $A_{\underline{m}}$ and $\zeta^{\hat{I}\hat{J}}$ should take their values in the centralizer $\mathbf{h}_{\mathrm{L}}^C$ of the subalgebra $\mathbf{h}_{\mathrm{L}}$ in $\mathbf{g}_{\mathrm{L}}$, while $C_{\underline{m}}$ should be valued in the centraiizer $\mathbf{h}_{\mathrm{R}}^C$ of the subalgebra $\mathbf{h}_{\mathrm{R}}$ in $\mathbf{g}_{\mathrm{R}}$.

In general, the coupling to some other massless scalars via the $(1,0)$ supersymmetric non-Abelian 'lefton-righton Thirring model' [456] should be added to (2.101). This would require slight modification of the argument given below, as regards the absence of the Siegel anomaly, when one uses the first-order action [456]. Since the main line of our reasoning remains the same, we prove the theorem only in the simpler case just described.

To cancel the anomalies in the left and right Siegel symmetries for the heterotic NLSM described by the sum of (2.99), (2.100) and (2.101), we have to impose the *same* restrictions (2.84), as in the free case, and nothing more [323]. In other words, if a free $(1,0)$ supersymmetric heterotic string theory on the coset $\mathbf{G}/\mathbf{H}$ does not have Siegel anomalies, this is also true for the corresponding heterotic NLSM!

As regards the righton sector, we introduce the scalar superfields $Y(z)$ valued in the centralizer $\mathbf{H}_\mathrm{R}^C$ of the subgroup $\mathbf{H}_\mathrm{R}$ of the Lie algebra $\mathbf{G}_\mathrm{R}$. They satisfy the equation

$$A_+ = Y^{-1}(\nabla_+ Y) \ . \tag{2.103}$$

After a change of variables,

$$g_\mathrm{R} \to \bar{g}_\mathrm{R} = Y g_\mathrm{R} \ , \tag{2.104}$$

in the action $S_\mathrm{R}(g_\mathrm{R}, \Xi_+, \Xi_-; A_+)$, the new righton superfields $\bar{g}_\mathrm{R}$ vary under the transformation (2.23) as

$$\delta \bar{g}_\mathrm{R} \cdot \bar{g}_\mathrm{R}^{-1} = \mathrm{i}\epsilon_-(\ - \bar{R}_+^2 \ + \ \nabla_+ \bar{R}_+ \ ) \ + \ \frac{\mathrm{i}}{2}(\nabla_+ \epsilon_-)\bar{R}_+ \ . \tag{2.105}$$

In terms of the new variables, the action takes the form

$$S_\mathrm{R}(g_\mathrm{R}, \Xi_+, \Xi_-; A_+) = S_\mathrm{R}(\bar{g}_\mathrm{R}, \Xi_+, \Xi_-; 0) - S(Y) \ , \tag{2.106}$$

where the Polyakov-Wiegmann identity (2.4) has been used. The second term in (2.106) is now the only one that depends upon the background superfield A_+, while it is obviously invariant under the Siegel symmetry. At the same time, the first term in (2.106) coincides with the free heterotic coset righton action, while the transformations (2.105) are apparently the same as that of the free theory. Finally, the change of variables in (2.104) does not contribute to the anomaly. Stated differently, the associated determinant in the quantum measure does not depend upon $\mathbf{g}_\mathrm{R}$ and, hence, it has nothing to do with the Siegel anomaly. The proof of a similar statement in the lefton case is completely analogous.

Since the Siegel symmetry was shown to be non-anomalous in quantum theory, we can switch off the associated Lagrange multipliers , $\Lambda_+{}^=$ and $\Lambda_-{}^{\ne}$, and all notons. The classical heterotic NLSM action,

$$S_\mathrm{HS} = S_\mathrm{string} \ + \ S_\mathrm{YM} \ + \ k_\mathrm{L} S_\mathrm{L}\big|_{\Lambda_+{}^= = 0} \ + \ k_\mathrm{R} S_\mathrm{R}\big|_{\Lambda_-{}^{\ne} = 0} \ , \tag{2.107}$$

may be rewritten in the form of the usual heterotic NLSM defined on the *extended* space $M = M_4 \otimes (\mathbf{G}_\mathrm{R}/\mathbf{H}_\mathrm{R}) \otimes (\mathbf{G}_\mathrm{L}/\mathbf{H}_\mathrm{L})$, where M_4 is Minkowskian spacetime. This would allow one to use the quantum results of Sect. 7.1 about the structure of the low-energy heterotic string effective action referred to the extended space. After dimensional reduction of M down to M_4, the four-dimensional ($D = 4$) heterotic string effective action appears. It can be considered as a useful prerequisite for phenomenological studies of four-dimensional heterotic strings.

8. LEEA in 4d, $N = 2$ Gauge Field Theories

Non-perturbative solutions to the Low-Energy Effective Action (LEEA) in four-dimensional (4d), $N = 2$ supersymmetric gauge field theories can be obtained either by QFT methods (e.g., using a strong-weak coupling duality), or by the Type-IIA superstring/M-theory methods of brane technology (Sect. 8.4). In this chapter we review the field-theoretical results about the $N = 2$ gauge (Seiberg-Witten) LEEA (Sect. 8.2) and the hypermultiplet LEEA (Sect. 8.3). We also introduce the alternative methods of brane technology (Sect. 8.4). The latter are based on the exact solutions to the 11d and 10d type-IIA supergravities, which describe classical configurations of intersecting BPS branes with *eight* conserved supercharges (the same number as in 4d, $N = 2$ supersymmetry). The crucial role of M-theory in providing the classical resolution of singularities in the 10-dimensional (Type-IIA superstring) brane picture is made manifest. The existing methods of derivation of the exact LEEA in 4d, $N = 2$ gauge field theories are shown to be complementary to each other.

The hypermultiplet LEEA in 4d, $N = 2$ gauge field theories can also receive both perturbative and non-perturbative (due to instantons) quantum corrections. The manifestly $N = 2$ supersymmetric Feynman rules in Harmonic Superspace (HSS) can be used to calculate the 4d hypermultiplet LEEA in the Coulomb branch, which results in the $N = 2$ supersymmetric NLSM with the Taub-NUT metric. The HSS approach also yields the simple form of non-perturbative contributions to the hypermultiplet LEEA in the Higgs branch, whose geometrical meaning can be best understood after projection from $N = 2$ HSS to $N = 2$ Projective Superspace (PSS), in terms of an $O(4)$ projective $N = 2$ multiplet. The most general $SU(2)_{\mathrm{R}}$-invariant *Ansatz* for the LEEA of a single (charged) hypermultiplet takes the form of the $N = 2$ NLSM whose metric is given by a two-parametric deformation of the Atiyah-Hitchin (AH) metric, while the corresponding action in $N = 2$ PSS can be encoded in terms of an auxiliary elliptic curve. This is very similar to the Seiberg-Witten construction [16, 17] of the exact 4d, $N = 2$ vector multiplet LEEA in the Coulomb branch, while it also agrees with their proposals [463] about the 3d hypermultiplet quantum moduli space, related to one of the 3d, $N = 4$ gauge field theories via the c-map. Some general remarks about supersymmetry breaking and confinement are given in Sect. 8.5.

8.1 Motivation and Setup

The standard textbook description of gauge QFT is often limited to *perturbative* considerations, whereas many physical phenomena (e.g., confinement) are essentially non-perturbative. It is usually straightforward (although, it may be quite non-trivial!) to develop a quantum perturbation theory with all the fundamental (non-anomalous) symmetries to be manifestly (i.e. linearly) realized. Unfortunately, a perturbative expansion usually does not make sense when the field coupling becomes strong. The path integral representing the quantum generating functional of QFT is supposed to be defined in practical terms. Generally speaking, it may be done in *many* ways beyond perturbation theory (cf. e.g., lattice regularization, instantons, duality). Because of this reasoning, it was common to believe that a non-perturbative gauge QFT is not well defined to allow one to do non-perturbative calculations from first principles and thus make certain predictions.

Since the recent discovery of *exact* non-perturbative QFT solutions to the LEEA in quantum 4d, $N = 2$ supersymmetric gauge field theories, pioneered by Seiberg and Witten [16, 17], and subsequent advances in non-perturbative M-theory 'formerly known as the theory of superstrings', initiated in another of Witten's papers [464], [1] the conventional wisdom may have to be revised. Though the non-trivial exact solutions were only found in a certain class of $N = 2$ supersymmetric gauge QFT having no immediate phenomenological applications, they are, nevertheless, of great theoretical value. The $N = 2$ extended supersymmetric gauge field theories in 4d cannot directly serve for phenomenological applications at (low) energies of order 100 GeV, because $N = 2$ supersymmetric matter can only be defined in *real* representations of the gauge group, and with conserved parity. The $N = 2$ gauge theories may, nevertheless, appear as effective theories at some intermediate energies provided that the ultimate (yet unknown at the microscopic level) unified theory of Nature (e.g. M-theory!) lives in higher spacetime dimensions and has even more supersymmetries. A solvable (in the LEEA sense) gauge QFT may be a good starting point for further symmetry breaking towards the phenomenologically applicable LEEA, with the underlying integrability properties being preserved and the non-perturbative structure under control.

The exact solutions to the LEEA of the 4d, $N = 2$ supersymmetric gauge field theories can be obtained either by conventional QFT methods, after taking into account S-duality in the Seiberg-Witten approach [16, 17, 467, 468, 469, 470, 471], or by the alternative Type-IIA superstring/M-theory methods of brane technology [472, 473, 474, 475, 476, 477, 478]. We restrict ourselves to the case of four (uncompactified) spacetime dimensions (4d) with $N = 2$ extended supersymmetry.

---

[1] See e.g., [465, 466] for an elementary introduction.

Four-dimensional (4d), $N = 2$ supersymmetric gauge field theories are not integrable, either classically or quantum mechanically. [2] The full quantum effective action Γ in those theories is highly non-local and intractable. Nevertheless, it can be decomposed into a sum of local terms, in powers of spacetime derivatives or momenta divided by some dynamically generated scale Λ (in components). The leading kinetic terms of this expansion are called the Low-Energy Effective Action (LEEA). Calculation of the exact LEEA is a great achievement because it provides information about the non-perturbative spectrum and exact static couplings in the full quantum theory at energies well below Λ. Since we are only interested in the 4d, $N = 2$ gauge field theories with spontaneously broken gauge symmetry via the Higgs mechanism, the effective low-energy field theory may include only Abelian massless vector particles. All the massive fields (like the charged W-bosons) are supposed to be integrated out. This very general concept of LEEA is sometimes called the *Wilsonian* LEEA since it is familiar from statistical mechanics. There is a difference between the quantum effective action Γ defined as the quantum generating functional of the 1-Particle-Irreducible (1PI) Green's functions, and the Wilsonian effective action defined above, as far as the gauge theories with *massless* particles are concerned. In this chapter we use the Wilsonian action since it is well defined.

$N = 2$ supersymmetry severely restricts the form of LEEA. The very presence of $N = 2$ supersymmetry in the full non-perturbatively defined gauge QFT follows from the fact that its Witten index [481] does not vanish, $\Delta_{\mathrm{W}} = \mathrm{tr}(-1)^F \neq 0$. There are only two basic supermultiplets (modulo classical duality transformations) in rigid 4d, $N = 2$ supersymmetry: an $N = 2$ vector multiplet and a hypermultiplet (Chap. 4). The $N = 2$ vector multiplet field components (in the WZ-gauge) are

$$\{\, a\,, \quad \lambda^i_\alpha\,, \quad V_\mu\,, \quad D^{(ij)}\,\}\,, \tag{1.1}$$

where a is a complex Higgs scalar, λ^i is a chiral spinor ('gaugino') $SU(2)_{\mathrm{R}}$ doublet, V_μ is a real vector gauge field, and D^{ij} is an auxiliary scalar $SU(2)_{\mathrm{R}}$ triplet. Similarly, the on-shell physical components of the FS hypermultiplet are

$$\mathrm{FS}: \quad \{\, q^i\,, \quad \psi_\alpha\,, \quad \bar\psi_{\dot\alpha}\,\}\,, \tag{1.2}$$

where q^i is a complex scalar $SU(2)_{\mathrm{R}}$ doublet, and ψ is a Dirac spinor. There exists another (dual) HST hypermultiplet, whose on-shell physical components are

$$\mathrm{HST}: \quad \{\, \omega\,, \quad \omega^{(ij)}\,, \quad \chi^i_\alpha\,\}\,, \tag{1.3}$$

where ω is a real scalar, $\omega^{(ij)}$ is a scalar $SU(2)_{\mathrm{R}}$ triplet, and χ^i is a chiral spinor $SU(2)_{\mathrm{R}}$ doublet. The hypermultiplet spinors are sometimes referred

---

[2] It is the (Euclidean) SDYM sector that is integrable in the classical sense [479, 480].

to as 'quarks', even though $N = 2$ supersymmetry implies (apparently absent in experiment) extra 'mirror' particle for each 'true' quark in $N = 2$ QCD.

The general *Ansatz* for the $N = 2$ supersymmetric LEEA, in terms of Abelian $N = 2$ vector multiplets, reads in $N = 2$ superspace as

$$\Gamma_V[W, \bar{W}] = \int_{\text{chiral}} \mathcal{F}(W) + \text{h.c.} + \int_{\text{full}} \mathcal{H}(W, \bar{W}) + \ldots , \qquad (1.4)$$

where we have used the fact that the Abelian $N = 2$ superfield strength W is the $N = 2$ chiral and gauge-invariant superfield. The leading term in (1.4) is given by the chiral $N = 2$ superspace integral of a holomorphic function $\mathcal{F}$ of the gauge superfield strength W that is valued in the Cartan subalgebra of the gauge group. The next-to-leading order term is given by the full $N = 2$ superspace integral of the real function $\mathcal{H}$ of W *and* $\bar{W}$. The dots in (1.4) stand for the higher-order terms containing the derivatives of W and $\bar{W}$.

Similarly, the leading term in the hypermultiplet LEEA reads

$$\Gamma_H[q^+, \overset{*}{\bar{q}}{}^+; \omega] = \int_{\text{analytic}} \mathcal{K}^{(+4)}(q^+, \overset{*}{\bar{q}}{}^+; \omega; u_i^{\pm}) + \ldots , \qquad (1.5)$$

where $\mathcal{K}^{(+4)}$ is an analytic function of the FS-type superfields q^+, their conjugates $\overset{*}{\bar{q}}{}^+$, the HST-type superfields ω and, perhaps, the harmonics $u_i^{\pm}$ too. The action (1.5) is supposed to be added to the standard kinetic hypermultiplet action whose analytic Lagrangian is quadratic in q^+ and ω, while it has $U(1)$-charge $(+4)$ (see Subsect. 4.4.2). The function $\mathcal{K}$ is known as a hyper-Kähler potential. An arbitrary choice of this function in (1.5) automatically leads to the $N = 2$ supersymmetric NLSM with a hyper-Kähler metric, due to manifest $N = 2$ supersymmetry of the HSS construction.

After being expanded in components, the first term in (1.4) leads to the special Kähler NLSM in the scalar (Higgs) sector $(a, \bar{a})$. The corresponding NLSM Kähler potential $K_{\mathcal{F}}(a, \bar{a})$ is $K_{\mathcal{F}} = \text{Im}[\bar{a}\mathcal{F}'(a)]$, where the holomorphic function $\mathcal{F}$ plays the role of the potential of the special Kähler (but not hyper-Kähler) geometry described by $K_{\mathcal{F}}(a, \bar{a})$.

As regards the hypermultiplet NLSM of (1.5), the relation between the hyper-Kähler potential $\mathcal{K}$ and the corresponding Kähler potential $K_{\mathcal{K}}$ of the same NLSM is much more involved (Sect. 4.4). The HSS approach offers the formal 'solution' (1.5) to the hyper-Kähler constraints, in terms of a single analytic potential $\mathcal{K}$. Of course, the real problem is now translated into a derivation of the precise relation between the hyper-Kähler potential $\mathcal{K}$ and the corresponding Kähler potential (or the Kähler metric) in components, which requires a solution to an infinite number of auxiliary fields. Nevertheless, the HSS notion of a hyper-Kähler potential appears to be very useful in dealing with the hypermultiplet LEEA (Sect. 8.3).

The LEEA gauge-invariant functions $\mathcal{F}(W)$ and $\mathcal{H}(W, \bar{W})$ generically receive both perturbative and non-perturbative contributions,

$$\mathcal{F} = \mathcal{F}_{\text{per.}} + \mathcal{F}_{\text{inst.}} \,, \qquad \mathcal{H} = \mathcal{H}_{\text{per.}} + \mathcal{H}_{\text{non-per.}} \,, \qquad (1.6)$$

while the non-perturbative corrections to the holomorphic function $\mathcal{F}$ are entirely due to instantons. This feature is in contrast to the situation in (bosonic) non-perturbative QCD whose LEEA is known to be dominated by mixed (instanton/anti-instanton) contributions. The latter may, however, contribute to the hypermultiplet LEEA (Subsect. 8.3.3).

It is remarkable that the perturbative contributions to the leading and (in some cases) even subleading terms of the $N = 2$ supersymmetric gauge LEEA entirely come from one loop only. Indeed, supersymmetry puts the trace of the energy-momentum tensor $T_\mu{}^\mu$ and the anomaly $\partial_\mu j_R^\mu$ of the Abelian R-symmetry into a single $N = 2$ supermultiplet. The trace $T_\mu{}^\mu$ is essentially determined by the perturbative renormalization group β-function, [3] $T_\mu{}^\mu \sim \beta(g)FF$, whereas the one-loop contribution to the R-anomaly, $\partial \cdot j_R \sim C_{1-\text{loop}} F^* F$, is known to saturate the exact solution to the Wess-Zumino consistency condition for the same anomaly (e.g. by the use of index theorems). Hence, $\beta_{\text{per.}}(g) = \beta_{1-\text{loop}}(g)$ by $N = 2$ supersymmetry. Finally, since $\beta_{\text{per.}}(g)$ is effectively determined by the second derivative of $\mathcal{F}_{\text{per.}}$, one concludes that $\mathcal{F}_{\text{per.}} = \mathcal{F}_{1-\text{loop}}$. This naive component argument can be extended to a proof [482] by using quantum perturbation theory in HSS, where the whole chiral contribution $\int_{\text{chiral}} \mathcal{F}_{\text{per.}}(W)$ arises as an anomaly. The non-vanishing central charges of $N = 2$ supersymmetry algebra in the underlying $N = 2$ gauge field theory turn out to be primarily responsible for the non-vanishing leading holomorphic contribution to the LEEA. Therefore, the perturbative part of the gauge LEEA takes the form $\mathcal{F}_{\text{per.}}(W) \sim W^2 \log(W^2/M^2)$, where M is the renormalization scale, with the coefficient being fixed by the one-loop β-function of the renormalization group (Sect. 8.2).

The usual strategy of calculating the exact gauge LEEA is based on exploiting the exact symmetries of the underlying $N = 2$ quantum gauge theory together with certain physical input. The 4d, $N = 2$ supersymmetric QCD with the gauge group $G_c = SU(N_c)$, and $N = 2$ matter described by some number (N_f) of hypermultiplets in the fundamental representation $(N_c + N_c^*)$ of the gauge group $SU(N_c)$, gives a basic example. Quantum consistency of the non-Abelian gauge theory requires asymptotic freedom that implies $N_f < 2N_c$ in the $N = 2$ QCD.

All possible $N = 2$ supersymmetric vacua can be classified as follows:

- *Coulomb branch:* $\langle q \rangle = \langle \omega \rangle = 0$, whereas $\langle a \rangle \neq 0$; the gauge group G_c is broken to its Abelian subgroup $U(1)^{\text{rank } G_c}$; non-vanishing 'quark' masses are allowed;
- *Higgs branch:* $\langle q \rangle \neq 0$ or $\langle \omega \rangle \neq 0$ for some hypermultiplets, whereas $\langle a \rangle = 0$; all 'quark' masses vanish; the gauge group G_c is completely broken;
- *mixed (Coulomb-Higgs) branch:* some $\langle q \rangle \neq 0$ *and* $\langle a \rangle \neq 0$; it requires $N_c > 2$.

---

[3] Here and in what follows g denotes the gauge coupling constant.

In addition, there may be less symmetric vacua where e.g., a non-vanishing Fayet-Iliopoulos (FI) term is present, i.e. $\langle D^{ij} \rangle = \xi^{ij} \neq 0$. Though the FI term is usually associated with spontaneous or soft breaking of supersymmetry, it does not automatically imply the supersymmetry breaking. We only consider the FI terms associated with fictitious (i.e. non-dynamical) $N = 2$ vector multiplets playing the role of Lagrange multipliers (Subsect. 8.3.5).

8.2 4d, $N = 2$ Gauge LEEA in the Coulomb Branch

Seiberg and Witten [16, 17] gave a full solution to the holomorphic function $\mathcal{F}(W)$ by using certain physical assumptions about the global structure of the quantum moduli space $\mathcal{M}_{\mathrm{qu}}$ of vacua *and* electric-magnetic duality, i.e. not from first principles. Their main assumption was the precise value of the Witten index, [4] $\Delta_{\mathrm{W}} = 2$, which implies only two physical singularities in $\mathcal{M}_{\mathrm{qu}}$. The electric-magnetic duality (also known as S-duality) was used in [16, 17] to connect the weak and strong coupling regions of $\mathcal{M}_{\mathrm{qu}}$.

The Seiberg-Witten solution in the simplest case of the $SU(2)$ gauge group (no fundamental $N = 2$ matter) reads [16]

$$a_D(u) = \frac{\sqrt{2}}{\pi} \int_1^u \frac{\mathrm{d}x \sqrt{x-u}}{\sqrt{x^2-1}} \ , \quad a(u) = \frac{\sqrt{2}}{\pi} \int_{-1}^1 \frac{\mathrm{d}x \sqrt{x-u}}{\sqrt{x^2-1}} \ , \quad (2.1)$$

where the renormalization-group independent (Seiberg-Witten) scale $\Lambda^2 = 1$, and

$$a_D \equiv \frac{\mathrm{d}\mathcal{F}(a)}{\mathrm{d}a} \ . \quad (2.2)$$

The holomorphic parameter u can be identified with the second Casimir eivenvalue, $u = \langle \mathrm{tr}\, a^2 \rangle$, that parameterizes $\mathcal{M}_{\mathrm{qu}}$. We find it convenient to use the same lower-case letter (a) to denote the leading component (scalar field) of the $N = 2$ vector multiplet and its (constant) expectation value simultaneously. The holomorphic function $\mathcal{F}$ is locally defined over the quantum moduli space of vacua, while the S-duality can be identified with the action of the modular group $SL(2, \mathbf{Z})$ [16, 17]. The monodromies of the multi-valued function $\mathcal{F}$ around the singularities are supplied by the perturbative β-functions, whereas the whole function $\mathcal{F}$ is a (unique) solution to the corresponding Riemann-Hilbert problem [467, 468, 469, 470, 471].

In order to make contact with our general discussion in Sect. 8.1, let us consider an expansion of the SW solution in the semiclassical region, $|W| \gg \Lambda$. One finds

$$\mathcal{F}(W) = \frac{\mathrm{i}}{2\pi} W^2 \log \frac{W^2}{\Lambda^2} + \frac{1}{4\pi \mathrm{i}} W^2 \sum_{m=1}^{\infty} c_m \left(\frac{\Lambda^2}{W^2} \right)^{2m} \ , \quad (2.3)$$

---

[4] A formal derivation of Witten's index Δ_{W} from the path integral is plagued with ambiguities.

for the interacting terms, where the Λ-dependence has been restored. The first term in (2.3) represents the perturbative (one-loop) contribution, whereas the rest is the sum over non-perturbative instanton contributions (see Subsect. 8.2.1). It is straightforward to calculate the numerical coefficients $\{c_m\}$ from the explicit solution (2.1) and (2.2), with the results [483, 484]

$$\begin{array}{c|cccccc} m & 1 & 2 & 3 & 4 & 5 & \cdots \\ \hline c_m & 1/2^5 & 5/2^{14} & 3/2^{18} & 1469/2^{31} & 4471/5 \cdot 2^{34} & \cdots \end{array} \qquad (2.4)$$

8.2.1 On Instanton Calculations

The SW solution predicts that the non-perturbative holomorphic contributions to the 4d, $N = 2$ gauge LEEA in the Coulomb branch are entirely due to instantons. It is therefore desirable to reproduce them 'from first principles', e.g. by the use of a path integral. The $N = 2$ supersymmetric instantons are solutions to the classical self-duality equations,

$$F = {}^*F , \qquad i\gamma^\mu D_\mu \lambda = 0 , \qquad D^\mu D_\mu a = [\bar\lambda, \lambda] , \qquad (2.5)$$

whose Higgs scalar a approaches a non-vanishing constant at spatial infinity, so that the whole configuration has a non-vanishing topological charge $m \in \mathbf{Z}$.

From the path-integral point of view, the sum over instantons should be of the form

$$\mathcal{F}_{\text{inst.}} = \sum_{m=1}^{\infty} \mathcal{F}_m , \qquad \text{where} \quad \mathcal{F}_m = \int d\mu_{\text{inst.}}^{(m)} \exp\left[-S_{(m)-\text{inst.}}\right] . \qquad (2.6)$$

Each term $\mathcal{F}_m$ in this sum can be interpreted as the partition function in the multi(m)-instanton background. The non-trivial measure $d\mu_{\text{inst.}}^{(m)}$ in (2.6) appears as the result of changing variables from the original fields to the collective instanton coordinates in the path integral, whereas the action $S_{(m)-\text{inst.}}$ is just the Euclidean action of an $N = 2$ superinstanton configuration of charge m [485]. One usually assumes that the scalar surface term $(\sim \text{tr} \int dS^\mu \bar a D_\mu a)$ is the only relevant term in the action $S_{(m)-\text{inst.}}$. The bosonic and fermionic determinants, which always appear in the saddle-point expansion and describe small fluctuations of the fields, actually cancel in a supersymmetric self-dual gauge background [486]. Hence, supersymmetry is also in charge for the absence of infrared divergences present in the determinants.

The functional dependence $\mathcal{F}_m(a)$ easily follows from the integrated RG equation for the one-loop β-function,

$$\exp\left(-\frac{8\pi^2 m}{g^2}\right) = \left(\frac{\Lambda}{a}\right)^{4m} , \qquad (2.7)$$

and dimensional reasons as follows:

$$\mathcal{F}_m(a) = \frac{a^2}{4\pi i} \left(\frac{\Lambda}{a} \right)^{4m} c_m \; , \tag{2.8}$$

as should have been expected, up to a numerical coefficient c_m. It is, therefore, the exact values of the coefficients $\{c_m\}$ that is the real problem. Their evaluation is reduced to the problem of calculating the finite-dimensional multi-instanton measure $\{d\mu_{\text{inst.}}^{(m)}\}$.

A straightforward computation of the measure naively amounts to an explicit solution of the $N = 2$ supersymmetric self-duality equations in terms of the collective $N = 2$ instanton coordinates for any positive integer instanton charge. As is well known, the Yang-Mills self-duality differential equations of motion (as well as their supersymmetric counterparts) can be reduced to a purely algebraic (though highly non-trivial) set of equations when using the standard ADHMN construction [487, 488]. Unfortunately, an explicit solution to the algebraic ADHMN equations is only known for $m = 1$ [489] and $m = 2$ [490]. Nevertheless, the correct multi-instanton measure for any instanton number can be fixed indirectly [491], by imposing $N = 2$ supersymmetry and the cluster decomposition requirements, without using the electric-magnetic duality! In the Seiberg-Witten model with the $SU(2)$ gauge group, the instanton solution for $\{c_m\}$ is known in quadratures [491]. It was demonstrated in [492, 493, 494, 495] that the leading instanton corrections ($m = 1, 2$) agree with the exact Seiberg-Witten solution (2.4).

8.2.2 Seiberg-Witten Curve

From the mathematical point of view, the Seiberg-Witten exact solution (2.1) in the case of two colours ($N_c = 2$) or the $SU(2)$ gauge group is a solution to the standard Riemann-Hilbert problem of fixing a holomorphic multi-valued function $\mathcal{F}$ by its given monodromy and singularities. The number (and nature) of the singularities is the physical input: they are identified with the appearance of massless non-perturbative BPS-like physical states (dyons) like the famous t'Hooft-Polyakov magnetic monopole. The monodromies are supplied by perturbative RG β-functions and S-duality.

The $SU(2)$ solution (2.1) can be encoded in terms of the auxiliary (Seiberg-Witten) elliptic curve (or torus) Σ_{SW} defined by the algebraic equation [16]

$$\Sigma_{\text{SW}} : \qquad y^2 = (v^2 - u)^2 - \Lambda^4 \; . \tag{2.9}$$

The multi-valued functions $a_D(u)$ and $a(u)$ then appear by integration of a certain Abelian (Seiberg-Witten) differential λ_{SW} (of the third kind) over the torus periods A and B of Σ_{SW},

$$a(u) = \oint_A \lambda_{\text{SW}} \; , \qquad a_D(u) = \oint_B \lambda_{\text{SW}} \; . \tag{2.10}$$

The SW differential λ_{SW} is simply related to the unique holomorphic 1-form ω on the torus Σ_{SW},

$$\frac{\partial \lambda_{\text{SW}}}{\partial u} = \omega , \qquad \omega \equiv \frac{dv}{y(v,u)} . \tag{2.11}$$

In the case (2.9) one easily finds that $\lambda_{\text{SW}} = v^2 dv/y(v,u)$ up to a total derivative.

The relation to the theory of Riemann surfaces [496] can be further generalized to more general simply laced gauge groups and $N = 2$ super-QCD [484, 497, 498, 499, 500, 501]. For instance, a solution to the LEEA of the purely $N = 2$ gauge theory with the gauge group $SU(N_c)$ can be encoded in terms of a *hyperelliptic* curve of genus $(N_c - 1)$, whose algebraic equation reads

$$\Sigma_{\text{SW}} : \qquad y^2 = W^2_{A_{N_c-1}}(v,u) - \Lambda^{2N_c} . \tag{2.12}$$

The polynomial $W_{A_{N_c-1}}(v,\boldsymbol{u})$ is known in the mathematical literature [502] as the *simple singularity* associated with $A_{N_c-1} \sim SU(N_c)$. In two-dimensional $N = 2$ superconformal field theory, the same polynomial is known as the Landau-Ginzburg potential [45]. Its explicit form is given by

$$W_{A_{N_c-1}}(v,\boldsymbol{u}) = \sum_{l=1}^{N_c}(v - \boldsymbol{\lambda}_l \cdot \boldsymbol{a}) = v^{N_c} - \sum_{l=0}^{N_c-2} u_{l+2}(\boldsymbol{a})v^{N_c-2-l} , \tag{2.13}$$

where $\boldsymbol{\lambda}_l$ are the weights of $SU(N_c)$ in the fundamental representation, and $\boldsymbol{u}$ are the Casimir eigenvalues, i.e. the Weyl group-invariant polynomials in $\boldsymbol{a}$ (they are usually constructed by a Miura transformation [471]). The simple singularity seems to be the only remnant of the fundamental non-Abelian gauge symmetry in the Coulomb branch.

Adding fundamental $N = 2$ matter does not represent a problem in calculating the corresponding Seiberg-Witten curve. The result reads [500]

$$\Sigma_{\text{SW}} : \qquad y^2 = W^2_{A_{N_c-1}}(v,\boldsymbol{u}) - \Lambda^{2N_c-N_f}\prod_{j=1}^{N_f}(v - m_j) , \tag{2.14}$$

where $\{m_j\}$ are bare hypermultiplet masses of N_f hypermultiplets ($N_f < N_c$), in the fundamental representation of the gauge group $SU(N_c)$.

In the canonical first homology basis (A_α, B^β), $\alpha, \beta = 1, \ldots, g$, of the genus-$g$ Riemann surface Σ_{SW}, the multi-valued sections $a(u)$ and $a_{\text{D}}(u)$ are determined by the equations

$$\frac{\partial a_\alpha}{\partial u_\beta} = \oint_{A_\alpha} \omega^\beta , \qquad \frac{\partial a_{\text{D}}^\alpha}{\partial u_\beta} = \oint_{B^\alpha} \omega^\beta , \tag{2.15}$$

in terms of g independent holomorphic 1-forms ω^α on Σ_{SW}. Equation (2.15) is quite similar to (2.10) after taking into account that the Seiberg-Witten differential λ_{SW} is defined similarly to (2.11), namely,

$$\frac{\partial \lambda_{\text{SW}}}{\partial u_\alpha} = \omega^\alpha . \tag{2.16}$$

The minimal data $(\Sigma_{\mathrm{SW}}, \lambda_{\mathrm{SW}})$ needed to reproduce the Seiberg-Witten exact solution to the four-dimensional gauge LEEA can be associated with a *two-dimensional* integrable system [503, 504]. In particular, the SW potential $\mathcal{F}$ appears to be a solution [505] to the Dijkgraaf-Verlinde-Verlinde-Witten type (DVVW) [506, 507] non-linear differential equations known in 2d (conformal) topological field theory,

$$\mathcal{F}_i \mathcal{F}_k^{-1} \mathcal{F}_j = \mathcal{F}_j \mathcal{F}_k^{-1} \mathcal{F}_i \ , \quad \text{where} \quad (\mathcal{F}_i)_{jk} \equiv \frac{\partial^3 \mathcal{F}}{\partial a_i \partial a_j \partial a_k} \ . \tag{2.17}$$

There also exists yet another non-trivial equation on $\mathcal{F}$, which follows from the anomalous (chiral) $N = 2$ superconformal Ward identities in 4d [508, 509].

Though the mathematical relevance of the Seiberg-Witten curve is already clear, its geometrical origin and physical interpretation are still obscure at this point. These issues can be naturally understood by using the brane technology in M-theory (Sect. 8.4).

8.2.3 Next-to-Leading Order Correction to the SW Action

The next-to-leading order correction to the 4d, $N = 2$ gauge (Seiberg-Witten) LEEA in the Coulomb branch is determined by a real function $\mathcal{H}$ of W *and* $\overline{W}$, which is supposed to be integrated over the whole $N = 2$ superspace. Since the full $N = 2$ superspace measure $\mathrm{d}^4 x \mathrm{d}^8 \theta$ is dimensionless, the function $\mathcal{H}(W, \overline{W})$ should also be dimensionless, i.e. without any $N = 2$ superspace derivatives [510]. Moreover, the exact function

$$\mathcal{H}(W, \overline{W}) = \mathcal{H}_{\mathrm{per.}}(W, \overline{W}) + \mathcal{H}_{\mathrm{non-per.}}(W, \overline{W}) \tag{2.18}$$

has to be S-duality invariant [511]. It is also worth mentioning that the non-holomorphic function $\mathcal{H}$ is only defined modulo Kähler gauge transformations

$$\mathcal{H}(W, \overline{W}) \to \mathcal{H}(W, \overline{W}) + f(W) + \bar{f}(\overline{W}) \ , \tag{2.19}$$

with an arbitrary holomorphic function $f(W)$ as the gauge parameter.

In the manifestly $N = 2$ supersymmetric background-field approach in $N = 2$ HSS [512], the one-loop contribution to $\mathcal{H}_{\mathrm{per.}}$ is given by a sum of the $N = 2$ HSS graphs schematically pictured in Fig. 8.1.

The sum in Fig. 8.1 goes over the external V^{++}-legs, whereas the loop consists of the $N = 2$ matter (and $N = 2$ superghost) superpropagators (Sect. 8.3). The $N = 2$ ghost contributions are similar to $N = 2$ matter contributions since the ghost supermultiplets are also given by the FS- and HST-type hypermultiplets (with the opposite statistics of components) [283, 512, 513]. Because of the (Abelian) gauge invariance, the result can only depend upon the Abelian $N = 2$ superfield strength W and its conjugate $\overline{W}$, as in (2.18). In fact, Fig. 8.1 simultaneously determines the one-loop perturbative contribution to the leading holomorphic gauge LEEA, which arises

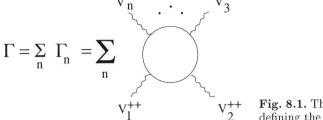

$$\Gamma = \sum_n \Gamma_n = \sum_n$$

Fig. 8.1. The HSS one-loop series defining the $N = 2$ gauge LEEA

as the *anomaly* associated with the non-vanishing central charges [283]. The self-energy HSS supergraph with only two external legs in Fig. 8.1 is the only one that is UV-divergent. The IR divergences of all the HSS graphs in Fig. 8.1 can be regularized by introducing the IR-cutoff Λ which is proportional to the Seiberg-Witten scale Λ_{SW} defined in Subsect. 8.2.2, with the relative co-efficient being dependent upon the actual renormalization scheme [513]. In the case of a single q^+-type (FS) matter hypermultiplet in an Abelian $N = 2$ gauge theory, calculation of the infinite HSS series depicted in Fig. 8.1 yields the holomorphic contribution [283, 512]

$$\mathcal{F}_q(W) = -\frac{1}{32\pi^2} W^2 \ln \frac{W^2}{M^2} \ , \qquad (2.20)$$

with the renormalization scale M being fixed by the condition $\mathcal{F}_q(M) = 0$. The non-holomorphic contribution is given by [478, 513]

$$\mathcal{H}_q(W, \overline{W}) = \frac{1}{(16\pi)^2} \sum_{k=1}^{\infty} \frac{(-1)^{k+1}}{k^2} \left(\frac{W\overline{W}}{\Lambda^2} \right)^k = \frac{1}{(16\pi)^2} \int_0^{W\overline{W}/\Lambda^2} \frac{d\xi}{\xi} \ln(1+\xi) \ , \qquad (2.21)$$

where we have used the standard integral representation for the dilogarithm function. It is not difficult to verify that the asymptotic perturbation series (2.21) can be rewritten as

$$\mathcal{H}_q(W, \overline{W}) = \frac{1}{(16\pi)^2} \ln \left(\frac{W}{\Lambda} \right) \ln \left(\frac{\overline{W}}{\Lambda} \right) \ , \qquad (2.22)$$

or, equivalently,

$$\mathcal{H}_q(W, \overline{W}) = \frac{1}{2(16\pi)^2} \ln^2 \left(\frac{W\overline{W}}{\Lambda^2} \right) \ , \qquad (2.23)$$

modulo irrelevant Kähler gauge terms, see (2.19). After being integrated over the whole $N = 2$ superspace, the non-holomorphic contribution (2.22) or (2.23) does not depend upon the scale Λ at all, because of the Kähler gauge invariance (2.19) [514].

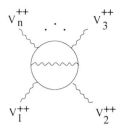

Fig. 8.2. A two-loop HSS graph

The HSS result (2.20) for the perturbative part of the SW gauge $N = 2$ LEEA agrees with the standard (Seiberg) argument [515] based on the perturbative $U(1)_R$ symmetry and the integration of the associated chiral anomaly. As is obvious from (2.22), the next-to-leading-order non-holomorphic contribution to the SW gauge LEEA satisfies the simple differential equation [513]

$$W\overline{W}\partial_W\partial_{\overline{W}}\mathcal{H}_q(W,\overline{W}) = \text{const.} \tag{2.24}$$

which can be considered as a direct consequence of the scale and $U(1)_R$ invariances [514].

In the more general case of N_f FS-type hypermultiplets in the fundamental representation of the gauge group $SU(N_c)$, i.e. in the $N = 2$ super-QCD, the extra coefficient in front of the holomorphic contribution $\mathcal{F}$ is proportional to the one-loop RG β-function $(N_f - 2N_c)$, whereas the extra coefficient in front of the non-holomorphic contribution $\mathcal{H}$ is proportional to $(2N_f - N_c)$, in the $N = 2$ super-Feynman gauge [513]. In another interesting case of the N=4 super-Yang-Mills theory, whose $N = 2$ matter content is given by a single HST-type hypermultiplet in the *adjoint* representation of the gauge group, the numerical coefficient in front of the holomorphic function $\mathcal{F}$ vanishes together with the RG β-function, whereas the numerical coefficient in front of the non-holomorphic contribution $\mathcal{H}$ always appears to be positive [516, 517]. In the case of finite and $N = 2$ *superconformally* invariant gauge field theories $(N_f = 2N_c)$, the leading non-holomorphic contribution to the LEEA is given by (2.22) multiplied by $3N_c$, so that it never vanishes.

It is also straightforward to verify in the HSS approach that there are no two-loop contributions to $\mathcal{F}_{\text{per.}}(W)$ *and* $\mathcal{H}_{\text{per.}}(W,\overline{W})$, since the two-loop HSS graph shown in Fig. 8.2 does not contribute in the local limit. This conclusion agrees with other calculations in terms of $N = 1$ superfields [518], while it is also consistent with the general perturbative structure of $N = 2$ supersymmetric gauge field theories in the manifestly $N = 2$ supersymmetric background-field method (in $N = 2$ HSS) [512]. It is thus conceivable that *all* higher-loop (perturbative) contributions to $\mathcal{H}_{\text{per.}}(W,\overline{W})$ may be absent too, at least in the case of $N = 2$ superconformally invariant gauge field theories. As regards non-perturbative (instanton) contributions to $\mathcal{H}(W,\overline{W})$, they are also expected to vanish in the 4d gauge field theories with $N = 2$ extended superconformal symmetry, since the dynamically generated instanton scale is apparently incompatible with scale invariance. The absence of instanton corrections in $N = 2$ supersymmetric, UV-finite, gauge field theories was also argued in [519, 520] by using instanton calculus. Stated differently, the one-loop non-holomorphic contribution to the function $\mathcal{H}(W, \bar{W})$ in the 4d, $N = 2$ superconformally invariant gauge field theories seems to be exact. Of course, this does not exclude non-holomorphic quantum corrections having the form of full $N = 2$ superspace integrals over functions depending upon the $N = 2$ superspace covariant derivatives of W and $\bar{W}$.

The exact function $\mathcal{H}(W, \bar{W})$ in a quantized (non-conformal) 4d, $N = 2$ supersymmetric gauge field theory is unknown. It was suggested in [521] that $\mathcal{H}(W, \bar{W})$ may satisfy a non-linear differential equation,

$$\partial_{\overline{W}} \partial_W \ln \left[\mathcal{H} \partial_W \partial_{\overline{W}} \ln \mathcal{H} \right] = 0 \,, \qquad (2.25)$$

which may be interpreted as the fully non-perturbative (non-chiral) $N = 2$ superconformal Ward identity. The leading non-holomorphic one-instanton contribution in the pure $N = 2$ gauge (SW) theory was already calculated in [522, 523], while it does not vanish. The full non-perturbative contribution is unlikely to be given by a sum over instanton contributions alone, since it can also include (multi)anti-instanton and mixed (instanton-anti-instanton) contributions that are allowed in a non-conformal QFT.

Brane technology (Sect. 8.4) may offer an alternative way of calculating the exact next-to-leading-order non-holomorphic correction, e.g. by using the covariant M-5-brane action describing the classical M-theory 5-brane dynamics, simply by expanding it further in powers of the spacetime derivatives up to fourth-order terms in components. This opportunity was investigated in [524], where it was demonstrated that the actual results obtained from brane technology qualitatively differ from those of the $N = 2$ quantum gauge field theory, as regards the non-holomorphic terms in the LEEA. There is only one dynamically generated dimensional scale Λ in the four-dimensional LEEA of the QFT under consideration, whereas there are at least two dimensional parameters in M-theory, namely, the radius R of the compactified 11th dimension and the typical scale L of the brane configuration (Sect. 8.4). The non-holomorphic contributions, derived from the M5-brane effective action, are dependent upon the radius R in a highly non-trivial way, while this dependence does not decouple in any simple limit [524].

8.3 Hypermultiplet LEEA in 4d

The previous section was essentially devoted to the holomorphic function $\mathcal{F}$ appearing in the gauge (Seiberg-Witten) LEEA (1.4), in the Coulomb branch. In this section we consider another *analytic* function $\mathcal{K}$ dictating the exact hypermultiplet LEEA (1.5) and called a hyper-Kähler potential. Its role in the hypermultiplet LEEA is similar to the role of $\mathcal{F}$ in the Seiberg-Witten theory.

8.3.1 Hypermultiplet LEEA in the Coulomb Branch

Since the HSS formulation of $N = 2$ vector multiplets *and* hypermultiplets has manifest off-shell $N = 2$ supersymmetry, it is perfectly suitable for a discussion of the induced hypermultipet self-interactions that are highly restricted by $N = 2$ supersymmetry and its $SU(2)_R$ automorphisms. The manifestly $N = 2$ supersymmetric Feynman rules in HSS are also suitable for calculating the perturbative hypermultiplet LEEA in 4d.

To illustrate the power of HSS, let us consider a single FS hypermultiplet for simplicity. We find it convenient to use pseudo-real notation for the hypermultiplet, combining q^+ and $\overset{*}{\tilde{q}}{}^+$ into a doublet of the Pauli-Gürsey $Sp(1)$ group, $q_a^+ = (\overset{*}{\tilde{q}}{}^+, q^+)$, with the central charge being realized as $Z_c^b = a(\tau_3)_c^b$, where a is a complex modulus and $\tau_3 = \text{diag}(1, -1)$ is the third Pauli matrix. An Abelian $N = 2$ vector superfield $\tilde{V}^{++}$ also receives a τ_3 factor, $(\tilde{V}^{++})_a^b = \tilde{V}^{++}(\tau_3)_a^b$, so that the full harmonic covariant derivative of the hypermultiplet superfield reads

$$\nabla^{++} q_a^+ = D^{++} q_a^+ + (v^{++} + \tilde{V}^{++})_a^b q_b^+ \equiv [\mathcal{D}^{++} + \tilde{V}^{++}]_a^b q_b^+ , \qquad (3.1)$$

where the central charge operator is represented by the background vector superfield v^{++} (see Subsect. 4.4.2). We also introduce the following bookkeeping notation for various covariantly constant superfields that are going to be relevant for our purposes,

$$v_b^c = \nu(\tau_3)_b^c, \quad (v^{\pm\pm})_b^c = \nu^{\pm\pm}(\tau_3)_b^c , \qquad (3.2)$$

$$\nu = i[\bar{a}(\theta^+\theta^-) + a(\bar{\theta}^+\bar{\theta}^-)], \quad \nu^{\pm\pm} = D^{\pm\pm}\nu , \qquad (3.3)$$

$$\mathcal{D}^{\pm\pm} = D^{\pm\pm} + \nu^{\pm\pm}\tau_3 = e^{-\nu\tau_3} D^{\pm\pm} e^{\nu\tau_3} . \qquad (3.4)$$

The $U(1)$ gauge harmonic superfield $\tilde{V}^{++}$ is considered to be massless, whereas the (bare) hypermultiplet mass squared is equal to $m^2 = a\bar{a} = |a|^2$ which is just the BPS mass.

The free hypermultiplet HSS action (4.4.61) in pseudo-real notation,

$$S[q] = \text{tr} \int_{\text{analytic}} q^+ \mathcal{D}^{++} q^+ , \qquad (3.5)$$

has the manifest internal symmetry

$$SU(2)_R \otimes Sp(1)_{\text{PG}} , \qquad (3.6)$$

where the $SU(2)_R$ is the automorphism symmetry of the $N = 2$ supersymmetry algebra. [5] Adding the minimal coupling to an Abelian $N = 2$ vector superfield $\tilde{V}^{++}$ breaks the internal symmetry (3.6) down to its subgroup

---

[5] It is easy to keep track of the $SU(2)_R$ symmetry in HSS, where it simply amounts to the absence of an explicit dependence of the HSS Lagrangian upon harmonics.

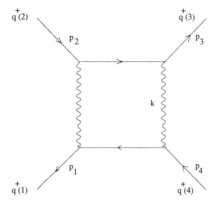

Fig. 8.3. The one-loop HSS graph contributing to the perturbative hypermultiplet LEEA, in the presence of central charges

$$SU(2)_{\mathrm{R}} \otimes U(1) \,. \tag{3.7}$$

The *unique* FS hypermultiplet self-interaction, consistent with the symmetry (3.7), is given by the hyper-Kähler potential

$$\mathcal{K}^{(+4)} = \frac{\lambda}{2} \left(\overset{*}{\widetilde{q}}{}^{+} q^{+} \right)^{2} , \tag{3.8}$$

since this is the only admissible term of $U(1)$-charge $(+4)$ that can be added to the free Lagrangian of (3.5). We thus arrive at the exact LEEA of the single charged matter hypermultiplet in the Coulomb branch, up to the induced NLSM coupling constant λ. The latter may be fixed by perturbative calculations (see below).

Similarly, the unique FS hypermultiplet self-interaction in the $N = 2$ super-QCD with N_{c} colours and N_{f} flavours, which is consistent with the $SU(N_{\mathrm{f}}) \otimes SU(2)_{\mathrm{R}} \otimes U(1)^{N_{\mathrm{c}}-1}$ symmetry, reads [525]

$$\mathcal{K}^{(+4)}_{\mathrm{QCD}} = \frac{\lambda}{2} \sum_{i,j=1}^{N_{\mathrm{f}}} \left(\overset{*}{\widetilde{q}}{}^{i+} \cdot q_{j}^{+} \right) \left(\overset{*}{\widetilde{q}}{}^{j+} \cdot q_{i}^{+} \right) , \tag{3.9}$$

where the dots stand for contractions of colour indices.

The induced coupling constant λ of (3.8) is determined (in the one-loop approximation) by the HSS graph shown in Fig. 8.3. Since the result vanishes ($\lambda = 0$) in the absence of central charges [526], let us assume that $\langle a \rangle \neq 0$, i.e. we are in the Coulomb branch. The free HSS actions of an $N = 2$ vector multiplet and of a hypermultiplet (Subsect. 4.4.2) can be used to compute the corresponding $N = 2$ superpropagators in HSS. For example, the $N = 2$ (Abelian) vector multiplet action takes the particularly simple form in the $N = 2$ super-Feynman gauge,

$$S[\tilde{V}]_{\text{Feynman}} = \tfrac{1}{2} \int_{\text{analytic}} \tilde{V}^{++} \Box \tilde{V}^{++} . \tag{3.10}$$

The corresponding analytic HSS propagator (the wave lines in Fig. 8.3) is therefore given by

$$i \left\langle \tilde{V}^{++}(1) \tilde{V}^{++}(2) \right\rangle = \frac{1}{\Box_1} (D_1^+)^4 \delta^{12}(\mathcal{Z}_1 - \mathcal{Z}_2) \delta^{(-2,2)}(u_1, u_2) , \tag{3.11}$$

where the harmonic delta-function $\delta^{(-2,2)}(u_1, u_2)$ has been introduced [221], while $\mathcal{Z}$ stands for the ordinary $N = 2$ superspace coordinates. The FS hypermultiplet HSS propagator (the solid lines in Fig. 8.3) with non-vanishing central charges reads [527, 282]

$$i \left\langle q^+(1) q^+(2) \right\rangle = \frac{-1}{\Box_1 + a\bar{a}} \frac{(D_1^+)^4 (D_2^+)^4}{(u_1^+ u_2^+)^3} e^{\tau_3(\nu_2 - \nu_1)} \delta^{12}(\mathcal{Z}_1 - \mathcal{Z}_2) . \tag{3.12}$$

The rest of the $N = 2$ supersymmetric Feynman rules in HSS is very similar to those of ordinary ($N = 0$) QED.

According to the general strategy of handling HSS graphs [221], one should first restore the *full* Grassmann integration measures at the vertices. This can be done by taking the factors $(D_1^+)^4 (D_2^+)^4$ off the hypermultiplet propagators. Then one integrates over two sets of Grassmann and harmonic coordinates by using the corresponding delta-functions in the integrand. There still remain $(D^+)^4$ factors in two $\tilde{V}^{++}$ propagators. After integrating by parts with respect to one of these factors, the only non-vanishing contribution comes from the term in which both such factors hit one of the two remaining Grassmann δ functions. Then the integral over one more set of the remaining Grassmann coordinates can be done by the use of the identity [221]

$$\delta^8(\theta_1 - \theta_2)(D_1^+)^4 (D_2^+)^4 \delta^8(\theta_1 - \theta_2) = (u_1^{+i} u_{2i}^+)^4 \delta^8(\theta_1 - \theta_2) . \tag{3.13}$$

As a result, the single Grassmann integration over $d^8\theta$ remains. The supergraph integral then takes the form [282]

$$\Gamma_4 = - \frac{ig^4}{(2\pi)^{16}} \int d^4 p_1 d^4 p_2 d^4 p_3 d^4 p_4 \int d^8\theta \int du_1 du_2$$

$$\times \frac{1}{(u_1^+ u_2^+)^2} F(p_1, u_1 | p_2, u_2) F(p_3, u_1 | p_4, u_2) \tag{3.14}$$

$$\times \int \frac{d^4 k \, \delta(p_1 - p_2 + p_3 - p_4)}{k^2 (k + p_1 - p_4)^2 [(k - p_3)^2 + m^2][(k - p_4)^2 + m^2]} ,$$

where we have introduced the notation $F(p_1, u_1 | p_2, u_2) \equiv F(1|2)$ (the θ-dependence is implicit) with

$$F(1|2) \equiv q^+(1) \exp\{\tau_3[\nu(2) - \nu(1)]\} q^+(2) . \tag{3.15}$$

According to the definition of the Wilsonian LEEA (at energies $\ll \Lambda$), we are supposed to integrate over all massive fields as well as over momenta squared $\geq \Lambda^2$ for all massless fields present in the fundamental (microscopic) Lagrangian, with the dynamically generated scale Λ, $\Lambda < m_{\text{lightest}} \neq 0$. In our case, we simply drop the massive gauge fields since they do not contribute to the hypermultiplet LEEA. Being only interested in calculating the leading contribution to the Wilsonian hypermultiplet LEEA, we are also allowed to omit in Γ_4 any terms that may only contribute to the higher-order terms in the momentum expansion of the quantum LEEA. In order to extract the relevant low-energy contribution out of (3.14), it is convenient to employ the covariant harmonic derivatives defined in (3.4). This makes integrations by parts with respect to them simpler since the 'bridge' e^{-v} is covariantly constant, $\mathcal{D}^{\pm\pm}e^{-v} = 0$. This also implies that $\mathcal{D}^{\pm\pm}e^{-v} = e^{-v}\mathcal{D}^{\pm\pm}$. The covariant derivatives satisfy the algebra

$$[\mathcal{D}^{++}, \mathcal{D}^{--}] = D^0 , \tag{3.16}$$

where the flat derivative D^0 measures the harmonic $U(1)$-charge.

We now insert (3.16) into the harmonic integral in (3.14) by rewriting

$$F(1|2)F(1|2) = \frac{1}{2}[\mathcal{D}_1^{++}, \mathcal{D}_1^{--}]F(1|2)F(1|2) . \tag{3.17}$$

It is easy to see that only the first term $\sim \mathcal{D}^{++}\mathcal{D}^{--}$ of (3.17) contributes to the leading low-energy contribution in the local limit of (3.14). Integrating by parts, one can cancel the harmonic distribution $1/(u_1^+ u_2^+)^2$ by the use of another identity [221]

$$\mathcal{D}_1^{++}\frac{1}{(u_1^+ u_2^+)^2} = \mathcal{D}_1^{--}\delta^{(2,-2)}(u_1, u_2) . \tag{3.18}$$

The harmonic delta-function on the right-hand side of (3.18) removes one of the two remaining harmonic integrals, while it allows us to rewrite (3.14) in the form

$$\Gamma_4 \rightarrow -ig^4 \int \frac{\mathrm{d}^4 k}{(2\pi)^4} \frac{1}{k^4(k^2 + m^2)^2} S_4 , \tag{3.19}$$

where we have ignored the dependence upon all the external momenta p_i (the terms omitted do not contribute to the hypermultiplet LEEA), and we have introduced the notation

$$S_4 = \int \mathrm{d}^4 x \mathrm{d}^8\theta \mathrm{d}u \, (\mathcal{D}^{--}q^+ \cdot q^+)^2 . \tag{3.20}$$

The S_4 is gauge-invariant as it should, while it has the form of the full superspace integral, in formal agreement with the non-renormalization 'theorem' in superspace. However, the connection v^{--} in $\mathcal{D}^{--}$ at $Z \neq 0$ is quadratic in θ^- because of (3.3) and (3.4). This results in a $(\theta^-)^4$-dependent term in the integrand of S_4, and, therefore, it yields a non-vanishing *analytic* contribution to the induced hypermultiplet effective action. In other words, it is the non-vanishing central charge Z that is responsible for the appearance of the leading analytic term in the hypermultiplet LEEA, already at the one-loop level of quantum perturbation theory. The analytic part in S_4 is given by

$$(S_4) \to \int d^{12} Z du \, (\nu^{--} q^+ \tau_3 q^+)^2 = -8|a|^2 \int d\zeta^{(-4)} du (\overset{*}{\tilde{q}}{}^+ q^+)^2 \equiv (S_4)_{\text{analytic}}$$

(3.21)

The terms omitted, in particular those with $(\mathcal{D}^{--})^2 q^+$ or $\mathcal{D}^{++} q^+$, contribute to the higher-order terms in the momentum expansion of the effective action. Hence, the one-loop induced low-energy hypermultiplet self-interaction takes the form [525, 282]

$$\int d\zeta^{(-4)} du \, \mathcal{K}_{\text{ind.}}^{(+4)} = \tfrac{\lambda}{2} \int d\zeta^{(-4)} du (\overset{*}{\tilde{q}}{}^+ q^+)^2 .$$

(3.22)

The one-loop induced coupling constant λ is given by

$$\begin{aligned}
\lambda_{1-\text{loop}} &= (2g)^4 |a|^2 \int_{k^2 \geq \Lambda^2} \frac{d_E^4 k}{(2\pi)^4} \frac{1}{k^4 (k^2 + m^2)^2} \\
&= \frac{g^4}{\pi^2} \left[\frac{1}{m^2} \ln\left(1 + \frac{m^2}{\Lambda^2}\right) - \frac{1}{\Lambda^2 + m^2} \right] ,
\end{aligned}$$

(3.23)

where we have used the BPS mass relation. The induced effective coupling λ is going to be renormalized in higher orders of perturbation theory, unlike the form of the effective self-interaction because of the symmetry (3.7). The dependence of λ upon the IR cutoff is expected to disappear after summing up contributions from higher loops.

Since the Taub-NUT metric is the only regular and asymptotically locally flat (ALF) hyper-Kähler metric having the $SU(2) \times U(1)$ isometry, it is not very surprising that the hypermultiplet LEEA (3.8) has the Taub-NUT effective metric. The proof amounts to decoding the HSS result (3.22) in component form. The general procedure of getting the bosonic NLSM (in components) from a hypermultiplet self-interaction in HSS consists of the following steps:

- expand the equations of motion in Grassmann (anticommuting) coordinates, and ignore all the fermionic field components;
- solve the kinematic linear differential equations for all the auxiliary fields, thus eliminating the infinite tower of them in the harmonic expansion of the hypermultiplet HSS analytic superfields;
- substitute the solution back into the HSS hypermultiplet action, and integrate over all the anitcommuting and harmonic HSS coordinates.

Doing the second step amounts to solving infinitely many linear differential equations on the sphere. In the case of (3.8), an explicit solution is possible by using the parametrization

$$q^+\big|_{\theta=0} = f^i(x)u_i^+ \exp\left[\lambda f^{(j}(x)\bar{f}^{k)}(x)u_j^+ u_k^-\right] . \tag{3.24}$$

Then one arrives at the bosonic 4d NLSM action,

$$S_{\text{NLSM}} = \int d^4x \left\{ g_{ij}\partial_m f^i \partial^m f^j + \bar{g}^{ij}\partial_m \bar{f}_i \partial^m \bar{f}_j + h^i{}_j \partial_m f^j \partial^m \bar{f}_i - V(f) \right\} , \tag{3.25}$$

whose metric is given by [285]

$$g_{ij} = \frac{\lambda(2 + \lambda f\bar{f})}{4(1 + \lambda f\bar{f})}\bar{f}_i\bar{f}_j , \qquad \bar{g}^{ij} = \frac{\lambda(2 + \lambda f\bar{f})}{4(1 + \lambda f\bar{f})}f^i f^j ,$$

$$h^i{}_j = \delta^i{}_j(1 + \lambda f\bar{f}) - \frac{\lambda(2 + \lambda f\bar{f})}{2(1 + \lambda f\bar{f})}f^i \bar{f}_j , \qquad f\bar{f} \equiv f^i \bar{f}_i . \tag{3.26}$$

In addition, one finds the induced scalar potential [282, 284]

$$V(f) = |Z|^2 \frac{f\bar{f}}{1 + \lambda f\bar{f}} . \tag{3.27}$$

In the form (3.26) the induced metric is apparently free from singularities.

It is usually non-trivial to compare a given hyper-Kähler metric with any standard metric since the metrics themselves are defined modulo field redefinitions, i.e. modulo four-dimensional diffeomorphisms. Nevertheless, it is known how to transform the metric (3.26) into the standard (Euclidean) Taub-NUT form (Subsect. 4.4.1),

$$ds^2 = \frac{r + M}{2(r - M)}dr^2 + \tfrac{1}{2}(r^2 - M^2)(d\vartheta^2 + \sin^2 \vartheta d\varphi^2)$$
$$+ 2M^2 \left(\frac{r - M}{r + M}\right)(d\psi + \cos \vartheta d\varphi)^2 , \tag{3.28}$$

by using the following change of variables [285]:

$$f^1 = \sqrt{2M(r - M)}\cos\frac{\vartheta}{2}\exp\frac{i}{2}(\psi + \varphi) ,$$

$$f^2 = \sqrt{2M(r - M)}\sin\frac{\vartheta}{2}\exp\frac{i}{2}(\psi - \varphi) , \tag{3.29}$$

$$f\bar{f} = 2M(r - M) , \qquad r \geq M = \frac{1}{2\sqrt{\lambda}} ,$$

where $M = \tfrac{1}{2}\lambda^{-1/2} \sim g^{-2}$ is the Taub-NUT mass parameter. The hyper-Kähler metric (3.28) is also known as the (gravitational) *Kaluza-Klein monopole* [528, 529] (see Sect. 8.4).

As a simple application, let us consider the Seiberg-Witten model [16], whose initial (microscopic) action describes the purely gauge $N = 2$ SYM theory with the $SU(2)$ gauge group spontaneously broken to its $U(1)$ subgroup

(Fig. 8.4). In the non-perturbative region (the Coulomb branch) near the singularity of the quantum moduli space where a BPS-like (t'Hooft-Polyakov) monopole becomes massless, the Seiberg-Witten model is equivalent to the *dual* $N = 2$ supersymmetric QED, $V^{++} \to V_D^{++}$ and $a \to a_D$. The t'Hooft-Polyakov monopole belongs to a hypermultiplet q_{HP}^+ [16] that represents the non-perturbative degrees of freedom in the SW theory. Our results imply that the induced HP-hypermultiplet self-interaction in the vicinity of the monopole singularity is regular, and it is given by

$$\mathcal{L}_{\text{Taub-NUT}}^{(+4)}(q_{HP}) = \frac{\lambda_D}{2}(\bar{q}_{HP}^+ q_{HP}^+)^2 \, , \qquad (3.30)$$

whose one-loop induced coupling constant λ_D is given by (3.23) in terms of the dual (magnetic) coupling constant g_D and the dual modulus a_D.

The relation between the HSS results and brane technology (Sect. 8.4) is provided by S-duality. From the type-IIA superstring (or M-theory) point of view, the HP-hypermultiplet components are zero modes of open superstring stretching between a magnetically charged D-6-brane and a D-4-brane. The magnetically charged (HP) hypermultiplet is the only $N = 2$ matter that survives in the effective four-dimensional $N = 2$ gauge theory (given by dual $N = 2$ super-QED) near the HP singularity, after taking the proper LEEA limit of the brane configuration where supergravity decouples. Therefore, in full agreement with the results of Sect. 8.4, the NLSM target space geometry governing the induced HP hypermultiplet self-interaction is given by the Taub-NUT (or KK-monopole) metric!

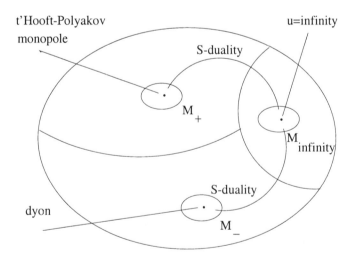

Fig. 8.4. The quantum moduli space in the Seiberg-Witten model

8.3.2 $N = 2$ Reduction from HSS to PSS

To understand the hyper-Kähler geometry associated with the induced hyper-Kähler potential (3.8), one can also perform an $N = 2$ supersymmetric reduction of the FS hypermultiplet to an $N = 2$ tensor multiplet, and then rewrite the corresponding dual HSS action into the projective superspace (PSS) (see Sect. 4.3). Unlike the FS hypermultiplet, the $N = 2$ tensor multiplet has a finite number of auxiliary fields. Let us recall that the $N = 2$ tensor superfield $L^{ij}(\mathcal{Z})$ is defined by the off-shell $N = 2$ superspace constraints (4.3.1) and (4.3.2). In HSS they read

$$D^{++}L^{++} = 0 \quad \text{and} \quad \overset{*}{L}{}^{++} = L^{++} \;, \qquad (3.31)$$

where $L^{++} = u_i^+ u_j^+ L^{ij}(\mathcal{Z})$. Let us substitute (we temporalily set $\lambda = 1$)

$$\mathcal{K}_{\text{TN}}^{(+4)} = \tfrac{1}{2}(\overset{*}{q}{}^+ + q^+)^2 = -2(L^{++})^2 \quad \text{or, equivalently,} \quad \overset{*}{q}{}^+ + q^+ = 2iL^{++} \qquad (3.32)$$

in (3.22). This is certainly allowed because of (4.4.100). The constraints (3.31) can be incorporated off-shell by using the extra real analytic superfield ω as the Lagrange multiplier. Changing the variables from $(\overset{*}{q}{}^+, q^+)$ to (L^{++}, ω) then amounts to an $N = 2$ duality transformation in HSS. An explicit solution to (3.32) is given by [239, 250]

$$q^+ = -i\left(2u_1^+ + if^{++}u_1^-\right)e^{-i\omega/2} \;, \quad \overset{*}{q}{}^+ = i\left(2u_2^+ - if^{++}u_2^-\right)e^{i\omega/2} \;, \quad (3.33)$$

where the function f^{++} has the form

$$f^{++}(L, u) = \frac{2(L^{++} - 2iu_1^+u_2^+)}{1 + \sqrt{1 - 4u_1^+u_2^+u_1^-u_2^- - 2iL^{++}u_1^-u_2^-}} \;. \qquad (3.34)$$

It is straightforward to rewrite the free (massless) HSS action (3.5) in terms of the new variables. This results in the so-called *improved* ($N = 2$ superconformally invariant) tensor multiplet action [250] (see Subsect. 4.3.3)

$$S_{\text{impr.}} = \tfrac{1}{2} \int d\zeta^{(-4)} du (f^{++})^2 \;. \qquad (3.35)$$

The action dual to the NLSM action defined by the sum of (3.5) and (3.22) is, therefore, given by the sum of the non-improved (quadratic) and improved (non-polynomial) HSS actions for the $N = 2$ tensor multiplet [114, 239],

$$S_{\text{TN}}[L; \omega] = S_{\text{impr.}} + \tfrac{1}{2} \int d\zeta^{(-4)} du \left[(L^{++})^2 + \omega D^{++}L^{++}\right] \;. \qquad (3.36)$$

To understand the peculiar structure of the improved $N = 2$ tensor multiplet action defined by (3.34) and (3.35), one may extract a constant piece c^{ij} out of L^{ij} by rewriting it in the form [250]

$$L^{++}(\zeta, u) = c^{++} + l^{++}(\zeta, u) , \qquad (3.37)$$

where

$$c^{\pm\pm} = c^{ij} u_i^\pm u_j^\pm , \quad \overline{(c^{ij})} = \varepsilon_{ik}\varepsilon_{jl}c^{kl} , \quad c^2 = \tfrac{1}{2}c_{ij}c^{ij} \neq 0 ,$$

$$f^{++}(L, u) \equiv l^{++} f(y) , \quad y = l^{++} c^{--} . \qquad (3.38)$$

The structural function f is a solution to the hypergeometric equation

$$z(1 - z)f_{zz} + \tfrac{1}{2}(6 - 7z)f_z - \tfrac{3}{2}f = 0 , \qquad z = -y , \qquad (3.39)$$

which comes from requiring rigid $N = 2$ superconformal invariance of the improved action [251]. The *action* (3.35) does not depend upon the 'Dirac-type string' c^{ij} because of the $SU(2)_{\mathrm{conf.}}$ internal symmetry that is part of the $N = 2$ superconformal symmetry [250].

It is straightforward to rewrite (3.36) into $N = 2$ PSS, where it takes the form of (4.3.7) with the particular holomorphic PSS potential G and the integration contour C given by (4.3.55). The $N = 2$ PSS action gives rise to the Taub-NUT NLSM metric after the $N = 1$ superspace Legendre transform [114]. This observation confirms once more that (3.22) describes the hyper-Kähler potential of the Taub-NUT metric .

It is instructive to investigate realizations of the $U(2)$ internal symmetry in different $N = 2$ superspace formulations of the Taub-NUT $N = 2$ NLSM. The hyper-Kähler potential (3.8), in terms of the Fayet-Sohnius hypermultiplet q^+, gives the manifestly $SU(2)_{\mathrm{R}}$ invariant formulation. In the dual HSS form, in terms of (L^{++}, ω), the non-Abelian symmetry $SU(2)$ is represented by the $SU(2)_{\mathrm{conf.}}$, whereas the Abelian symmetry $U(1)$ is realized by constant shifts of ω. The $SU(2)$ transformations act in PSS in the form of projective (or fractional) transformations (4.3.10) and (4.3.12).

Equation (4.3.11) implies that the $SU(2)$-invariant PSS potential $G(Q_2)$ should be 'almost' linear in Q_2, like the second term of (4.3.55). Imposing the off-shell constraint (3.31) together with $u_i \to \xi_i = (1, \xi)$ allow us to connect HSS to PSS. The analytic HSS superfield $L^{++}(\zeta, u)$ is replaced in PSS by the holomorphic (with respect to ξ) section $Q_2(L, \xi)$ of the line bundle $O(2)$ whose fibre is parametrized by constrained $N = 2$ superfields.

8.3.3 Exact Hypermultiplet LEEA in the Higgs Branch

In an Abelian QFT there are no instantons, so that the results of the pre-ceeding subsection are, in fact, exact in this case. If, however, the underlying $N = 2$ gauge field theory has a non-Abelian gauge group of rank higher than one, one may expect non-perturbative contributions to the hypermultiplet LEEA of a single (magnetically charged) hypermultiplet from instantons and anti-instantons. This happens, for example, in the Higgs branch where the gauge symmetry is completely broken [290].

The most general $SU(2)_R$-invariant hyper-Kähler potential in HSS is given by (4.4.99). Let us make the substitution [290],

$$\frac{\lambda}{2}(\overset{*}{q}{}^+)^2(q^+)^2 + \left[\gamma\,\overline{(q^+)}^{\,4} + \beta\,\overline{(q^+)}^{\,3}q^+ + \text{h.c.}\right] = L^{++++}(\zeta, u) , \quad (3.40)$$

where the left-hand side satisfies the conservation law (4.4.100). Hence, the real analytic superfield L^{++++} satisfies the HSS constraint

$$D^{++}L^{++++} = 0 . \qquad (3.41)$$

Equation (3.41) can be recognized as the *off-shell* $N = 2$ superspace constraints (4.3.1) and (4.3.2), defining the projective $O(4)$ supermultiplet, where $L^{++++} = u_i^+ u_j^+ u_k^+ u_l^+ L^{ijkl}(\mathcal{Z})$. Unlike the $O(2)$ tensor multiplet, the $O(4)$ multiplet does not have a conserved vector (or a gauge antisymmetric tensor) amongst its field components (Subsect. 4.3.1). This implies the absence of the related $U(1)$ triholomorphic isometry in the $N = 2$ NLSM to be constructed in terms of L^{++++}.

The most straightforward procedure of calculating the dependence $q(L)$, as well as performing an explicit $N = 2$ transformation of the unconstrained HSS action (3.5) into the PSS action in terms of the constrained $N = 2$ superfield L^{++++}, uses roots of the quartic polynomial. Though it is possible to calculate the roots, the results are not very illuminating. Remarkably, the equivalent (or dual) $N = 2$ PSS action can be fixed on the general grounds of $SU(2)_R$ symmetry and regularity of the NLSM metric. The one real and two complex constants, (λ, β, γ), respectively, parametrizing the hyper-Kähler potential in (4.4.99) and (3.40), can be naturally united into an $SU(2)$ 5-plet c^{ijkl} subject to the reality condition (4.3.2). After extracting a constant piece (VEV) out of q^+, say, $q_a^+ = u_a^+ + \tilde{q}_a^+$ and $u_a = (1, \xi)$, and collecting all constant pieces on the left-hand side of (3.40), we can identify their sum with the spacetime independent part $c^{++++} = c^{ijkl}u_i^+ u_j^+ u_k^+ u_l^+$ of L^{++++} on the right-hand side of (3.40). The c^{++++} thus represents the constant VEV of the superfield components of L^{++++}, defined by (4.3.42), i.e.

$$\lambda = \langle V \rangle , \quad \beta = \langle W \rangle , \quad \gamma = \langle \chi \rangle . \qquad (3.42)$$

The $SU(2)_R$ transformations in PSS are the projective transformations (4.3.10) and (4.3.12). Hence, the PSS potential G in the 'improved' $O(4)$ multiplet action having the general form (4.3.7) should be proportional to $\sqrt{Q_4}$, where $Q_4(\xi) = \xi_i\xi_j\xi_k\xi_l L^{ijkl}$ and $\xi_i = (1, \xi)$, because of (4.3.11) and (4.3.12) which imply (cf. [251])

$$G(Q_4{}'(\xi'), \xi') = \frac{1}{(a + b\xi)^2}G(Q_4(\xi), \xi) \quad \text{and} \quad Q_4{}'(\xi') = \frac{1}{(a + b\xi)^4}Q_4(\xi) . \qquad (3.43)$$

The most general non-trivial contour C_r in the complex ξ-plane, whose definition is compatible with the projective $SU(2)$ symmetry, encircles the roots of the quartic

$$Q_4(\xi)| = p + \xi q + \xi^2 r - \xi^3 \bar{q} + \xi^4 \bar{p} , \qquad (3.44)$$

with one real (r) and two complex (p, q) additional parameters belonging to yet another 5-plet of $SU(2)$. The projective $SU(2)$ invariance of the PSS action can be used to reduce the number of independent parameters in the corresponding family of hyper-Kähler metrics from five to two, which is consistent with the HSS predictions of Sect. 4.4. [6] Therefore, the PSS potential of the dual $N = 2$ action in question is given by [252, 290, 251]

$$\frac{1}{2\pi i} \oint G = -\frac{1}{2\pi i} \oint_{C_0} \frac{Q_4^2}{\xi} + \oint_{C_r} \sqrt{Q_4} , \qquad (3.45)$$

where the contour C_0 encircles the origin in the complex ξ-plane, whereas the contour C_r is defined by (3.44). The most natural non-trivial choice [252],

$$Q_4(\xi)| = 0 , \qquad (3.46)$$

which was already considered in Subsect. 4.3.3, gives rise to the *regular* Atiyah-Hitchin (AH) solution (see Subsect. 8.3.4). The generalizations of the AH metric, described by (3.44) and (3.45), lead to the hyper-Kähler metrics with singularities (Subsect. 8.3.4).

The AH space M_2 was originally introduced as the moduli space of two BPS $SU(2)$ monopoles [204]. The metric of the AH space is known to be the *only* regular hyper-Kähler metric with the entirely non-triholomorphic $SO(3)$ symmetry rotating the hyper-Kähler complex structures [204]. In the Donaldson description [530] of the AH space, M_2 is described by the quotient of an algebraic curve in $\mathbf{C}^3$,

$$x^2 - zy^2 = 1 , \quad \text{where} \quad x, y, z \in \mathbf{C} , \qquad (3.47)$$

under $Z_2 : (x, y, z) \equiv (-x, -y, z)$. Equation (3.47) describes the $SU(2)$-symmetric universal (2-fold) covering $\tilde{M}_2$ of the AH space.

The standard parametrization of the AH metric is given by (4.4.39) and (4.4.40), while it appears to be exponentially close to the Taub-NUT metric in the limit $k \to 1$ (Subsect. 4.4.1). The extra $U(1)$ symmetry of the Taub-NUT metric is a direct consequence of the relation $a^2 = b^2$ in (4.4.19) which arises from the AH metric only in the asymptotic limit described by (4.4.42) and (4.4.43). It is the vicinity of $k' \approx 0^+$ that describes the region of the hypermultiplet moduli space where quantum perturbation theory applies, with the exponentially small AH corrections (to the Taub-NUT metric) being interpreted as the non-perturbative (mixed) instanton and anti-instanton contributions to the hypermultiplet LEEA. The AH metric, as the metric in the hypermultiplet quantum moduli space, was also proposed by Seiberg and

---

[6] A similar generalization in the case of the improved $N = 2$ tensor multiplet action (4.3.55) does not lead to a more general action since the quadratic polynomial $c_2(\xi) = p + \xi r - \xi^2 \bar{p}$ can always be removed by an $SU(2)$ transformation.

Witten [463] in the context of 3d, $N = 4$ supersymmetric gauge field theories, where it can be related to the (Seiberg-Witten) gauge LEEA via the c-map [248, 247].

From the $N = 2$ PSS viewpoint, the transition from the perturbative hypermultiplet LEEA (in the Coulomb branch) to the non-perturbative LEEA (in the Higgs branch) corresponds to the transition from the $O(2)$ holomorphic line bundle, associated with the standard $N = 2$ tensor supermultiplet, to the $O(4)$ holomorphic line bundle, associated with the $O(4)$ $N = 2$ supermultiplet. The two holomorphic bundles are topologically different: with respect to the standard covering of CP^1 by two open affine sets, the $O(2)$ bundle has transition functions ξ^{-2}, whereas the $O(4)$ bundle has transition functions ξ^{-4}.

8.3.4 Atiyah-Hitchin Metric and Elliptic Curve

The quadratic dependence of the $N = 2$ tensor multiplet PSS superfield $Q_2 \equiv y^2$ upon ξ in (3.3.6) and (3.3.29) allows us to globally interpret $Q_2(\xi)$ as a holomorphic (of degree 2) section of PSS, fibred by the superfields (χ, H) and topologically equivalent to a complex line (or Riemann sphere of genus 0). Similarly, the quartic dependence of Q_4 upon ξ in (3.3.6) and (3.3.47) allows us to globally interpret it as a holomorphic (of degree 4) section of PSS, fibred by the superfields (χ, W, V) and topologically equivalent to an elliptic curve $\Sigma_{\text{hyper.}}$ (or a torus) of genus one. The non-perturbative hypermultiplet LEEA in the Higgs branch can, therefore, be encoded in terms of the genus-one Riemann surface $\Sigma_{\text{hyper.}}$ in close analogy to the exact $N = 2$ gauge LEEA in terms of the elliptic curve Σ_{SW} of Seiberg and Witten [16, 17].

The classical twistor construction of hyper-Kähler metrics [204] is known to be closely related to the Hurtubise elliptic curve Σ_{H} [531]. The spectral curve Σ_{H} naturally arises in the process of uniformization of the algebraic curve (3.47) in the Donaldson description of the AH space [204]. The curve Σ_{H} can be identified with $\Sigma_{\text{hyper.}}$ which carries the same information and whose defining equation (3.3.6) can be put into the normal (Hurtubise) form,

$$\tilde{Q}_4(\tilde{\xi}) = K^2(k)\tilde{\xi}\left[kk'(\tilde{\xi}^2 - 1) + (k^2 - k'^2)\tilde{\xi}\right] , \qquad (3.48)$$

by a projective $SU(2)$ transformation [532]. Equation (3.48) is simply related to another standard (Weierstrass) form, $y^2 = 4x^3 - g_2x - g_3$ [496]. Therefore, in accordance with [204, 496], the real period ω of Σ_{H} is

$$\omega \equiv 4k_1 , \quad \text{where} \quad 4k_1^2 = kk'K^2(k) , \qquad (3.49)$$

whereas the complex period matrix of Σ_{H} is given by

$$\tau = \frac{iK(k')}{K(k)} . \qquad (3.50)$$

At generic values of the AH modulus k, $0 < k < 1$, the roots of the Weierstrass form are all different from each other, while they all lie on the real axis, say, at $e_3 < e_2 < e_1 < \infty = (e_4)$. Accordingly, the branch cuts run from e_3 to e_2 and from e_1 to ∞. The C_r integration contour in the PSS formulation of the exact hypermultiplet LEEA in (3.45) can now be interpreted as the contour integral over the non-contractible α-cycle of the elliptic curve Σ_H [532], in a similar way to how the Seiberg-Witten solution to the $SU(2)$-based $N = 2$ gauge LEEA is written down in terms of the Abelian differential λ_{SW} integrated over the periods of Σ_{SW} [16, 17]. The most general (non-trivial) integration contour C_r in (3.44) is given by a linear combination of the non-contractible α and β cycles of S_H. Since the alternative integration over β leads to a singularity [204], the choice of the AH metric gives the only regular solution.

The perturbative (Taub-NUT) limit $k \to 1$ corresponds to the situation when $e_2 \to e_1$, so that the β-cycle of Σ_H degenerates. The curve (3.48) then asymptotically approaches a complex line, $\tilde{Q}_4 \sim \pm K\tilde{\xi}$. Another limit, $k \to 0$, leads to a coordinate bolt-type singularity of the AH metric in the standard parameterization (4.4.39) [204]. In the context of monopole physics, this corresponds to the coincidence limit of two centred monopoles. In the context of the hypermultiplet LEEA, $k \to 0$ implies $e_2 \to e_3$, so that the α-cycle of Σ_H degenerates, as well as the whole hypermultiplet action associated with (3.45). The two limits, $k \to 1$ and $k \to 0$, are related by the modular transformation exchanging k with k', and α-cycle with β-cycle [532]. The non-perturbative corrections to the hypermultiplet LEEA are therefore dictated by the hidden (in 4d) elliptic curve parametrizing the exact solution.

The 4d twistor approach considered above apparently has many similarities with the inverse scattering method in the theory of (lower-dimensional) *integrable systems*. In turn, the integrable systems are known to be closely connected with hyper-Kähler geometry and the theory of Riemann surfaces [503, 504]. In our approach to the construction of the 4d hypermultiplet LEEA, it was the passage from $N = 2$ HSS to $N = 2$ PSS that provided the link of harmonic analysis to complex analysis, and thus it allowed us to introduce the holomorphic quantities interpreted in terms of the elliptic curve associated with the AH metric. The uniformization of the algebraic curve (3.47) describing the AH space is closely related to the continuous 3d Toda equation [532], while the latter naturally arises in the large N limit of 2d conformal field theories [45].

It is also remarkable that the very simple (quartic) hyper-Kähler potential on the left-hand side of (3.40) provides an exhaustive description of the highly non-trivial (two-parametric) deformation of the standard Atiyah-Hitchin metric, with both $N = 2$ supersymmetry and $SU(2)$ internal symmetry being manifest. The corresponding $N = 2$ NLSM in HSS have natural holomorphic projections to $N = 2$ PSS, where they are related to the holomorphic $O(4)$ line bundle. A bare hypermultiplet mass, or non-vanishing

$N = 2$ central charges, gives rise to a non-trivial (yet unknown) scalar potential, whose form is entirely dictated by the Atiyah-Hitchin NLSM metric.

8.3.5 Hypermultiplet LEEA in the Presence of FI Term

The Higgs branch in $N = 2$ super-QCD can be divided into two phases called 'baryonic' and 'non-baryonic' [533]. In the baryonic phase, some Fayet-Iliopoulos (FI) terms are present. In the non-baryonic phase, all FI terms vanish. In this subsection we briefly discuss the baryonic phase. It intersects with the Coulomb branch at a single point in the quantum moduli space [533].

The most natural and manifestly $N = 2$ supersymmetric description of *neutral* hypermultiplets in the baryonic Higgs branch is also provided by HSS, in terms of the analytic (HST) superfields ω of vanishing $U(1)$ charge. The free action of a single ω superfield reads (Subsect. 4.4.2)

$$S[\omega] = -\tfrac{1}{2} \int_{\text{analytic}} \left(D_A^{++}\omega \right)^2 . \tag{3.51}$$

Similarly to the free action (3.5) for a massless FS analytic superfield q^+, the action (3.51) possesses the extended internal symmetry

$$SU(2)_{\text{R}} \otimes SU(2)_{\text{PG}} , \tag{3.52}$$

where $SU(2)_{\text{R}}$ is the automorphism symmetry of the $N = 2$ supersymmetry algebra. The $SU(2)_{\text{PG}} = Sp(1)$ symmetry of (3.51) is less obvious [282],

$$\delta\omega = c^{--} D^{++}\omega - c^{+-}\omega , \tag{3.53}$$

where $c^{--} = c^{(ij)}u_i^- u_j^-$ and $c^{+-} = c^{(ij)}u_i^+ u_j^-$, while $c^{(ij)}$ are the infinitesimal parameters of $SU(2)_{\text{PG}}$.

It is clearly not possible to construct any non-trivial self-interaction of the $U(1)$-chargeless superfield ω alone, without breaking the $SU(2)_{\text{R}}$ symmetry, simply because a hyper-Kähler potential has the $U(1)$ charge $(+4)$. Therefore, if $N = 2$ supersymmetry and the $SU(2)_{\text{R}}$ internal symmetry are not broken, one has [16, 17]

$$\mathcal{K}^{(+4)}{}_{\text{Higgs}}(\omega) = 0 , \tag{3.54}$$

i.e. there are no quantum corrections to the hyper-Kähler metric of neutral hypermultiplets in the $N = 2$ supersymmetric (non-baryonic) Higgs branch.

It is, however, possible to break the internal symmetry (3.52) down to

$$U(1)_{\text{R}} \otimes SU(2)_{\text{PG}} , \tag{3.55}$$

by introducing an FI term

$$\langle D^{ij} \rangle = \xi^{ij} = \text{const.} \neq 0 \tag{3.56}$$

that may softly break $N = 2$ supersymmetry too [534]. This way of symmetry breaking allows us to maintain control over the $N = 2$ supersymmetric hypermultipet LEEA because of the non-Abelian internal symmetry (3.55). Indeed, the only non-trivial hyper-Kähler potential, which is invariant under the symmetry (3.55), is given by [282]

$$\mathcal{K}^{(+4)}{}_{\mathrm{EH}}(\omega) = - \frac{(\xi^{++})^2}{\omega^2} \, , \qquad (3.57)$$

where $\xi^{++} = \xi^{ij} u_i^+ u_j^+$. The hyper-Kähler metric associated with (3.57) is known [286, 284], being equivalent to the Eguchi-Hanson (EH) ALE metric.

To understand the origin of the EH metric in HSS, it is useful to apply the *gauging* procedure known as 'hyper-Kähler quotient construction' [114]. The useful resources for generating new hyper-Kähler potentials in HSS are given by (i) gauging isometries, and (ii) adding FI terms. For instance, given two FS-type hypermultiplets $q_A^+ \in \mathbf{2}$ of $SU(2)_f$, we can gauge a $U(1)$ subgroup of $SU(2)_f$ and simultaneously add a FI term as follows [286]:

$$S_{\mathrm{EH}} = \int_{\mathrm{analytic}} \left\{ q_A^{a+} D^{++} q_{aA}^+ + V_{\mathrm{L}}^{++} \left(\tfrac{1}{2} \varepsilon^{AB} q_A^{a+} q_{aB}^+ + \xi^{++} \right) \right\} \, , \qquad (3.58)$$

where V_{L}^{++} is the $N = 2$ vector gauge potential without a kinetic term (i.e. the HSS Lagrange multiplier), $\xi^{++} = \xi^{(ij)} u_i^+ u_j^+$, and $A, B = 1, 2$. The action (3.58) has the manifest $Sp(1)$ internal symmetry rotating lower-case Latin indices only. The $SU(2)_{\mathrm{R}}$ symmetry is explicitly broken down to its $U(1)_{\mathrm{R}}$ subgroup by the non-vanishing FI term ξ^{++}. After some algebra, the Lagrange multiplier superfield V_{L}^{++} and one of the hypermultiplet superfields can be eliminated, while the resulting equivalent action takes the form [286]

$$S_{\mathrm{EH}} = \int_{\mathrm{analytic}} \left\{ q^{a+} D^{++} q_a^+ - \frac{(\xi^{++})^2}{(q^{a+} u_a^-)^2} \right\} \, . \qquad (3.59)$$

Changing the variables $q_a^+ = u_a^+ \omega + u_a^- f^{++}$, in terms of the dual ω-type hypermultiplet and yet another Lagrange multiplier f^{++}, gives us the HSS action with the hyper-Kähler potential (3.57) after eliminating f^{++} according to its algebraic equations of motion in HSS. It should be noted that the hyper-Kähler potential (3.57) already implies that $\langle \omega \rangle \neq 0$, so that we are indeed in the baryonic Higgs branch.

It is useful to imagine how a FI term could be non-perturbatively generated. First, we can slightly generalize this problem by allowing non-vanishing vacuum expectation values for all bosonic components of the Abelian $N = 2$ gauge superfield strength W, [282]

$$\langle W \rangle = \{ \ \langle a \rangle = Z \, , \quad \langle F_{\mu\nu} \rangle = n_{\mu\nu} \, , \quad \langle D \rangle = \boldsymbol{\xi} \ \} \, , \qquad (3.60)$$

where all the parameters $(Z, n_{\mu\nu}, \boldsymbol{\xi})$ are constants. It may also imply *soft* $N = 2$ supersymmetry breaking [534]. The physical meaning of Z is already known — it is the complex central charge in the $N = 2$ supersymmetry algebra. The related gauge-invariant quantity $u \sim \langle \mathrm{tr}\, a^2 \rangle$ parametrizes the quantum moduli space of vacua in the Coulomb branch. The central charge Z can be naturally generated via the standard Scherk-Schwarz mechanism of dimensional reduction from six dimensions [282]. Similarly, $n_{\mu\nu} \neq 0$ can be interpreted as a *toron* background after replacing the effective spacetime R^{1+3} by a hypertorus T^{1+3} and imposing t'Hooft's twisted boundary conditions [535]. The $\xi \neq 0$ is just the FI term.

The brane technology (Sect. 8.4) can help us to address the question of the *dynamical* generation of both $n_{\mu\nu}$ and ξ in a geometrical way [282]. Let us deform the brane configuration of Fig. 8.5 by allowing some of the NS5- or D4-branes to intersect at angles instead of being parallel! The vector $w = (x^7, x^8, x^9)$ is the same for both solitonic 5-branes depicted in Fig. 8.5. Its non-vanishing value,

$$\xi = w_1 - w_2 \neq 0 \,, \tag{3.61}$$

effectively generates the FI term. Similarly, when allowing the D4-branes to intersect at angles, some non-trivial values of $\langle F_{\mu\nu} \rangle = n_{\mu\nu} \neq 0$ are generated [536].

Since the spacetime LEEA of BPS branes is governed by a gauge field theory, it may not be surprising that torons can also be understood as the BPS bound states of certain D-branes in the field theory limit $M_{\mathrm{Planck}} \to \infty$ [536]. Moreover, the torons are known to generate gluino condensate [537],

$$\langle \lambda^i \lambda^j \rangle = \Lambda^3 (\xi^2)^{ij} \,, \quad \xi^{ij} \sim \delta^{ij} \exp\left(-\frac{2\pi^2}{g^2} \right) \,, \tag{3.62}$$

where $\xi \sim \{\xi^{ij}\}$ have to be constants [538], so that they can be identified with the FI term by $N = 2$ supersymmetry.

It is worth mentioning here that D4-branes can also end on D6-branes (in the type-IIA picture), while those D4-branes support hypermultiplets, not $N = 2$ vector multiplets [476]. This leads to the hyper-Kähler manifold (Subsect. 8.4.2) of topology $\sim S^3/Z_2$ in its spatial infinity. The EH-instanton is the only hyper-Kähler manifold that possesses this topology amongst the four-dimensional ALE spaces (Subsect. 4.3.3).

8.4 Brane Technology

The exact LEEA solutions to 4d, $N = 2$ supersymmetric *quantum* gauge field theories can also be obtained from the *classical* M-theory brane dynamics [476, 539, 540]. Though very little is known about microscopic M-theory, we really need only two of its general features:

- M-theory is the strong coupling limit of the type-IIA superstring theory;
- the low-energy limit of M-theory is 11d supergravity.

The first statement is not very constructive, while the second statement implies that solitonic (stable and unique) BPS states of 11d supergravity can be used to extract non-perturbative information about the spectrum and static couplings of M-theory. This information can then be applied to study the effective supersymmetric gauge field theories in the BPS brane world-volumes. These 'built-in' effective field theories are not quite the same as those we studied in the preceding sections of this chapter. Nevertheless, they may share the same LEEA under certain circumstances to be discussed below [541, 542, 543]).

8.4.1 11d Supergravity and Its BPS Solutions

Eleven is believed to be the maximum dimension of spacetime (with Lorentz signature), where a consistent interacting supersymmetric field theory exists [544, 545]. It is essentially a consequence of the fact that the massless physical particles mediating long-range forces in 4d spacetime can have spin two at most, while there is only one type of particle of spin two (i.e. gravitons). A supersymmetry charge (a component of a 4d spinor) changes the helicity λ of a massless particle by a half, so that the maximal non-vanishing product of N supersymmetry charges changes λ by $N/2$. Hence, in a massless representation of the 4d, N-extended supersymmetry, the helicity varies from λ to $\lambda + \frac{1}{2}N$. This immediately implies $N \leq 8$ provided that $|\lambda| \leq 2$. The maximal $N = 8$ supersymmetry in 4d has $8 \times 4 = 32$ real component charges, while the maximal spacetime dimension (with the Lorentz signature), where the minimal spinor representation also has 32 real components, is eleven (11d). The simplest supersymmetry algebra in 11d reads

$$\{Q_\alpha, Q_\beta\} = (C\Gamma^M)_{\alpha\beta}P_M \;, \tag{4.1}$$

where Q_α is the Majorana spinor supersymmetry charge, $\alpha = 1, 2, \ldots, 32$, P_M is the 11d (flat) spacetime momentum operator, Γ^M are real gamma matrices in 11d, and C is the 11d charge conjugation matrix, $M = 0, 1, 2, \ldots, 10$.

The 11d supergravity [545] is described in terms of three fields: a metric g_{MN}, a gravitino $\psi_{M\alpha}$ and a 3-form gauge potential A_{MNP} subject to gauge transformations $\delta A_{(3)} = d\Lambda_{(2)}$ and having the field strength $F_{(4)} = dA_{(3)}$. The 11d supergravity theory has $128_\text{B} + 128_\text{F}$ on-shell physical components. The bosonic part of the 11d supergravity action reads

$$S_\text{bosonic} = \tfrac{1}{2} \int d^{11}x \left\{ \sqrt{-g}(R - \tfrac{1}{48}F^2) + \tfrac{1}{6}F_{(4)} \wedge F_{(4)} \wedge A_{(3)} \right\} \;, \tag{4.2}$$

where we have taken the gravitational coupling constant to be equal to one. The last (Chern-Simons) term in (4.2) is, in fact, required by 11d supersymmetry of the total action including the gravitino-dependent terms.

Quantized 11d supergravity is not expected to be a consistent theory, e.g., because of its apparent non-renormalizability [546]. It should rather be interpreted as the *effective* low-energy approximation (i.e. the LEEA) to presumably consistent M-theory. This most basic and natural assumption, in fact, implies many deep consequences. Though we are unable to describe the underlying (microscopic) dynamics of M-theory from its LEEA alone, knowing the latter allows us to determine the spectrum of BPS states in M-theory, by constructing classical solitonic solutions to the 11d supergravity equations of motion, which preserve some part (ν) of 11d supersymmetry and, hence, some part of the translational invariance in 11d as well. Since the supersymmetry variations of the bosonic fields of 11d supergravity are all proportional to the gravitino field, the latter should vanish in a supersymmetric solution, $\psi_{M\alpha} = 0$. The vanishing supersymmetry variation of the gravitino field itself,

$$\delta\psi_M\big|_{\psi=0} = \widetilde{D}_M\varepsilon \equiv \left(D_M - \tfrac{1}{288}[\Gamma_M{}^{NPQS} - 8\delta_M^N\Gamma^{PQS}]F_{NPQS}\right)\varepsilon = 0 , \tag{4.3}$$

where $D_M\varepsilon = (\partial_M + \tfrac{1}{4}\omega_M{}^{BC}\Gamma_{BC})\varepsilon$, then implies the existence of a Majorana *Killing spinor* field ε satisfying the first-order differential equation (4.3). In (4.3) we denote by $\Gamma^{M_1\cdots M_p}$ the antisymmetric products of the gamma matrices with unit weight.

Assuming (as usual) the existence of asymptotic states with a supersymmetric vacuum, and requiring the 11d metric to be asymptotically Minkowskian, it is easy to see that the only BPS states with respect to the supersymmetry algebra (4.1) are just massless particles, since [7]

$$0 = \det\langle\{Q_\alpha, Q_\beta\}\rangle = \langle\det(\Gamma \cdot P)\rangle = \langle(P^2)^{16}\rangle , \tag{4.4}$$

while the matrix $\{Q_\alpha, Q_\beta\}$ has 16 independent zero eigenvalues ($\nu = \tfrac{1}{2}$). This means that a massless representation of 11d supersymmetry is $\tfrac{1}{2}$-shorter than the massive one, as is well known in supersymmetry *without* central charges. The corresponding asymptotically flat classical BPS solution of 11d supergravity with $P^2 = 0$ (called M-wave) was found, e.g., in [547].

It is one of the lessons of modern QFT that the massless particles appearing in its perturbative spectrum may not be the only BPS states. Nonperturbative (massive) BPS states in extended 4d supersymmetry carry electric and magnetic charges saturating the Bogomolnyi bound, whereas linear combinations of those charges appear as the central charges on the right-hand side of the 4d supersymmetry algebra. The symmetric matrix on the left-hand side of (4.1) belongs to the adjoint representation **528** of the Lie algebra of $Sp(32)$, which is decomposed with respect to its (Lorentz) subgroup $SO(1, 10)$ as

$$\mathbf{528} \to \mathbf{11} \oplus \mathbf{55} \oplus \mathbf{462} . \tag{4.5}$$

The **11** is apparently associated with P_M in (4.1), whereas the rest has to be associated with some additional 'central' charges commuting with supersymmetry charges and monenta, but not commuting with Lorentz rotations.

---

[7] It is the asymptotic form of the local 11d supersymmetry algebra that is given by the rigid superalgebra (4.1). The Killing spinor $\varepsilon(x)$ should also be asymptotically constant.

The 11d Lorentz representations **55** and **462** are associated with a 2-form $Z_{(2)}$ and a 5-form $Y_{(5)}$, respectively, so that the maximal 11d supersymmetry algebra reads [548]

$$\{Q_\alpha, Q_\beta\} = (C\Gamma^M)_{\alpha\beta} P_M + \tfrac{1}{2}(\Gamma^{MN}C)_{\alpha\beta} Z_{MN} + \tfrac{1}{5!}(\Gamma^{MNPQS}C)_{\alpha\beta} Y_{MNPQS} , \tag{4.6}$$

where Z_{MN} represent the 'electric' charges and Y_{MNPQS} are the 'magnetic' ones. The BPS object carrying non-vanishing electric charges is known as a *supermembrane* or an electric *M2-brane* [549]. Associated with the 11d spacetime symmetries broken by the supermembrane are the Nambu-Goldstone (NG) modes. The three-dimensional LEEA action describing the dynamics of small fluctuations of the NG fields about the supermembrane in 11d supergravity background was constructed in [550]. [8]

The BPS object magnetically *dual* to the M2-brane in 11 dimensions is a magnetically charged 5-brane called an *M5-brane*. According to Gauss's law, the electric charge of a particle (a 0-brane) in some number (D) of spacetime dimensions is measured by the dual gauge field strength according to the integral $Q_{\text{electric}} = \int_{S^{D-2}} {}^*F$ over the sphere S^{D-2} surrounding the particle, where $F_{(2)} = dA_{(1)}$ is the Abelian field strength of a $U(1)$ gauge field $A_{(1)}$, and ${}^*F_{(D-2)}$ is the Hodge dual to $F_{(2)}$ in D dimensions. In the case of an 'electric' p-brane charged with respect to a gauge $(p+1)$-form $A_{(p+1)}$ in D dimensions, the field strength is $F_{(p+2)} = dA_{(p+1)}$, and its dual is ${}^*F_{(D-p-2)}$. For magnetically charged objects the roles of F and ${}^*F$ are supposed to be interchanged. For example, the object carrying a magnetic charge in $D = 4$ is again a 0-brane (i.e. particle or monopole) since the dual potential $\tilde{A}$ defined by ${}^*F_{(2)} = d\tilde{A}$ is a 1-form, whereas the charge of the $D = 4$ monopole is measured by $F_{(2)}$ as $Q_{\text{magnetic}} = \int_{S^2} F$. Similarly, since the potential $\tilde{A}$ of the dual field strength $*F_{(D-p-2)}$ is a $(D-p-3)$-form, ${}^*F_{(D-p-2)} = d\tilde{A}_{(D-p-3)}$, it is a $D(p-4)$-brane that can support magnetic charges. The general rule for an electrically charged p-brane and its dual, magnetically charged q-brane reads

$$p + q = D - 4 . \tag{4.7}$$

Given $D = 11$ and $p = 2$, one has $q = 5$. The magnetic charge of an M5-brane is proportional to the integral $\int_{S^4} F_{(4)}$ over the sphere S^4 surrounding the brane at spatial infinity in five directions transverse to its six-dimensional world-volume. The integral is obviously topological (i.e. homotopy invariant) due to the Bianchi identity $dF_{(4)} = 0$.

The explicit forms of electric (M2-brane) and magnetic (M5-brane) BPS solutions to 11d supergravity are known [541, 542, 543]. For our purposes we only need a solitonic 5-brane solution found in [551],

$$ds^2 = H^{-1/3}(y)dx^\mu dx^\nu \eta_{\mu\nu} + H^{2/3}(y)dy^m dy^n \delta_{mn} , \qquad F_{(4)} = {}^*_5 dH , \tag{4.8}$$

---

[8] In fact, any NG-type action is essentially unique, being entirely determined by the spontaneously broken symmetries.

where the 11d spacetime coordinates have been split into the 'world-volume' coordinates labelled by $\mu, \nu = 0, 1, 2, 3, 4, 5$ and the 'transverse to the world-volume' coordinates labelled by $m, n = 6, 7, 8, 9, 10$, according to the space-time decomposition $R^{1,10} = R^{1,5} \times R^5$. In (4.8), the Hodge dual ($\ast$) in the five transverse dimensions has been introduced, whereas $H(y)$ is supposed to be a harmonic function in R^5, i.e. $\nabla^2 H(y) \equiv \Delta H(y) = 0$. All the other components of $F_{(4)}$ vanish. For a single M-5-brane of magnetic charge k, the harmonic function $H(y)$ is given by

$$H(y) = 1 + \frac{|k|}{r^3} , \quad \text{where} \quad r^2 = y^m y_m . \tag{4.9}$$

This M5-brane solution is completely regular (i.e. truly solitonic). In fact, it interpolates between the two maximally supersymmetric 11d 'vacua', with the one being asymptotically flat in the limit $r \to \infty$ and another approaching $(AdS)_7 \times S^4$ in the limit $r \to 0$ [552, 553].

When using the harmonic function

$$H(y) = 1 + \sum_{s=1}^{n} \frac{|k|}{|y - y_s|^3} , \tag{4.10}$$

one arrives at the classical configuration of n *parallel* and similarly oriented M5-branes, each having magnetic charge k and located at y_s in R^5-space. This multi-centre BPS solution also admits 16 Killing spinor fields by construction, so that it also preserves $\nu = \frac{1}{2}$ of 11d supersymmetry. The existence of the multi-centre brane solutions can be physically interpreted as a result of cancellation of gravitational and antigravitational (due to the antisymmetric tensor) forces between equally charged BPS branes, which is quite similar to the well-known 'no force condition' (zero binding energy) in the four-dimensional physics of monopoles.

8.4.2 NS and D Branes in Ten Dimensions

In 11 dimensions there are only M-waves, M2- and M5-branes as the 'elementary' BPS states preserving just half of 11d supersymmetry. In order to make contact with the type-IIA superstring theory in 10d (Chap. 6), let us assume that one of the transverse (to the brane world-volume) dimensions is compactified on a circle S^1 of radius $R_{[11]}$. An M5-brane can now be either (i) Kaluza-Klein-like 'reduced' to 10 dimensions, which results in a solitonic NS5-brane, or (ii) it can be 'wrapped' around the circle S^1, which results in a D4-brane.

By construction, the NS5-brane is magnetically charged with respect to a gauge NS-NS 2-form (Kalb-Ramond field) $B_{(2)}^{[10]}$ descending from the gauge 3-form $A_{(3)}^{[11]}$ in 11 dimensions. In accordance with (4.7), the NS5-brane is magnetically dual to the 'fundamental' 10d superstring. Since the NS5-brane still depends upon the compactified (periodic) coordinate ϱ of S^1, it contains all the associated Kaluza-Klein (KK) physical modes. In order to become a BPS solution to the 10d, Type-IIA supergravity, the NS5-brane solution should therefore be 'averaged' over the compactified coordinate ϱ, which just amounts to dropping all the massive KK modes. Though the latter is fully legitimate for a small compactification radius $R_{[11]}$ of S^1 (i.e for weakly coupled superstrings — see (4.15) below), it becomes illegitimate for large $R_{[11]}$ (i.e. for strongly coupled superstrings) when some massive KK modes become light. From the viewpoint of 10-dimensional type-IIA superstring theory, all the KK modes appear as non-perturbative states.

The wrapped M5-brane (=D4-brane) is (RR) charged with respect to the gauge 3-form $A_{(3)}^{[10]}$ of type-IIA supergravity, which is also descending from $A_{(3)}^{[11]}$, so that it is indeed a Dirichlet 4-brane. [9]

The KK *Ansatz* for the bosonic fields of 11d supergravity, which leads to the 10d type-IIA action (in the so-called string frame), reads [541, 542, 543]

$$
\begin{aligned}
ds_{[11]}^2 &= e^{-\frac{2}{3}\phi} ds_{[10]}^2 + e^{\frac{4}{3}\phi}(d\varrho + C_M dx^M)^2 \ , \\
A_{(3)}^{[11]} &= A_{(3)}^{[10]} + B_{(2)}^{[10]} \wedge d\varrho \ , \quad M = 0,1,2,\ldots,9 \ ,
\end{aligned}
\tag{4.11}
$$

where the S^1 coordinate ϱ is supposed to be periodic (with period 2π), and the 10d dilaton ϕ and KK vector C_M have been introduced.

The 10d bosonic action descending from (4.2) is

$$
\begin{aligned}
S_{\text{IIA}} = \tfrac{1}{2} \int d^{10}x \sqrt{-g} \Big\{ & e^{-2\phi} \Big[R + 4\nabla_M \phi \nabla^M \phi - \tfrac{1}{12} F_{MNP} F^{MNP} \Big] \\
& - \tfrac{1}{48} F_{MNPQ} F^{MNPQ} - \tfrac{1}{4} F_{MN} F^{MN} \Big\} + (\text{Chern} - \text{Simons terms}) \ ,
\end{aligned}
\tag{4.12}
$$

where $F_{(2)} = dC_{(1)}$, $F_{(3)} = dB_{(2)}$ and $F_{(4)} = dA_{(3)}$. The kinetic terms of the type-IIA superstring NS-NS fields (g_{MN}, B_{MN}, ϕ) in the first line of (4.12) are uniformly coupled to the dilaton factor $e^{-2\phi}$ (cf. the familiar factor g^{-2} in front of the Yang-Mills action), whereas the field strengths of the RR-fields (C_M, A_{MNP}) in the second line of (4.12) do *not* couple to the dilaton at all. Therefore, the superstring coupling constant g_{string} is given by the asymptotical value of e^ϕ,

$$
g_{\text{string}} = \langle e^\phi \rangle \ ,
\tag{4.13}
$$

while the RR-field couplings to the D-branes in type-IIA supergravity should contain non-perturbative information about type-IIA superstring or M-theory.

---

[9] The D-branes have a simple interpretation in the perturbative superstring theory, where they appear as the spacetime topological defects on which the open type-I superstrings with Dirichlet boundary conditions can end. The charges carried by the D-branes are known to be the Ramond-Ramond (RR) charges in superstring theory [349].

It follows from the KK *Ansatz* (4.11) that the compactification radius $R_{[11]}$ is also related to the dilaton as

$$R_{[11]} = \left\langle e^{\frac{2}{3}\phi} \right\rangle .$$

(4.14)

Combining (4.13) and (4.14) results in the remarkable relation [464]

$$R_{[11]} = g_{\text{string}}^{2/3} .$$

(4.15)

The strong coupling limit of the type-IIA superstring theory is, therefore, *eleven*-dimensional [464]! Since there is only one supersymmetric field theory (of second-order in spacetime derivatives) in 11d, namely, 11d supergravity, it should thus be interpreted as the LEEA of some 11-dimensional M-theory that is also supposed to be supersymmetric by consistency.

The solitonic (i.e. regular) NS5-brane BPS solution to 10d, type-IIA supergravity, which is obtained by plain Dimensional Reduction (DR) of the M5-brane solution (4.8) down to $R^{1,5} \times R^4$, is given by [541, 542]

$$ds_{[10]}^2 = dx^\mu dx^\nu \eta_{\mu\nu} + H(y) dy^m dy^n \delta_{mn} , \qquad F_{(3)} = {}^*_4 dH , \qquad e^{2\phi} = H ,$$

(4.16)

where x^μ ($\mu = 0, 1, 2, 3, 4, 5$) parametrize $R^{1,5}$ and y^m ($m = 6, 7, 8, 9$) parametrize R^4, $({}^*_4)$ is the Hodge dual in R^4, and $H(y)$ is a harmonic function in R^4. A BPS configuration of n parallel and similarly oriented NS5-branes is obtained with the 'multi-centred' harmonic function $H(y)$, as in (4.10).

Similarly, the D4-brane solution to the 10d, type-IIA supergravity reads [541, 542]

$$ds_{[10]}^2 = H^{-\frac{1}{2}} dx^\mu dx^\nu \eta_{\mu\nu} + H^{\frac{1}{2}} dy^m dy^n \delta_{mn} , \qquad F_{(3)} = 0 , \qquad e^{-4\phi} = H ,$$

(4.17)

where x^μ ($\mu = 0, 1, 2, 3, 4$) parametrize $R^{1,4}$, whereas y^m ($m = 5, 6, 7, 8, 9$) parametrize R^5, and $H(y)$ is a harmonic function in R^5. Given the choice (4.10) for the harmonic H-function in (4.17), one arrives at a BPS configuration of n parallel and similarly oriented D4-branes in static equilibrium. The 'no force condition' in this case can again be physically interpreted as the result of mutual cancellation of the gravitational (NS-NS) and antigravitational (R-R) forces between the D4-branes. However, unlike the similar NS5-brane solution (4.16), the solution (4.17) has isolated singularities at the positions $\{y_s\}$ of the D4-branes in R^5.

In the case of n parallel and similarly oriented D4-branes, there are n Abelian gauge fields in their common world-volume, which originate as the zero modes of the open superstrings stretched between the D4-branes [349]. In the coincidence limit, where all n D4-branes collapse, i.e. when they are 'on top of each other', the gauge symmetry $U(1)^n$ enhances to $U(n)$ [554]. This gauge symmetry enhancement can be understood from the viewpoint of the perturbative open (T-dual, or subject to Dirichlet boundary conditions) superstring theory [349] due to the appearance of extra massless vector bosons in the coincidence limit.

However, it is not yet the end of the 10d BPS brane story. The KK massive particles associated with the compactification circle S^1 can be naturally interpreted in 10d as the D0-branes charged with respect to the RR (and KK) gauge field $C_{(1)}$ [554]. Indeed, the eleventh component of the spacetime momentum in the 11-dimensional supersymmetry algebra (4.1) plays the role of an Abelian central charge in the compactified theory, whereas this central charge in 10d originates from the RR charges of D0-branes. From the viewpoint of the type-IIA superstring, all these BPS states are truly nonperturbative. [10]

According to the rule (4.7), magnetically dual to the D0-branes in 10d are D6-branes that are, therefore, also of KK origin in type-IIA supergravity. The corresponding classical BPS solution to the type-IIA supergravity equations of motion in 10d reads [555]

$$ds_{[10]}^2 = H^{-\frac{1}{2}} dx^\mu dx^\nu \eta_{\mu\nu} + H^{\frac{1}{2}} dy^m dy^n \delta_{mn}, \quad F_{(2)} = {}^*_3 dH, \quad e^{-4\phi} = H^3,$$
$$(4.18)$$

where x^μ ($\mu = 0,1,2,3,4,5,6$) parametrize $R^{1,6}$ and y^m ($m = 7,8,9$) parametrize R^3, $\binom{*}{3}$ is the Hodge dual in R^3, and $H(y)$ is a harmonic function in R^3. A BPS configuration of n parallel D6-branes is described by the harmonic function $H(y)$ similar to (4.10) but with the different power (-1) instead of (-3) there. Like a D4-brane, a single D6-brane is singular in 10d at the position of the D6-brane in R^3. The M-theory resolution of the D4-brane singularity will be discussed in Subsect. 8.4.3. As regards the D6-brane singularity in 10d, its M-theory resolution in 11 dimensions is provided by the following *non-singular* 11d supergravity solution [556]:

$$ds_{[11]}^2 = dx^\mu dx^\nu \eta_{\mu\nu} + H dy \cdot dy + H^{-1}(d\varrho + C \cdot dy)^2 , \quad (4.19)$$

where $H = 1 + \frac{1}{2} r^{-1}$, $r = |y|$, and $\nabla \times C = \nabla H$ (cf. (4.4.2) and (4.4.3)!). The 11d spacetime (4.19) is given by the product of the flat space $R^{1,6}$ with the Euclidean Taub-NUT space. The Taub-NUT space can be thought of as a non-trivial bundle (Hopf fibration) with the base R^3 and the fibre S^1. After dimensional reduction to 10 dimensions, (4.19) results in a single D6-brane located at the origin of R^3. Though the Taub-NUT metric seems to be singular at $r = 0$, this is merely a coordinate singularity provided that ϱ is periodic with period 2π. [11]

Therefore, the M-theory interpretation of a D6-brane is given by the Taub-NUT (or Kaluza-Klein (KK) [528, 529]) monopole that interpolates between the two maximally supersymmetric M-theory 'vacua': the flat 11d spacetime

---

[10] There are no RR charged states in the perturbative superstring spectrum.

[11] The Taub-NUT metric near the singularity of H is diffeomorphism-equivalent to a flat metric.

near $r = 0$ and the KK spacetime $R^{1,9} \times S^1$ near $r \to \infty$ [556]. It is straightforward to generalize this result to a system of n parallel and similarly oriented D6-branes in 10d, whose M-theory interpretation is given by the Euclidean multi-Taub-NUT monopole (Subsect. 4.4.1) described by the multi-centred harmonic function (cf. eq. (4.4.5))

$$H_{\text{multi}-\text{Taub}-\text{NUT}}(y) = 1 + \sum_{i=1}^{n} \frac{|k|}{2 |y - y_i|} . \tag{4.20}$$

This solution is non-singular in 11d provided that no two centres coincide, i.e. $y_i \neq y_j$ for all $i \neq j$. In the coincidence limit of parallel and similarly oriented D6-branes with equal RR charges ($= 1$), non-isolated singularities appear.

To investigate the coincidence limit in some more detail, one may have to generalize the harmonic function H of (4.20) to the form (4.4.5). For instance, in the case of two centres with equal charges, one has [541]

$$H(y) = \frac{\lambda}{2} + \frac{1}{2 |y - \epsilon y_0|} + \frac{1}{2 |y + \epsilon y_0|} , \tag{4.21}$$

where $\{\lambda, \epsilon\}$ are some positive constants. The double Taub-NUT metric is then obtained by choosing $\lambda/2 = 1$ in (4.21), whereas the limit $\lambda \to 0$ results in the Eguchi-Hanson (EH) metric. In the coincidence limit $|\epsilon| \to 0$, near the singularity of H, the value of λ is obviously irrelevant, whereas the gauge symmetry in the common world-volume of the D6-branes is *enhanced* from $U(1) \times U(1)$ to $U(2)$ [464, 554]. From the M-theory perspective, the M2-branes can wrap about the compactification circle S^1, being simultaneously stretched between the D6-branes. It is the massless modes of these M2-branes that play the role of the additional (non-Abelian) massless vector particles in the coincidence limit [541]. In order to make this symmetry enhancement manifest, it is useful to employ the non-singular HSS description of the mixed Taub-NUT-EH metric (Sect. 9.1).

The 10d singularity of the single D6-brane solution can be physically interpreted as the result of an illegitimate neglect of the KK particles that become *massless* at the D6-brane core and whose inclusion resolves the singularity in 11d [556]. This phenomenon is known as the *M-Theory resolution* of short-distance singularities in 10d, type-IIA supergravity by relating them (via the strong-weak coupling duality) to the long-distance effects of the massless modes of the M2-brane wrapped around S^1 [541]. The M-theory resolution is one of the cornerstones of brane technology in non-perturbative 4d, $N = 2$ supersymmetric gauge field theories [476] (see Subsect. 8.4.5).

8.4.3 Intersecting Branes

Each of the BPS (single or parallel) 'elementary' brane solutions to 11d or 10d (type-IIA) supergravity considered in the preceding subsections breaks exactly 1/2 of the maximal supersymmetry having 32 supercharges, and it is governed by a single harmonic function of transverse spatial coordinates. More BPS brane solutions preserving some part $\nu \leq 1/2$ of the maximal supersymmetry can be obtained by a superposition of the (intersecting) 'elementary' branes. The construction procedure of the corresponding classical supergravity solutions depending upon several harmonic functions is outlined in [557], and it is known as the 'harmonic function rule' (see e.g., [543] and references therein for details).

Since our goal in this section is a description of the brane technology towards the 4d gauge field theories with $N = 2$ supersymmetry (i.e. having $2 \times 4 = 8$ supercharges), we only need 'marginal' (i.e. of vanishing binding energy) BPS brane configurations preserving exactly $8/32 = 1/4$ of the maximal supersymmetry and having (uncompactified) flat spacetime $R^{1,3}$ as their intersection. This limits our discussion (e.g. in 10 dimensions) to *orthogonally* intersecting NS-branes, D4-branes and D6-branes, all having $R^{1,3}$ as their common world-volume, where the effective four-dimensional $N = 2$ supersymmetric field theory lives. From the M-theory perspective (i.e. in 11d), we want the NS- and D4-branes to be represented by a *single* M5-brane described by a single harmonic function, perhaps, in the background of a multi-Taub-NUT monopole described by yet another harmonic function or the corresponding hyper-Kähler potential. The relevant 1/4-supersymmetric BPS brane solution is thus going to be parametrized by two functions, like in $N = 2$ gauge field theory with hypermultiplet matter. Unlike the 10d configuration of the intersecting BPS branes (which is singular), the corresponding M-theory brane configuration in 11d is non-singular, while it carries some non-perturbavive information about the 4d, $N = 2$ effective gauge theory. Our immediate tasks are, therefore, (i) to establish a correspondence (dictionary) between the (classical) brane- and (quantum) field theory quantities, and (ii) to fix the form of the M5-brane. Both problems were solved in [476]. The earlier work on brane technology was reported in [472, 473, 474, 475].

8.4.4 LEEA in the World-Volume of Type-IIA BPS Branes

The exact solutions to the LEEA of 4d, $N = 2$ supersymmetric gauge field theories (say, for definiteness, in $N = 2$ super-QCD with N_c colours and N_f flavours) can be interpreted (and, in fact, derived) in a nice geometrical way, when considering the effective field theory in the common world-volume (to be identified with 4d spacetime $R^{1,3}$) of the magnetically (or RR) charged BPS branes of the type-IIA superstring theory after resolving the classical singularities of the 10d BPS branes in 11d M-theory [476].

The relevant 1/4-supersymmetric configuration of the orthogonally intersecting branes in 10 dimensions $R^{1,9}$ parametrized by

$$(x^0, x^1, x^2, x^3, x^4, x^5, x^6, x^7, x^8, x^9)$$

is schematically given in Fig. 8.5. It consists of two parallel (magnetically charged) NS5-branes, N_c parallel Dirichlet 4-branes orthogonally stretching between the NS5-branes, and N_f Dirichlet 6-branes that are orthogonal to both NS- and D-branes.

The two parallel 5-branes are located at $w = (x^7, x^8, x^9) = 0$, while they have classically fixed x^6 values. Being parallel to each other but orthogonal to the 5-branes, the 4-branes have their world-volumes parametrized by $(x^0, x^1, x^2, x^3) \in R^{1,3}$ and x^6. Being orthogonal to both 5-branes and 4-branes, the 6-branes are located at fixed values of (x^4, x^5, x^6), while their world-volumes are parametrized by $(x^0, x^1, x^2, x^3) \in R^{1,3}$ and $w \in R^3$.

After 'blowing-up' the intersecting NS5- and D4-brane configuration depicted in Fig. 8.5, its two-dimensional projection (in the directions orthogonal to the 4d effective spacetime) looks like as a *hyperelliptic* curve Σ of genus $g = N_c - 1$, depicted in Fig. 8.6. Indeed, as is well known in the theory of Riemann surfaces [496], a hyperelliptic (compact) Riemann surface of genus g can be defined by taking two Riemann spheres, cutting each of them between $2g + 2$ ramification (Weierstrass) points e_i, and then identifying the cuts, as shown in Fig. 8.6. The corresponding algebraic (complex) equation reads $y^2 = \prod_{i=1}^{2g+2}(z - e_i)$, with $e_i \neq e_j$ for $i \neq j$. In other words, a two-sheeted cover of the sphere branched over $2g + 2$ points is a useful realization of a hyperelliptic surface of genus $g \geq 1$. Though the surface obtained by the projection of the M5-brane world-volume is actually non-compact (it goes through infinity), nevertheless, one may formally apply the theory of compact Riemann surfaces to this case too.

Back to 10 dimensions, the 5-brane world-volumes are thus given by the *local* product $R^{1,3} \times \Sigma_0$, where $R^{1,3}$ is the 4d spacetime parametrized by the coordinates (x^0, x^1, x^2, x^3), whereas Σ_0 is the singular, of genus $g = N_c - 1$, Riemann surface parametrized by two real coordinates (x^4, x^5) or, equivalently, by a complex variable $v \equiv x^4 + ix^5$ (Fig. 8.5).

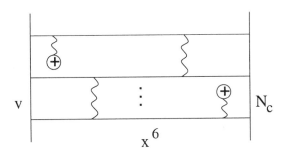

Fig. 8.5. A configuration of NS-5-branes (the two vertical lines), N_c of D4-branes (the horizontal lines and dots), and N_f of D6-branes (the encircled crosses) in the type-IIA picture

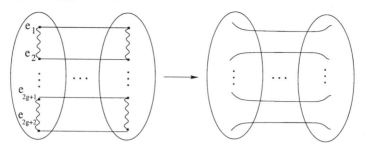

Fig. 8.6. A hyperelliptic curve of genus g

The type-IIA brane interaction in 10 dimensions can be visualized by the exchange of open superstrings, even though the ultimate force between some static branes may vanish. Associated with the zero modes of such open superstrings are BPS multiplets of a supersymmetric field theory. In particular, a gauge $N = 2$ vector multiplet in the effective 4d spacetime $R^{1,3}$ can be identified with massless modes of an open $(4 - 4)$ superstring carrying Chan-Paton factors at its ends and stretching between two D4-branes (Fig. 8.5), whereas the spacetime matter hypermultiplets are just zero modes of open $(6 - 4)$ superstrings connecting the D6-branes to the D4-branes. The BPS mass of a hypermultiplet is determined by the distance (in $x^{4,5}$ directions) between the corresponding D-6-brane and D-4-brane. [12]

For physical reasons, the effective gauge coupling constant g_{gauge} of the $N = 2$ supersymmetric gauge field theory in the effective 4d spacetime should be proportional to the distance between two NS-5-branes [476],

$$\frac{1}{g_{\text{gauge}}^2} = \frac{x_1^{[6]} - x_2^{[6]}}{g_{\text{string}}} , \tag{4.22}$$

where g_{string} is the type-IIA superstring coupling constant. Indeed, according to [476], the 5-brane x^6-coordinate should be thought of as a function of v by minimizing the total (BPS!) 5-brane world-volume. The BPS condition for large v is given by a two-dimensional Laplace equation on x^6, whose solution has a logarithmic dependence upon v for large values of v. Having interpreted $|v|$ as the mass scale in our theory, (4.22) is apparently consistent with the standard logarithmic behaviour of the four-dimensional effective gauge coupling at high energies in an asymptotically free gauge QFT. The presence of the superstring coupling g_{string} in (4.22) can be justified by the way it appears in the D-brane effective action induced by open superstrings (ending on a D-brane) in the brane world-volume. This effective action (or, at least, its bosonic part) is calculable, e.g. by the use of the standard sigma-model approach (Chap. 6) to open string theory [558]. As a leading contribution,

---

[12] The $(6 - 6)$ open superstrings decouple from the NS5- and D4-branes in the field theory LEEA.

one finds the Born-Infeld type effective action (Sect. 9.2) [559],

$$S_{\text{BI}} = T \int_{\text{world-volume}} e^{-\phi} \sqrt{\det(g_{\mu\nu} + B_{\mu\nu} + 2\pi F_{\mu\nu})} \,, \qquad (4.23)$$

where T is the constant brane tension, ϕ is the dilaton field, $g_{\mu\nu}$ is the induced metric in the world-volume, $B_{\mu\nu}$ and $F_{\mu\nu}$ are the pull-backs of the 2-form B and the Abelian field strength F, respectively. The factor $\langle e^{-\phi} \rangle = 1/g_{\text{string}}$ contributing to the effective brane tension in (4.23) is dictated by the disk topology of the relevant open superstring tree diagram. Extracting from (4.23) the term quadratic in F leads to the denominator on the right-hand side of (4.22).

At this end, let us summarize the most relevant properties of the brane configuration depicted in Fig. 8.5, from the 10-dimensional type-IIA point of view:

- its common world-volume is $(1+3)$-dimensional, being infinite in all directions;
- it is the BPS (and, hence, stable and unique) supersymmetric solution to the type-IIA supergravity equations of motion;
- it is invariant under $1/2 \times 1/2 = 1/4$ of the maximal supersymmetry, with the first $1/2$ factor being due to the parallel NS5-branes and the second $1/2$ factor due to the parallel D4-branes orthogonal to the NS5-branes; this results in $32/4 = 8$ conserved supercharges of $N = 2$ extended supersymmetry in the effective 4d spacetime;
- the D6-branes, orthogonal to both NS5- and D4-branes, do not break the $N = 2$ supersymmetry; they are the origin of matter hypermultiplets in 4d,
- the 10-dimensional Lorentz group $SO(1,9)$ is broken down to

$$SO(1,3) \otimes SU(2)_{\text{R}} \otimes U(1)_{\text{c.c.}} \,, \qquad (4.24)$$

in accordance with the local decomposition $R^{1,9} = R^{1,3} \times R_v^2 \times R_{x^6}^1 \times R^3$, respectively; the $SO(1,3)$ factor in (4.24) can be identified with the Lorentz group of the 4d spacetime $R^{1,3}$, the rotational symmetry $SO(3)$ of R^3 implies the $N = 2$ supersymmetry automorphism symmetry $SU(2)_{\text{R}} = SO(3) \times Z_2$ in 4d, whereas the $U(1) = SO(2)$ factor in (4.24) can be identified with the central charge transformations [282].

These basic properties, in fact, uniquely determine the most general BPS brane configuration of Fig. 8.5. One gets the effective gauge field theory in the common brane world-volume identified with the effective macroscopic 4d spacetime with $N = 2$ extended supersymmetry, while the field theory LEEA is under control. In order to accommodate non-perturbative quantum gauge field theory dynamics, the classical brane configuration of Fig. 8.5 should be blown-up. In the type-IIA picture considered in Subsect. 8.3, the corresponding BPS solution to the type-IIA supergravity suffers from singularities.

The singularities cannot be described semiclassically in 10d, but they can be resolved in 11d after reinterpreting the brane configuration of Fig. 8.5 in M-theory (see Subsect. 8.4.5).

8.4.5 M-Theory Resolution

It follows from (4.22) that one can keep the effective 4d gauge coupling constant g_{gauge} fixed while increasing the distance $L = x_1^6 - x_2^6$ between the two NS5-branes and, simultaneously, the type-IIA superstring coupling constant g_{string} accordingly. As we already know, at strong coupling the type-IIA superstring should be replaced by M-theory. This means that the additional compact dimension (x^{10}) represented by the circle S^1 of radius $R \sim g_{\text{string}}^{2/3}$, can no longer be ignored. Associated with the S^1-rotations is a non-perturbative $U(1)_M$ gauge symmetry.

The classical low-energy description of M-theory and its BPS branes turns out to be sufficient (see Subsect. 8.4.6) for a purely geometrical derivation of exact solutions to the LEEA of the effective 4d, $N = 2$ supersymmetric gauge field theories in the M-theory 5-brane world-volume, just because *all* the relevant distances in the non-perturbative 11-dimensional BPS brane configuration become large while no singularity appears, unlike that in the type-IIA picture considered in Subsects. 8.4.2 and 8.4.3. In particular, the D4-branes and NS5-branes in the type-IIA picture are replaced in M-theory by a *single* and *smooth* M5-brane whose world-volume is given by the *local* product $R^{1,3} \times \Sigma$, where Σ is the genus $g = N_c - 1$ hyperelliptic Riemann surface *holomorphically* embedded into a four-dimensional hyper-Kähler manifold Q given by the local product of R^3 and S^1. [13] The manifold Q is thus topologically a bundle $Q \sim R^3 \times S^1$ parametrized by the coordinates (x^4, x^5, x^6) and ϱ, whose base R^3 can also be interpreted as part of the D6-brane world-volume in the type-IIA picture and whose fibre S^1 is the eleventh dimension of M-theory [556].

The hyperelliptic curve Σ intersects with R^3 at a single point. After unifying the real coordinates x^6 and ϱ ($R = 1$) into a single complex coordinate $s = x^6 + i\varrho$ as in [476], the analytic equation defining the Riemann surface Σ should be of the form

$$F(s, v) = 0 \qquad (4.25)$$

with a holomorphic function F. Given a finite number of branes, the function F has to be a polynomial in v and $t = e^{-s}$ [476]. This polynomial can be fixed in terms of the non-Abelian $N = 2$ gauge field theory data $(SU(N_c), N_f, m_i)$ by using standard techniques of singularity theory [502]. Equation (4.25) then takes the form of the Seiberg-Witten curve Σ_{SW} described in Subsect. 8.2.2 [476].

---

[13] The holomorphicity of the embedding and the hyper-Kähler nature of the manifold Q are required by the BPS condition and unbroken supersymmetry, respectively.

In the absence of $N = 2$ matter, the hyper-Kähler manifold Q is flat. In the presence of (magnetically charged) hypermultiplets, the manifold Q is given by a multi-Taub-NUT (KK) monopole. The BPS bound for any Riemann surface Σ embedded into a hyper-Kähler manifold Q is given by [539]

$$\text{Area}(\Sigma) \geq \left| \int_\Sigma \Omega \right| , \qquad (4.26)$$

where the Kähler form Ω of Q [205] has been introduced. The bound (4.26) becomes saturated if and only if Σ is holomorphically embedded into Q, i.e. if the holomorphic description (4.25) of Σ is valid [504].

The origin of the Abelian $N = 2$ vector multiplets in the Coulomb branch of the effective 4d gauge field theory also becomes more transparent from the M-theory point of view [472, 473, 474]. The effective field theory in the six-dimensional world-volume of an M5-brane should have chiral six-dimensional $(2,0)$ supersymmetry [560]. The only admissible $(2,0)$ chiral supermultiplet in 6d is given by the tensor $(2,0)$ supermultiplet having a two-form $B_{(2)}$ with the *self-dual* field strength $T_{(3)}$ [429]. Indeed, being invariant under 16 linearly realized supersymmetries and having $11 - 6 = 5$ scalar fields describing transverse fluctuations to the 5-brane, the 6d supermultiplet in question should have $\frac{1}{2} \cdot 16 - 5 = 3$ additional bosonic on-shell degrees of freedom that can only be delivered by a bosonic gauge 2-form with the self-dual field strength belonging to the $(2,0)$ chiral tensor multiplet. Since, in our case, the M5-brane is wrapped around the Riemann surface, we can decompose the self-dual 3-form $T_{(3)}$ as

$$T_{(3)} = F_{(2)} \wedge \omega_{(1)} + {}^*_4 F_{(2)} \wedge {}^*_2 \omega_{(1)} , \qquad (4.27)$$

where $F_{(2)}$ is the two-form in $R^{1,3}$, whereas $\omega_{(1)}$ is the one-form on the Riemann surface Σ_{N_c-1} of genus $N_c - 1$. The equations of motion $\mathrm{d}T = 0$ then imply

$$\mathrm{d}F = \mathrm{d}^* F = 0 , \qquad (4.28)$$

and

$$\mathrm{d}\omega = \mathrm{d}^*\omega = 0 . \qquad (4.29)$$

Equation (4.29) means that the one-form ω is harmonic on Σ_{N_c-1}. Since the number of independent harmonic one-forms on a Riemann surface exactly equals its genus [496], one also has $(N_c - 1)$ two-forms F, while each of them satisfies (4.28). Since (4.28) is nothing but the Maxwell equations for an electromagnetic field strength F, this explains the origin of the Abelian gauge group $U(1)^{N_c-1}$ in the Coulomb branch of the effective 4d gauge field theory.

We conclude that the geometrical M-theory interpretation of the 4d, $N = 2$ gauge LEEA in the Coulomb branch is given by the identification [472, 473, 474]

$$\Sigma_{N_c-1} = \Sigma_{\text{SW}} . \qquad (4.30)$$

To understand the origin of the hypermultiplet LEEA in the Coulomb branch (Subsect. 8.3.1) in a similar way, let us notice that the D6-branes are magnetically charged with respect to the non-perturbative $U(1)_M$ gauge symmetry. Hence, the fibre S^1 of Q has to be non-trivial (i.e. of non-vanishing magnetic charge $m \neq 0$). After taking into account the $U(1)$ isometry of the hypermultiplet NLSM target space $Q_{(m)}$ in the Coulomb branch, it follows that $Q_{(m)}$ has to be the (Euclidean) multi-Taub-NUT space or a multi-KK monopole, whose metric (in the case of a single hypermultiplet) was already described in Subsects. 4.4.1 and 4.4.3.

The very existence of the 11d M-theory interpretation of the BPS multiplets is remarkable since there are no strings in 11d, whereas 4d, $N = 2$ multiplets are just identified with the open superstring zero modes in 10d. There are, however, M2-branes in 11d. They are also of the BPS-type, while they can end on an M5-brane. Having considered those M2-branes as the BPS deformations of the M5-brane, one can identify the BPS states in the effective $N = 2$ supersymmetric 4d field theory with zero modes of the M2-branes. The type of an $N = 2$ supermultiplet is determined by the static M2-brane topology: the cylinder yields an $N = 2$ vector multiplet, whereas the disk gives rise to a hypermultiplet [539].

8.4.6 SW Solution and M5-Brane Dynamics

We are now in a position to ask an educated question about the derivation of the Seiberg-Witten exact solution (Sect. 8.2) from the brane technology [524, 539, 540]. We should consider a single M5-brane in 11 dimensions, whose world-volume is given by the local product of flat spacetime $R^{1,3}$ with the hyperelliptic curve Σ of genus $g = N_{\mathrm{c}} - 1$ (Subsect. 8.4.4), where Σ is supposed to vary in the effective 4d spacetime $R^{1,3}$. Moreover, in order to be consistent with rigid $N = 2$ supersymmertry in 4d, the Riemann surface Σ should be holomorphically embedded into a hyper-Kähler four-dimensional manifold Q (say, with the Taub-NUT metric) which is part of the whole 11d spacetime given by the local product $R^{1,6} \times Q$. Since the Nambu-Goldstone type effective action of the M5-brane is *uniquely* determined by its symmetries, it is, in principle, straightforward to calculate it explicitly (see [561, 562] for the fully supersymmetric and covariant form of the M5-brane action). The KK reduction of the six-dimensional M-5-brane action on the complex curve Σ then gives rise to the effective $N = 2$ supersymmetric gauge field theory action which is nothing but the Seiberg-Witten effective action! This derivation of the SW exact result may also be interpreted in the theory of integrable systems [503, 504]. At the end of this subsection, we discuss its simplest technical realization (without hypermultiplets and with a flat hyper-Kähler manifold Q, for simplicity).

The fully covariant and supersymmetric M5-brane action of [561, 562] is not needed for our purposes. Even its bosonic part in a flat 11d supergravity background is too complicated, partly because it accommodates off-shell the self-duality condition on the field strength $T_{(3)} \equiv dB_{(2)}$ of the 2-form $B_{(2)}$ present in the six-dimensional chiral $(2,0)$ tensor multiplet. The bosonic part of the M5-brane action has the structure

$$S_{M-5,\,\text{bosonic}} = S_{\text{NG}} + S_{\text{self}-\text{dual}} + S_{\text{WZ}} , \qquad (4.31)$$

where S_{NG} is the standard 5-brane Nambu-Goto (NG) action, $S_{\text{self}-\text{dual}}$ stands for the naive world-volume integral over T^2 subject to the (implicit) self-duality constraint, whereas S_{WZ} is a higher-derivative Wess-Zumino (WZ) term that can be ignored here, since we are only interested in the leading contribution to the M5-brane LEEA with two spacetime derivatives at most. We restrict ourselves to the calculation of the *scalar* sector of the $N = 2$ supersymmetric LEEA in 4d (the rest of the field theory LEEA depending upon the vector and fermionic components of $N = 2$ vector supermultiplets is entirely determined by the special geometry of the scalar field components by $N = 2$ extended supersymmetry). Hence, we can ignore even the second term in the bosonic M-5-brane effective action (4.31).

The NG part of the action (4.31) reads

$$S_{\text{NG}} = T \int \text{Vol}(g) = T \int d^6\xi \sqrt{-\det(g_{\mu\nu})} , \qquad (4.32)$$

in terms of the 5-brane tension T and the induced metric

$$g_{\mu\nu} = \eta_{MN}\partial_\mu x^M \partial_\nu x^N \qquad (4.33)$$

in the M5-brane world-volume parametrized by the coordinates ξ^μ, where the functions $x^M(\xi)$ describe the embedding of the six-dimensional M5-brane world-volume into flat 11-dimensional spacetime, $M = 0, 1, \ldots, 10$, and $\mu = 0, 1, \ldots, 5$.

To simplify the form of the NG action (4.32) before its KK reduction down to four dimensions, we make use of (i) the reparametrizational invariance of this action in six world-volume dimensions, and (ii) the geometrical information about the M5-brane configuration collected in Subsects. 8.4.4 and 8.4.5. The local symmetry (i) allows us to choose a *static gauge*

$$x^{\underline{\mu}} = \xi^{\underline{\mu}} , \qquad (4.34)$$

where we have introduced the notation

$$\mu = \{\underline{\mu}, 4, 5\} , \quad \text{with} \quad \underline{\mu} = 0, 1, 2, 3 . \qquad (4.35)$$

We remind the reader that the M5-brane has $w = (x^7, x^8, x^9) = \mathbf{0}$, the Riemann surface Σ is parametrized by the coordinates $(v, \bar{v})$, whereas the flat (hyper-Kähler) manifold Q is parametrized by the four coordinates $(s, v, \bar{s}, \bar{v})$.

Since we are interested in the M-theory limit, where the supergravity decouples and the central charge $v = x^4 + i x^5$ of the 4d, $N = 2$ supersymmetry

algebra is constant, we assume in what follows that v is 4d spacetime independent. Moreover, since Σ is supposed to be holomorphically embedded into
Q, while Σ is also holomorphically dependent upon its complex moduli u_α,
$\alpha = 1, \ldots, g$, the actual dependence of the only remaining non-trivial function
s amongst the M-5-brane embedding functions $x^M(\xi)$ should be holomorphic,
i.e.

$$s = s(v, u_\alpha(x^{\underline{\mu}})) , \qquad (4.36)$$

where we have taken into account that $u_\alpha = u_\alpha(x^{\underline{\mu}})$. From the viewpoint
of the effective 4d gauge theory, the complex moduli $u_\alpha(x)$ of the Riemann
surface $\Sigma(x)$ are just the scalar field components of $N = 2$ vector multiplets
in 4d spacetime.

The induced metric (4.33) now takes the form

$$g_{\mu\nu} = \eta_{\mu\nu} + \tfrac{1}{2} \left(\partial_\mu s \partial_\nu \bar{s} + \partial_\nu s \partial_\mu \bar{s} \right) . \qquad (4.37)$$

We follow [540] here. Let us substitute (4.37) into (4.32), and keep terms of
second order in the derivatives ∂_μ. We then arrive at a free scalar 6d action,

$$S[s] = \frac{T}{2} \int d^6 \xi \, \eta^{\mu\nu} \partial_\mu s \partial_\nu \bar{s} \; \rightarrow \; S[u] = \frac{T}{2} \int d^6 \xi \, \partial_\mu s \partial^{\underline{\mu}} \bar{s} . \qquad (4.38)$$

The KK-reduction of the action (4.38) on Σ gives rise to the four-
dimensional NLSM action [540]

$$S[u] = \frac{T}{4i} \int d^4 x \, \partial_\mu u_\alpha \partial^{\underline{\mu}} \bar{u}_\beta \int_\Sigma \omega^\alpha \wedge \overline{\omega}^\beta , \qquad (4.39)$$

where the holomorphic 1-forms ω_α on Σ,

$$\omega^\alpha = \frac{\partial s}{\partial u_\alpha} dv , \qquad (4.40)$$

have been introduced.

The NLSM metric in (4.39) can be put into another equivalent form, by
using the Riemann bilinear identity [496]

$$\int_\Sigma \omega^\alpha \wedge \overline{\omega}^\beta = \sum_{\gamma=1}^g \left(\int_{A_\gamma} \omega^\alpha \int_{B^\gamma} \overline{\omega}^\beta - \int_{A_\gamma} \overline{\omega}^\beta \int_{B^\gamma} \omega^\alpha \right) , \qquad (4.41)$$

where a canonical (symplectic) basis (A_α, B^β) of the first homology class has
been introduced on the Riemann surface Σ_g of genus g, $\alpha, \beta = 1, \ldots, g$, and
$g = N_c - 1$. Substituting (4.41) into (4.39) and using the definitions (2.15) of
the multi-valued functions $a_\alpha(u)$ and their duals $a_{D\alpha}(u)$ yields [524, 539, 540]

$$\begin{aligned}
S[u] &= \frac{T}{4i} \int d^4 x \sum_{\alpha=1}^{N_c-1} \left(\partial_\mu a_\alpha \partial^{\underline{\mu}} \bar{a}_D^\alpha - \partial_\mu \bar{a}_\alpha \partial^{\underline{\mu}} a_D^\alpha \right) \\
&= -\frac{T}{2} \, \mathrm{Im} \left[\int d^4 x \, \partial_\mu \bar{a}_\alpha \partial^{\underline{\mu}} a_\beta \, \tau^{\alpha\beta} \right] ,
\end{aligned} \qquad (4.42)$$

where the period matrix $\hat{\tau}$ of Σ,

$$\tau^{\alpha\beta} = \frac{\partial a_D^\alpha}{\partial a_\beta} = \frac{\partial^2 \mathcal{F}}{\partial a_\alpha \partial a_\beta} \ , \tag{4.43}$$

has been introduced.

Equation (4.42) gives the scalar part of the full SW effective action in 4d, $N = 2$ super-QCD [16, 17]. Because of the NLSM special geometry described by (4.43), the unique $N = 2$ supersymmetric extension of (4.42), including all fermionic- and vector-dependent terms, is given by the superfield function $\mathcal{F}(W)$ integrated over the $N = 2$ chiral superspace (Sect. 8.1).

It is straightforward to generalize this derivation of the exact gauge (Seiberg-Witten) LEEA to more general cases with hypermultiplet matter, e.g., after replacing the flat background space Q by a curved hyper-Kähler manifold Q [524].

When being applied to the derivation of the hypermultiplet LEEA in 4d, brane technology suggests a dimensional reduction of the D6-brane effective action (see Subsects. 8.4.2 and 8.4.4) down to four spacetime dimensions. The D6-brane in M-theory is just the KK monopole described by the non-singular 11-dimensional metric (4.19). Hence, the induced metric in the D6-brane world-volume (in the static gauge) is given by

$$g_{\mu\nu} = \eta_{\mu\nu} + G_{ij}(y)\partial_\mu y^i \partial_\nu y^j \ , \tag{4.44}$$

where G_{ij} is the four-dimensional Taub-NUT metric, $\mu, \nu = 0, 1, \ldots, 6$ and $i, j = 1, 2, 3, 4$. Substituting (4.44) into the LEEA (Nambu-Goto) part of the D-6-brane effective action,

$$S[y] = \int \mathrm{d}^7\xi \sqrt{-\det(g_{\mu\nu})} \ , \tag{4.45}$$

expanding it up to second order in derivatives,

$$\sqrt{-\det(g_{\mu\nu})} = \mathrm{const.} - \tfrac{1}{2}\eta^{\mu\nu}g_{\mu\nu} + \ldots \ , \tag{4.46}$$

where the dots stand for higher-derivative terms, and performing plain dimensional reduction from seven to four dimensions (in fact, our fields do not depend upon three irrelevant coordinates) result in the NLSM action with the Taub-NUT metric,

$$S[y] = -\tfrac{1}{2} \int \mathrm{d}^4x \, G_{ij}(y)\partial_{\underline{\mu}} y^i \partial^{\underline{\mu}} y^j \ , \tag{4.47}$$

in full agreement with the perturbative QFT results obtained in 4d, $N = 2$ harmonic superspace (Subsect. 8.3.1). Therefore, as regards the *leading* (i.e. of second order in spacetime derivatives, in components) contributions to the 4d, $N = 2$ supersymmetric LEEA, brane technology offers rather simple classical tools for their derivation, when compared to conventional QFT methods. Being either holomorphic in the case of $N = 2$ vector multiplets or analytic in the case of hypermultiplets, the leading contributions are given by the 'protected' integrals over half of the $N = 2$ superspace anticommuting coordinates. The higher-order terms in the LEEA are given by the 'unprotected' integrals over the whole $N = 2$ superspace, which casts some doubt on the applicability of brane technology to their derivation.

8.5 On Supersymmetry Breaking and Confinement

Though being very different, (i) instanton calculus, (ii) the Seiberg-Witten approach and brane technology, and (iii) harmonic superspace lead to consistent results, as regards the *leading* terms in the LEEA of the four-dimensional $N = 2$ supersymmetric gauge field theories. No single universal method apparently exists for handling all problems associated with supersymmetric gauge field theories, since each approach has its own advantages and disadvantages. For example, in the Seiberg-Witten approach, the physical information is encoded in terms of functions defined over the quantum moduli space whose modular group is identified with the S-duality group. The SW approach is based on knowing exact perturbative limits of the non-Abelian $N = 2$ gauge field theory under consideration, whereas the HSS approach is the most efficient one in quantum perturbative calculations. The HSS description of non-perturbative hypermultiplet LEEA is natural, while HSS is not necessary for addressing the $N = 2$ gauge (SW) LEEA. The instanton calculus is very much dependent upon the applicability of its own basic assumptions, while it is not manifestly supersymmetric at all. Moreover, instanton methods usually need additional input. Being geometrically very transparent, the M-theory brane technology has limited analytic support, whereas its successful applications are so far limited to those terms in the LEEA that are protected by non-anomalous symmetries, i.e. they are either holomorphic or analytic.

We conclude this chapter with a few brief comments on supersymmetry breaking and colour confinement in 4d QCD. It is quite natural to take advantage of the existence of exact solutions to the LEEA in 4d, $N = 2$ supersymmetric gauge field theories, and apply them to the traditionally intractable, strongly coupled theories like QCD. This was, perhaps, the main motivation in the original work of Seiberg and Witten [16, 17]. The most attractive mechanism of colour confinement is the *dual* Meissner effect or *dual* (Type II) superconductivity [563, 564, 565]. It takes three major steps to connect a BCS superconductor to the Seiberg-Witten theory: first, one defines a relativistic version of the superconductor known as the Abelian Higgs model in field theory; second, one introduces the non-Abelian Higgs theory known as the Georgi-Glashow model; third, one extends the Georgi-Glashow model by $N = 2$ supersymmetry to the Seiberg-Witten (SW) model. Since they latter have (BPS) monopoles, it is quite natural to explain QCD confinement as the result of monopole condensation (= the dual Meissner effect as a consequence of the dual Higgs effect), due to the non-vanishing vacuum expectation value of the magnetically charged (dual Higgs) scalar belonging to the HP hypermultiplet. When monopoles condense, the electric flux is confined in dual (Abrikosov-Nielsen-Olesen) vortex tubes. Of course, this is only possible after $N = 2$ supersymmetry breaking in the SW theory.

It is worth mentioning that a derivation of the exact (SW) LEEA requires eight conserved supercharges or, equivalently, $N = 2$ extended supersymmetry in 4d. Accordingly, most of our results in this chapter are not applicable to the more phenomenologically interesting case of 4d gauge field theories with only $N = 1$ supersymmetry. [14] Therefore, on the one hand, it is the $N = 2$ extended supersymmetry that crucially simplifies the calculation of the 4d QFT LEEA. On the other hand, it is the same 4d, $N = 2$ supersymmetry that is incompatible with phenomenology, e.g., because of equal masses of bosons and fermions inside $N = 2$ supermultiplets (this is applicable, in fact, to any $N \geq 1$ supersymmetry), and the non-chiral nature of $N = 2$ supersymmetry (e.g. $N = 2$ 'quarks' belong to *real* representations of the gauge group). Therefore, if one believes in the fundamental role of $N = 2$ supersymmetry in high-energy physics, one has to find a way of judicious $N = 2$ supersymmetry breaking. The associated dual Higgs mechanism may then be responsible for chiral symmetry breaking and the appearance of the pion effective Lagrangian (Sect. 2.7) as well, provided that the dual Higgs field has flavour charges [16, 17]. In fact, Seiberg and Witten [16, 17] used a mass term of the $N = 1$ chiral multiplet that is part of the $N = 2$ vector multiplet, in order to *softly* break $N = 2$ supersymmetry to $N = 1$ supersymmetry 'by hand'. As a result, they found a non-trivial vacuum solution with a monopole condensation, i.e. a confinement. The weak point of their approach is the *ad hoc* assumption about the existence of the mass gap, i.e. the mass term itself.

The 4d, $N = 2$ supersymmetry can be broken either softly or spontaneously, if one wants to preserve the benefits of its presence (e.g. for maintaining control over the LEEA) at high energies. A detailed investigation of soft supersymmetry breaking in $N = 2$ supersymmetric QCD was done in [534] by using FI terms. [15] Though being pragmatic and more efficient when compared to the $N = 1$ case, soft $N = 2$ supersymmetry breaking still requires extra parameters. Spontaneous $N = 2$ supersymmetry breaking, by using a dynamically generated scalar potential, is possible too [284], while it does not need new parameters.

Dual (type II) superconductivity results in the creation of colour-electric fluxes (or strings) having quarks at their ends [563, 564, 565]. The usual type-II superconductivity (i.e. the confinement of magnetic charges) is a so-

---

[14] See e.g., [566] for a review of the known field theory results, and [477] for a review of brane technology, with less than eight supercharges.

[15] See [567] for a similar analysis in $N = 1$ supersymmetric gauge field theories.

lution to the standard Landau-Ginzburg theory, whereas QCD confinement is supposed to be a non-perturbative solution to a (1+3)-dimensional quantum $SU(N_c)$ gauge field theory with $N_c = 3$. A formal proof of the colour confinement amounts to a derivation of the area law for a Wilson loop $W[C]$. It may be based on the so-called 'string' *Ansatz* [1, 568]

$$W[C] \sim \int_{\substack{\text{surfaces } \Sigma, \\ \partial\Sigma = C}} \exp\left(-S_{\text{string}}\right) \ . \tag{5.1}$$

This formal equation clearly shows that the effective degrees of freedom (or collective coordinates) in QCD at strong coupling (in the infrared) are the *strings* whose world-sheets are given by the surfaces Σ, and whose dynamics is governed by the string action S_{string}. The fundamental (Schwinger-Dyson) equations of QCD can be reformulated in an equivalent infinite chain of equations for the Wilson loops [569]. The chain of loop equations drastically simplifies at *large* number of colours N_c to a *single* closed equation known as the Makeenko-Migdal (MM) loop equation [570]. [16]

The main problems with the MM loop equation are: (i) taking into account quantum renormalization, and (ii) determining the string action S_{string}. The first problem may be solved by replacing QCD by the 4d, $N = 4$ super-Yang-Mills (SYM) theory, and by treating the $N = 4$ supersymmetric MM-type loop equation instead of the original ($N = 0$) one. Indeed, being a scale-invariant QFT in 4d, the $N = 4$ SYM theory does not renormalize at all. It was conjectured by Maldacena [571] that the $N = 4$ SYM theory is dual to the IIB superstring theory in the AdS$_5 \times S^5$ background. The Maldacena conjecture may, therefore, be interpreted as the particular *Ansatz* $S_{\text{string}} = S_{\text{IIB/AdS}_5 \times S^5}$ for a solution to the $N = 4$ super-MM loop equation in the form (5.1). According to [571], the (1+3)-dimensional spacetime is supposed to be identified with the *boundary* of the anti-de-Sitter space AdS$_5$, where

$$\text{AdS}_5 = \frac{SO(4,2)}{SO(4,1)} \quad \text{and} \quad S^5 = \frac{SO(6)}{SO(5)} \ , \tag{5.2}$$

while the coupling constants are related to the AdS$_5$ radius R_{AdS} as [571]

$$(\alpha')^{-2} R_{\text{AdS}}^4 \sim g_{\text{YM}}^2 N_c \ , \quad \text{and} \quad g_{\text{string}} \sim g_{\text{YM}}^2 \ . \tag{5.3}$$

The proposed (Maldacena) duality is a strong-weak coupling duality since

- for small $\lambda = g_{\text{YM}}^2 N_c$ a perturbative SYM description applies;
- for large λ a perturbative IIB string/AdS supergravity description applies.

It is also in agreement with the *holography* proposal [572] since physics in the AdS$_5$ bulk is supposed to be encoded in terms of the field theory defined on the AdS$_5$ boundary. The quantum 4d, $N = 4$ SYM theory is conformally invariant, while its rigid symmetry is given by the supergroup $SU(2,2|4)$

---

[16] Only *planar* Feynman graphs survive in the large N_c limit.

that contains 32 supercharges. The isometries of $AdS_5 \times S^5$ form the group $SO(4,2) \times SO(6) \cong SU(2,2) \times SU(4)$ whose extension in AdS supergravity is also given by $SU(2,2|4)$. In addition, both the $N = 4$ SYM and type-IIB superstrings are believed to be self-dual under the S-duality group $SL(2, \mathbf{Z})$. In more practical terms, this CFT/AdS correspondence implies a one-to-one correspondence [573] between the $N = 4$ SYM correlators and the correlators of the certain superstring theory whose action S_{string} is known and whose correlators can be computed, in principle, by the methods of 2d conformal field theory [45]. Quantum corrections in powers of $(\alpha' \times \text{curvature})$ on the string theory side correspond to corrections in powers of $(g_{\text{YM}}^2 N_c)^{-1/2}$ on the gauge field theory side, while the string loop corrections are suppressed by powers of N_c^{-2} (see e.g., [574, 575] for more details).

The CFT/AdS correspondence provides simple mechanisms for simulating confinement, generating a mass gap, and breaking conformal invariance and supersymmetry [576], by considering 'finite-temperature' versions of anti-de Sitter spaces [577]. For example, by solving the supergravity wave equation in anti-de Sitter black hole geometry [577], one can calculate glueball masses in QCD. The glueball mass ratios found in this way [578] are in remarkable agreement with the available lattice results. These encouraging theoretical developments show promise for solving long-standing problems in QCD.

9. Generalizations of NLSM

In this chapter we give several different generalizations of the NLSM concept, by relaxing the assumptions made about the NLSM in the Introduction (Chap. 1). In Sect. 9.1 we consider the 4d, $N = 2$ NLSM with the ALE target spaces, and demonstrate that the Eguchi-Hanson 4d, $N = 2$ NLSM with a non-vanishing $N = 2$ central charge gives rise to dynamical generation of the composite $N = 2$ vector gauge multiplet, in one-loop perturbation theory [579, 580]. The composite $N = 2$ vector multiplet is identified with the zero modes of the superstring ending on a D6-brane. In Sect. 9.2 we relax another assumption that the NLSM fields should be represented by scalars or supersymmetric scalar multiplets. The NLSM (e.g., with an n-sphere as the target space) is often related to spontaneous breaking of internal symmetry (e.g., the $O(n)$ rotational symmetry). A spontaneous breakdown of internal symmetry can be accomplished by starting with a linear realization of the symmetry (e.g. in terms of the scalar n-field) and then imposing a non-linear constraint (e.g., by defining the $O(n)$ NLSM of the n-field by assigning its values to the n-sphere). This concept can be further generalized to *partial* (1/2) spontaneous supersymmetry breaking with a vector supermultiplet of Goldstone modes [581]. The corresponding Goldstone action can be interpreted as the gauge-fixed D-brane action that can be put into NLSM-type form by using a non-linear superfield constraint (Sect. 9.2). Section 9.3 is devoted to another natural generalization of NLSM in 4d, with higher space-time derivatives (of fourth order). Though the higher derivatives are usually accompanied by ghosts in QFT, the fourth-order NLSM and its ($N = 1$) supersymmetric generalizations are formally renormalizable and asymptotically free [582]. Finally, in Sect. 9.4 we briefly discuss the structure of the 4d NLSM originating in the scalar sector of $N \geq 4$ extended supergravitities, and their relation to U-duality [583].

9.1 Dynamical Generation of Particles in 4d, $N = 2$ ALE NLSM

The idea that some of the 'elementary' particles, like a photon, Higgs or W bosons, may be composite has been known in theoretical high-energy physics for many years. It was proposed as a possible solution to many different problems in QFT. For instance, the compositeness of photons was suggested a long time ago, in order to resolve the ultraviolet problems of QED related to the existence of a Landau pole and the divergence of the effective coupling at high energies [584]. If the Higgs particles are to be interpreted as bound states, this would simply explain the experimental failure to observe them. The compositeness of some of the vector bosons mediating weak or strong interactions was also proposed to accommodate the phenomenologically required gauge group $SU(3) \times SU(2) \times U(1)$ of the Standard Model in the maximally extended four-dimensional $N = 8$ supergravity [585]. Gauging the internal symmetry of the $N = 8$ supergravity merely produces the $SO(8)$ gauge group that does not contain the SM gauge group as a subgroup [586]. However, since the scalar sector of the $N = 8$ supergravity (Sect. 9.4) can be described by a non-compact NLSM over the coset $E_7/SU(8)$ [139], assuming that its auxiliary gauge fields become dynamical in quantum theory would give rise to the gauge group $SU(8)$ that is big enough. Though $N = 8$ supergravity is no longer considered to be the unifying quantum field theory because of its apparent non-renormalizability, its modern successor called M-theory [464] (Sect. 8.4) has 11d supergravity as the LEEA, whose dimensional reduction down to four spacetime dimensions yields N=8 supergravity. The bound states arising in a system of BPS-type extended classical solutions to 11d supergravity (called branes) play an important role in M-theory (Sects. 8.4 and 9.2).

The quantum field-theoretical mechanisms of dynamical generation of composite particles are known in two or three spacetime dimensions [1, 587, 588]. Unfortunately, little is known about the formation of bound states in quantized four-dimensional field theories (see, however, [589]) or in M-theory (see, however, [590]). In this section we present a dynamical mechanism of generation of composite $N = 2$ vector multiplets in the 4d, $N = 2$ supersymmetric NLSM with the ALE target space. Our approach is based on rewriting the classical ALE-based NLSM in renormalizable form given by the gauged 'linear' (non-compact) NLSM, and then taking into account the one-loop quantum corrections due to the quantized hypermultiplets comprising fields of both positive and negative norm. $N = 2$ extended supersymmetry in 4d spacetime, with a non-vanishing central charge, plays a very important role in our model. First, $N = 2$ susy protects the ALE hyper-Kähler geometry in the NLSM target space and the very particular form of the associated scalar potential. Second, $N = 2$ susy automatically gives rise to the (UV) divergence cancellations that, otherwise, could destroy the consistency of the proposed theory. The technical power of $N = 2$ harmonic superspace (Subsect. 4.4.2) allows us to take advantage of having manifest $N = 2$ extended supersymmetry in quantum perturbation theory, which significantly simplifies our calculations and makes them very transparent.

9.1.1 4d, $N = 2$ NLSM with ALE Metric

Our starting point is the harmonic potential (4.4.5) describing a two-centred ($n = 2$) monopole solution with equal charges ($m = 1/2$, $\boldsymbol{y}_1 = \boldsymbol{0}$, $\boldsymbol{y}_2 = \boldsymbol{\xi}$), and a constant $\lambda > 0$,

$$H(\boldsymbol{y}) = \lambda + \frac{1}{2} \left\{ \frac{1}{|\boldsymbol{y} - \boldsymbol{0}|} + \frac{1}{|\boldsymbol{y} - \boldsymbol{\xi}|} \right\} . \tag{1.1}$$

The real vector $\boldsymbol{\xi}$ can be equally represented as the $SU(2)$ triplet $\xi^{ij} = \mathrm{i}\boldsymbol{\xi} \cdot \boldsymbol{\tau}^{ij}$ satisfying the reality condition

$$(\xi^{ij})^{\dagger} \equiv \xi_{ij}^{\dagger} = \varepsilon_{il}\varepsilon_{jm}\xi^{lm} = \xi_{ij} , \tag{1.2}$$

where $\boldsymbol{\tau}$ are the usual 2×2 Pauli matrices. The hyper-Kähler metric defined by (4.4.2), (4.4.3) and (1.1) is the *double* Taub-NUT metric with a constant potential λ at infinity. According to Subsect. 4.4.3, the 4d, $N = 2$ NLSM with the double Taub-NUT target space metric is described by the HSS Lagrangian

$$\mathcal{L}^{(+4)} = - \overset{*}{\bar{q}}{}^{+}_{A} D^{++} q^{+}_{A} - V^{++} \left(\varepsilon^{AB} \overset{*}{\bar{q}}{}^{+}_{A} q^{+}_{B} + \xi^{++} \right) - \lambda \left(\sum_{A=1}^{2} \overset{*}{\bar{q}}{}^{+}_{A} q^{+}_{A} \right)^{2} , \tag{1.3}$$

where the $N = 2$ vector gauge superfield V^{++} has been introduced as the Lagrange multiplier, and $\xi^{++} = \xi^{ij} u^{+}_{i} u^{+}_{j}$. As is clear from (1.3), the NLSM is invariant under the local $U(1)$ gauge symmetry

$$\delta q^{+}_{1} = \Lambda q^{+}_{2}, \quad \delta q^{+}_{2} = -\Lambda q^{+}_{1}, \quad \delta V^{++} = D^{++}\Lambda , \tag{1.4}$$

with the analytic HSS superfield parameter $\Lambda(\zeta, u)$. The rigid $SU(2)$ automorphisms of $N = 2$ supersymmetry algebra are broken in (1.3) to its Abelian subgroup that leaves ξ^{++} invariant. The (Pauli-Gürsey) symmetry $Sp(1) \cong SU(2)_{\mathrm{PG}}$, rotating q^{+} and $\overset{*}{\bar{q}}{}^{+}$, is also broken in (1.3), as long as $\lambda \neq 0$.

Equation (1.3) takes a simple form in the limit $\xi \to 0$ where it reduces (after a superfield redefinition) to the Taub-NUT NLSM action in HSS (Subsect. 4.4.3). Similarly, in another limit $\lambda \to 0$, (1.3) yields the $N = 2$ NLSM with the Eguchi-Hanson (EH) metric (Subsect. 4.4.3). In other words, the double Taub-NUT metric interpolates between the Taub-NUT and Eguchi-Hanson metrics [287]. In both limits (Taub-NUT and Eguchi-Hanson), the metric has $U(2)$ isometry, whereas only the $U(1)$ isometry is left when both $\lambda \neq 0$ and $\xi \neq 0$.

Equation (1.3) at $\lambda = 0$ takes the form of the $SU(2)_{\mathrm{PG}}$-invariant minimal coupling between the two 'matter' FS-type hypermultiplets q^{+}_{A} and the Abelian $N = 2$ vector gauge multiplet V^{++}, in the presence of the gauge-invariant FI term linear in V^{++},

$$\mathcal{L}^{(+4)}(q_A, V) = -\tfrac{1}{2}q_A^{a+} D^{++} q_{aA}^+ - V^{++} \left(\tfrac{1}{2}\varepsilon^{AB} q_A^{a+} q_{Ba}^+ + \xi^{++}\right) . \tag{1.5}$$

Let us consider the following gauge-invariant HSS action, in terms of another two FS-type hypermultiplet superfields and an $N = 2$ vector gauge V^{++} superfield:

$$S_{\text{EH}}[q_1, q_2, V] = \int d\zeta^{(-4)} du \left[-\overset{*}{q}\,^+_1 \mathcal{D}^{++} q_1^+ + \overset{*}{q}\,^+_2 \mathcal{D}^{++} q_2^+ + V^{++} \xi^{++} \right] , \tag{1.6}$$

where we have returned to canonical dimensions for all the superfields involved, [1] and we have introduced the gauge-covariant harmonic derivative $\mathcal{D}^{++} = D^{++} + iV^{++}$, thus extending the rigid $U(1)$ symmetry (4.4.85) of a free hypermultiplet action to the local analytic symmetry. It is not difficult to check that the classical theory (1.6) is equivalent to (1.5), e.g., by considering a gauge $q_2^+ = 0$ in (1.6) and a gauge $q_2^+ = iq_1^+$ in (1.5), up to rescaling by a factor of 2. However, in the form (1.6), the $SU(2)_{\text{PG}}$ invariance is no longer manifest. Moreover, the action (1.6) has the wrong sign in front of the kinetic term for the q_2 hypermultiplet, which indicates its non-physical (ghost) nature. This also implies its anticausal propagation and the wrong (negative) sign of the residue in the propagator of q_2 superfield. It does not, however, make our theory (1.6) non-unitary since the q_2^+ hypermultiplet is a gauge degree of freedom, while the classical action (1.6) itself is dual to any of the manifestly unitary NLSM actions with the ALE (Eguchi-Hanson) target space (see (4.4.95) and (4.4.96) in Subsect. 4.4.3) The action (1.6) has the form of the $N = 2$ supersymmetric non-compact (gauged) $SU(1,1)/U(1)$ NLSM parametrized by the FS-type hypermultiplets q_A in the fundamental representation of $SU(1,1)$ whose $U(1)$ subgroup is gauged in HSS (cf. Sect. 6.2).

We now exploit the freedom of choosing a classical HSS Lagrangian with the on-shell EH metric, and take (1.6) as our starting point for quantizaton. It is worth mentioning that the minimal gauge interaction of hypermultiplets with $N = 2$ vector multiplets is the *only* renormalizable type of $N = 2$ supersymmetric field-theoretical interaction in four-dimensional spacetime [238]. The classical equivalence to the formally unitary (but non-renormalizable) NLSM actions (4.4.95) and (4.4.96) ensures unitarity in our theory (1.6), whereas the non-anomalous gauge Ward identities are supposed to take care of the gauge invariance after quantization. Our approach may be compared to the standard bosonic string theory (Sect. 6.1) where the 2d Nambu-Goto classical string action is replaced by the 2d Polyakov string action (Sect. 6.1). The non-polynomial Nambu-Goto action has a clear geometrical interpretation as the area of a string world-sheet but it is formally non-renormalizable. One defines a quantized bosonic string theory (in the critical dimension) after replacing the Nambu-Goto action by the classically equivalent Polyakov action that has a 2d auxiliary metric as the Lagrange multiplier. In our case,

---

[1] In units of mass one has $[q] = 1$, $[V] = 0$ and $[\xi] = +2$.

however, we will not integrate over the Lagrange multiplier given by the $N = 2$ vector gauge superfield in 4d. It is the ghost hypermultiplet that is going to be integrated out in quantum theory (Subsect. 9.1.2).

The quantized theory (1.6) is, however, of little interest unless it is supplemented by an $N = 2$ central charge $\hat{Z}$ giving BPS masses to hypermultiplets. A hypermultiplet of mass m can be described in HSS via the extension (Subsect. 4.4.2)

$$\hat{D}^{++} \equiv D^{++} + i(\theta^{\alpha+}\theta^+_\alpha)\hat{\bar{Z}} + i(\bar{\theta}^+_{\dot{\beta}}\bar{\theta}^{\dot{\beta}+})\hat{Z} \tag{1.7}$$

of the flat harmonic derivative D^{++}, with $\hat{Z}$ being an operator. It is is not difficult to verify that the free hypermultiplet equation of motion $\hat{D}^{++}q^+ = 0$ implies

$$\left(\Box + \hat{Z}\hat{\bar{Z}}\right) q^+ = 0 , \tag{1.8}$$

which allows us to identify $\hat{Z}\hat{\bar{Z}}q^+ = m^2q^+$. In this section we prefer to attach the non-vanishing central charge to D^{++}, not V^{++}, since we want to introduce *different* masses for the hypermultiplets q^+_1 and q^+_2 in (1.6) via (1.7).

The $N = 2$ central charges in 4d HSS can be generated from a 6d HSS [282] by the use of the standard (Scherk-Schwarz) mechanism of dimensional reduction [591], where the derivatives with respect to extra space coordinates play the role of the central charge operators. The six-dimensional notation simplifies the equations with implicit (4d) central charges. For example, the bosonic kinetic terms of the NLSM (1.6) to be rewritten to 6d, after an elimination of the HSS auxiliary fields in components, are given by [286, 284]

$$S_{\text{bosonic}}[\phi^{ai}_1, \phi^{ai}_2, V_\mu] = \tfrac{1}{2} \int d^6x \left\{ (D^\mu \phi^{ia}_1)(D_\mu \phi_{ia1}) - (D^\mu \phi^{ia}_2)(D_\mu \phi_{ia2}) \right.$$
$$\left. + \tfrac{1}{2} D_{ij} \left(\phi^{ia}_1 \phi^j_{1a} - \phi^{ia}_2 \phi^j_{2a} + \xi^{ij} \right) \right\}, \tag{1.9}$$

where $\mu = 0, 1, 2, 3, 4, 5$, $D_\mu = \partial_\mu + iV_\mu$, and D_{ij} is the scalar triplet of the auxiliary field components of the $N = 2$ vector superfield V^{++} in the WZ-gauge.

We are now in a position to formulate our model by the following HSS action:

$$S_{\text{ALE}}[q_1, q_2, V] = \int d\zeta^{(-4)}du \left\{ - \overset{*}{\bar{q}}{}^+_1(\hat{D}^{++}_1 + iV^{++})q^+_1 \right.$$
$$\left. + \overset{*}{\bar{q}}{}^+_2(\hat{D}^{++}_2 + iV^{++})q^+_2 + V^{++}\xi^{++} \right\}, \tag{1.10}$$

where

$$\hat{Z}\hat{\bar{Z}}q^+_1 = m^2_1 q^+_1 , \quad \hat{Z}\hat{\bar{Z}}q^+_2 = m^2_2 q^+_2 . \tag{1.11}$$

It is worth mentioning that the mass parameters m_1^2 and m_2^2 introduced in (1.11) do not represent physical masses. As is clear from (1.9), the classical on-shell physical significance has only their difference,

$$m_2^2 - m_1^2 \equiv m^2 , \qquad (1.12)$$

which can be identified with the classical mass of the single physical hyper-multiplet in the NLSM under consideration, after taking into account the constraint imposed by the Lagrange multiplier D_{ij}. Moreover, because of the presence of the FI term (linear in D_{ij}) in the action (1.9), the auxiliary triplet D^{ij} of V^{++} may develop a non-trivial vacuum expectation value in quantum theory after taking into account quantum corrections due to quantized hy-permultiplets. It affects the physical mass values to be defined with respect to the 'true' vacuum. In Subsect. 9.1.2 we examine whether the auxiliary field components D_{ij} get non-trivial vacuum expectation values. The latter have to be constant in order to maintain 4d Lorentz invariance. A constant solution $\langle D_{ij} \rangle \neq 0$ is consistent with Abelian gauge invariance in components.

9.1.2 Quantized ALE $N = 2$ NLSM and Composite Particles

To quantize both hypermultiplets of the theory (1.6) in the manifestly $N = 2$ supersymmetric way, we use HSS quantum perturbation theory in terms of analytic superfields in four spacetime dimensions (Subsect. 8.3.1). For our purposes in this section, the HSS propagator (8.3.12) of a massive physical hypermultiplet should be generalized in the background of an FI term [282]. Equation (8.3.12) can be written down in the form

$$i \left\langle q^+(1) \overset{*}{\bar{q}}{}^+(2) \right\rangle_{\text{phys.}} = \frac{-1}{\Box + m^2 - i0}(D_1^+)^4(D_2^+)^4 \frac{e^{v_2-v_1}}{(u_1^+ u_2^+)^3} \delta^{12}(\mathcal{Z}_1 - \mathcal{Z}_2) , \qquad (1.13)$$

where the 'bridge' v is defined by the relation

$$\mathcal{D} = e^{-v} D e^v \qquad (1.14)$$

between the manifestly analytic HSS derivatives, $\mathcal{D}$, and the covariantly an-alytic ones, D. In the case of the central charge background (1.7) we easily find

$$v = i(\theta^+\theta^-)\hat{\bar{Z}} + i(\bar{\theta}^+\bar{\theta}^-)\hat{Z} . \qquad (1.15)$$

The Green function $G^{(1,1)}(1|2)_{\text{phys.}} \equiv i \left\langle q^+(1) \overset{*}{\bar{q}}{}^+(2) \right\rangle_{\text{phys.}}$ satisfies the equation

$$\hat{D}_1^{++} G_{\text{phys.}}^{(1,1)}(1|2) = \delta_{\text{A}}^{(3,1)}(1|2) , \qquad (1.16)$$

where the analytic HSS delta-function $\delta_{\text{A}}^{(3,1)}(1|2)$ has been introduced [221].

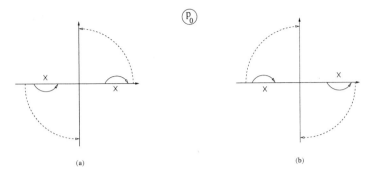

Fig. 9.1. The Wick rotations for the physical (a) and non-physical (b) fields

A causal (unitary) propagation is ensured in QFT by adding a small negative imaginary part to the mass squared, $m^2 \to m^2 - i\epsilon$, in the propagator
(1.13) [27]. The same prescription automatically takes care of (i) the convergence of the path integral defining the generating functional of quantum
Green's functions in Minkowski spacetime and (ii) free interchange of integrations. A propagator of the non-physical hypermultiplet entering the action
(1.6) with the wrong sign (and, hence, formally leading to negative norms of
the corresponding 'states') is also of the form (1.13) but with the negative
residue *and* the anticausal $i\epsilon$-prescription (Fig. 9.1),

$$i \left\langle q^+(1)\, \overset{*}{\bar{q}}{}^+(2) \right\rangle_{\text{non-phys.}} = \frac{1}{\Box + m^2 + i0}(D_1^+)^4(D_2^+)^4 \frac{e^{v_2 - v_1}}{(u_1^+ u_2^+)^3}\delta^{12}(\mathcal{Z}_1 - \mathcal{Z}_2) .$$

(1.17)

It can only occur as an internal line inside Feynman graphs.

Gauge couplings of physical hypermultiplets to $N = 2$ vector superfields
also differ by a minus sign from those of non-physical hypermultiplets. Hence,
a Feynman graph for non-physical fields has extra minus signs for every internal line and every vertex, when being compared to the same graph for the
physical fields. Though all these signs mutually cancel in loop diagrams with
the same number of vertices and internal lines, the difference in $i\epsilon$ prescription remains. It forces the non-physical poles in the complex p_0-plane to be
on the other side of the real axis [589]. This amounts to the appearance of a
relative minus sign for every non-physical loop compared to the same physical loop, because of the opposite Wick rotation in momentum space for the
non-physical fields (Fig. 9.1). In this respect, the quantized non-physical (or
of negative-norm) fields behave like fermions or Pauli-Villars regulators, so
that one may already expect UV-divergence cancellations in Feynman graphs
between physical and non-physical loops. This indeed happens to be the case
(see below).

We are now in a position to discuss the $N = 2$ gauge LEEA defined by
Gaussian integration over both hypermultiplets in (1.6), after expanding the

result in powers of external momenta. The quantum effective action $\Gamma(V^{++})$ is formally defined in HSS by the one-loop formula

$$\Gamma(V^{++}) = i\text{Tr}\ln\mathcal{D}^{++}_{\text{phys.}} - i\text{Tr}\ln\mathcal{D}^{++}_{\text{non-phys.}} , \qquad (1.18a)$$

or, in terms of the Green functions (1.16), as

$$\Gamma(V^{++}) = i\text{Tr}\ln\frac{\delta^{(3,1)}_A + iV^{++}G^{(1,1)}_{\text{phys.}}}{\delta^{(3,1)}_A + iV^{++}G^{(1,1)}_{\text{non-phys.}}} . \qquad (1.18b)$$

The supergraph calculation of the gauge LEEA $\Gamma[V]$ obtained by integrating over a single physical hypermultiplet in HSS was already described in Subsect. 8.2.3,

$$\Gamma[V^{++}] = \left[\int d^4x d^4\theta\, \mathcal{F}(W) + \text{h.c.}\right] + \int d^4x d^4\theta d^4\bar\theta\, \mathcal{H}(W,\bar W) , \qquad (1.19)$$

where the leading holomorphic term is given by the (perturbative) Seiberg-Witten LEEA [16, 17] or the integrated $N = 2$ supersymmetric (chiral) $U(1)_\text{R}$ anomaly [592], whereas the second term is the non-holomorphic (perturbative) next-to-leading order correction [513], with the real function $\mathcal{H}(W,\bar W)$ being subject to the Kähler gauge transformations (8.2.19). According to Sect. 8.2, we have

$$\mathcal{F}(W)_{\text{phys.}} = -\frac{1}{(8\pi)^2}W^2\ln\frac{W^2}{m^2} , \qquad (1.20)$$

and

$$\mathcal{H}(W,\bar W)_{\text{phys.}} = \frac{1}{(16\pi)^2}\left(\ln\frac{W}{\Lambda}\right)\left(\ln\frac{\bar W}{\Lambda}\right) , \qquad (1.21)$$

where Λ is an irrelevant parameter [514]. Equations (1.20) and (1.21) in our case (1.18) imply

$$\Gamma[V^{++}]_{\text{LEEA}} = -\frac{1}{32\pi^2}\ln\left(\frac{m_2^2}{m_1^2}\right)\int d^4x d^4\theta\, W^2 \equiv -\frac{1}{2e_0^2}\int d^4x d^4\theta\, W^2 \qquad (1.22)$$

that is just the free action of the $N = 2$ vector gauge superfield! Equation (1.22) implies, in particular, the dynamical generation of a term quadratic in the auxiliary field D_{ij}, which accompanies the standard kinetic terms of the $N = 2$ vector multiplet in components. Together with the FI term in (1.6), this gives rise to a non-vanishing vacuum expectation value $\langle D_{ij}\rangle \neq 0$.

Unlike the $N = 2$ gauge LEEA for a single hypermultiplet (Sect. 8.3), our LEEA (1.18) is both infrared (IR) *and* ultraviolet (UV) finite. IR divergences are obviously absent due to non-vanishing hypermultiplet masses acting as IR regulators. As regards the UV divergences of the $N = 2$ gauge LEEA for a physical hypermultiplet, the leading 2-point contribution in Fig. 8.1

is the only divergent one (all the higher n-point contributions in Fig. 8.1 are automatically UV finite for dimensional reasons) [283]. The holomorphic 2-point contribution to the $N = 2$ gauge LEEA due to a single physical hypermultiplet reads [283]

$$\Gamma^{(2)}_{\text{phys.}}[V] = -\frac{i}{2}\frac{1}{(2\pi)^4}\int d^4p d^8\theta du\, V^{++}(p,\theta,u,)\Pi_{\text{phys.}}(-p^2)V^{--}(-p,\theta,u)\ ,$$
(1.23)

where a (dimensionally regularized) one-loop structure function $\Pi_{\text{phys.}}(-p^2)$ has been introduced (with the renormalization scale μ):

$$\Pi_{\text{phys.}}(-p^2) = \mu^{2\varepsilon}\int_E \frac{d^{4-2\varepsilon}l}{(2\pi)^{4-2\varepsilon}}\frac{1}{[l^2+m^2][(l-p)^2+m^2]}\ .$$
(1.24)

Equation (1.24) is logarithmically UV-divergent in four spacetime dimensions ($\varepsilon \to +0$). This UV divergence is simultaneously the origin of the renormalization scale dependence of the renormalized low-energy effective action (1.19) via its holomorphic (anomalous) conribution. In our case (1.18), the UV divergence of the self-energy integral (1.24) cancels against the opposite UV divergence of the similar contribution to the LEEA due to the non-physical hypermultiplet,

$$\Pi(-p^2) \equiv \Pi_{\text{phys.}}(-p^2) + \Pi_{\text{non-phys.}}(-p^2)$$

$$= \int_E \frac{d^4l}{(2\pi)^4}\left\{\frac{1}{[l^2+m_1^2][(l-p)^2+m_1^2]} - \frac{1}{[l^2+m_2^2][(l-p)^2+m_2^2]}\right\}$$

$$= \frac{1}{16\pi^2}\int_0^1 dx \ln\frac{m_2^2+p^2x(1-x)}{m_1^2+p^2x(1-x)}\ ,$$
(1.25)

where Feynman parameterization has been used to evaluate the momentum integral in the Euclidean domain (we assume that $m_2^2 > m_1^2$). Continuation to Minkowski space entails changing the sign of p^2 — this explains our notation, $\Pi(s)$ and $s = -p^2$, above. The function $\Pi(s)$ is analytic in the cut s plane whose analytic structure is best exhibited by dispersion relations [593].

One gets (1.22) from (1.23) and (1.25) in the low-energy limit $p^2 \to 0$. We took vanishing momenta since we are interested in finding a Poincaré- and gauge-invariant vacuum background solution. It can only be represented by the spacetime-independent $N = 2$ vector gauge superfield strength $\langle W\rangle$ having the form

$$\langle W\rangle = \langle a\rangle + \tfrac{1}{2}(\theta_i^\alpha\theta_{\alpha j})\langle D^{ij}\rangle\ ,$$
(1.26)

where merely constant vacuum expectation values of the bosonic scalar components of W have been kept. We can assume that $\langle a\rangle = 0$ without a loss of generality since: (i) there is no equation on $\langle a\rangle$ at all, and (ii) a constant $\langle a\rangle$ would simply amount to an *equal* shift of both hypermultiplet masses in the theory (1.6). Hence, we are left with the induced scalar potential

$$V(D) = \xi \cdot D - \frac{1}{2e_0^2} D^2 \ , \tag{1.27}$$

in components, which has the only vacuum solution

$$\langle D \rangle = e_0^2 \xi \neq 0 \ . \tag{1.28}$$

It also follows from (1.12), (1.22) and (1.28) that

$$m_1^2 = \frac{m^2}{e^{16\pi^2/e_0^2} - 1} \ , \quad m_2^2 = \frac{m^2}{1 - e^{-16\pi^2/e_0^2}} \ . \tag{1.29}$$

This simple exercise can also be repeated in HSS. Varying the $N = 2$ gauge effective action with respect to the Abelian $N = 2$ vector gauge superfield V^{++} in HSS yields the equation of motion (in vacuum)

$$\frac{1}{e_0^2}(D^+)^4 \langle A^{--} \rangle = \xi^{++} \ , \tag{1.30}$$

while the relation between the HSS potentials A^{--} and V^{++} is given by (4.4.67) and (4.4.70). A Poincaré-invariant solution to (1.30) reads

$$\langle V^{++} \rangle = (\theta^+)^2 (\bar{\theta}^+)^2 e_0^2 \xi^{--} \ , \tag{1.31}$$

while it is equivalent to (1.28) because of (4.4.75).

Any other non-trivial $N = 2$ gauge LEEA (1.19) having a form different from (1.22), i.e. with a non-quadratic holomorphic function $\mathcal{F}(W)$, does not admit a constant non-vanishing solution for D, because of the appearance of the extra equation $\partial^3 \mathcal{F}/\partial W^3 \big|_{W=a} D^2 = 0$ [594]. A non-vanishing value of $\langle D \rangle$ implies the appearance of Goldstone fermions that inhomogeneously transform under on-shell $N = 2$ supersymmetry (cf. Subsect. 4.2.2). In other words, $N = 2$ supersymmetry is spontaneously broken in our theory (1.6).

We conclude that quantum effects due to hypermultiplets lead to the appearance of a propagating (physical) Abelian $N = 2$ vector multiplet V^{++}. In the classical theory (1.6), V^{++} is merely present as a (non-propagating) Lagrange multiplier. Because of (1.23) and (1.25), the induced gauge coupling constant is momentum-dependent,

$$\frac{1}{e_{\text{ind.}}^2} = \frac{1}{16\pi^2} \int_0^1 dx \ \ln \frac{m_2^2 + p^2 x(1-x)}{m_1^2 + p^2 x(1-x)} = \frac{1}{e_0^2} + O(p^2/m^2) \ . \tag{1.32}$$

Notably, the UV finiteness enjoyed by our theory in four spacetime dimensions is also necessary for its consistency: if there were UV-divergent contributions to $e_{\text{ind.}}^2$, they would have to be removed by the corresponding counterterm proportional to the $N = 2$ gauge action. The latter must, however, be absent in the bare action (1.6) since, otherwise, it would contradict the classical nature of V^{++} as the Lagrange multiplier.

In order to calculate the full gauge LEEA, one has to repeat the calculation of the HSS graphs depicted in Fig. 8.1, in terms of the *new* hypermultiplet propagators defined with respect to the 'true' vacuum with the non-vanishing FI term (1.28). The Green function of a physical hypermultiplet in a generic $N = 2$ vector superfield background $\hat{V}^{++}$ satisfies the defining equation

$$\mathcal{D}_1^{++} G_{\text{phys.},\hat{V}}^{(1,1)}(1|2) = \delta_A^{(3,1)}(1|2) \,, \tag{1.33}$$

whose solution can be formally written down in the form

$$G_{\text{phys.},\hat{V}}^{(1,1)}(1|2) = \frac{-1}{\Box_{\text{cov.}} - i0}(D_1^+)^4(D_2^+)^4 e^{\mathcal{V}_2 - \mathcal{V}_1}\delta^{12}(\mathcal{Z}_1 - \mathcal{Z}_2)\frac{1}{(u_1^+ u_2^+)^3} \,, \tag{1.34}$$

where the covariantly constant 'bridge' $e^{-\mathcal{V}}$ and the covariant d'Alambertian $\Box_{\text{cov.}}$ in the analytic HSS have been introduced [283, 282]. The defining equation for the 'bridge' reads

$$\mathcal{D}^{++}e^{-\mathcal{V}} = (D^{++} + i\hat{V}^{++})e^{-\mathcal{V}} = 0 \,, \tag{1.35}$$

whereas the defining equation for the covariant d'Alambertian is given by

$$-\tfrac{1}{2}(D^+)^4(D^{--})^2\Phi^{(p)} = \Box_{\text{cov.}}\Phi^{(p)} \,, \tag{1.36}$$

where $\Phi^{(p)}$ is a HSS analytic superfield of (positive) $U(1)$ charge p. The definition (1.36) obviously implies that

$$[D_\alpha^+, \Box_{\text{cov.}}] = [\bar{D}_{\dot\alpha}^+, \Box_{\text{cov.}}] = 0 \,. \tag{1.37}$$

The explicit form of the operator $\Box_{\text{cov.}}$ in a generic background $\hat{V}^{++}$ was calculated in [283], in the covariantly analytic form

$$\Box_{\text{cov. analytic}} = \mathcal{D}^\mu\mathcal{D}_\mu + \tfrac{i}{2}(\mathcal{D}^{\alpha+}W)\mathcal{D}_\alpha^- + \tfrac{i}{2}(\bar{D}_{\dot\alpha}^+\bar{W})\bar{D}^{\dot\alpha-}$$
$$- \tfrac{i}{4}(\bar{D}_{\dot\alpha}^+\bar{D}^{\dot\alpha+}\bar{W})\mathcal{D}^{--} + \tfrac{i}{4}(\mathcal{D}^{\alpha-}\mathcal{D}_\alpha^+ W) + \bar{W}W \,, \tag{1.38}$$

which is related to $\Box_{\text{cov.}}$ via the 'bridge' transform,

$$\Box_{\text{cov.}} = e^{-\mathcal{V}}\Box_{\text{cov. analytic}}e^{\mathcal{V}} \,. \tag{1.39}$$

The particular form of the operator $\Box_{\text{cov.}}$ in the spacetime-constant gauge-invariant backround (1.26) reads

$$\Box_{\text{c.c.}} = \Box + \hat{Z}\hat{\bar{Z}} + \tfrac{1}{2}\xi^{+-} - \tfrac{i}{2}\xi^{--}\left[(\theta^+\theta^+)\hat{\bar{Z}} + (\bar{\theta}^+\bar{\theta}^+)\hat{Z} + (\theta^+)^2(\bar{\theta}^+)^2 e_0^2\xi^{--}\right]$$
$$+ \left[\tfrac{i}{2}\xi^{++}(\theta^-\partial\bar{\theta}^-) + \tfrac{i}{2}\xi^{--}(\theta^+\partial\bar{\theta}^+) - \xi^{+-}(\theta^+\partial\bar{\theta}^-) + \text{h.c.}\right]$$
$$+ \left[\tfrac{1}{2}\xi^{++}(\theta^-D^-) - \tfrac{1}{2}\xi^{+-}(\theta^+D^-) + \text{h.c.}\right] + \tfrac{1}{4}\xi^{++}D^{--} \,. \tag{1.40}$$

The 'bridge' itself is given by

$$\mathcal{V}_{\text{const}} = \tfrac{1}{2}\xi^{--}(\theta^{+}\theta^{-})(\bar{\theta}^{+})^2 + \tfrac{1}{4}\xi^{++}(\theta^{-})^2(\bar{\theta}^{+}\bar{\theta}^{-})$$
$$- \tfrac{1}{4}\xi^{+-}\left[2(\theta^{+}\theta^{-})(\bar{\theta}^{+}\bar{\theta}^{-}) + (\theta^{+})^2(\bar{\theta}^{-})^2\right] - \text{h.c.} \tag{1.41}$$

The non-physical hypermultiplet propagator is similar to (1.34) after obvious modifications.

The hypermultiplet propagator defined by (1.34), (1.40) and (1.41) is rather complicated. Nevertheless, the qualitative picture remains the same as above [579]: the kinetic term of the $N = 2$ vector gauge superfield is dynamically generated, with the induced (dimensionless and momentum-dependent) gauge coupling constant being given by

$$e^2_{\text{ind.}}(p) = e^2 + O(p^2/m^2) . \tag{1.42}$$

The momentum dependence of $e^2_{\text{ind.}}$ is calculable, while its low-energy value e^2 is non-vanishing, being a function of the dimensionless ratio m^2/ξ. Indeed, the modification (1.40) of the box operator in the low-energy limit essentially amounts to a shift of the hypermultiplet mass, which is clearly *the same* for both (i.e. physical and non-physical) hypermultiplet propagators.

The dynamical generation of the whole $N = 2$ vector multiplet implies, of course, the dynamical deneration of all of its physical components, i.e. a complex scalar which can be interpreted as a 'Higgs' particle, a chiral spinor doublet representing a complex 'photino', and a real 'photon'. In Subsect. 9.1.3 we interpret the composite $N = 2$ vector multiplet components as the zero modes of the superstring ending on a Dirichlet (D) 6-brane (Sect. 8.4).

9.1.3 Relation to M-Theory and Brane Technology

The exact solution to the LEEA of $N = 2$ super-QCD with N_c colours in spacetime $R^{1,3}$ can be identified with the LEEA of the effective $N = 2$ MQCD defined in the single M5-brane (Sect. 8.4). The M5-brane world-volume is given by the local product of $R^{1,3}$ with the hyperelliptic curve Σ_g of genus $g = N_c - 1$. The hyperelliptic curve Σ_g is holomorphically embedded into the hyper-Kähler four-dimensional multi-centre Taub-NUT space Q_{mTN} associated with a multiple KK monopole (Subsect. 8.4.2). The identification of the LEEA in the two apparently very different field theories (namely, the $N = 2$ super-QCD in the Coulomb branch, on the one hand, and the $N = 2$ MQCD defined in the M5-brane world-volume, on the other hand), is highly non-trivial, since the former is defined as the leading contribution to the *quantum* LEEA in the gauge field theory, whereas the latter is determined by the *classical* M5-brane dynamics or the 11d supergravity equations of motion whose extended BPS solutions preserving some part of 11d supersymmetry are the M-theory branes.

The multiple KK monopole is a non-singular (solitonic) BPS solution to the classical equations of motion of 11d supergravity, with 11d spacetime being the product of the seven-dimensional (flat) Minkowski spacetime $R^{1,6}$ and the four-dimensional Euclidean multi-centre Taub-NUT space Q_{mTN} — see (8.4.19). The eleventh coordinate in 11d is identified with the periodic coordinate of the multi-Taub-NUT space whose harmonic potential H is given by (8.4.20). The moduli (k, y_i) in (8.4.20) are interpreted as charges and locations of KK monopoles. For instance, the simple Taub-NUT space ($n = 1$) can be thought of as the non-trivial bundle (Hopf fibration) with base R^3 and fibre S^1 of magnetic charge k. In general ($n \geq 1$), there exist n linearly independent normalizable self-dual harmonic 2-forms ω_A in Q_{mTN}, which satisfy the orthogonality condition [595]

$$\frac{1}{(2\pi k)^2} \int_{Q_{\mathrm{mTN}}} \omega_A \wedge \omega_B = \delta_{AB} \ . \qquad (1.43)$$

Two adjacent KK monopoles are connected by a homology 2-sphere having poles at the positions of the monopoles. Near a singularity of H, the KK circle S^1 contracts to a point. A *holomorphic* embedding of the Seiberg-Witten spectral curve Σ_g into the hyper-Kähler manifold Q_{mTN} is a consequence of the BPS condition (Subsect. 8.4.5),

$$\mathrm{Area}_{\Sigma} = \left| \int_{\Sigma} \Omega_{\Sigma} \right| \ , \qquad (1.44)$$

where Ω_{Σ} is the pullback of the Kähler $(1,1)$ form Ω of Q_{mTN} on Σ_g. In fact, any four-dimensional hyper-Kähler manifold possesses the holomorphic $(2,0)$ form ω that is simply related to the Kähler form Ω [205],

$$\Omega^2 = \omega \wedge \bar{\omega} \ . \qquad (1.45)$$

The M-theory BPS states, whose zero modes appear in the effective field theory defined in the M5-brane world-volume, correspond to the supermembranes (or M2-branes) having minimal area (BPS!) and ending on the M5-brane. The *spatial* topology of such an M2-brane determines the type of the corresponding $N = 2$ supermultiplet in the effective (macroscopic) spacetime $R^{1,3}$: a cylinder (Y) leads to an $N = 2$ vector multiplet, whereas a disk (D) gives rise to a hypermultiplet [539, 596]. Since the pullback ω_Y on Y is closed, there exists a meromorphic differential λ_{SW} satisfying the relations $\omega_Y = d\lambda_{\mathrm{SW}}$ and

$$Z = \int_Y \omega_Y = \oint_{\partial Y} \lambda_{\mathrm{SW}} \ , \qquad (1.46)$$

where Z is the central charge and $\partial Y \in \Sigma$ [539, 596, 597]. Hence, λ_{SW} can be identified with the Seiberg-Witten differential that determines the $N = 2$ gauge LEEA in $R^{1,3}$ (Subsect. 8.2.2).

We are now in a position to discuss the *symmetry enhancement* in the case of two nearly coincident D6-branes. The non-singular interpretation of D6-branes in M-theory is based on the fact that the isolated singularities of the harmonic function (8.4.20) are merely the *coordinate* singularities of the 11-dimensional metric (8.4.19), though they are truly singular with respect to the dimensionally reduced 10-dimensional metric that is associated with the D6-branes in the type-IIA picture (Subsect. 8.4.2). The physical significance of the 10-dimensional metric singularities is understood due to the illegitimate neglect of the KK modes related to the compactification circle S^1, since the KK particles (also called D0-branes) become massless near the D6-brane core [556]. Their inclusion is equivalent to accounting for instanton corrections in the four-dimensional $N = 2$ supersymmetric gauge field theory.

When some parallel and similarly oriented D-branes coincide (this may happen in some special points of the moduli space of M-theory), it is accompanied by gauge symmetry enhancement [554, 583]. Since the brane singularities become non-isolated in the coincidence limit, first, they have to be resolved by considering the branes separated by some distance ξ, as in (1.1). Then one takes the limit of small ξ. In the case of two parallel D6-branes, the double-centred Taub-NUT metric with a constant potential λ at infinity is just described by (1.1). Equation (8.4.20) with the potential (1.1) yields two parallel and similarly oriented M-theory KK monopoles, with their centres on the line $\boldsymbol{\xi}$ in the sixth direction. This 11d KK monopole configuration dimensionally reduces to the double D6-brane configuration in 10d. The homology 2-sphere connecting two KK monopoles contracts to a point in the limit $\xi \to 0$, which gives rise to a curvature singularity of the dimensionally reduced metric in 10 dimensions. From the 11-dimensional perspective, M2-branes can wrap about the 2-sphere connecting the KK monopoles, while the energy of the wrapped M2-brane is proportional to the area of the sphere [583]. When the sphere collapses, its area vanishes and, hence, the zero modes of the wrapped M2-brane become massless. This gives rise to an extra massless vector supermultiplet in the LEEA and, hence, the gauge symmetry enhancement

$$U(1) \times U(1) \ \to \ U(2) \tag{1.47}$$

assiciated with the A_1-type singularity. The non-perturbative phenomenon of gauge symmetry enhancement was first observed in the K3-compactification of M-theory due to collapsing 2-cycles of K3 on the basis of duality with the heterotic string compactifications [598, 474]. It is worth mentioning here that the geometry near a collapsing 2-cycle of K3 is the same as the geometry near two almost coincident parallel KK monopoles in M-theory [599], i.e. the corresponding hyper-Kähler metric is governed by the harmonic function (1.1) in the limit $\xi \to 0$.

From the 10-dimensional viewpoint, the wrapped M2-branes are just the 6–6 superstrings stretched between two D6-branes, so that it is the zero modes

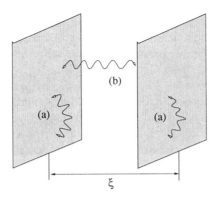

Fig. 9.2. The 6–6 superstrings ending on D-6-branes: with both of their ends on the same brane (a), and on the different branes (b)

of these 6-6 superstrings that become massless in the coincidence limit for D-6-branes (Fig. 9.2).

Each of the $U(1)$ factors on the left-hand side of (1.47) is associated with a single D6-brane, being related to the 6–6 superstring both ends of whose are on this brane (i.e. of type (a) in Fig. 9.2). The massless zero modes of this 6–6 superstring define a $U(1)$ gauge vector supermultiplet in the field theory LEEA describing the dynamics of small fluctuations about the D-6-brane. We can therefore identify this Abelian vector supermultiplet with the composite vector supermultiplet dynamically generated from the hypermultiplet LEEA in the D6-brane world-volume (Subsect. 9.1.2).

Unlike the 6–6 superstrings of type (a) in Fig. 9.2, the 6–6 superstrings of type (b), with their ends on the different D-6-branes, have a truly non-perturbative origin. Indeed, our QFT (1.6) becomes singular in the limit $\xi \to 0$. Accordingly, the *non-Abelian* gauge symmetry enhancement (1.47) is beyond the scope of the hypermultiplet LEEA approach described in Sect. 8.3, which is apparently limited to a single D-brane world-volume.

The Abelian gauge symmetry in the hypermultiplet LEEA is the gauged isometry in the corresponding NLSM target space, while the gauging itself can be made manifest in HSS (Subsect. 9.1.1). The $N = 2$ supersymmetric NLSM with the double Taub-NUT metric (two KK monopoles) is equivalent to the one with the mixed (Eguchi-Hanson-Taub-NUT) metric [287]. The corresponding HSS action is given by (1.3). It follows from (1.3) that the mixed hyper-Kähler NLSM metric interpolates between the Eguchi-Hanson metric ($\lambda = 0$) and the Taub-NUT metric ($\xi = 0$), both having the maximal isometry group $U(2)$. The action of the $U(2)$ isometry is linear in both limiting cases, while it is even holomorphic in the second case. Within the HSS approach, the internal symmetry enhancement $U(1) \to SU(2)$ can be understood either as the restoration of the $SU(2)_R$ automorphism symmetry of the $N = 2$ supersymmetry algebra in the Taub-NUT limit, or as the restoration of the $SU(2)_{\mathrm{PG}}$ symmetry in the Eguchi-Hanson limit [478]. Therefore, on the one hand, the geometry of two almost coinciding D6-branes near the origin $y = 0$ can be approximated by the Eguchi-Hanson metric in M-theory since a finite asymptotical potential λ in (1.1) becomes irrelevant near the singularity $y = 0$. The corresponding hypermultiplet LEEA (1.3) then reduces to our model (1.6) whose one-loop quantum fluctuations were investigated in Subsect. 9.1.2. On the other hand, a D6-brane possesses in its world-volume a massless Abelian vector supermultiplet that can be understood as the Nambu-Goldstone mode associated with the 11d symmetries broken by the D6-brane (BPS!) classical solution (Sect. 8.4). Therefore, the dynamical generation of an Abelian $N = 2$ vector multiplet in the quantized 4d field theory (1.6) is consistent with the effective classical dynamics of two nearly coincident D6-branes.

9.2 D-Brane Actions as NLSM

In this section we construct the supersymmetric generalizations of the Born-Infeld-Nambu-Goto (BING) actions, describing the gauge-fixed D3-branes in ambient four-dimensional (4d) or six-dimensional (6d) spacetime, both in components and in superspace [581]. The actions found can be interpreted as the Goldstone-Maxwell actions associated with partial (1/2) spontaneous breaking of extended supersymmetry. We show that those actions can be put into NLSM form by using certain non-linear superfield constraints [581].

9.2.1 D-Brane Actions in Components

First we briefly discuss some general features of the gauge-invariant and gauge-fixed D-brane actions in components [600, 601, 602, 603, 604]. This allows us to specify our motivation.

A good starting point is provided by the type-II Dp-brane embedded into flat (10d) spacetime. The gauge-invariant Dp-brane action is usually written down in terms of the world-volume fields $(X^m(\xi), \theta_{\alpha A}(\xi), A_\mu(\xi))$ depending upon world-volume coordinates ξ^μ, where $(X^m, \theta_{\alpha A})$ themselves can be considered as the coordinates of 10d, $N = 2$ superspace ($m = 0, 1, \ldots, 9$, $\alpha = 1, \ldots, 16$, $A = 1, 2$), and A_μ is an Abelian gauge field, $\mu = 0, 1, \ldots, p$. The gauge symmetries of the action comprise (i) world-volume diffeomorphisms, (ii) a fermionic κ-symmetry, and (iii) a $U(1)$ gauge invariance, whereas the global or rigid invariances are given by 10d, $N = 2$ super-Poincaré symmetry. The gauge-invariant Dp-brane action is the sum of the BING and WZ terms, [2]

$$S_p = -\int \mathrm{d}^{p+1}\xi \sqrt{-\det(\mathcal{G}_{\mu\nu} + \mathcal{F}_{\mu\nu})} + \int \Omega_{p+1} \,, \qquad (2.1)$$

[2] The D-brane torsion coefficient is chosen to be one.

where $\mathcal{G}_{\mu\nu}$ is the supersymmetric induced metric in the world-volume,

$$\mathcal{G}_{\mu\nu} = \eta_{mn}\Pi^m_\mu\Pi^n_\nu \ , \quad \Pi^m_\mu = \partial_\mu X^m - \bar{\theta}\Gamma^m\partial_\mu\theta \ , \tag{2.2}$$

$\mathcal{F}_{\mu\nu}$ is the supersymmetric Abelian field strength,

$$\mathcal{F}_{\mu\nu} = \left[\partial_\mu A_\nu - \bar{\theta}\hat{\Gamma}\Gamma_m\partial_\mu\theta\left(\partial_\nu X^m - \tfrac{1}{2}\bar{\theta}\Gamma^m\partial_\nu\theta\right)\right] - (\mu \leftrightarrow \nu) \ , \tag{2.3}$$

and

$$\hat{\Gamma} = \begin{cases} I \otimes \tau_3, & p \text{ odd}, \\ \Gamma_{11} \otimes I, & p \text{ even}, \end{cases} \tag{2.4}$$

with respect to the (αA) indices. The WZ term in (2.1) describes a coupling of the D-brane to the background Ramond-Ramond (RR) gauge fields [349], while its explicit form is fixed by the κ-symmetry of the whole action (2.1),

$$\delta_\kappa X^m = \bar{\theta}\Gamma^m\delta\theta \ , \quad \delta_\kappa\theta = \tfrac{1}{2}(1+\Gamma)\kappa \ , \tag{2.5}$$

where Γ is a (field-dependent) projector [600]. The world-volume diffeomorphisms ensure that only the $(9-p)$ coordinates $\{X^i\}$, $i = p+1,\ldots,9$, transverse to the D-brane world-volume are physical, whereas the κ-symmetry effectively eliminates half of the fermionic θ's in accordance with the BPS nature of the D-brane that breaks just half of spacetime supersymmetry. The rigid 10d, $N = 2$ supersymmetry transformations are

$$\delta_\varepsilon X^m = \bar{\varepsilon}\Gamma^m\theta \ , \quad \delta_\varepsilon\theta = \varepsilon \ . \tag{2.6}$$

All physical fields in the D-brane world-volume can be interpreted as the Goldstone fields associated with the symmetries broken by the D-brane [349, 605]. The spontaneously broken symmetries (including broken supersymmetry) are non-linearly realized in the gauge-fixed D-brane action that is obtained by fixing the local symmetries and removing unphysical degrees of freedom. A covariant physical gauge for the world-volume general coordinate transformations is given by the static gauge, in which the first $(p+1)$ spacetime coordinates are identified with the D-brane world-volume coordinates, i.e. $X^\mu = \xi^\mu$. The remaining scalars X^i representing transverse excitations of the D-brane can then be identified with the Goldstone bosons (collective modes) ϕ^i associated with spontaneously broken translations [349, 605]. The bosonic part of the induced metric in the static gauge reads

$$G_{\mu\nu} = \eta_{\mu\nu} + \partial_\mu\phi^i\partial_\nu\phi^i \ , \quad \text{where} \quad i = 1,\ldots,9-p. \tag{2.7}$$

A covariant gauge-fixing of the κ-symmetry is also possible (e.g. taking either $\theta_{\alpha 1} = 0$ and $\theta_{\alpha 2} = \psi$ in the type-IIB case, or just the opposite, $\theta_{\alpha 1} = \psi$ and $\theta_{\alpha 2} = 0$, in the type-IIA case), while the WZ term vanishes in this gauge [600]. The covariant gauge-fixed Dp-brane action can therefore be identified with a supersymmetric extension of the following BING (or Goldstone)-type action:

$$S_{\text{bosonic}} = - \int d^{p+1}\xi \sqrt{-\det\left(\eta_{\mu\nu} + F_{\mu\nu} + \partial_\mu\phi^i\partial_\nu\phi^i\right)} . \tag{2.8}$$

This action depends upon the Abelian gauge field A_μ only via its field strength $F_{\mu\nu}$ so that the $U(1)$ gauge invariance is kept.

The number $(p-1)+(9-p) = 8$ of the bosonic physical degrees of freedom in the action (2.8) matches with the number of fermionic degrees of freedom $16/2 = 8$ associated with the 10d Majorana-Weyl (MW) spinor ψ, and it does not depend upon p. It is not, therefore, surprising that supersymmetric extensions of all the gauge-fixed Dp-brane actions (2.8) can be deduced by dimensional reduction from a single master 10d action [600],

$$S_{\text{master}} = - \int d^{10}\xi \sqrt{-\det\left[\eta_{\mu\nu} + F_{\mu\nu} - 2\bar\psi\Gamma_\mu\partial_\nu\psi + (\bar\psi\Gamma^\rho\partial_\mu\psi)(\bar\psi\Gamma_\rho\partial_\nu\psi)\right]} ,$$

$$\tag{2.9}$$

associated with the top value $p = 9$ of the 10d 'spacetime-filling' D9-brane.

By construction [600], the component 10d Super-Born-Infeld (SBI) action (2.9) is invariant under two 10d MW supersymmetries (one unbroken susy and another spontaneously broken susy), with the 10d Maxwell supermultiplet (A_μ, ψ_α) being the Goldstone vector supermultiplet associated with the second non-linearly realized supersymmetry. In particular, the spinor superpartner ψ_α of the BI vector is a Goldstone fermion [606]. The unbroken supersymmetry of the action (2.9) is not the same as the original rigid supersymmetry (2.6) since it has to be supplemented by the compensating gauge transformation needed to preserve the gauge. Neither of the supersymmetries is manifest in the action (2.9).

Our goal is to rewrite some of the supersymmetric gauge-fixed Dp-brane actions in superspace in the NLSM-type form, in order to make their unbroken supersymmetries manifest. The superfield formulation is also useful in deciphering the unique non-trivial geometry underlying the complicated Goldstone actions associated with partial supersymmetry breaking in various spacetime dimensions [607, 608]. The superspace formulation becomes indispensable if one wants to address quantum properties of D-branes [609].

Supersymmetrization of the BI actions in various spacetime dimensions represents a challenge in supersymmetry since one has to deal with a non-polynomial field theory containing higher derivatives of all orders [610]. Causal propagation of the physical fields has to be maintained, while the auxiliary fields needed to close the off-shell supersymmetry algebra are to be kept non-propagating (this property is sometimes called 'auxiliary freedom' [611]). This is, nevertheless, possible in the superfield formulation of the SBI actions due to their special (Goldstone) nature [581].

Yet another important asset of the BI action in 4d [389] is its *electric-magnetic* (e.-m.) self-duality (see, e.g., [612, 613]). The self-duality and causal propagation [614] together are responsible for the characteristic ('square root of a determinant') non-polynomial structure of the 4d BI action [615]. It is worth mentioning here that the fundamental motivation in favour of the non-linear BI generalization of Maxwell electrodymanics is the well-known BI taming of the Coulomb self-energy, i.e. the existence of a non-singular charged soliton with finite self-energy [389, 614].

The leading term in the expansion of the BI action with respect to the gauge field strength is the Maxwell action. As is well known, even a covariant off-shell manifestly N-extended supersymmetrization of the 4d free(!) Maxwell theory is a difficult problem once the number (N) of supersymmetries exceeds two. An infinite number of auxiliary fields beyond the $N = 2$ (or eight supercharges) barrier is, in fact, required. In this section we restrict ourselves to the cases of $N = 1$ and $N = 2$ supersymmetry in 4d, where an off-shell formulation of the SBI theory is still possible in conventional superspace with a finite number of auxiliary fields. Similar reasoning also restricts the number of real Goldstone bosons ϕ^i in the 4d (gauge-fixed) super-BING action (2.8), if one wants to achieve its off-shell superspace reformulation by using a finite set of auxiliary fields. In 4d (i.e. for a D3-brane) with $N = 2$ unbroken supersymmetry we are led to restrict $i = 1, 2$, which implies a six-dimensional ambient spacetime for propagation of the D3-brane [616, 617]. It is worth mentioning that the initial motivation towards a supersymmetrization of the BI action came from the fact that it is the relevant part of the 10d open superstring effective action [618, 619]. We adopt the superspace approach for this purpose. The alternative coset construction, underlying non-linear realizations of internal and spacetime symmetries, leads in practice to highly involved calulations of Goldstone actions. Moreover, it appears to be incomplete in the case of non-linearly realized supersymmetry [13, 620, 621], since the irreducibility constraints on the Goldstone superfields are not apparent in the coset construction [607, 608].

In Subsect. 9.2.2 we review the 4d bosonic BI action in 4d. Its $N = 1$ supersymmetric generalization is constructed in Subsect. 9.2.3, where it is recognized as the 4d Goldstone-Maxwell action related to $N = 2$ supersymmetry spontaneously broken to $N = 1$, with a massless $N = 1$ vector superfield being the Goldstone-Maxwell superfield [622, 623]. The 4d Goldstone-Maxwell action, related to $N = 4$ supersymmetry spontaneously broken to $N = 2$, and a massless $N = 2$ vector superfield as the Goldstone superfield, is constructed in Subsect. 9.2.4 [624]. Both SBI actions (after non-linear field redefinitions) can be interpreted as the gauge-fixed actions of a D3-brane that either 'fills in' 4d spacetime or propagates in six-dimensional spacetime, respectively. The manifestly 6d Lorentz invariant and (1,0) supersymmetric Goldstone-Maxwell action, related to partial breaking of (2,0) supersymmetry down to (1,0) supersymmetry in 6d, with a massless (1,0) vector superfield being the Goldstone superfield, was constructed in [581].

9.2.2 Born-Infeld Action

The BI action in flat four-dimensional (4d) spacetime with Minkowski metric $\eta_{\mu\nu} = \mathrm{diag}(+, -, -, -)$,

$$S_{\mathrm{BI}} = -\frac{1}{b^2} \int \mathrm{d}^4 x \sqrt{-\det(\eta_{\mu\nu} + bF_{\mu\nu})} \, , \tag{2.10}$$

was introduced [389] as a non-linear generalization of Maxwell electrody-namics. This action also naturally arises (i) as the bosonic part of the 4d low-energy effective action of open superstrings (together with other mass-less superstring modes), and (ii) as the bosonic 4d spacetime-filling D3-brane action as well (Subsect. 9.2.1). In string/brane theory $b = 2\pi\alpha'$, whereas we choose $b = 1$ for notational simplicity.

The BI action (2.10) is manifestly Lorentz-invariant, it depends upon the gauge field A_μ only via its field strength $F_{\mu\nu} = \partial_\mu A_\nu - \partial_\nu A_\mu$, it contains no spacetime derivatives of F, and, after being expanded in powers of F, it gives the Maxwell action as the leading contribution. In fact, the BI action shares with the Maxwell action some other physically important properties, such as

- causal propagation (no ghosts),
- positive energy density,
- electric-magnetic self-duality,

which are non-trivial in the BI case [614, 615]. Unlike the Maxwell action, the BI action provides a natural taming of the Coulomb self-energy, which is yet another argument in favour of quantum consistency of superstring theory!

Taking advantage of the Lorentz invariance of the BI action, it is always possible to simplify the calculation of its expansion in powers of the gauge field strength by putting $F_{\mu\nu}$ into the particular form

$$F_{\mu\nu} = \begin{pmatrix} 0 & \lambda_1 & 0 & 0 \\ -\lambda_1 & 0 & 0 & 0 \\ 0 & 0 & 0 & \lambda_2 \\ 0 & 0 & -\lambda_2 & 0 \end{pmatrix} \tag{2.11}$$

in terms of real 'eigenvalues' (λ_1, λ_2). Equation (2.11) is just a manifestation of the fact that the Lorentz group $SO(1,3)$ has two independent Casimir operators. In other words, it suffices to pick up two independent Lorentz-invariant F-products in order to parametrize any Lorentz-invariant function of $F_{\mu\nu}$. For example,

$$\det(\eta_{\mu\nu} + F_{\mu\nu}) = -1 - \tfrac{1}{2}F^2 + \det(F_{\mu\nu}) = -1 - \tfrac{1}{2}F^2 + \tfrac{1}{4}\left[F^4 - \tfrac{1}{2}(F^2)^2\right] \, , \tag{2.12}$$

where we have introduced two real independent Lorentz invariants,

$$F^2 \equiv F^{\mu\nu}F_{\mu\nu} \quad \text{and} \quad F^4 \equiv F_{\mu\nu}F^{\nu\lambda}F_{\lambda\rho}F^{\rho\mu} \, . \tag{2.13}$$

The choice (2.13) is, of course, not unique, and it is not really the most convenient one in 4d supersymmetry. Let us introduce the 4d dual of $F_{\mu\nu}$,

$$\tilde{F}^{\mu\nu} = \tfrac{1}{2}\varepsilon^{\mu\nu\lambda\rho}F_{\lambda\rho} \, , \tag{2.14}$$

and form (anti)self-dual linear combinations,

$$F^{\pm}{}_{\mu\nu} = \tfrac{1}{2}\left(F \pm i\tilde{F}\right)_{\mu\nu} ,\tag{2.15}$$

that satisfy the identities

$$(F^{\pm})^2 = \tfrac{1}{2}(F^2 \pm iF\tilde{F}) , \quad (F^2)^2 + (F\tilde{F})^2 = 4(F^+)^2(F^-)^2 .\tag{2.16}$$

Note that

$$\det(F_{\mu\nu}) = \tfrac{1}{16}(F\tilde{F})^2 .\tag{2.17}$$

Using yet another identity

$$F^4 = \tfrac{1}{2}(F^2)^2 + \tfrac{1}{4}(F\tilde{F})^2 ,\tag{2.18}$$

it is possible to simplify (2.12) as

$$-\det(\eta_{\mu\nu} + F_{\mu\nu}) = 1 + \tfrac{1}{2}F^2 - \tfrac{1}{16}(F\tilde{F})^2 .\tag{2.19}$$

It follows that

$$\begin{aligned}
L_{\mathrm{BI}} \equiv 1 - \sqrt{-\det(\eta_{\mu\nu} + F_{\mu\nu})} &= -\tfrac{1}{4}F^2 - \tfrac{1}{8}\left[\tfrac{1}{4}(F^2)^2 - F^4\right] + O(F^6)\\
&= -\tfrac{1}{4}F^2 + \tfrac{1}{32}\left[(F^2)^2 + (F\tilde{F})^2\right] + O(F^6)\\
&= -\tfrac{1}{4}F^2 + \tfrac{1}{8}(F^+)^2(F^-)^2 + O(F^6) .
\end{aligned}\tag{2.20}$$

By a complex 'rotation' of the Lie algebra of $SO(1,3)$ to that of $SL(2,\mathbf{C})$, it is sometimes useful (in supersymmetry) to replace $F^+_{\mu\nu}$ by the 2×2 matrix

$$\hat{F}_\alpha{}^\beta = (\sigma^{\mu\nu})_\alpha{}^\beta F_{\mu\nu} , \quad \alpha, \beta = 1, 2 ,\tag{2.21}$$

where we have introduced the two-component spinor notation,

$$(\sigma^{\mu\nu}) = \tfrac{1}{4}\left(\sigma^\mu\tilde{\sigma}^\nu - \sigma^\nu\tilde{\sigma}^\mu\right) , \quad \sigma^\mu = (1, \boldsymbol{\sigma}), \quad \tilde{\sigma}^\mu = (1, -\boldsymbol{\sigma}) ,\tag{2.22}$$

in terms of Pauli matrices $\boldsymbol{\sigma}$. We find in addition that

$$\tfrac{1}{4}\left|\det\hat{F}\right|^2 = 4(F^+)^2(F^-)^2 = (F^2)^2 + (F\tilde{F})^2 ,\tag{2.23}$$

where we have introduced the chiral (2×2) determinant, $\det\hat{F}$, on the left-hand side. The right-hand side of the identity (2.23) is called the Euler-Heisenberg (EuH) Lagrangian. It also arises as the bosonic part of the one-loop effective action in $N = 1$ supersymmetric scalar electrodynamics (= the supersymmetric quantum field theory of a massive $N = 1$ scalar multiplet minimally coupled to an $N = 1$ vector multiplet in 4d) with the parameter $b^2 = e^4/(24\pi^2 m^4)$.

The single complex Lorentz invariant

$$\tfrac{1}{16}\mathrm{tr}(\hat{F}^2) = -\tfrac{1}{4}F^2 - \tfrac{i}{4}F\tilde{F} \equiv A + iB \tag{2.24}$$

is another natural variable for an expansion of the BI action in terms of the field strength F (it will be used in Subsect. 9.2.3). Yet another natural choice of variables is given by the Maxwell-Lagrangian and the associated energy-momentum tensor squared (= the EuH Lagrangian!),

$$-\tfrac{1}{4}F^2 = A \quad \text{and} \quad \tfrac{1}{32}\left[(F^2)^2 + (F\tilde{F})^2\right] \equiv E . \tag{2.25}$$

The Lorentz invariants (2.25) have natural supersymmetric extensions (Subsects. 9.2.3 and 9.2.4), with the first one having the form of a chiral superspace integral and the second one being a full superspace integral. This justifies our choice (2.25). We find

$$-\det(\eta_{\mu\nu} + F_{\mu\nu}) = 1 - 2A - B^2 = (1 - A)^2 - 2E . \tag{2.26}$$

This allows us to rewrite the BI Lagrangian in the form

$$L_{\mathrm{BI}}(F) = A + E + \ldots = A + EY(A, E) , \tag{2.27}$$

where the function $Y(A, E)$ has been introduced. It is not difficult to check that $Y(A, E)$ is just a solution to the quadratic equation

$$Ey^2 + 2(A - 1)y + 2 = 0 . \tag{2.28}$$

Similarly, it is straightforward to calculate $L_{\mathrm{BI}}(F)$ as a function of A and B (see Subsect. 9.2.3), e.g., by using the identity $A^2 + B^2 = 2E$, (2.27) and (2.28).

The Lagrangian 'magnetically dual' to the BI one is obtained via a first-order action

$$L_1 = L_{\mathrm{BI}}(F) + \tfrac{1}{2}\tilde{A}_\mu \varepsilon^{\mu\nu\lambda\rho} \partial_\nu F_{\lambda\rho} , \tag{2.29}$$

where $\tilde{A}_\mu$ is the (dual) magnetic vector potential. $\tilde{A}_\mu$ enters (2.29) as the Lagrange multiplier enforcing the Bianchi identity $\varepsilon^{\mu\nu\lambda\rho}\partial_\nu F_{\lambda\rho} = 0$. Varying (2.29) with respect to $F_{\mu\nu}$ instead, solving the arising algebraic equation on $F_{\mu\nu}$ as a function of the magnetically dual gauge field strength $^*F_{\mu\nu} = \partial_\mu \tilde{A}_\nu - \partial_\nu \tilde{A}_\mu$ (use the representation (2.11) for $F_{\mu\nu}$ and similarly for $^*F_{\mu\nu}$!), and substituting the solution back into (2.29) yields the magnetically dual action in terms of $^*F_{\mu\nu}$, which has *the same* form as the original BI action (2.10) in terms of $F_{\mu\nu}$. This is known as electric-magnetic (e.-m.) self-duality [612, 615], while it is connected to the classical $SL(2, \mathbf{R})$ symmetry of IIB superstrings [625]. The non-Gaussian BI Lagrangian (2.10) is *uniquely* fixed by the requirements of causal propagation and classical e.-m. self-duality if, in addition, one insists on the Maxwell low-energy limit, i.e. the (strong) correspondence principle. In general, there exists a family of e.-m. self-dual Lagrangians parametrized by one variable, with all of them being solutions to a first-order Hamilton-Jacobi partial differential equation [615].

9.2.3 $N = 1$ Supersymmetric Born-Infeld Action

The manifestly $N = 1$ supersymmetric 4d Born-Infeld (or Goldstone-Maxwell) action associated with partial spontaneous breaking of rigid $N = 2$ supersymmetry in terms of the Goldstone-Maxwell $N = 1$ supermultiplet (A_μ, ψ_α, D) was constructed in superspace in [626, 622, 623]. Amongst the superpartners of the Maxwell gauge field A_μ are the Goldstone (Majorana) fermion ψ_α and the real auxiliary scalar D. In this subsection we follow [622].

The standard 4d, $N = 1$ superspace is parametrized by the coordinates $Z^M = (x^\mu, \theta_\alpha, \bar\theta_{\dot\alpha})$, where θ_α and $\bar\theta_{\dot\alpha}$ are (Majorana) spinor anticommuting coordinates in the 2-component notation, $(\theta_\alpha)^* = \bar\theta_{\dot\alpha}$ and $\alpha = 1, 2$. An Abelian vector $N = 1$ supermultiplet is described in $N = 1$ superspace by the irreducible chiral spinor superfield W_α satisfying the off-shell constraints [141]

$$\bar{D}_{\dot\alpha} W_\alpha = 0 , \quad D^\alpha W_\alpha = \bar{D}_{\dot\alpha} \bar{W}^{\dot\alpha} . \tag{2.30}$$

As a result of the constraints, the bosonic components of the $N = 1$ superfield strength W_α can be introduced as follows [141]:

$$D^\alpha W_\beta| = (\sigma^{\mu\nu})_\beta{}^\alpha F_{\mu\nu} + i\delta_\beta^\alpha D , \tag{2.31}$$

where $F_{\mu\nu}$ is the Maxwell field strength of the gauge field A_μ.

The superfield constraints (2.30) can be solved in terms of a real gauge superfield pre-potential $V(x, \theta, \bar\theta)$ as [141]

$$W_\alpha = \bar{D}^2 D_\alpha V , \tag{2.32}$$

subject to Abelian gauge transformations $\delta V = i(\Lambda - \bar\Lambda)$ where Λ is a chiral superfield gauge parameter, $\bar{D}_{\dot\alpha}\Lambda = 0$. This gives the necessary input for a superfield quantization in terms of the unconstrained superfield V.

The 4d, $N = 2$ supersymmetry algebra can be decomposed with respect to unbroken $N = 1$ supersymmetry as [622]

$$\{Q_\alpha, \bar{Q}_{\dot\alpha}\} = 2\sigma^\mu_{\alpha\dot\alpha} P_\mu , \qquad \{S_\alpha, \bar{S}_{\dot\alpha}\} = 2\sigma^\mu_{\alpha\dot\alpha} P_\mu ,$$
$$\{Q_\alpha, S_\beta\} = 0 , \qquad \{Q_\alpha, \bar{S}_{\dot\alpha}\} = 0 , \tag{2.33}$$

where the Q's stand for the unbroken ($N = 1$) supersymmetry generators, S's stand for the broken ($N = 1$) supersymmetry generators, while P_μ are 4d translation generators. It is worth mentioning that a non-vanishing central charge does not appear in the $N = 2$ algebra (2.33). The vanishing central charge is, in fact, required for the consistency of the Goldstone-Maxwell action. In particular, the BPS nature of this action also implies the vanishing vacuum expectation values for composites of the physical fields, while the auxiliary field D should vanish on-shell too. In general, vanishing vacuum expectations for the physical (Goldstone) composites protect the auxiliary fields from becoming propagating due to interacting terms in all supersymmetric BI actions provided that pure kinetic terms for the auxiliary fields do not appear. The latter is the case for the manifestly supersymmetric (superfield) BING actions.

A generalization of the $N = 1$ supersymmetric constraints (2.30), which would be invariant under the second (S) non-linearly realized supersymmetry, is possible, in principle, by using the standard perturbative approach of non-linear realizations [13], though the full answer in closed form is still unknown in this case [622]. It is, nevertheless, possible to determine the full and manifestly $N = 1$ supersymmetric Goldstone-Maxwell action by a direct $N = 1$ supersymmetrization of the BI action. The result is given by the SBI action [626, 622, 623]

$$
\begin{aligned}
S_{\text{N=1 GM}} &= \left[\tfrac{1}{4} \int \mathrm{d}^4 x \mathrm{d}^2 \theta \, W^2 + \text{h.c.} \right] + \tfrac{1}{8} \int \mathrm{d}^4 x \mathrm{d}^2 \theta \mathrm{d}^2 \bar{\theta} \, f(A, B) W^2 \bar{W}^2 \\
&= \tfrac{1}{4} \int \mathrm{d}^4 x \mathrm{d}^2 \theta \left\{ W^2 + \tfrac{1}{4} \bar{D}^2 \left[f(A, B) W^2 \bar{W}^2 \right] \right\} + \text{h.c.} \\
&\equiv \tfrac{1}{4} \int \mathrm{d}^4 x \mathrm{d}^2 \theta \, W^2_{\text{improved}} + \text{h.c.} ,
\end{aligned}
$$
(2.34)

where the structure function $f(A, B)$ is given by

$$
f(A, B) = \frac{1}{1 - A + \sqrt{1 - 2A - B^2}} , \tag{2.35}
$$

whereas A and B stand for the $N = 1$ superfields

$$
\begin{aligned}
A &= \tfrac{1}{4} D^2 W^2 + \text{h.c.} , \\
iB &= \tfrac{1}{4} D^2 W^2 - \text{h.c.} ,
\end{aligned}
\tag{2.36}
$$

respectively, whose leading (F-dependent) components (at $\theta_\alpha = \bar{\theta}_{\dot{\alpha}} = 0$) are just given by A and B of (2.24). [3]

The Goldstone-Maxwell action (2.34) is the sum of the chiral $N = 1$ superspace integral (= super-Maxwell or super-A invariant) and the full $N = 1$ superspace integral (= super-Euler-Heisenberg or super-E invariant), with the latter being modified by the 'form factor' $f(D^2 W^2, \bar{D}^2 \bar{W}^2)$. The only quartic (higher derivative) combination, $\tfrac{1}{4}(F^2)^2 - F^4$, that can be supersymmetrized up to the full (EuH) $N = 1$ superinvariant, was also identified in [627] by using helicity conservation of four-particle scattering amplitudes in $N = 1$ supersymmetric scalar QED.

In terms of our 'smart' variables (2.25) the bosonic BI Lagrangian in the form (2.27) can be immediately supersymmetrized to the form (2.34).

---

[3] It is customary (in supersymmetry) to denote both a superfield and its first component by the same letter.

The $N = 1$ supersymmetric Goldstone-Maxwell action is therefore given by the $N = 1$ supersymmetric Born-Infeld action. The *same* SBI action emerges from the $N = 2$ supersymmetric non-linear Antoniadis-Partouche-Taylor (APT) model [628] where $N = 2$ supersymmetry is partially broken to $N = 1$ supersymmetry due to the non-linearity of the Seiberg-Witten type action for an $N = 2$ (Abelian) vector supermultiplet in the presence of 'electric' *and* 'magnetic' Fayet-Iliopoulos (FI) terms, after 'integrating out' (or decoupling) the massive $N = 1$ scalar superfield component of the $N = 2$ vector superfield [623].

It is worth mentioning that positivity of the 'discriminant' (under the square root in the denominator of (2.35)) is ensured by positivity of the BI determinant on the left-hand side of (2.26). A causal (no ghosts) propagation of the physical fields in the SBI theory is achieved due to the Goldstone nature of the whole $N = 1$ vector multiplet and its irreducibility with respect to unbroken supersymmetry. The auxiliary field D does not propagate, with $D = 0$ being an on-shell solution to its equation of motion.

The whole non-linear structure of the $N = 1$ SBI a ction (2.34) is dictated by the hidden non-linearly realized S-supersymmetry [622]. It is, therefore, not very surprising that the same action (2.34) can be nicely represented as the 'non-linear sigma-model' [622]

$$S_{N=1\ \mathrm{GM}} = \tfrac{1}{4} \int \mathrm{d}^4 x \mathrm{d}^2 \theta\, X + \mathrm{h.c.}\ , \qquad (2.37)$$

where the chiral $N = 1$ superspace Lagrangian X obeys the non-linear $N = 1$ superfield constraint [622]

$$X = \tfrac{1}{4} X \bar{D}^2 \bar{X} + W^2\ . \qquad (2.38)$$

The uniqueness of the $N = 1$ Goldstone-Maxwell action (2.37) now becomes apparent because of the identity $X^2 = 0$. The $N = 1$ chiral superfield X can be identified with the chiral $N = 1$ superfield constituent of an $N = 2$ vector superfield W after its decomposition, $W = \{X, W_\alpha\}$, in $N = 1$ superspace [623].

The e.-m. self-duality of the BI action is also naturally generalized to the $N = 1$ supersymmetric e.-m. self-duality of the $N = 1$ SBI action, when using the $N = 1$ supersymmetric analogue

$$S_{N=1} = S_{N=1\ \mathrm{GM}} + \left[\tfrac{i}{2} \int \mathrm{d}^4 x \mathrm{d}^2 \theta\, \tilde{W}^\alpha W_\alpha + \mathrm{h.c.} \right] \qquad (2.39)$$

of the bosonic first-order action (2.29). In (2.39) the $N = 1$ chiral Lagrange multiplier superfield $\tilde{W}^\alpha$ has been introduced to enforce the $N = 1$ Bianchi identity given by the second equation (2.30) on W_α that is merely an $N = 1$ chiral superfield in (2.39). Hence, on the one hand, varying the action (2.39) with respect to $\tilde{W}^\alpha$ gives us back the action (2.34), whereas, on the other hand, varying the action (2.39) with respect to W_α instead, solving the arising equation on W_α in terms of $\tilde{W}^\alpha$, and substituting the result back into the action (2.39), yield *the same* SBI action (2.34) in terms of $\tilde{W}^\alpha$. This is the $N = 1$ e.-m. self-duality in terms of $N = 1$ superfiels [622, 623, 613].

Other known Goldstone actions, associated with partial supersymmetry breaking $N = 2 \to N = 1$ in 4d, are related to different massless Goldstone (chiral or tensor) $N = 1$ supermultiplets [629].

9.2.4 $N = 2$ Supersymmetric Born-Infeld Action

A manifestly $N = 2$ supersymmetric and e.-m. self-dual extension of the 4d BING action (2.8) with two real scalars can be constructed as the $N = 2$ superspace extension of the bosonic BI action (2.10) [624, 613]. Two massless Goldstone bosons and Maxwell vector can be unified into a single massless $N = 2$ vector supermultiplet. The $N = 2$ SBI action [624] may be considered either as the Goldstone action related to partial breaking of $N = 4$ supersymmetry down to $N = 2$ in 4d, with the Goldstone-Maxwell $N = 2$ multiplet with respect to unbroken $N = 2$ supersymmetry, or, equivalently, as the gauge-fixed $N = 2$ superfield action of a D3-brane in flat six-dimensional ambient spacetime (Subsect. 9.2.1), up to a field redefinition. In the standard $N = 2$ superspace (Subsect. 4.2.1) the Goldstone-Maxwell $N = 2$ multiplet is described by the restricted chiral $N = 2$ superfield W subject to the off-shell constraints (4.2.26) and (4.2.27).

The $N = 2$ supersymmetric extension of the Maxwell Lagrangian $A = -\frac{1}{4}F_{\mu\nu}^2$ is given by

$$\frac{1}{2}\int d^4\theta\, W^2 = -a\Box\bar{a} - \frac{i}{2}\psi_j^\alpha \partial_{\alpha\dot{\alpha}}\bar{\psi}^{\dot{\alpha}j} - \frac{1}{2}(F^+)^2 + \frac{1}{2}\boldsymbol{D}^2 \ . \tag{2.40}$$

The Maxwell energy-momentum tensor squared or, equivalently, the EuH Lagrangian $E = \frac{1}{8}(F^+)^2(F^-)^2$, is also easily extended in $N = 2$ superspace,

$$\int d^4\theta d^4\bar{\theta}\, W^2\bar{W}^2 = (F^+)^2(F^-)^2 + (\boldsymbol{D}^2)^2 - \boldsymbol{D}^2 F^2 + \dots \ . \tag{2.41}$$

Equation (2.41) also arises as the leading (one-loop) non-holomorphic (non-BPS) contribution to the $N = 2$ gauge LEEA in the interacting $N = 2$ supersymmetric QFT of a charged hypermultipet minimally coupled to an $N = 2$ Maxwell supermultiplet (Sect. 8.2).

The gauge-invariant $N = 2$ superfield strength squared, W^2, is an $N = 2$ chiral but not a restricted $N = 2$ chiral superfield. As is clear from (2.40), the first component of the $N = 2$ anti-chiral superfield $K \equiv D^4 W^2$ takes the form

$$K \equiv D^4 W^2 = -2a\Box\bar{a} - (F^+)^2 + \boldsymbol{D}^2 + \dots \ . \tag{2.42}$$

It is now straightforward to $N = 2$ supersymmetrize the BI Lagrangian (2.20) by engineering the proper $N = 2$ superspace invariant,

$$L = \tfrac{1}{2} \int \mathrm{d}^4\theta \, W^2 + \tfrac{1}{8} \int \mathrm{d}^4\theta \mathrm{d}^4\bar{\theta} \, \mathcal{Y}(K,\bar{K}) W^2 \bar{W}^2 \,, \qquad (2.43)$$

whose 'form factor' $\mathcal{Y}(K,\bar{K})$ is dictated by the known bosonic structure function $Y(A,E)$ in (2.27). Note that the vector-dependent contributions to the first scalar components of the $N = 2$ superfields K and $\bar{K}$ are simply related to A und E as

$$K + \bar{K} = 4A \,, \qquad K\bar{K} = 8E \,, \qquad (2.44)$$

i.e. they are just the roots of the quadratic equation

$$k^2 - 4Ak + 8E = 0 \,. \qquad (2.45)$$

We find

$$\mathcal{Y}(K,\bar{K}) = \frac{1 - \tfrac{1}{4}(K + \bar{K}) - \sqrt{(1 - \tfrac{1}{4}K - \tfrac{1}{4}\bar{K})^2 - \tfrac{1}{4}K\bar{K}}}{K\bar{K}} \qquad (2.46)$$

$$= 1 + \tfrac{1}{4}(K + \bar{K}) + O(K^2) \,.$$

Since the BI action itself is defined modulo spacetime derivatives of the Maxwell field-strength, its $N = 2$ supersymmetrization is also not unique [613]. The proposed $N = 2$ SBI action [624]

$$S[W,\bar{W}]_{\mathrm{restr.}} = \tfrac{1}{2} \int \mathrm{d}^4 x \mathrm{d}^4\theta \, W^2 + \tfrac{1}{8} \int \mathrm{d}^4 x \mathrm{d}^4\theta \mathrm{d}^4\bar{\theta} \, \mathcal{Y}(K,\bar{K}) W^2 \bar{W}^2$$

$$= \tfrac{1}{2} \int \mathrm{d}^4 x \mathrm{d}^4\theta \, \{ W^2 + \tfrac{1}{4}\bar{D}^4 \left[\mathcal{Y}(K,\bar{K}) W^2 \bar{W}^2 \right] \} \qquad (2.47)$$

$$\equiv \tfrac{1}{2} \int \mathrm{d}^4 x \mathrm{d}^4\theta \, W^2_{\mathrm{restr.}} \,,$$

nevertheless captures the most relevant terms (modulo spacetime derivatives of the Maxwell $N = 2$ superfield strength). The $N = 2$ supersymmetric *and* e.-m. self-dual extension of the BI action is given by the 'non-linear sigma-model' [624, 613]

$$S[W,\bar{W}] = \tfrac{1}{2} \int \mathrm{d}^4 x \mathrm{d}^4\theta \, X \,, \qquad (2.48)$$

where the $N = 2$ chiral superfield $X \equiv W^2_{\mathrm{improved}} = W^2_{\mathrm{restr.}} + O(\partial W)$ has been introduced as a solution to the non-linear $N = 2$ superfield constraint [624]

$$X = \tfrac{1}{4}X\bar{D}^4\bar{X} + W^2 \,. \qquad (2.49)$$

Equation (2.48) is the 'improved' non-linear extension of the $N = 2$ Maxwell Lagrangian in $N = 2$ chiral superspace. The existence of the non-linear sigma-model form of the $N = 2$ SBI action implies its uniqueness and supports its interpretation as the Goldstone action associated with partial breaking of $N = 4$ supersymmetry down to $N = 2$, with the $N = 2$ vector multiplet as the

Goldstone multiplet, in remarkable similarity with the $N = 1$ supersymmetric Goldstone-Maxwell theory discussed in Subsect. 9.2.3. In particular, (2.49) can be considered as the $N = 2$ superfield generalization of the $N = 1$ superfield non-linear constraint (2.38).

Like the $N = 1$ SBI action, the $N = 2$ SBI action (2.48) does not lead to the propagating auxiliary fields D, despite of the presence of higher derivatives to all orders, with $D = 0$ being an on-shell solution. Non-vanishing expectation values for fermionic and scalar composite operators in front of the 'dangerous' interacting terms, which could lead to propagation of the auxiliary fields, are also forbidden because of the unbroken Lorentz symmetry and $N = 2$ supersymmetry with vanishing central charge.

To verify that (2.47) is indeed an $N = 2$ supersymmetric extension of the $N = 1$ SBI action, it is useful to rewrite it in terms of $N = 1$ superfields by integrating over a half of the $N = 2$ superspace anticommuting coordinates. The standard identification of the $N = 1$ superspace anticommuting coordinates, [4]

$$\theta^\alpha{}_{\underline{1}} = \theta^\alpha , \quad \text{and} \quad \bar{\theta}_{\dot{\alpha}}{}^{\underline{1}} = \bar{\theta}_{\dot{\alpha}} , \tag{2.50}$$

implies the $N = 1$ superfield projection rule

$$G = G(Z)| , \tag{2.51}$$

where $|$ means taking the $(\theta^\alpha{}_{\underline{2}}, \bar{\theta}_{\dot{\alpha}}{}^{\underline{2}})$-independent part of an $N = 2$ superfield $G(Z)$. As regards the $N = 2$ restricted chiral superfield W, its $N = 1$ superspace constituents are given by $N = 1$ complex superfields Φ and W_α,

$$W| = \Phi , \quad D^2_{\underline{\alpha}} W| = W_\alpha , \quad \tfrac{1}{2}(D^{\underline{2}})^\alpha (D^{\underline{2}})_\alpha W| = \bar{D}^2 \bar{\Phi} , \tag{2.52}$$

which follow from the constraints (4.2.26) and (4.2.27). In particular, (4.2.27) implies the $N = 1$ superfield Bianchi identity

$$D^\alpha W_\alpha = \bar{D}_{\dot{\alpha}} \bar{W}^{\dot{\alpha}} , \tag{2.53}$$

as well as the relations

$$K| = D^2 \left(W^\alpha W_\alpha + 2\Phi \bar{D}^2 \bar{\Phi} \right) .$$

$$(\bar{D}_{\underline{2}})^{\dot{\alpha}} K| = 2\mathrm{i} D^2 \partial^{\dot{\alpha}\beta} (W_\beta \Phi) , \tag{2.54}$$

$$(\bar{D}_{\underline{2}})_{\dot{\alpha}} (\bar{D}_{\underline{2}})^{\dot{\alpha}} K| = -4 D^2 \partial_\mu (\Phi \partial^\mu \Phi) ,$$

together with their conjugates. Equations (2.52), (2.53) and (2.54) are enough to perform a reduction of any $N = 2$ superspace action depending upon W and $\bar{W}$ into $N = 1$ superspace by differentiation,

---

[4] We underline particular values $i = \underline{1}, \underline{2}$ of the internal $SU(2)$ indices, and use the $N = 1$ notation $D^2 = \tfrac{1}{2}(D^{\underline{1}})^\alpha (D^{\underline{1}})_\alpha$ and $\bar{D}^2 = \tfrac{1}{2}(\bar{D}_{\underline{1}})_{\dot{\alpha}} (\bar{D}_{\underline{1}})^{\dot{\alpha}}$ here.

$$\int d^4\theta \ \rightarrow \ \int d^2\theta \, \tfrac{1}{2}(D^{\underline{2}})^\alpha (D^{\underline{2}})_\alpha \ ,$$

$$\int d^4\theta d^4\bar\theta \ \rightarrow \ \int d^2\theta d^2\bar\theta \, \tfrac{1}{2}(D^{\underline{2}})^\alpha (D^{\underline{2}})_\alpha \tfrac{1}{2}(\bar D_{\underline{2}})_{\dot\alpha}(\bar D_{\underline{2}})^{\dot\alpha} \ . \tag{2.55}$$

It is now straightforward to calculate the $N = 1$ superfield form of the $N = 2$ action (2.47). For our purposes, it is enough to notice that the first term in (2.47) gives rise to the kinetic terms for the $N = 1$ chiral superfields Φ and W_α,

$$\mathrm{Re} \int d^2\theta \, (\tfrac{1}{2}W^\alpha W_\alpha + \Phi \bar D^2 \bar\Phi) \ , \tag{2.56}$$

whereas the $N = 1$ vector multiplet contribution arising from the second term in (2.47) is given by

$$\tfrac{1}{8}\int d^2\theta d^2\bar\theta \, \mathcal{Y}(K|,\bar K \,|)W^\alpha W_\alpha \bar W_{\dot\alpha}\bar W^{\dot\alpha} + \ldots \, , \tag{2.57}$$

where the dots stand for Φ-dependent terms. The W-dependent contributions of (2.56) and (2.57) exactly coincide with the $N = 1$ supersymmetric extension (2.34) of the BI action after taking into account that the vector field dependence in the first component of the $N = 1$ superfield $K|$ is given by

$$K| = D^4 W^2| = 2D^2(\tfrac{1}{2}W^\alpha W_\alpha + \Phi\bar D^2\bar\Phi)| = -(F^+)^2 + D^2 + \ldots, \tag{2.58}$$

and similarly for $\bar K \,|$.

The dependence of the $N = 2$ SBI action upon the $N = 1$ chiral part Φ of the $N = 2$ vector multiplet is clearly of most interest, since it is dictated by $N = 2$ extended supersymmetry and electric-magnetic self-duality. Let us now take $W_\alpha = 0$ in the action (2.48), and calculate merely the leading terms depending upon Φ and $\bar\Phi$ there. After some algebra one gets the following $N = 1$ superspace action:

$$S[\Phi,\bar\Phi] = \int d^4x d^2\theta d^2\bar\theta \left[\Phi\bar\Phi - 4(\Phi\partial^\mu\Phi)(\bar\Phi\partial_\mu\bar\Phi) + 4\partial^\mu(\Phi\bar\Phi)\partial_\mu(\Phi\bar\Phi)\right] + \ldots \, , \tag{2.59}$$

where the dots stand for the higher-order terms depending upon the derivatives of $\mathcal{Y}$. The field components of the $N = 1$ chiral superfield Φ are conveniently defined by the projections

$$\Phi| = \frac{1}{\sqrt{2}}\phi \equiv \frac{1}{\sqrt{2}}(P + iQ) \ , \quad D_\alpha \Phi| = \psi_\alpha \ , \quad D^2 \Phi| = F \ , \tag{2.60}$$

where P is a real physical scalar, Q is a real physical pseudo-scalar, ψ_α is a chiral physical spinor, and F is a complex auxiliary field. It is not difficult to check that the kinetic terms for the auxiliary field components F and $\bar F$ cancel in (2.59), as they should. This allows us to simplify the calculation of the quartic term in (2.59) even further by going on-shell, i.e. assuming that $\Box\phi = F = 0$ there, even though this is not really necessary. A simple calculation now yields

$$S[\phi, \bar\phi] = \int d^4x \left\{ \partial^\mu\phi\partial_\mu\bar\phi - 2(\partial_{(\mu}\phi\partial_{\nu)}\bar\phi)^2 + (\partial_\mu\phi\partial^\mu\bar\phi)^2 \right\} , \qquad (2.61)$$

which exactly coincides with the leading terms in the derivative expansion of the Nambu-Goto (NG) action

$$S = -\int d^4x \sqrt{-\det(\eta_{\mu\nu} + \partial_\mu P\partial_\nu P + \partial_\mu Q\partial_\nu Q)} . \qquad (2.62)$$

To verify the correspondence in all orders, without going on-shell, non-linear redefinitions of field components are necessary to remove higher derivatives.

Equation (2.62) yields the effective action of a (static-gauge) 3-brane in flat six-dimensional ambient spacetime, with the Goldstone scalars (P, Q) being two collective coordinates corresponding to a 'transverse' motion of the 3-brane. The $N = 1$ superspace description of two transverse 3-brane coordinates in terms of the $N = 1$ chiral, complex linear and real linear Goldstone superfields was obtained in [630]. A 3-brane solution to (1,0), 6d super-Maxwell theory coupled to chiral (scalar) multiplets was constructed in [616, 617]. The solution of [616, 617] breaks translational invariance in two spatial directions and half of the 6d supersymmetry. This observation implies a six-dimensional origin of the four-dimensional $N = 2$ supersymmetric BI action. The latter is obtained by DR from a supersymmetric BI action in six spacetime dimensions after identifying the extra two components of a six-dimensional Abelian vector potential with the scalars P and Q [581]. The existence of the 6d SBI action guarantees the Goldstone nature of the scalars in (2.52) and (2.53), as well as the *off-shell* invariance of our 4d, $N = 2$ action under constant shifts of the scalars. We refer the reader to [581] for the explicit form of the 6d SBI action in superspace. Unlike the 4d bosonic BI action and the 4d, $N = 1$ supersymmetric BING action that merely contains the *first* derivatives of the transverse (scalar) collective coordinates, as in (2.62), the manifestly $N = 2$ supersymmetric BING action in 4d contains *higher* derivatives of the scalars. These higher derivatives may, however, be eliminated by field redefinitions at the expense of making $N = 2$ extended supersymmetry non-manifest [631].

To indicate the possibility of adding some additional structure given by a magnetic FI term into our 4d, $N = 2$ theory, it is worth mentioning here that the $N = 2$ superspace constraints (4.2.26) and (4.2.27) imply

$$\Box\left(D^{ij}W - \bar D^{ij}\bar W\right) = 0 . \qquad (2.63)$$

This means that the function $\mathrm{Im}\,(D^{ij}W)$ is harmonic and therefore it should be constant, [5] i.e.

---

[5] We assume that all components of the superfield W are regular in spacetime.

$$D^{ij}W - \bar{D}^{ij}\bar{W} = 4\mathrm{i}M^{ij} \ . \tag{2.64}$$

Taking into account the constant (FI) vector $\boldsymbol{M}$ in the constraint (2.64) is equivalent to adding a 'magnetic' Fayet-Iliopoulos (FI) term to the dual action [628]. The FI term can be formally removed from the constraint (2.64) by a field redefinition of W, i.e. at the expense of adding a constant imaginary part to the auxiliary scalar triplet $\boldsymbol{D}$ of the $N = 2$ vector multiplet. The APT model [628] is defined by adding the usual (electric) and magnetic FI terms to the general (Seiberg-Witten type) $N = 2$ chiral action in terms of W [628].

The 4d, $N = 2$ 'Bianchi identity' can be enforced by introducing a real unconstrained (Lagrange multiplier) $N = 2$ superfield (known as the *Mezincescu pre-potential* [632]) $\boldsymbol{L} = \frac{1}{2}(\boldsymbol{\tau})^i{}_j L^j{}_i \equiv \frac{1}{2}\mathrm{tr}(\boldsymbol{\tau}L)$ into the $N = 2$ SBI action (2.48) by rewriting it in the first-order form in $N = 2$ superspace,

$$
\begin{aligned}
S[W,\bar{W}] \to S[W,\bar{W};L] &= S[W,\bar{W}] + \mathrm{i}\int \mathrm{d}^4x \mathrm{d}^4\theta \mathrm{d}^4\bar{\theta}\, L_{ij}\left(D^{ij}W - \bar{D}^{ij}\bar{W}\right) \ , \\
&= S[W,\bar{W}] + \left[\mathrm{i}\int \mathrm{d}^4x \mathrm{d}^4\theta\, WW_{\mathrm{magn.}} + \mathrm{h.c.}\right] \ ,
\end{aligned}
\tag{2.65}
$$

where the $N = 2$ superfield W is now a chiral (unrestricted) $N = 2$ superfield, while

$$W_{\mathrm{magn.}} \equiv \bar{D}^4 D^{ij} L_{ij} \tag{2.66}$$

is the dual or 'magnetic' $N = 2$ superfield strength that automatically satisfies the $N = 2$ constraints (4.2.26) and (4.2.27) due to the defining equation (2.66). Varying the action (2.65) with respect to W, solving the resulting algebraic equation for W in terms of $W_{\mathrm{magn.}}$, and substituting the result back into the action (2.65), results in the dual $N = 2$ action $S[W_{\mathrm{magn.}}, \bar{W}_{\mathrm{magn.}}]$ that takes exactly *the same* form as (2.48) [613]. In other words, it is self-dual with respect to the $N = 2$ supersymmetric electric-magnetic duality.

The invariance of the action (2.48) under constant shifts of the $N = 2$ superfield strength W is, of course, a necessary condition for its Goldstone interpretation. It is easy to verify this symmetry if W is subject to the on-shell condition $\Box W = 0$. The 6d super-BI action of [581] dimensionally reduced down to four dimensions automatically implies this symmetry off-shell.

9.3 Fourth-Order NLSM

Since any non-trivial 4d NLSM (of second order in spacetime derivatives) is not renormalizable, it may be useful to investigate its possible generalizations that would be renormalizable in 4d. The renormalizability (by divergence index) can be easily achieved by introducing higher spacetime derivatives of quantum fields. The basic examples are provided by (manifestly renormalizable) scalar QFT models whose kinetic terms are of fourth-order in spacetime derivatives, because those models are formulated in terms of dimensionless scalar fields and dimensionless (NLSM) coupling constants. We call them *fourth-order* NLSM [582, 633, 634]. Our philosophy in this section is entirely different from the one adopted in Chap. 8 where similar NLSM actions were considered as the next-to-leading-order corrections to the 4d LEEA. Here we treat the fourth-order NLSM as a QFT model, i.e. we quantize it.

The standard argument against quantization of the fourth-order NLSM is its apparent non-unitarity as a fundamental QFT, i.e. the presence of ghosts in its classical action. This situation is very similar to the relationship between Einstein gravity and R^2-gravity: the former is formally unitary but non-renormalizable, whereas the latter is formally renormalizable but apparently non-unitary [635, 636, 637, 638]. The status of unitarity in R^2 gravity is, however, still far from an ultimate solution, while the asymptotical freedom of R^2-gravity [639, 640] apparently favours unitarity [641, 642, 643, 644].

In this section we formulate the general geometrical action of the fourth-order bosonic NLSM in 4d [582], as well as its $N = 1$ and $N = 2$ supersymmetric generalizations [633, 634]. We calculate the one-loop counterterms, and demonstrate asymptotic freedom in the quantized fourth-order NLSM whose target space is a sphere. The crucial role of supersymmetry in providing stability at high energies is emphasized. The mixed 4d NLSM, having both second- and fourth-order spacetime derivatives, was considered in [645] with similar results.

9.3.1 Bosonic 4d NLSM with Higher Derivatives

Let us consider flat 4d (Euclidean) spacetime and a set of scalar fields $\{\Phi^i\}$. The most general action with fourth-order spacetime derivatives reads [582]

$$
I[\Phi] = \frac{1}{2} \int \mathrm{d}^4x \left[G_{ij}^{(0)} \Box\Phi^i \Box\Phi^j + A_{ijk}^{(0)} \partial_\mu \partial_\nu \Phi^i \partial_\mu \Phi^j \partial_\nu \Phi^k \right.
$$
$$
\left. + T_{ijkl}^{(0)} \partial_\mu \Phi^i \partial_\mu \Phi^j \partial_\nu \Phi^k \partial_\nu \Phi^l + H_{ijkl}^{(0)} \varepsilon_{\mu\nu\lambda\rho} \partial_\mu \Phi^i \partial_\nu \Phi^j \partial_\lambda \Phi^k \partial_\rho \Phi^l \right] ,
$$
$$
(3.1)
$$

with some functions $G_{ij}^{(0)}(\Phi)$, $A_{ijk}^{(0)}(\Phi)$, $T_{ijkl}^{(0)}(\Phi)$ and $H_{ijkl}^{(0)}(\Phi)$. All the other possible terms can be reduced to those written down in (3.1) after integration by parts. It is convenient to rewrite (3.1) in the equivalent (NLSM-covariant) form

$$
I[\Phi] = \frac{1}{2} \int \mathrm{d}^4x \left[G_{ij} \ominus\Phi^i \ominus\Phi^j + A_{ijk} D_\mu \partial_\nu \Phi^i \partial_\mu \Phi^j \partial_\nu \Phi^k \right.
$$
$$
\left. + T_{ijkl} \partial_\mu \Phi^i \partial_\mu \Phi^j \partial_\nu \Phi^k \partial_\nu \Phi^l + H_{ijkl} \varepsilon_{\mu\nu\lambda\rho} \partial_\mu \Phi^i \partial_\nu \Phi^j \partial_\lambda \Phi^k \partial_\rho \Phi^l \right] ,
$$
$$
(3.2)
$$

where we have introduced the NLSM-covariant derivatives with respect to the metric G_{ij},

$$\hat{\Box} \equiv (D_\mu \partial_\mu \Phi)^i = \left[\delta^{ij}\partial_\mu + \Gamma^i_{jk}\partial_\mu\Phi^k\right]\partial_\mu\Phi^j = \Box\Phi^i + \Gamma^i_{jk}\partial_\mu\Phi^k\partial_\mu\Phi^j \ ,$$

$$(D_\mu\partial_\nu\Phi)^i \equiv \partial_\mu\partial_\nu\Gamma^i + \Gamma^i_{jk}\partial_\mu\Phi^k\partial_\nu\Phi^j \ , \tag{3.3}$$

with $\Gamma^i_{jk} = \frac{1}{2}G^{kl}\left(\partial_i G_{jl} + \partial_j G_{il} - \partial_l G_{ij}\right)$. It is easy to establish a one-to-one correspondence between the coefficient functions in (3.1) and (3.2). The action (3.2) can be geometrically interpreted as the (fourth-order) NLSM with the NLSM *metric* G_{ij}, and the additional *tensors* (A, T, H) defined in the NLSM target space,

$$G_{ij} = G_{(ij)} \ , \quad A_{ijk} = A_{i(jk)} \ , \quad T_{ijkl} = T_{(ij)(kl)} \ , \quad H_{ijkl} = H_{[ijkl]} \ . \tag{3.4}$$

In particular, the last term in (3.2) stands for a (generalised) WZ term in 4d. It is worth mentioning that the WZ term in the fourth-order NLSM is on an equal footing with the other terms in the action since they are all of fourth-order in spacetime derivatives. The renormalizability of the 4d action (3.1) or (3.2) (in the generalized NLSM sense – see Chap. 2) follows from the observation that it is the most general fourth-order NLSM action all of whose coupling constants are dimensionless.

The action (3.2) can be used as the starting point for the covariant background-field expansion (Sect. 2.2). The quantum fields ξ^i can be identified with the normal coordinates in the fourth-order NLSM target space $\mathcal{M}$ equipped with the metric G_{ij}. In the one-loop approximation we only need the quadratic terms of the expansion with respect to ξ. We find [582]

$$\frac{1}{2}\frac{d^2}{ds^2}I[\rho(s)]\Big|_{s=0} = \frac{1}{2}\int d^4x\, \xi^a \hat{F}_{ab}(\nabla)\xi^b \ , \tag{3.5}$$

where the kinetic operator $\hat{F}(\nabla)$ has the structure

$$\hat{F}(\nabla) = \hat{\Box}^2 + \hat{\Omega}_{\rho\mu\nu}\nabla_\rho\nabla_\mu\nabla_\nu + \hat{D}_{\mu\nu}\nabla_\mu\nabla_\nu + \hat{H}_\mu\nabla_\mu + \hat{P} \ , \tag{3.6}$$

with certain symmetric coefficient functions, $\hat{\Omega}_{\rho\mu\nu}$, $\hat{D}_{\mu\nu}$, $\hat{H}_\mu$ and $\hat{P}$, of the arguments (3.4). In what follows we only consider the case with $A_{ijk} = 0$ for simplicity (see [646] for the most general case). Then we find that $\hat{\Omega}_{\rho\mu\nu} = 0$ also, whereas

$$\hat{D}_{\mu\nu ab} = 2\delta_{\mu\nu}\partial_\rho\Phi^m\partial_\rho\Phi^p(R_{bmap} - 2T_{abmp}) - 4\partial_{(\mu}\Phi^m\partial_{\nu)}\Phi^p T_{ambp} \ ,$$

$$\hat{D}_{ab} = \hat{D}_{\mu\mu ab} = \partial_\rho\Phi^m\partial_\rho\Phi^p(8R_{bmap} - 16T_{abmp} - 4T_{ambp}) \ ,$$

$$\hat{H}_{\mu ab} = \partial_\mu\Phi^m(D_\nu\partial_\nu\Phi)^j(4R_{jbam} - T_{ajbm}) + 2\partial_\mu\Phi^m\partial_\nu\Phi^j\partial_\nu\Phi^p$$
$$\times [2D_a T_{bmjp} - D_m T_{abjp} - 2D_j T_{apbm}] - 4(D_\mu\partial_\nu\Phi)^j\partial_\nu\Phi^p$$
$$\times [T_{abjp} + T_{apbj}] + 10S_{abjpm}\varepsilon_{\mu\nu\lambda\sigma}\partial_\nu\Phi^j\partial_\lambda\Phi^p\partial_\sigma\Phi^m \ ,$$

$$\hat{P}_{ab} = \partial_\mu \Phi^m \partial_\mu \Phi^p (D_\nu \partial_\nu \Phi)^j [D_a R_{jmbp} + D_m R_{jabp}]$$

$$+ (D_\mu \partial_\mu \Phi)^p (D_\nu \partial_\nu \Phi)^j R_{jabp} + \partial_\mu \Phi^m \partial_\mu \Phi^p \partial_\nu \Phi^q \partial_\nu \Phi^t$$

$$\times \left[R_{jmap} R^j{}_{qbt} + \tfrac{1}{2} D_b D_a T_{mpqt} + 2T_{jpqt} R^j{}_{bam} - 2T_{apjm} R^j{}_{bmt} \right]$$

$$+ \tfrac{5}{2} D_b S_{aijkl} \varepsilon_{\mu\nu\lambda\sigma} \partial_\mu \Phi^i \partial_\nu \Phi^j \partial_\lambda \Phi^k \partial_\sigma \Phi^l \ ,$$

$$(3.7)$$

where we have introduced the curvature tensor R_{ijkl} in terms of the metric G_{ij}, and the WZ 'field-strength' tensor

$$S_{nijkl} = \partial_{[n} H_{ijkl]} \quad \text{or, just} \quad S = dH \ . \tag{3.8}$$

Calculation of the UV divergences of the one-loop effective action of the fourth-order NLSM is thus reduced to calculation of the UV divergences in the determinant of the fourth-order minimal differential operator $\hat{F}(\nabla)$ defined by (3.6),

$$i\Gamma_{\text{div}}^{(1)} = -\tfrac{1}{2} \text{Tr} \ln \hat{F}(\nabla) \Big|_{\text{div}} \ . \tag{3.9}$$

A solution to the last problem is provided by the generalized Schwinger-De Witt techniques [647]. By using dimensional regularization (DR) we find

$$-\tfrac{1}{2} \text{Tr} \ln \hat{F}(\nabla) \Big|_{\text{div}} = \frac{i}{(4-d)16\pi^2} \int d^4 x \, \text{tr} \left\{ \tfrac{1}{6} \hat{R}_{\mu\nu} \hat{R}_{\mu\nu} \right.$$

$$\left. -\hat{P} + \tfrac{1}{48} \hat{D}^2 + \tfrac{1}{24} \hat{D}_{\mu\nu} \hat{D}_{\mu\nu} \right\} \ , \tag{3.10}$$

where we have introduced the notation

$$\hat{R}_{\mu\nu}{}^{ab} = \partial_\mu \Phi^i \partial_\nu \Phi^j R_{ij}{}^{ab} \ , \quad R_{ij}{}^{ab} = V^{ak} V^{bl} R_{ijkl} \ , \tag{3.11}$$

and the 'vielbein' V^{ai} associated with the metric G_{ij}. A derivation of (3.10) can be found in [647]. In our case specified by (3.7) and (3.8), we get

$$-\Gamma_{\text{div}}^{(1)} = \frac{1}{16\pi^2(d-4)} \int d^4 x \, L(x) \ , \tag{3.12}$$

and

$$L(x) = L_1(x) + L_2(x) + L_3(x) + L_4(x) \ , \tag{3.13}$$

where we have separated the terms contributing to the one-loop (NLSM) renormalization of the (G, A, T, S) tensors, respectively, with

$$L_1(x) = (D_\mu \partial_\mu \Phi)^i (D_\nu \partial_\nu \Phi)^j R_{ij} \ ,$$

$$L_2(x) = (D_\mu \partial_\nu \Phi)^i \partial_\mu \Phi^j \partial_\nu \Phi^k (D^p R_{ijkp} - 2D_{(i} R_{k)j}) \ ,$$

$$L_3(x) = \partial_\mu \Phi^i \partial_\nu \Phi^j \partial_\nu \Phi^k \partial_\nu \Phi^l \left[R_{ipqj} R_k{}^{pq}{}_l - \frac{1}{6} R_{ikpq} R^{pq}{}_{jl} - D_k D_p R^p{}_{ijl} \right.$$

$$- D_k D_i R_{jl} + 8 T_{ijpq} T_{kl}{}^{pq} + 4 T_{ijpq} T_k{}^{pq}{}_l + 4 T_{(i}{}^{pq}{}_{k)} T_{jpql}$$

$$+ \frac{1}{3} T_{ipqj} T_k{}^{pq}{}_l + 2 R_i^p T_{pjkl} - 2 R_{ik}{}^{pq} T_{jpql} + 2 R_k{}^{pq}{}_l (4 T_{ijpq} + T_{ipqj})$$

$$\left. - \frac{1}{2} D^p D_p T_{ijkl} \right] ,$$

$$L_4(x) = -\frac{5}{2} \varepsilon_{\mu\nu\lambda\rho} \partial_\mu \Phi^i \partial_\nu \Phi^j \partial_\lambda \Phi^k \partial_\rho \Phi^l D^p S_{pijkl} ,$$

$$(3.14)$$

and we have introduced the Ricci tensor R_{ij} of the metric G_{ij}.

Subject to $A_{ijk} = 0$, the one-loop renormalizability and the second line of (3.14) imply the additional restrictions

$$D_{(i} R_{jk)} = 0 , \quad D^p R_{i(jk)p} = 0 . \tag{3.15}$$

We restrict ourselves to the Ricci-flat spaces ($R_{ij} = 0$) and the locally symmetric spaces ($D^p R_{ijkl} = 0$) that both satisfy (3.15) due to the Bianchi identities for the Riemann curvature.

It is worth mentioning that the first two terms (quadratic in the curvature) in L_3, governing the renormalization of the T-tensor, are proportional to the UV divergence of the standard (second-order) 4d NLSM with metric G_{ij} [58]. The one-loop renormalization of the metric G_{ij} in (3.14) is governed by the Ricci-tensor, as in 2d NLSM (Chap. 2). However, unlike the 2d NLSM models with a torsion, a WZ term (or torsion) in the fourth-order 4d NLSM does not affect the metric renormalization. The one-loop renormalization of the WZ term in (3.14) is absent if its field-strength is co-closed,

$$D^p S_{pijkl} = 0 , \quad \text{or, equivalently,} \quad \delta S = 0 , \tag{3.16}$$

in terms of forms. The condition (3.16) is clearly sufficient but is not necessary. According to (3.14), the total one-loop counterterm (3.12) vanishes for the Ricci-flat NLSM target spaces $\mathcal{M}$ provided that

$$D^p D_p T_{ijkl} - 16 T_{ij}{}^{pq} T_{klpq} - 8 T_{ijpq} T_k{}^{pq}{}_l - \frac{8}{3} T_{(i}{}^{pq}{}_{k)} T_{jpql}$$

$$-\frac{2}{3} T_{ipqj} T_k{}^{pq}{}_l - 4 R_k{}^{pq}{}_l (4 T_{ijkl} + T_{ipqj}) + 4 R_{ik}{}^{pq} T_{jpql} - 2 R_{ipqj} R_k{}^{pq}{}_i$$

$$+\frac{1}{3} R_{ikpq} R^{pq}{}_{jl} + [\text{symmetrization in } (ij), (kl) , \text{ and } ((ij)(kl))] = 0 , \tag{3.17}$$

and $\delta S = dQ$ for some 3-form Q.

Since (3.17) is still very complicated, let us consider the simpler case of the *multiplicatively* renormalizable fourth-order NLSM whose metric G_{ij} is the only independent tensorial structure in the action (without a WZ term). The simplest non-trivial choice of the symmetric NLSM target space $\mathcal{M}$ is given by an n-sphere of radius r, with

$$R_{ijkl} = r^{-2}\left(G_{il}G_{kj} - G_{ik}G_{jl}\right), \quad R_{ij} = \frac{(n-1)}{r^2}G_{ij}, \quad R = \frac{n(n-1)}{r^2}.$$
(3.18)

The most general (metric-dependent) form of T-tensor is parametrized by two coupling constants, f_1 and f_2,

$$T_{ijkl}(G) = \frac{f_1}{2r^2}\left(G_{il}G_{jk} + G_{ik}G_{jl}\right) + \frac{f_2}{r^2}G_{ij}G_{kl}.$$
(3.19)

Stated differently, we now consider the fourth-order NLSM action

$$I[\varPhi] = \frac{1}{2\lambda^2}\int d^4x\left[G_{ij}(D_\mu\partial_\mu\varPhi)^i(D_\nu\partial_\nu\varPhi)^j + T_{ijkl}(G)\partial_\mu\varPhi^i\partial_\mu\varPhi^j\partial_\nu\varPhi^k\partial_\nu\varPhi^l\right]$$
(3.20)

with three coupling constants $(\lambda, \phi_1, \phi_2)$. The radius value is at our disposal. We choose $8\pi^2 r^2 = \pm 1$.

It is easy to see that the restrictions (3.15) are satisfied in the case (3.18), while the one-loop renormalization of the QFT (3.14) is multiplicative (i.e. it amounts to renormalization of the coupling constants). Having obtained the one-loop counterterm, we can write down the corresponding renormalization group (RG) equations for the effective charges $\lambda^2(t)$, $f_1(t)$ and $f_2(t)$, in accordance with the general relation (Sect. 2.4)

$$\frac{1}{\lambda_B^2}G_{ij} = \mu^{d-4}\left[\frac{1}{\lambda^2(\mu)}G_{ij} + \frac{1}{d-4}G_{ij}^{(1)} + \dots\right],$$
(3.21)

where μ is the RG scale and $t = \ln\mu$. It follows from (3.21) that

$$\mu\frac{d}{d\mu}\left[\frac{1}{\lambda^2(\mu)}G_{ij}\right] = -G_{ij}^{(1)} = \frac{1}{8\pi^2}R_{ij} = \pm(n-1)G_{ij}.$$
(3.22)

Similarly, we find

$$\mu\frac{d}{d\mu}\left[\frac{1}{\lambda^2(\mu)}T_{ijkl}(G)\right] = \frac{1}{8\pi^2}\left[R_{(i}{}^{pq}{}_{j)}R_{kpql} + \dots\right].$$
(3.23)

Equations (3.14), (3.18) and (3.19) together yield

$$\frac{d}{dt}\left(\frac{1}{\lambda^2}\right) = \pm(n-1),$$
(3.24a)

and

$$\frac{d}{dt}\left(\frac{f_1}{\lambda^2}\right) = \pm\left[\frac{(4n+131)}{12}f_1^2 + f_2^2 + \frac{19}{3}f_1f_2 + 2(n-5)f_1 - 4f_2 + \frac{4}{3}\right],$$

$$\frac{d}{dt}\left(\frac{f_2}{\lambda^2}\right) = \pm\left[\frac{(n+28)}{12}f_1^2 + \frac{2(12n+7)}{3}f_2^2 + (2n+19)f_1f_2\right.$$

$$\left. + (n+7)f_1 + 2(5n-3)f_2 + n - \frac{7}{3}\right].$$
(3.24b)

Equation (3.24a) is easily solved,

$$\lambda^2(t) = \frac{\lambda_0^2}{1 \pm (n-1)\lambda_0^2 t} \ , \tag{3.25}$$

whereas the remaining two equations (3.24b), after a change of variables,

$$\frac{d\tau}{dt} = \pm\lambda^2(t) \ , \tag{3.26}$$

take the form

$$\frac{df_1}{d\tau} = \frac{(4n+131)}{12}f_1^2 + f_2^2 + \frac{19}{3}f_1 f_2 + (n-9)f_1 - 4f_2 + \frac{4}{3} \ ,$$

$$\frac{df_2}{d\tau} = \frac{(n+28)}{12}f_1^2 + \frac{2(12n+7)}{3}f_2^2 + (2n+19)f_1 f_2 \tag{3.27}$$

$$+ (n+7)f_1 + (9n-5)f_2 + n - \frac{7}{3} \ .$$

Equation (3.25) allows the standard interpretation (cf. [14, 15]):

- the choice of the lower sign ($-$) in (3.25) results in increasing the effective NLSM coupling $\lambda^2(t)$ with energy, $t \to \infty$, at $n > 1$; it gives rise to a violation of applicability of quantum perturbation theory in the NLSM under consideration, which is similar to the behaviour of the effective coupling in abelian gauge theories like QED;
- the choice of the upper sign in (3.25) for $n > 1$ results in decreasing the effective NLSM coupling $\lambda^2(t)$ up to zero when the energy increases, $t \to \infty$, which is similar to the behaviour of the effective coupling in non-Abelian gauge theories like QCD.

In other words, the upper sign choice in (3.25) gives rise to asymptotic freedom with respect to the coupling constant λ^2 in the fourth-order 4d NLSM (3.20). The action (3.20) has, however, two additional coupling constants, whose one-loop RG flow is governed by the differential equations (3.27). The QFT (3.20) would be asymptotically free with respect to *all* of its coupling constants if $f_1(t)$ and $f_2(t)$ were finite in the limit $\tau \to \infty$. In turn, it is determined by zeros of the NLSM β-functions defined by the right-hand sides of (3.27). Numerical calculations on a computer [582] show the existence of real fixed points for the β-functions of (3.27) at $n \geq 21$, although all of them turn out to be *unstable*. Hence, the asymptotic freedom in the NLSM (3.20) is realized only if the initial values of the coupling constants f_1 and f_2 were chosen in exact correspondence with the RG fixed points, i.e. on the special solutions of the RG equations.

Stable solutions are obtained by considering the multiplicatively renormalizable fourth-order NLSM with a *single* coupling constant on a group manifold [648]. In Subsect. 9.3.2 we demonstrate that introducing supersymmetry into the fourth-order NLSM also leads to stable RG solutions.

9.3.2 4d, $N = 1$ NLSM with Higher Derivatives

The $N = 1$ supersymmetrization of the 4d bosonic fourth-order NLSM in 4d (Subsect. 9.3.1) is straightforward [633], while the relation between the standard (second-order) 4d, $N = 1$ NLSM and its fourth-order supersymmetric counterpart is similar to the relation between $N = 1$ (Poincaré) supergravity and R^2 supergravity in 4d [649].

Similarly to the standard 4d, $N = 1$ NLSM whose $N = 1$ supersymmetry requires a Kählerian NLSM metric (Subsect. 3.4.2), the $N = 1$ supersymmetry in fourth-order 4d NLSM poses certain restrictions on the allowed tensors (G, A, T, H). In particular, a supersymmetric WZ term enters the fourth-order action in combination with some other terms, i.e. T and H tensors are no longer independent [633]. The supersymmetric WZ term is needed, for example, for a stabilization of solitonic solutions in the (standard) 4d supersymmetric CP^1 NLSM [650] generalizing the Skyrme model (Sect. 2.7). We show in this subsection that the one-loop renormalization of the 4d, $N = 1$ fourth-order NLSM leads to the appearance of Chern-Simons-like terms in its quantum effective action [633].

The most general 4d, $N = 1$ supersymmetric fourth-order NLSM Lagrangian, in terms of dimensionless $N = 1$ chiral superfields $\Phi^i(x_\mu, \theta_\alpha, \bar{\theta}_{\dot{\alpha}})$ and their conjugates $\bar{\Phi}^{\bar{i}}(x_\mu, \theta_\alpha, \bar{\theta}_{\dot{\alpha}})$ reads as follows:

$$
L = G_{i\bar{j}} D^2 \Phi^i \bar{D}^2 \bar{\Phi}^{\bar{j}} + \left[A^{(0)}_{i(\bar{j}\bar{k})} D^2 \Phi^i \bar{D}^{\dot{\alpha}} \bar{\Phi}^{\bar{j}} \bar{D}_{\dot{\alpha}} \bar{\Phi}^{\bar{k}} + \text{h.c.} \right]
$$

$$
+ T^{(0)}_{i\bar{j}k\bar{n}} D^\alpha \Phi^i D_\alpha \Phi^j \bar{D}^{\dot{\alpha}} \bar{\Phi}^{\bar{k}} \bar{D}_{\dot{\alpha}} \bar{\Phi}^{\bar{n}} + \left[-H^{(0)}_{i[\bar{j}\bar{k}]} D^\alpha \Phi^i D_\alpha \Phi^j \bar{D}^{\dot{\alpha}} \bar{\Phi}^{\bar{j}} \bar{D}_{\dot{\alpha}} \bar{\Phi}^{\bar{k}} + \text{h.c.} \right]
$$
$$(3.28)$$

where we have introduced some arbitrary functions $(G, A^{(0)}, T^{(0)}, H)$ of the superfields Φ and $\bar{\Phi}$, and the $N = 1$ superspace (flat) covariant derivatives D_α and $\bar{D}^{\dot{\alpha}}$. It is straightforward to calculate the leading bosonic terms in components from the $N = 1$ superspace NLSM action

$$
I = \int \mathrm{d}^4 x \mathrm{d}^2 \theta \mathrm{d}^2 \bar{\theta} \, L \, .
\tag{3.29}
$$

We find the following $\Phi(x)$-dependent terms:

$$
I = \int \mathrm{d}^4 x \left\{ G_{i\bar{j}} \Box \Phi^i \Box \bar{\Phi}^{\bar{j}} + \left[A^{(0)}_{i(\bar{j}\bar{k})} \Box \Phi^i \partial_\mu \bar{\Phi}^{\bar{j}} \partial_\mu \bar{\Phi}^{\bar{k}} + \text{h.c.} \right] \right.
$$

$$
+ T^{(0)}_{i\bar{j}k\bar{n}} \partial_\mu \Phi^i \partial_\mu \Phi^j \partial_\nu \bar{\Phi}^{\bar{k}} \partial_\nu \bar{\Phi}^{\bar{n}} + \left[H_{i[\bar{k}\bar{n}],j} \left(-\frac{i}{2} \varepsilon_{\mu\nu\lambda\rho} \partial_\mu \Phi^i \partial_\nu \Phi^j \partial_\lambda \bar{\Phi}^{\bar{k}} \partial_\rho \bar{\Phi}^{\bar{n}} \right. \right.
$$

$$
\left. \left. - \partial_\mu \Phi^j \partial_\mu \bar{\Phi}^{\bar{k}} \partial_\nu \Phi^i \partial_\nu \bar{\Phi}^{\bar{n}} \right) + \text{h.c.} \right] + \ldots \right\} \, .
$$
$$(3.30)$$

The 'auxiliary' fields $F^i(x) = D^2\Phi^i(x_\mu, \theta_\alpha, \bar{\theta}_{\overset{\bullet}{\alpha}})\big|$ are, of course, propagating in the action (3.29), i.e. they lead to additional bosonic degrees of freedom in the $N = 1$ fourth-order NLSM.

As in Subsect. 9.3.1, let us rewrite the $N = 1$ fourth-order action (3.29) in covariant NLSM form by interpreting $G_{i\bar{j}}$ as the Hermitian NLSM metric and introducing complex connections

$$\Gamma_{jk}{}^i = \tfrac{1}{2}G^{i\bar{j}}\left(\partial_j G_{\bar{j}k} + \partial_k G_{\bar{j}j}\right) , \quad \Gamma_{j\bar{k}}{}^i = 0 ,$$

$$\Gamma_{\bar{j}\bar{k}}{}^{\bar{i}} = \tfrac{1}{2}G^{\bar{i}j}\left(\partial_{\bar{j}} G_{j\bar{k}} + \partial_{\bar{k}} G_{j\bar{j}}\right) , \quad \Gamma_{\bar{n}j}{}^{\bar{k}} = 0 . \tag{3.31}$$

The $N = 1$ superspace covariant derivatives should also be covariantized with respect to the NLSM target space connection (3.31) (cf. Sect. 3.1),

$$(D_\alpha^c \Phi)^i = D_\alpha \Phi^i + \Gamma_{jk}{}^i D_\alpha \Phi^k \Phi^j , \quad (\bar{D}_{\overset{\bullet}{\alpha}}^c \bar{\Phi})^{\bar{i}} = \bar{D}_{\overset{\bullet}{\alpha}} \bar{\Phi}^{\bar{i}} + \Gamma_{\bar{j}\bar{k}}{}^{\bar{i}} \bar{D}_{\overset{\bullet}{\alpha}} \bar{\Phi}^{\bar{k}} \bar{\Phi}^{\bar{j}} ,$$

$$\bar{D}^c \Phi = D^c \bar{\Phi} = 0 , \quad D^c D\Phi \equiv (D^c)^2 \Phi . \tag{3.32}$$

The 4d, $N = 1$ supersymmetric fourth-order NLSM action reads [633]

$$I = \int d^4x d^4\theta \left\{ G_{i\bar{j}}(D^c D\Phi)^i (\bar{D}^c \bar{D}\bar{\Phi})^{\bar{j}} + \left[A_{i(\bar{j}\bar{k})}(D^c D\Phi)^i (\bar{D}\bar{\Phi}^{\bar{j}} \bar{D}\bar{\Phi}^{\bar{k}}) + \text{h.c.} \right] \right.$$

$$\left. + T_{ij\bar{k}\bar{n}}(D\Phi^i D\Phi^j)(\bar{D}\bar{\Phi}^{\bar{k}} \bar{D}\bar{\Phi}^{\bar{n}}) - \left[H_{i[\bar{j}\bar{k}]} D^\alpha \Phi^i (D_\alpha^c \bar{D}^{\overset{\bullet}{\alpha}} \bar{\Phi})^{\bar{j}} \bar{D}_{\overset{\bullet}{\alpha}} \bar{\Phi}^{\bar{k}} + \text{h.c.} \right] \right\} \tag{3.33}$$

where we have simply redefined the coefficient functions. The action (3.33) is invariant under the gauge transformations $\delta H_{i[\bar{j}\bar{k}]} = \partial_i \lambda_{[\bar{j}\bar{k}]}$ with arbitrary gauge parameters $\lambda_{[\bar{j}\bar{k}]}(\Phi, \bar{\Phi})$. Accordingly, it is useful to rewrite the WZ-like term in (3.33) in the manifestly gauge-invariant form, by introducing an extra bosonic coordinate η belonging to the interval $[0, 1]$, with the special dependence of the chiral superfields upon η,

$$\Phi(x, \theta, \bar{\theta}; \eta) = \eta \Phi(x, \theta, \bar{\theta}) , \tag{3.34}$$

and the 'torsion' field strength $H_{ki\bar{j}\bar{k}} = \partial_{[k} H_{i]\bar{j}\bar{k}}$. We find

$$\int d^4x d^4\theta \left[H_{i[\bar{j}\bar{k}]} D\Phi^i (D^c \bar{D}\bar{\Phi})^{\bar{j}} \bar{D}\bar{\Phi}^{\bar{k}} + \text{h.c.} \right]$$

$$= \int d^4x d^4\theta \left\{ \int_0^1 d\eta\, \eta \left[H_{ki[\bar{j}\bar{k}]} D\Phi^i (D^c \bar{D}\bar{\Phi})^{\bar{j}} (\bar{D}\bar{\Phi}^{\bar{k}})\Phi^k \right] + \text{h.c.} \right\} . \tag{3.35}$$

To compute the one-loop UV divergences of the $N = 1$ fourth-order NLSM, we use the covariant background-field method directly in 4d, $N = 1$ superspace (cf. Sect. 3.1). The geodesics $\rho(\tau)$ between the 'points' Φ and $\Phi + \Pi$ of the complex NLSM target space have an expansion

$$\rho^i(\tau) = \Phi^i + \xi^i\tau - \tfrac{1}{2}\Gamma_{jk}{}^i\xi^j\xi^k\tau^2 + \dots \, , \qquad (3.36)$$

where the dots stand for higher-order terms in the parameter τ. Hence, the covariant quantum superfields ξ^i introduced via (3.36) are not chiral,

$$0 = \bar{D}^{\dot\alpha}\Pi^i = \bar{D}^{\dot\alpha}\xi^i - \Gamma^i{}_{jk}\bar{D}^{\dot\alpha}\xi^j\xi^k - \tfrac{1}{2}R^i{}_{j\bar{k}l}\bar{D}^{\dot\alpha}\bar\Phi^{\bar{k}}\xi^j\xi^l + \dots \, . \qquad (3.37)$$

Nevertheless, it is still possible to treat ξ^i as $N = 1$ chiral superfields in one-loop perturbation theory since the terms violating the chirality in (3.37) do not contribute in the one-loop approximation. The covariant background field expansion of the action (3.33) has quadratic (in ξ) terms having the form (cf. Sect. 3.1),

$$I^{(2)} = \int d^4x d^4\theta \left\{ G_{i\bar{j}}\xi^i(D^{(c)2}\bar{D}^{(c)2}\bar\xi)^{\bar{j}} + \xi^I F_{IJ}\xi^J \right\} \, . \qquad (3.38)$$

Let us introduce a chiral 'vielbein' $V_i^a(\Phi)$ by the relations

$$G_{i\bar{j}}(\Phi,\bar\Phi) = V_i^a(\Phi)\bar{V}_{\bar{j}a}(\bar\Phi) \, , \quad D_\alpha \bar{V}_{\bar{j}a} = \bar{D}_{\dot\alpha} V_i^a = 0 \, , \qquad (3.39)$$

and transform the vector ξ^i into the tangential space, $\xi^a = V_i^a\xi^i$, as usual (Chaps. 2 and 3). We find

$$I^{(2)} = \int d^4x d^4\theta (\xi^a, \bar\xi^{\bar{a}}) \left[\begin{pmatrix} 0 & \tfrac{1}{2}\Box_- \\ \tfrac{1}{2}\Box_+ & 0 \end{pmatrix} + \begin{pmatrix} F_{ab} & \bar{F}_{a\bar{b}} \\ F_{\bar{a}b} & \bar{F}_{\bar{a}\bar{b}} \end{pmatrix} \right] \begin{pmatrix} \xi^b \\ \bar\xi^{\bar{b}} \end{pmatrix} \qquad (3.40)$$

where we have used the notation $\Box_- = \bar{D}^{(c)2}D^{(c)2}$ and $\Box_+ = D^{(c)2}\bar{D}^{(c)2}$.

As in Subsect. 9.3.1, we only consider the NLSM (3.33) with $A_{ij\bar{k}} = 0$ for simplicity — see [651] for the general case. The operator F of (3.40) is then given by [633]

$$F_{ab} = K_{ad}(D^{(c)\alpha}D_\alpha^c)^d{}_b + (H_{\alpha\dot\alpha})_{ad}(D^{(c)\dot\alpha\alpha})^d{}_b + (T_\alpha)_{ad}(D^{(c)\alpha})^d{}_b + H_{ab} \, ,$$

$$F_{\bar{a}b} = (P_{\alpha\dot\alpha})_{\bar{a}d}(D^{(c)\dot\alpha\alpha})^d{}_b + L_{\bar{a}b} \, ,$$

$$\qquad (3.41)$$

where we have introduced the notation

$$K_{ab} = (\bar{D}\bar\Phi^{\bar{k}}\bar{D}\bar\Phi^{\bar{n}})[R_{b\bar{k}a\bar{n}} - T_{ab\bar{k}\bar{n}}] \, ,$$

$$(H_{\alpha\dot\alpha})_{ab} = \bar{D}_{\dot\alpha}\bar\Phi^{\bar{k}}D_\alpha\Phi^n \left[D_a H_{\bar{k}bn} + \tfrac{1}{2}D_n H_{\bar{k}ab} \right] + \tfrac{1}{2}\partial_{\alpha\dot\alpha}\bar\Phi^{\bar{k}}H_{\bar{k}ba} \, ,$$

$$(T_\alpha)_{ab} = \tfrac{1}{2}D^c_{\alpha\dot\alpha}\bar{D}^{\dot\alpha}\bar\Phi^{\bar{k}}H_{\bar{k}ab} + 2D_\alpha\Phi^k(\bar{D}\bar\Phi^{\bar{k}}\bar{D}\bar\Phi^{\bar{n}})\left[D_a T_{bk\bar{k}\bar{n}} \right.$$

$$\left. - \tfrac{1}{2}D_k T_{ab\bar{k}\bar{n}} \right] + \partial_{\alpha\dot\alpha}\bar\Phi^{\bar{k}}\bar{D}^{\dot\alpha}\bar\Phi^{\bar{n}} \left[D_a H_{b\bar{k}\bar{n}} - 2T_{ab\bar{k}\bar{n}} + \tfrac{1}{2}D_{\bar{k}}H_{\bar{n}ab} \right]$$

$$+ \partial_{\alpha\dot\alpha}\Phi^k\bar{D}^{\dot\alpha}\bar\Phi^{\bar{n}}\left[-D_a H_{\bar{n}kb} + \tfrac{1}{2}D_k H_{\bar{n}ab} \right] \, ,$$

$$(P_{\alpha\dot\alpha})_{\bar{a}b} = \left(\bar{D}_{\dot\alpha}\bar\Phi^{\bar{k}}D_\alpha\Phi^n \right)[2T_{bn\bar{a}\bar{k}} + D_{\bar{a}}H_{\bar{k}bn}] \, ,$$

$$H_{AB} = \tfrac{1}{2}(D^{(c)2}\Phi)^k(\bar{D}\bar{\Phi}^{\bar{k}}\bar{D}\bar{\Phi}^{\bar{n}})D_{[A}R_{k\bar{k}]B\bar{n}}$$

$$+ \tfrac{1}{2}(D\Phi^k D\Phi^n)(\bar{D}\bar{\Phi}^{\bar{k}}\bar{D}\bar{\Phi}^{\bar{n}})\left[\tfrac{1}{2}D_A D_B T_{kn\bar{k}\bar{n}} + 2T_{pn\bar{k}\bar{n}}R^p{}_{ABk}\right.$$

$$\left. -4T_{pnA\bar{n}}R^p{}_{B\bar{k}k} + R_{\bar{p}kAn}R^{\bar{p}}{}_{\bar{k}Bn}\right] + \tfrac{1}{2}(D^{(c)2}\Phi)^k(\bar{D}^{(c)2}\bar{\Phi})^{\bar{n}}R_{kAB\bar{n}}$$

$$- \tfrac{1}{2}D_\alpha\Phi^k\partial^{\dot{\alpha}\alpha}\Phi^L\bar{D}_{\dot\alpha}\bar{\Phi}^{\bar{n}}H_{\bar{n}Ap}R^p{}_{BLk} + \tfrac{1}{2}D_\alpha\Phi^k\partial^{\dot{\alpha}\alpha}\bar{\Phi}^{\bar{k}}\bar{D}_{\dot\alpha}\bar{\Phi}^{\bar{n}}$$

$$\times \left[D_A D_B H_{k\bar{k}\bar{n}} + H_{p\bar{k}\bar{n}}R^p{}_{ABk} + H_{kp[\bar{n}}R^p{}_{\underline{AB}\bar{k}]} - 2H_{A\bar{p}[\bar{n}}R^{\bar{p}}{}_{\underline{B}k\bar{k}]}\right.$$

$$\left. - H_{\bar{n}pk}R^p{}_{AB\bar{k}}\right] + \text{h.c.} ,$$

$$L_{\bar{a}b} = D^{(c)\alpha}\left\{2D_\alpha\Phi^n(\bar{D}^{(c)2}\bar{\Phi})^{\bar{k}}\left[R_{\bar{k}b\bar{a}n} - T_{bn\bar{a}\bar{k}}\right]\right.$$

$$+ 2D_\alpha\Phi^k(\bar{D}\bar{\Phi}^{\bar{k}}\bar{D}\bar{\Phi}^{\bar{n}})D_{[\bar{a}}T_{\underline{bk\bar{k}}]n} + \partial_{\dot\alpha\alpha}\bar{\Phi}^{\bar{k}}\bar{D}^{\dot\alpha}\bar{\Phi}^{\bar{n}}D_a H_{b\bar{k}\bar{n}} \qquad (3.42)$$

$$\left. +\partial_{\dot\alpha\alpha}\Phi^k\bar{D}^{\dot\alpha}\bar{\Phi}^{\bar{n}}\left[D_{[\bar{n}}H_{\bar{a}]kb} - 2T_{bk\bar{a}\bar{n}}\right]\right\} + H_{\bar{a}b} .$$

To calculate the UV divergences of the supersymmetric NLSM one-loop effective action

$$W = \tfrac{i}{2}\text{Tr}\ln\left[\frac{\delta^2 I^{(2)}}{\delta\xi^A\delta\xi^B}\right] , \qquad (3.43)$$

we use the (generalized) Schwinger-De Witt techniques in flat $N = 1$ superspace [652], after expanding (3.43) in powers of F. The divergent part of the one-loop effective action W is given by [633]

$$W_{\text{div}} = -\frac{3i}{4}\text{Tr}\ln\left[\frac{1}{\Box_-}\bar{D}^{(c)2}\delta^8(1,2)\right]$$

$$+ i\text{Tr}\left[D^{(c)2}F_{\bar{a}b}\frac{1}{\Box_-^2}\bar{D}^{(c)2}\delta^8(1,2)\right]$$

$$- i\text{Tr}\left[D^{(c)2}\bar{F}_{\bar{a}\bar{b}}D^{(c)2}\bar{D}^{(c)2}F_{bc}\frac{1}{\Box_-^4}\bar{D}^{(c)2}\delta^8(1,2)\right] \qquad (3.44)$$

$$- i\text{Tr}\left[D^{(c)2}F_{\bar{a}b}\bar{D}^{(c)2}D^{(c)2}F_{\bar{b}c}\frac{1}{\Box_-^4}\bar{D}^{(c)2}\delta^8(1,2)\right] + \text{h.c.}$$

Our calculation thus reduces to computing the universal functional traces $\text{Tr}D_A^c\cdots D_B^c\Box_-^{-k}\bar{D}^{(c)2}\delta^8(1,2)$. The Schwinger-De Witt representation for the kernel of the operator $\Box_-^{-k}$ in flat superspace is quite natural for this purpose. Let us consider the equation for the Green function G of the operator $\Box_-$,

$$\Box_- G(1,2) = \bar{D}_1^2\delta^8(1,2) , \qquad (3.45)$$

and use the integral representation of the Green function,

$$G^a{}_b(1,2) = i \int_0^\infty ds\, U^a{}_b(1,2;s) \,, \tag{3.46}$$

whose kernel satisfies the equation

$$\frac{\partial U}{\partial s} = -\Box_- U \,, \tag{3.47}$$

with the initial condition $U(1,2;0) = \bar{D}_1^2 \delta^8(1,2)$. The superfield *Ansatz* for a solution to (3.47) is given by (cf. [653, 654])

$$U(1,2;s) = -\frac{i}{4\pi^2} \frac{1}{s^2} \exp\left[\frac{i\omega(1,2)}{2s}\right] \sum_{n=0}^\infty (is)^n a_n(1,2) \,, \tag{3.48}$$

where we have introduced the 2-point scalar ω representing the invariant interval between the superspace points 1 and 2,

$$\omega(1,2) = \tfrac{1}{2}\omega^\mu \omega_\mu \,, \quad \omega^\mu = \Delta x^\mu - i(\Delta\theta\sigma^\mu\bar\theta_2) + i(\theta_1\sigma^\mu\Delta\bar\theta) \,, \tag{3.49}$$

and the coefficients $a_n(1,2)$ to be determined.

The operator $\Box_-$ (acting on a chiral superfield) reads

$$\Box_- = -2\hat{D}^{\dot\alpha\alpha} \hat{D}_{\alpha\dot\alpha} + R^\alpha D_\alpha^{(c)} + R \,, \tag{3.50a}$$

where we have introduced the notation

$$(\hat{D}^{\dot\alpha\alpha})^a{}_b = (\bar{D}^{(c)\dot\alpha} D^{(c)\alpha})^a{}_b = \tfrac{1}{2}\tilde\sigma_\mu^{\dot\alpha\alpha} D_b^{(c)\mu a} + R^a{}_{b\bar{k}p}\bar{D}^{\dot\alpha}\bar\Phi^{\bar{k}} D^\alpha \Phi^p \,,$$

$$(D^{(c)\mu})^a{}_b = \delta^a{}_b \partial^\mu + \Gamma^a{}_{bp} \partial^\mu \Phi^p \,,$$

$$(R^\alpha)^a{}_b = 2\bar{D}^2(\Gamma^a{}_{bp}) D^\alpha \Phi^p = -2\left[2R^a{}_{b\bar{k}p}\bar{D}_{\dot\alpha}\bar\Phi^{\bar{k}}\partial^{\dot\alpha\alpha}\Phi^p \right.$$

$$\left. +\bar{D}^{(c)}_{\dot\alpha}(R^a{}_{b\bar{k}p}\bar{D}^{\dot\alpha}\bar\Phi^{\bar{k}}) D^\alpha \Phi^k\right] \,, \tag{3.50b}$$

$$-\tfrac{1}{2}R^a{}_b = \left(D_\alpha^{(c)} R^\alpha\right)^a{}_b \,.$$

Substituting (3.50) into (3.47) leads to the recursion relations on the coefficients $a_n(1,2)$. The results in the coincidence limit $1 \to 2$ are [633]

$$\bar{D}_{\dot\alpha}\omega = 0 \,, \quad D_\alpha\omega|_{1=2} = D^{(c)\dot\alpha\alpha}\omega\Big|_{1=2} = D_\alpha D_\beta\omega|_{1=2} = 0 \,,$$

$$D^{(c)\dot\alpha\alpha} D^{(c)}_{\dot\beta\beta}\omega\Big|_{1=2} = \tfrac{1}{2}\delta^{\dot\alpha}{}_{\dot\beta}\delta^\alpha{}_\beta \,, \quad a_0(1,1) = a_1(1,1) = 0 \,,$$

$$D_\alpha^{(c)} D_\beta^{(c)} a_0\Big|_{1=2} = \tfrac{1}{2}\varepsilon_{\alpha\beta} \,, \quad D_\alpha^{(c)} a_1\Big|_{1=2} = \tfrac{1}{2}R_\alpha \,,$$

$$D^{(c2)} D^{(c)\dot\alpha\alpha} a_0 \Big|_{1=2} = \frac{1}{2} \bar\sigma^{\dot\alpha\alpha}_\mu \left[R^a{}_{b\bar k d} \bar\sigma^{\mu\dot\beta\beta} \bar D_{\underset{\beta}{\bullet}} \bar\Phi^{\bar k} D_\beta \Phi^d \right] . \tag{3.51}$$

It is now straightforward to deduce the UV-divergent part of the one-loop effective action (3.44) by using the results above and DR ($d = 2n$, $\varepsilon = n - 2$). The pole originates from the identity [647]

$$\int_0^\infty \frac{ds}{s^{n-1}} f(s, n) = -\frac{1}{\varepsilon} f(0, 2) \tag{3.52}$$

valid for any function $f(s, n)$ regular at $s = 0$ and $n = 2$. We find [633]

$$\begin{aligned}
W_{\text{div}} = \; & -\frac{1}{16\pi^2\varepsilon} \int \mathrm{d}^4 x \mathrm{d}^4\theta \Big\{ -\frac{3}{4} R^a{}_{b\bar k [k} \Gamma^b{}_{\underline{a}n]} \bar D\bar\Phi^{\bar k} \bar D^{(c)} D\Phi^n D\Phi^k \\
& -\frac{3}{4} R^{\bar a}{}_{\bar b k [\bar k} \Gamma^{\bar b}{}_{\underline{\bar a}\bar n]} D\Phi^k D^{(c)} \bar D\bar\Phi^{\bar n} \bar D\bar\Phi^{\bar k} + \frac{3}{2} R^{\bar a}{}_{\bar b k \bar k} R^{\bar b}{}_{\bar a n \bar n} \\
& \times (D\Phi^k D\Phi^n)(\bar D\bar\Phi^{\bar k} \bar D\bar\Phi^{\bar n}) + (P_{\underset{\alpha\alpha}{\bullet}})_{\bar a b} R^{b\bar a}{}_{\bar k n} \bar D^{\dot\alpha}\bar\Phi^{\bar k} D^\alpha \Phi^n - L^a{}_a \\
& + (P_{\underset{\alpha\alpha}{\bullet}})_{a\bar b} R^{\bar b a}{}_{k\bar n} D^\alpha \Phi^k \bar D^{\dot\alpha}\bar\Phi^{\bar n} - L^{\bar a}{}_{\bar a} - \frac{1}{8} (P_{\underset{\alpha\alpha}{\bullet}})^a{}_b (P^{\dot\alpha\alpha})^b{}_a \\
& -\frac{1}{8} (P_{\underset{\alpha\alpha}{\bullet}})^{\bar a}{}_{\bar b} (P^{\dot\alpha\alpha})^{\bar b}{}_{\bar a} + \frac{16}{3} K^{\bar a}{}_b K^b{}_{\bar a} \Big\} .
\end{aligned} \tag{3.53}$$

The final result is obtained after substitution of (3.42) into (3.53). Remarkably, (3.53) contains apparently non-covariant (Chern-Simons-like) terms that are explicitly dependent upon the connection (3.31). The integral on the right-hand side of (3.53) is, nevertheless, truly covariant.

Having assumed (for simplicity) the absence of the $A_{i\bar j\bar k}$ structure in our classical NLSM action (3.33), we get extra restrictions from the one-loop renormalizability (Subsect. 9.3.1),

$$D^a R_{k(\bar k \bar n)a} + D_{(\bar k} R_{\bar n)k} = 0 , \quad D^{\bar a} R_{\bar k(kn)\bar a} + D_{(k} R_{n)\bar k} = 0 . \tag{3.54}$$

Under these restrictions, the one-loop counterterm of the 4d, $N = 1$ supersymmetric fourth-order NLSM takes the form

$$I_R = \frac{1}{2\varepsilon} [I_1 + I_3 + I_4] , \tag{3.55}$$

where we have separated the structurally different terms,

$$I_1 = -\frac{1}{8\pi^2} \int \mathrm{d}^4 x \mathrm{d}^4\theta (D^{(c)2}\Phi)^k (\bar D^{(c)2}\bar\Phi)^{\bar n} R_{k\bar n} ,$$

$$I_3 = \frac{1}{8\pi^2} \int d^4x d^4\theta (D\Phi^k D\Phi^n)(\bar{D}\bar{\Phi}^{\bar{k}}\bar{D}\bar{\Phi}^{\bar{n}}) \left\{ \frac{3}{2} R^a{}_{b\bar{k}k} R^b{}_{a\bar{n}n} \right.$$

$$+ \frac{13}{3} R_{\bar{a}k\bar{b}n} R^{\bar{b}}{}_{\bar{k}}{}^{\bar{a}}{}_{\bar{n}} - \frac{1}{2} D^A D_A T_{kn\bar{k}\bar{n}} + T_{bn\bar{a}\bar{k}} T^{\bar{a}}{}_k{}^b{}_{\bar{n}} + \frac{16}{3} T_{ab\bar{k}\bar{n}} T^{ba}{}_{kn}$$

$$- T_{\bar{p}\bar{n}kn} R^{\bar{p}}{}_{\bar{k}} - T_{pn\bar{k}\bar{n}} R^p{}_k - \frac{16}{3}\left(T_{ab\bar{k}\bar{n}} R^a{}_k{}^b{}_n + T_{\bar{a}\bar{b}kn} R^{\bar{a}}{}_{\bar{k}}{}^{\bar{b}}{}_{\bar{n}} \right)$$

$$+ D^a H_{\bar{n}bk} T_{an}{}^b{}_{\bar{k}} + \frac{1}{4} D^a H_{\bar{k}bn} D^b H_{\bar{n}ak} - (D_{\bar{a}} H_{\bar{n}bk} + D_b H_{k\bar{a}\bar{n}}) R^{b\bar{a}}{}_{\bar{k}n} \right\},$$

$$I_4 = \frac{1}{8\pi^2} \int d^4x d^4\theta D\Phi^k D\bar{D}\bar{\Phi}^{\bar{k}} \bar{D}\bar{\Phi}^{\bar{n}}) \left\{ -\frac{3}{4} R^a{}_{\bar{b}k[\bar{n}} \Gamma^{\bar{b}}{}_{\bar{a}\bar{k}]} - \frac{1}{4} D^A D_A H_{k\bar{k}\bar{n}} \right.$$

$$\left. - \frac{1}{2} H_{k\bar{p}[\bar{n}} R^{\bar{p}}{}_{\bar{k}]} - \frac{1}{4} H_{p\bar{k}\bar{n}} R^p{}_k + \frac{1}{2} H_{a\bar{p}[\bar{n}} R^{\bar{p}a}{}_{\underline{k}\bar{k}]} \right\} + \text{h.c.}$$

$$(3.56)$$

According to the first line of (3.56), the $N = 1$ fourth-order NLSM metric renormalization is governed by the Ricci tensor, as in the bosonic case (Subsect. 9.3.1). The renormalization of the WZ-like term is dictated by I_4 in (3.56), while the latter can be rewritten to the manifestly covariant form similar to (3.35),

$$I_4 = +\frac{1}{8\pi^2} \int d^4x d^4\theta \int_0^1 d\eta\eta\, \Phi^n D\Phi^k D\bar{D}\bar{\Phi}^{\bar{k}} \bar{D}\bar{\Phi}^{\bar{n}} \left\{ -\frac{3}{4} R^a{}_{\bar{b}k[\bar{n}} R^{\bar{b}}{}_{\underline{a}n\bar{k}]} \right.$$

$$\left. - \frac{1}{4} D_n D^A D_A H_{k\bar{k}\bar{n}} - \frac{1}{2} D_n H_{k\bar{p}[\bar{n}} R^{\bar{p}}{}_{\bar{k}]} - \frac{1}{4} D_n H_{p\bar{k}\bar{n}} R^p{}_k \right.$$

$$\left. + D_n H_{a\bar{p}[\bar{n}} R^{\bar{p}a}{}_{\underline{k}\bar{k}]} \right\} + \text{h.c.}$$

$$(3.57)$$

The leading term (inside the curly brackets) in (3.57) is proportional to the first Pontryagin class of the NLSM target manifold $\mathcal{M}$.

The one-loop NLSM metric renormalization is absent for a Ricci-flat manifold $\mathcal{M}$, when $R_{k\bar{n}} = 0$. By using Bianchi identities for the curvature tensor, it is not difficult to verify that the restrictions (3.54) are also satisfied for a Ricci-flat $\mathcal{M}$. The fourth-order $N = 1$ NLSM is one-loop finite only if I_3 and I_4 vanish too. Their vanishing implies a complicated equation on the T and H tensors, whose explicit form can be easily read off from (3.56), similarly to the bosonic case (3.17), though we did not attempt to find its general solution. Instead, we restrict ourselves to the locally symmetric target spaces satisfying $D_p R_{k\bar{k}\bar{n}a} = D_{\bar{p}} R_{\bar{k}kn\bar{a}} = 0$. Equation (3.54) is then automatically satisfied, whereas (3.56) is considerably simplified. Moreover, we now consider the T and H tensors to be dependent upon the NLSM metric (cf. the preceding subsection). Stated differently, we investigate the special fourth-order $N = 1$ NLSM whose $N = 1$ superfield action reads [633]

$$I = \frac{1}{\lambda^2} \int d^4x d^4\theta \left\{ G_{i\bar{j}} (D^c D\Phi)^i (\bar{D}^c \bar{D}\bar{\Phi})^{\bar{j}} + T_{ij\bar{k}\bar{n}} (D\Phi^i D\Phi^j)(\bar{D}\bar{\Phi}^{\bar{k}} \bar{D}\bar{\Phi}^{\bar{n}}) \right\}$$

$$+ \frac{1}{\lambda^2} \int d^4x d^4\theta \int_0^1 d\eta \, \eta H_{nk\bar{k}\bar{n}} \Phi^n D\Phi^k D^c \bar{D}\bar{\Phi}^{\bar{k}} \bar{D}\bar{\Phi}^{\bar{n}} + \text{h.c.} ,$$

$$(3.58)$$

where $T = T(G)$ and $H = H(G)$. The simplest consistent choice of $\mathcal{M}$ is given by a Kählerian space of constant holomorphic section curvature c [196] with

$$R_{k\bar{k}n\bar{n}} = \tfrac{1}{2}c \, (G_{k\bar{k}} G_{n\bar{n}} + G_{k\bar{n}} G_{n\bar{k}}) . \qquad (3.59)$$

One arrives at a multiplicatively renormalizable $N = 1$ fourth-order NLSM if one chooses [633]

$$T_{kn\bar{k}\bar{n}} = 8\pi^2\alpha \, (G_{k\bar{k}} G_{n\bar{n}} + G_{k\bar{n}} G_{n\bar{k}}) ,$$

$$H_{kn\bar{k}\bar{n}} = 16\pi^2\beta \, (G_{k\bar{k}} G_{n\bar{n}} - G_{k\bar{n}} G_{n\bar{k}}) ,$$

$$(3.60)$$

with some real dimensionless coupling constants α and β. Having chosen λ^2 as the NLSM coupling constant, the absolute value of c is at our disposal. We take $c = \pm 16\pi^2$. The positive sign of c leads to complex projective spaces CP^k of complex dimension k, while the corresponding metric $G_{i\bar{j}}$ is known as the Fubini-Study metric [196].

The 4d QFT (3.58) then has three coupling constants. The one-loop RG equations for the effctive charges $\lambda^2(t)$, $\alpha(t)$ and $\beta(t)$ follow from the known one-loop counterterms (3.56) after substituting (3.59) and (3.60). We find $(t = \ln\mu)$ [633]

$$\frac{d}{dt}\left(\frac{1}{\lambda^2(t)}\right) = \pm (k-1) ,$$

$$\frac{d}{dt}\left(\frac{\alpha(t)}{\lambda^2(t)}\right) = -\frac{1}{12}\left\{(12k+292)\alpha^2 - (48k+560)\alpha + 12(k-1)\alpha\beta\right.$$

$$\left. + 3(k-1)\beta^2 - 24(k-1)\beta + 18k + 262\right\} ,$$

$$\frac{d}{dt}\left(\frac{\beta(t)}{\lambda^2(t)}\right) = \tfrac{1}{4}(k+1)(3+\beta) .$$

$$(3.61)$$

The first line of (3.61) is solved by

$$\lambda^2(t) = \frac{\lambda_0^2}{1 \pm (k-1)\lambda_0^2 t} . \qquad (3.62)$$

Changing the variables as

$$\frac{d\tau}{dt} = \pm\lambda^2(t) \qquad (3.63)$$

allows us to rewrite the rest of the RG equations (3.61) in the form

$$\frac{\mathrm{d}\alpha}{\mathrm{d}\tau} = -\frac{1}{12}\left\{(12k+292)\alpha^2 - (36k+548)\alpha + 12(k-1)\alpha\beta\right.$$
$$\left. +3(k-1)\beta^2 - 24(k-1)\beta + 18k + 262\right\}, \tag{3.64}$$

$$\frac{\mathrm{d}\beta}{\mathrm{d}\tau} = \frac{3(k+1)}{4}(1-\beta) .$$

A solution to the last equation is given by

$$\beta(\tau) = 1 + (\beta_0 - 1)\exp\left[-\tfrac{3}{4}(k+1)\tau\right], \tag{3.65}$$

so that $\beta \to 1$ when $\tau \to \infty$. As regards $\alpha(t)$, its asymptotic behaviour at $\tau \to \infty$ is easily deduced from (3.64). We find

$$\begin{aligned} \alpha \to \alpha_1 \quad \text{when} \quad (\alpha_1 - \alpha_2) > 0 , \\ \alpha \to \alpha_2 \quad \text{when} \quad (\alpha_1 - \alpha_2) < 0 , \end{aligned} \tag{3.66}$$

where $\alpha_{1,2}$ are the RG fixed points given by the roots of the quadratic equation

$$\alpha_{1,2} = \frac{(16k+140) \pm \sqrt{75k^2 + 180k - 329}}{6k + 146} . \tag{3.67}$$

Given the choice of upper sign in the supersymmetric RG equations above, the effective charge $\lambda^2(t)$ vanishes when the energy increases, $\tau \to \infty$, which is quite similar to the behaviour of the effective charge in 4d QCD and 2d (standard) NLSM on CP^k spaces. Unlike the bosonic case (Subsect. 9.3.1), the 4d, $N = 1$ supersymmetric fourth-order NLSM on CP^k space is asymptotically free uniformly in all its couplings since both $\alpha(t)$ and $\beta(t)$ approach constant values in the limit $\tau \to \infty$. Therefore, we conclude that supersymmetry is crucial for stabilization of asymptotic freedom in the fourth-order 4d NLSM.

9.3.3 4d, $N = 2$ NLSM with Higher Derivatives

The most general 4d, manifestly $N = 2$ supersymmetric fourth-order NLSM in terms of hypermultiplets can be formulated in harmonic superspace (Subsects. 4.4.2 and 4.4.3). In this subsection we consider non-linear 4d, $N = 2$ supersymmetric QFT in terms of Abelian $N = 2$ vector multiplets with higher derivatives (of fourth-order in components) [634]. This QFT can be formulated in conventional 4d, $N = 2$ superspace (Subsect. 4.2.1) in terms of the restricted chiral $N = 2$ superfields $\{\Phi^a\}$ satisfying the off-shell constraints (4.2.26) and (4.2.27). The component content of the $N = 2$ Abelian vector multiplet is given by (4.2.29) with the Bianchi 'identity' (4.2.30), while the natural $N = 2$ supersymmetric fourth-order action in $N = 2$ superspace reads (cf. Subsect. 8.2.3)

$$I = \frac{1}{16} \int \mathrm{d}^4 x \mathrm{d}^4 \theta \mathrm{d}^4 \bar{\theta}\, \mathcal{H}(\Phi, \bar{\Phi}) \,, \tag{3.68}$$

with an arbitrary potential $\mathcal{H}(\Phi, \bar{\Phi})$. We call the QFT action (3.68) the 4d, $N = 2$ supersymmetric fourth-order NLSM action, even though it contains propagating vector fields in addition to propagating scalars, as is required by $N = 2$ supersymmetry. The bosonic terms of the action (3.68) in components are given by [634]

$$
\begin{aligned}
I_{\mathrm{bos.}} = \tfrac{1}{2} \int \mathrm{d}^4 x \Big\{ &\mathcal{H}_{a\bar{b}} \Big[\Box A^a \Box \bar{A}^b + \tfrac{1}{4} \left(F^a_{\mu\nu} - \mathrm{i} \tilde{F}^a_{\mu\nu} \right) \partial_\nu \partial_\rho \tilde{F}^b_{\rho\mu} \\
&- \tfrac{7}{24} \partial_\mu C^a_m \partial_\mu C^b_m \Big] + \mathcal{H}_{a\bar{b}\bar{c}} \Big[\Box A^a \partial_\mu \bar{A}^b \partial_\mu \bar{A}^c - \tfrac{7}{24} \partial_\mu \bar{A}^c C^b_m \partial_\mu C^a_m \\
&- \tfrac{1}{16} \Box \bar{A}^a C^b_m C^c_m + \tfrac{1}{32} \Box \bar{A} F^c_{\mu\nu} \left(F^b_{\mu\nu} - \mathrm{i} \tilde{F}^b_{\mu\nu} \right) \Big] \\
&+ \tfrac{1}{2} K_{a\bar{b}\bar{c}d} \Big[\partial_\mu A^a \partial_\mu A^d \partial_\nu \bar{A}^b \partial_\nu \bar{A}^c + \tfrac{1}{8} (C^a_m C^d_m)(C^b_n C^c_n) \Big] \\
&+ \tfrac{1}{32} F^d_{\mu\nu} \left(F^a_{\mu\nu} + \mathrm{i} \tilde{F}^a_{\mu\nu} \right) F^c_{\rho\lambda} F^b_{\rho\lambda} + \tfrac{1}{32} F^d_{\mu\nu} \left(F^a_{\mu\nu} - \mathrm{i} \tilde{F}^a_{\mu\nu} \right) F^c_{\rho\lambda} \tilde{F}^b_{\rho\lambda} \\
&- \tfrac{3}{4} \partial_\mu \bar{A}^c \partial_\mu \bar{A}^a (C^b_m C^d_m) + \tfrac{1}{16} (C^c_m C^b_m) \left(F^a_{\mu\nu} + \mathrm{i} \tilde{F}^a_{\mu\nu} \right) F^d_{\mu\nu} \\
&+ \tfrac{1}{16} (C^d_m C^a_m) F^c_{\mu\nu} \left(F^b_{\mu\nu} - \mathrm{i} \tilde{F}^b_{\mu\nu} \right) \\
&+ \tfrac{3}{4} \partial_\nu \bar{A}^c \partial_\rho A^d \left(F^a_{\rho\mu} + \mathrm{i} \tilde{F}^a_{\rho\mu} \right) \left(F^b_{\mu\nu} - \mathrm{i} \tilde{F}^b_{\mu\nu} \right) \Big] + \mathrm{h.c.} \Big\} \,,
\end{aligned}
\tag{3.69}
$$

where we have simplified our notation by removing bars on the field indices (but not on the indices of $\mathcal{H}$!), and we have defined

$$\mathcal{H}_{a\cdots\bar{b}\cdots} \equiv \frac{\partial^n \mathcal{H}(A, \bar{A})}{\partial A^a \cdots \partial \bar{A}^b \cdots} \,. \tag{3.70}$$

The NLSM (3.68) is formally renormalizable if one understands its renormalization in the generalized sense, namely, as a quantum deformation of the NLSM geometry defined by the potential $\mathcal{H}$, as usual (Chap. 2). The one-loop UV divergences of the (Euclidean) quantum effective action for the QFT (3.68) can be deduced from the generating functional of its (connected) Green functions,

$$
\begin{aligned}
Z[J] = \exp\{-W[J]\} = \int [\mathrm{d}\Phi \mathrm{d}\bar{\Phi}] \exp \Big\{ &- \int \mathrm{d}^4 x \mathrm{d}^8 \theta\, \mathcal{H}(\Phi, \bar{\Phi}) \\
&- \int \mathrm{d}^4 x \mathrm{d}^4 \theta\, \Phi^a J_a - \int \mathrm{d}^4 x \mathrm{d}^4 \bar{\theta}\, \bar{\Phi}^a \bar{J}_a \Big\} \,,
\end{aligned}
\tag{3.71}
$$

where we have introduced the $N = 2$ chiral sources J_a and their conjugates. Let us now define the mean fields

$$\phi^a = \frac{\delta W}{\delta J_a} , \quad \bar{\phi}^a = \frac{\delta W}{\delta \bar{J}_a} , \tag{3.72}$$

and introduce the generating functional of the one-particle-irreducible Green functions (cf. Sect. 2.2)

$$\Gamma(\phi, \bar{\phi}) = W(J, \bar{J}) - \phi^a J_a - \bar{\phi}^a \bar{J}_a . \tag{3.73}$$

The *linear* background-quantum field splitting is most convenient for the one-loop calculations,

$$\exp[-\Gamma(\phi, \bar{\phi})] = \int [\mathrm{d}\Phi \mathrm{d}\bar{\Phi}] \exp\left\{-\mathcal{H}(\phi + \Phi, \bar{\phi} + \bar{\Phi}) - \mathcal{H}_a \Phi^a - \mathcal{H}_{\bar{a}} \bar{\Phi}^a\right\} \tag{3.74}$$

in the condensed notation where all coordinate integrations are implicit. Expanding $\mathcal{H}(\phi + \Phi, \bar{\phi} + \bar{\Phi})$ up to second order in the quantum fields Φ, $\bar{\Phi}$ and integrating over the latter in the Gaussian (one-loop) approximation yield the one-loop effective action

$$\begin{aligned}
\Gamma^{(1)}(\phi, \bar{\phi}) = \ & I(\phi, \bar{\phi}) - \ln \int [\mathrm{d}\Phi \mathrm{d}\bar{\Phi}] \exp\left\{-\int \mathrm{d}^4 x \mathrm{d}^8 \theta \, \mathcal{H}_{a\bar{b}} \Phi^a \bar{\Phi}^b \right. \\
& \left. -\frac{1}{2} \int \mathrm{d}^4 x \mathrm{d}^4 \theta \, (\bar{D}^4 \mathcal{H}_{ab}) \Phi^a \Phi^b - \frac{1}{2} \int \mathrm{d}^4 x \mathrm{d}^4 \bar{\theta} \, (D^4 \mathcal{H}_{\bar{a}\bar{b}}) \bar{\Phi}^a \bar{\Phi}^b \right\} .
\end{aligned} \tag{3.75}$$

The last two terms in (3.75) do not contribute to the one-loop UV divergences. As regards the UV divergent part of the one-loop effective action, we have

$$\Gamma^{(1)}_{\mathrm{div.}}(\phi, \bar{\phi}) = \mathrm{Tr}\ln\left[\mathcal{H}_{a\bar{b}} D^4 \bar{D}^4 \delta^{12}(\mathcal{Z}_1 - \mathcal{Z}_2)\right]\Big|_{\mathrm{div.}} , \tag{3.76}$$

where

$$\delta^{12}(\mathcal{Z}_1 - \mathcal{Z}_2) = \int \mathrm{d}^4 p \exp[ip \cdot (x_1 - x_2)] \delta^8(\theta_1 - \theta_2) . \tag{3.77}$$

Equation (3.76) can be rewritten in the more explicit form,

$$\begin{aligned}
\Gamma^{(1)}_{\mathrm{div.}} = \ & \mathrm{Tr}\ln\left[\delta_{a\bar{b}} D^4 \bar{D}^4 + (\mathcal{H}_{a\bar{b}} - \delta_{a\bar{b}}) D^4 \bar{D}^4\right] \delta^{12}(\mathcal{Z}_1 - \mathcal{Z}_2) \\
= \ & \mathrm{Tr}\ln\left[\delta_{a\bar{b}} D^4 \bar{D}^4 \delta^{12}(\mathcal{Z}_1 - \mathcal{Z}_2)\right] + \mathrm{Tr}\left\{G_{ab} \frac{D^4 \bar{D}^4}{\Box^2} \delta^{12}(\mathcal{Z}_1 - \mathcal{Z}_2)\right. \\
& \left. -\frac{1}{2} G_{a\bar{b}} \frac{D_1^4 \bar{D}_1^4}{\Box_1^2} \delta^{12}(\mathcal{Z}_1 - \mathcal{Z}_2) G_{\bar{b}c} \frac{D_2^4 \bar{D}_2^4}{\Box_2^2} \delta^{12}(\mathcal{Z}_1 - \mathcal{Z}_2) + \ldots\right\} ,
\end{aligned} \tag{3.78}$$

where $G_{a\bar{b}} = \mathcal{H}_{a\bar{b}} - \delta_{a\bar{b}}$ and $\Box = \partial_\mu \partial^\mu$. We normalize the delta-function here by the condition $D^4 \bar{D}^4 \delta^8(\theta_1 - \theta_2)\big| = 1$.

The first term in (3.78) is background-independent, so that it can be ignored. Calculation of the remaining (one-loop) terms in (3.78) is straight-forward: one integrates by parts in order to release $\delta^8(\theta_1 - \theta_2)$, while $G_{a\bar{b}}$

should not be differentiated in the divergent terms for dimensional reasons. We find [634]

$$\mathrm{Tr}\, G_{a\bar{b}} \frac{D_1^4 \bar{D}_1^4}{\Box_1^2} \delta^{12}(\mathcal{Z}_1 - \mathcal{Z}_2) = \int \mathrm{d}^4 x \mathrm{d}^8\theta \, G_{a\bar{a}} \int \frac{\mathrm{d}^4 p}{(p^2 + m^2)^2} \ ,$$

$$\mathrm{Tr}\, G_{a\bar{b}} G_{\bar{b}c} \frac{D_1^4 \bar{D}_1^4}{\Box_1^2} \delta^{12}(\mathcal{Z}_1 - \mathcal{Z}_2) \frac{D_1^4 \bar{D}_1^4}{\Box_1^2} \delta^{12}(\mathcal{Z}_1 - \mathcal{Z}_2)$$

$$= -\frac{1}{2} \int \mathrm{d}^4 x \mathrm{d}^8\theta \, G_{a\bar{b}} G_{\bar{b}a} \int \frac{\mathrm{d}^4 p}{(p^2 + m^2)^2} \ , \quad \dots \tag{3.79}$$

where we have introduced the mass m as the IR regulator. It follows that

$$\Gamma^{(1)}_{\mathrm{div.}} = \int \mathrm{d}^4 x \mathrm{d}^8\theta \left[G_{a\bar{a}} - \tfrac{1}{2} G_{a\bar{b}} G_{\bar{b}a} + \tfrac{1}{3} G_{a\bar{b}} G_{\bar{b}c} G_{c\bar{a}} + \dots \right] \int \frac{\mathrm{d}^4 p}{(p^2 + m^2)^2}$$

$$= \mathrm{Tr}\ln[\delta_{a\bar{b}} - G_{a\bar{b}}] \int \frac{\mathrm{d}^4 p}{(p^2 + m^2)^2} \ ,$$

$$\tag{3.80}$$

or, equivalently,

$$\Gamma^{(1)}_{\mathrm{div.}} = -\frac{1}{16\pi^2(4-d)} \mathrm{Tr}\ln\left[\mathcal{H}_{a\bar{b}}(\phi, \bar{\phi}) \right] \ . \tag{3.81}$$

Equation (3.81) shows that the one-loop renormalization of the 4d, $N = 2$ NLSM (3.68) is a quantum deformation of the NLSM target space geometry described by the potential $\mathcal{H}$ [634].

At the end of this subsection, we would like to mention that the use of $N = 2$ extended supersymmetry and higher derivatives of fourth-order allows us to construct a UV-*finite* QFT to all orders of quantum perturbation theory in 4d [655]. The basic example is provided by the $N = 2$ superspace action (cf. Sect. 8.2)

$$I_f = -\tfrac{1}{2} \int \mathrm{d}^4 x \mathrm{d}^8\theta \, \bar{\Phi}\Phi + \left[\alpha^2 \int \mathrm{d}^4 x \mathrm{d}^4\theta \, \mathcal{F}(\Phi) + \mathrm{h.c.} \right] \ , \tag{3.82}$$

with arbitrary $N = 2$ chiral potential $\mathcal{F}(\Phi)$, and the coupling constant α of the dimension of mass [655]. The first term of (3.82) in components is given by

$$I_1 = -\tfrac{1}{2} \int \mathrm{d}^4 x \mathrm{d}^8\theta \, \bar{\Phi}\Phi = -\tfrac{1}{2} \int \mathrm{d}^4 x \mathrm{d}^4\theta \, \Phi\Box\Phi$$

$$= -\int \mathrm{d}^4 x \left[\bar{A}\Box^2 A - \mathrm{i}\psi^{\alpha i} \partial_{\alpha\dot{\beta}} \Box\bar{\psi}_i^{\dot{\beta}} + C_m \Box C_m + \tfrac{1}{2} F^{\mu\nu}\Box F_{\mu\nu} \right] \ . \tag{3.83}$$

The second (interaction) term in (3.82) was discussed (in a different context) in Sects. 4.2 and 8.2. The 4d, $N = 2$ superfield Feynman rules for the QFT (3.82) are similar to those found in Sect. 4.2, but with the $1/p^4$-factor in the superpropagator. It is not difficult to prove, along the lines of Subsect. 4.2.3, that the divergence index of any $N = 2$ supergraph in the 4d QFT (3.82) is positive. We are not aware of physical applications of this model.

9.4 Extended 4d Supergravities and U-Duality

The maximal 11d supergravity (Subsect. 8.4.1) can be compactified or dimensionally reduced to lower spacetime dimensions. As a result, one gets scalars amongst the physical field components, whose kinetic action takes the form of a NLSM, just because of the non-linearity of the 11d supergravity action. The classical structure of this NLSM is of great interest since it is entirely determined by the Einstein action and extended (local) supersymmetry.

For example, after being dimensionally reduced down to 4d, an elfbein of 11d supergravity gives rise to a vierbein, seven photons and 28 physical scalars in 4d, whereas the three-form of 11d supergravity yields 21 photons and 42 physical scalars in 4d. The 4d (graviton, gravitino, photon, photino, scalar) numbers $(1, 8, 28, 56, 70)$, respectively, in the 4d, $N = 8$ supergravity exactly correspond to the dimensions of antisymmetric tensor representations of $SU(8)$. This is not accidental since 4d, $N = 8$ supergravity has a large symmetry structure [545, 585]. For example, all its scalars can be united into a complex $SU(8)$ tensor $\phi^{ijkl} = A^{ijkl} + iB^{ijkl}$ satisfying the relation

$$(\phi^*)_{i_1 i_2 i_3 i_4} \equiv \phi_{i_1 i_2 i_3 i_4} = \varepsilon_{i_1 i_2 i_3 i_4 i_5 i_6 i_7 i_8} \phi^{i_5 i_6 i_7 i_8} \ . \tag{4.1}$$

If central charges in the supersymmetry algebra can be ignored on-shell, one should, therefore, expect the $SU(8)$ global symmetry of the 4d, $N = 8$ supergravity equations of motion. The actual on-shell symmetry may be even larger (see below).

Associated with dimensional reduction $11 \to 4$ is the local $SO(7)$ symmetry originating from a decomposition of the 11d Lorentz group, $SO(1, 10) \to SO(1, 3) \times SO(7)$. Since 4d, $N = 8$ supergravity has 28 propagating vector fields that may be assigned in the adjoint of $SO(8)$, an off-shell $SO(8)$ local symmetry is to be expected, whose $SO(7)$ part can be manifestly realized in the 4d, $N = 8$ supergravity Lagrangian. Remarkably, the 4d, $N = 8$ supergravity theory actually possesses a larger local $SU(8)$ symmetry, with the extra $(63 - 28 = 35)$ connections being represented by composite gauge fields [545, 585]. Given the local $SU(8)$ symmetry having 63 generators, and 70 physical scalars parametrizing the NLSM target manifold $\mathcal{M}$, the latter may be identified with a coset G/H where $H = SU(8)$. The dimension of G is therefore given by $63 + 70 = 133$ which exactly coincides with the dimension of a non-compact simple Lie group E_7. The full symmetry group of 4d, $N = 8$ supergravity is indeed given by the product of the local $SU(8)$ symmetry and the global E_7 symmetry [545, 585]. Since the *non-compact E_7* symmetry is *non-linearly* realized in 4d, $N = 8$ supergravity, it does not lead to ghosts. It is also important to note that the E_7 symmetry is an on-shell symmetry, while it rotates the gauge field strengths, not the gauge fields themselves. In other words, the E_7 symmetry should be interpreted as the non-Abelian generalization of the (strong-weak coupling) electric-magnetic (S) duality in field theory. The global $SU(8)$ symmetry mentioned above is the maximal compact subgroup of the non-compact E_7 symmetry.

To give the simplest example, let us consider the 4d, $N = 4$ supergravity [656] which has the lowest number of 4d supersymmetries needed for the presence of physical scalars in a supergravity multiplet, namely, a dilaton ϕ and an axion B. The appearance of the $SU(4)$ local symmetry in 4d, $N = 4$ supergravity is obvious since the latter can be derived by dimensional reduction of 10d, $N = 1$ supergravity [656], which implies a Lorentz group decomposition $SO(1,9) \rightarrow SO(1,3) \times SO(6)$, while $SO(6) \cong SU(4)$. The vector gauge fields of 4d, $N = 4$ supergravity transform in $\mathbf{6}$ of $SU(4)$. Two real scalars (ϕ, B) can be unified into a single complex scalar, $Y = e^{-2\phi} - 2iB$, while the scalar sector of the 4d, $N = 4$ supergravity Lagrangian is given by the NLSM [656]

$$\mathcal{L}(Y) = \frac{1}{(Y + \bar{Y})^2} \partial_\mu Y \partial^\mu \bar{Y} \ . \tag{4.2}$$

It is not difficult to verify that (4.2) is invariant under the $SU(1,1)$ transformations

$$Y \ \rightarrow \ \frac{aY + ib}{icY + d} \ , \tag{4.3}$$

whose real parameters (a, b, c, d) are subject to the relation $ad + bc = 1$. A discrete subgroup $SL(2, \mathbf{Z})$ of $SL(2, \mathbf{R}) \cong SU(1,1)$ can be identified with the (electric-magnetic) S-duality group [657]. Truncating the 4d, $N = 8$ supergravity down to 4d, $N = 4$ supergravity gives rise to the $SU(4) \times SU(1,1)$ symmetry of the latter as the remnant of the $SU(8) \times E_7$ symmetry of the former.

Since the 11d supergravity is the LEEA of M-theory (Sect. 8.4), extended supergravity theories in various dimensions lower than eleven should be considered as the LEEA of the corresponding compactified superstrings. Perturbative M-theory is supposed to reproduce the known perturbative theory of superstrings. Non-perturbative M-theory is supposed to have a very rich symmetry structure including all known S- and T-dualities of compactified superstrings, relating their strongly coupled phases to weakly coupled phases and momentum modes to winding modes, respectively. The S-duality relations imply non-trivial equivalences between perturbatively very different superstring theories. For instance, let us consider type-II superstrings toroidally compactified to d dimensions [583]. The LEEA describing the massless fields of the superstring is a d-dimensional supergravity with a discrete 'duality' group G. Since G must also be the symmetry of the supergravity equations of motions, its continuous (rigid) extension may be identified with the Lie group G in the coset G/H describing the scalar NLSM sector of the supergravity, where H is the maximal compact subgroup of G. One then conjectures [657] that the continuous Lie symmetry G of the d-dimensional supergravity is broken down to the discrete subgroup $G(\mathbf{Z})$ in d-dimensional (compactified) superstring theory. This conjecture is supported by the observation that the quantized charges of solitonic Dp-branes belong to a lattice, while the discrete duality group $G(\mathbf{Z})$ leaves the charge lattice invariant [583]. The discrete symmetry $G(\mathbf{Z})$ is called U-duality since it unifies S- and T-dualities of the superstring. The list of conjectured U-dualities is summarized in Table 9.1 [658].

Table 9.1. Duality symmetries of type-II superstrings compactified to d dimensions

Spacetime dimension d	Supergravity duality group G	String T-duality	U-duality
10A	$SO(1,1)/\mathbf{Z}_2$	1	1
10B	$SL(2,\mathbf{R})$	1	$SL(2,\mathbf{Z})$
9	$SL(2,\mathbf{R}) \times SO(1,1)$	1	$SL(2,\mathbf{Z})$
8	$SL(3,\mathbf{R}) \times SL(2,\mathbf{R})$	$O(2,2;\mathbf{Z})$	$SL(3,\mathbf{Z}) \times SL(2,\mathbf{Z})$
7	$SL(5,\mathbf{R})$	$SO(3,3;\mathbf{Z})$	$SL(5,\mathbf{Z})$
6	$O(5,5)$	$SO(4,4;\mathbf{Z})$	$SO(5,5;\mathbf{Z})$
5	E_6	$SO(5,5;\mathbf{Z})$	$E_6(\mathbf{Z})$
4	E_7	$SO(6,6;\mathbf{Z})$	$E_7(\mathbf{Z})$
3	E_8	$SO(7,7;\mathbf{Z})$	$E_8(\mathbf{Z})$
2	E_9	$SO(8,8;\mathbf{Z})$	$E_9(\mathbf{Z})$
1	E_{10}	$SO(9,9;\mathbf{Z})$	$E_{10}(\mathbf{Z})$

The Lie algebra of E_9 is actually the AKM algebra associated with E_8 [659], while the (hyperbolic) algebra E_{10} can be associated with the E_{10} Dynkin diagram [660, 661]. The duality symmetries arising in two dimensions ($d = 2$) include the infinite-dimensional *Geroch* symmetry groups of toroidally compactified general relativity [659]. So much can be learned just by studying the NLSM in the scalar sector of an extended supergravity in d spacetime dimensions!

References

1. A.M. Polyakov: *Gauge Fields and Strings* (Harwood Academic Publ., 1987)
2. T.H.R. Skyrme: Nucl. Phys. **31**, 556 (1962)
3. S.V. Ketov: *Non-Linear Sigma-Models in Quantum Field Theory and Strings* (Nauka Publ., 1992)
4. V. de Alfaro, S. Fubini, G. Furlan: Nuovo Cim. **48**, 485 (1978)
5. R. Rajaraman: *Solitons and Instantons* (Elsevier Publ., 1982)
6. W.J. Zakrzewski: *Low-Dimensional Sigma Models* (IOP Publ., 1989)
7. A.M. Perelomov: Phys. Rep. **174**, 229 (1989)
8. V.G. Makhankov, Y.P. Rybakov, V.I. Sanyuk: *The Skyrme Model: Fundamentals, Methods, Applications* (Springer, 1993)
9. C.N. Yang, R.L. Mills: Phys. Rev. **96**, 191 (1954)
10. P.W. Higgs: Phys. Rev. Lett. **13**, 508 (1964)
11. G. t'Hooft: 'The Confinement Phenomenon in Quantum Field Theory'. In: *Under the Spell of the Gauge Principle* (World Scientific, 1981) p. 514
12. G. t'Hooft: Nucl. Phys. **B190**, 455 (1981)
13. S. Coleman, J. Wess, B. Zumino: Phys. Rev. **177**, 2239 (1969)
    C.G. Callan Jr., S. Coleman, J. Wess, B. Zumino: Phys. Rev. **177**, 2247 (1969)
14. D. Gross, F. Wilczek: Phys. Rev. Lett. **30**, 1343 (1973)
15. H.D. Politzer: Phys. Rev. Lett. **30**, 1346 (1973)
16. N. Seiberg, E. Witten: Nucl. Phys. **B426**, 19 (1994)
17. N. Seiberg, E. Witten: Nucl. Phys. **B431**, 484 (1994)
18. *Three Hundred Years of Gravitation*, ed. by S.W. Hawking, W. Israel (Cambridge Univ. Press, 1987)
19. M.H. Goroff, A. Sagnotti: Phys. Lett. **B160**, 81 (1985)
20. S. Deser, J.H. Kay: Phys. Lett. **B76**, 400 (1978)
21. R.E. Kallosh: Phys. Lett. **B99**, 122 (1981)
22. P.S. Howe, U. Lindström: Nucl. Phys. **B181**, 487 (1981)
23. V.A. Novikov, M.A. Shifman, A.I. Vainstein, V.I. Zakharov: Phys. Rep. **116**, 103 (1984)
24. D.M. Haldane: Phys. Rev. Lett. **50**, 1153 (1983)
25. H. Levine, S. Libby, A. Pruisken: Phys. Rev. Lett. **51**, 1915 (1983)
26. G.E. Volovik: *Exotic Properties of Superfluid He3-A* (World Scientific, 1992)
27. S. Weinberg: *The Quantum Theory of Fields* (Cambridge Univ. Press, 1995)
28. C.M. Hull: 'Lectures on Non-Linear Sigma-Models and Strings'. In: *Super Field Theories*, ed. by H.C. Lee, et al. (Plenum Press, 1987) p. 77
29. A.A. Tseytlin: Int. J. Mod. Phys. **A6**, 1257 (1989)
30. S. Randjbar-Daemi, A. Salam, J. Strathdee: Int. J. Mod. Phys. **A2**, 667 (1987)
31. M. Lüscher, K. Pohlmeyer: Nucl. Phys. **B137**, 46 (1978)
32. A.B. Zamolodchikov, Al.B. Zamolodchikov: Ann. Phys. (N.Y.) **120**, 253 (1979)
33. V.L. Golo, A.M. Perelomov: Phys. Lett. **79B**, 112 (1978)

34. D. Friedan: Phys. Rev. Lett. **45**, 1057 (1980); Ann. Phys. (N.Y.) **163**, 318 (1985)
35. B.E. Fridling, A.E.M. van de Ven: Nucl. Phys. B**268**, 719 (1986)
36. R.D. Pisarski: Phys. Rev. D**20**, 3358 (1979)
37. F. Delduc, G. Valent: Nucl. Phys. B**253**, 494 (1985)
38. A.V. Bratchikov, I.V. Tyutin: Teor. Mat. Fiz. **66**, 360 (1986)
39. A. Blasi, F. Delduc, S.P. Sorella: The Background-Quantum Split Symmetry in Two-Dimensional σ-models, a Regularization-Independent Proof of its Renormalizability. CERN preprint TH. 5046 (1988)
40. C. Becchi, O. Piguet: Nucl. Phys. B**315**, 153 (1989)
41. H. Osborn: Nucl. Phys. B**294**, 595 (1987)
42. H. Osborn: Phys. Lett. B**214**, 555 (1988)
43. H. Osborn: Nucl. Phys. B**308**, 629 (1988)
44. E. Braaten, T.L. Curtright, C.K. Zachos: Nucl. Phys. B**260**, 630 (1985)
45. S.V. Ketov: *Conformal Field Theory* (World Scientific, 1995)
46. J. Wess, B. Zumino: Phys. Lett. B**37**, 95 (1971)
47. S.P. Novikov: Sov. Math. Dokl. **24**, 222 (1981)
48. E. Witten: Nucl. Phys. B**223**, 422 (1983)
49. A.M. Polyakov, P.B. Wiegmann: Phys. Lett. B**131**, 121 (1983)
50. A.M. Polyakov, P.B. Wiegmann: Phys. Lett. B**141**, 223 (1984)
51. A.A. Belavin, A.M. Polyakov, A.B. Zamolodchikov: Nucl. Phys. B**241**, 333 (1984)
52. V.G. Knizhnik, A.B. Zamolodchikov: Nucl. Phys. B**247**, 83 (1984)
53. E. Witten: Commun. Math. Phys. **92**, 455 (1984)
54. P. Di Vecchia: The Wess-Zumino Action in Two Dimensions and Non-Abelian Bosonization. Copenhagen preprint NBI-HE-02 (1984)
55. C. Itzykson, J-B. Zuber: *Quantum Field Theory* (McGraw-Hill, 1987)
56. M.E. Peskin, D. Schröder: *An Introduction to Quantum Field Theory* (Addison-Wesley, 1995)
57. L.F. Abbott: Nucl. Phys. B**185**, 189 (1981)
58. J. Honerkamp, G. Ecker: Nucl. Phys. B**35**, 481 (1971)
59. J. Honerkamp: Nucl. Phys. B**36**, 130 (1971)
60. S. Ichinose, M. Omote: Nucl. Phys. B**203**, 221 (1982)
61. I. Jack, H. Osborn: Nucl. Phys. B**234**, 331 (1984)
62. S. Mukhi: Nucl. Phys. B**264**, 640 (1986)
63. S.V. Ketov, A.I. Samolov: The Background-Field Method in the Two-Dimensional Non-Linear Sigma-Models. Tomsk preprint TF/SB/AS-2 (1988)
64. U. Müller, Ch. Schubert, A.E.M. van de Ven: Gen. Rel. Grav. **31**, 1759 (1999)
65. G. Leibrandt: Rev. Mod. Phys. **47**, 849 (1975)
66. G.A. Vilkovisky: Nucl. Phys. B**234**, 125 (1984)
67. P.S. Howe, G. Papadopoulos, K.S. Stelle: Nucl. Phys. B**296**, 26 (1986)
68. B.L. Voronov, I.V. Tyutin: Jadern. Fiz. **33**, 1137 (1981)
69. G. t'Hooft, M. Veltman: Nucl. Phys. B**44**, 189 (1971)
70. G. t'Hooft: Nucl. Phys. B**61**, 455 (1973)
71. M.T. Grisaru, A.E.M. van de Ven, D. Zanon: Nucl. Phys. B**277**, 388 (1986)
72. M.T. Grisaru, A.E.M. van de Ven, D. Zanon: Nucl. Phys. B**277**, 409 (1986)
73. M.T. Grisaru, D.I. Kazakov, D. Zanon: Nucl. Phys. B**287**, 189 (1987)
74. C. Nash: *Relativistic Quantum Fields* (Academic Press, 1978)
75. J.C. Collins: *Renormalization: An Introduction to Renormalization Group and Operator-Product Expansion* (Cambridge Univ. Press, 1984)
76. H.W. Braden, D.R.T. Jones: Phys. Rev. D**35**, 1519 (1987)
77. G. Curci, G. Paffuti: Nucl. Phys. B**312**, 227 (1989)
78. P. Breitenlohner, D. Maison: Commun. Math. Phys. **52**, 11 (1977)

79. M. Bos: Ann. Phys. (N.Y.) **181**, 177 (1988)
80. S.V. Ketov: The Three-Loop Beta-Function of the Two-Dimensional Non-Linear Sigma-Model with a Wess-Zumino-Novikov-Witten Term. Report at the Int. Workshop on Superstrings at ICTP, Trieste, Italy (1988), unpublished
81. D.M. Capper, D.R.T. Jones, P. van Nieuwenhuizen: Nucl. Phys. **B167**, 479 (1980)
82. S.V. Ketov: Nucl. Phys. **B294**, 813 (1987)
83. C.M. Hull, P.K. Townsend: Phys. Lett. **B191**, 115 (1987)
84. R.R. Metsaev, A.A. Tseytlin: Phys. Lett. **B191**, 354 (1987)
85. R.R. Metsaev, A.A. Tseytlin: Nucl. Phys. **B293**, 385 (1987)
86. D. Zanon: Phys. Lett. **B191**, 363 (1987)
87. A.B. Zamolodchikov: ZhETF **43**, 565 (1986)
88. T.L. Curtright, C.K. Zachos: Phys. Rev. Lett. **53**, 1799 (1984)
89. S.V. Ketov: Sov. Phys. J. **31**, 720 (1988)
90. C.G. Callan, I.R. Klebanov, M.J. Perry: Nucl. Phys. **B278**, 78 (1986)
91. C.G. Callan, D. Friedan, E.J. Martinec, M.J. Perry: Nucl. Phys. **B262**, 593 (1985)
92. S.V. Ketov, A.A. Deriglazov, Ya.S. Prager: Nucl. Phys. **B332**, 447 (1990)
93. S.J. Graham: Phys. Lett. **B197**, 543 (1987)
94. A.P. Foakes, N. Mohammedi: Phys. Lett. **B198**, 359 (1987)
95. S.V. Ketov: Sov. J. Nucl. Phys. **49**, 184 (1988)
96. R.R. Metsaev, A.A. Tseytlin: Phys. Lett. **B185**, 52 (1987)
97. I. Jack, D.R.T. Jones, D.A. Ross: Nucl. Phys. **B307**, 531 (1988)
98. S.V. Ketov: JETP Lett. **47**, 339 (1988)
99. Z.-M. Xi: Nucl. Phys. **B314**, 112 (1988)
100. I. Jack, D.R.T. Jones: Phys. Lett. **B193**, 449 (1987)
101. I. Jack, D.R.T. Jones, D.A. Ross: Nucl. Phys. **B307**, 130 (1988)
102. A.A. Tseytlin: Phys. Lett. **B194**, 63 (1987)
103. S.P. de Alwis: Phys. Lett. **B217**, 467 (1989)
104. S.V. Ketov, A.I. Samolov: The Kalb-Ramond Field in the Low-Energy Bosonic String Effective Action. Tomsk preprint TF/SB/AS-56 (1988), unpublished
105. S.V. Ketov, A.I. Samolov: JETP Lett. **48**, 558 (1988)
106. S.V. Ketov: Gen Rel. Grav. **22**, 193 (1988)
107. I. Jack, D.R.T. Jones, N. Mohammedi: Nucl. Phys. **B322**, 431 (1989)
108. A.A. Tseytlin: Yadern. Fiz. **40**, 1363 (1984)
109. T.H. Buscher: Phys. Lett. **B201**, 466 (1988)
110. D.Z. Freedman, P.K. Townsend: Nucl. Phys. **B177**, 282 (1981)
111. I.L. Buchbinder, S.M. Kuzenko: Nucl. Phys. **B308**, 162 (1988)
112. S.V. Ketov, K.E. Osetrin, Ya.S. Prager: JETP Lett. **50**, 343 (1989)
113. A.P. Demichev, A.Yu. Rodionov: Computer Phys. Commun. **38**, 441 (1985)
114. C.M. Hull, A. Karlhede, U. Lindström, M. Roček: Nucl. Phys. **B266**, 1 (1986)
115. S. Coleman: 'Secret Symmetry: an Introduction to Spontaneous Symmetry Breakdown and Gauge Theories. In: *Aspects of Symmetry: Selected Erice Lectures of Sidney Coleman* (Cambridge Univ. Press, 1985)
116. C.G. Callan, K. Hornbostel, I.R. Klebanov: Phys. Lett. **B202**, 269 (1988)
117. U.G. Meissner, I. Zahed: Adv. Nucl. Phys. **17**, 143 (1986)
118. T. Gisiger, M.B. Paranjape: Phys. Rep. **306**, 109 (1998)
119. G.S. Adkins, C.R. Nappi: Nucl. Phys. **B228**, 552 (1983)
120. N.S. Manton, P.J. Ruback: Phys. Lett. **B181**, 137 (1986)
121. Yu.A. Gol'fand, E.P. Lichtman: Pis'ma ZhETF **13**, 452 (1971)
122. D.V. Volkov, V.P. Akulov: Pis'ma ZhETF **16**, 621 (1972)
123. D.V. Volkov, V.A. Soroka: Pis'ma ZhETF **18**, 529 (1973)
124. J. Wess, B. Zumino: Phys. Lett. **B49**, 52 (1974)

402   References

125. J. Wess, B. Zumino: Nucl. Phys. B**70**, 39 (1974)
126. J. Wess, B. Zumino: Nucl. Phys. B**78**, 1 (1974)
127. J. Wess, B. Zumino: Nucl. Phys. B**79**, 413 (1974)
128. R. Haag, J.T. Lopuszanski, M.F. Sohnius: Nucl. Phys. B**88**, 257 (1975)
129. D.Z. Freedman, P. van Nieuwenhuizen, S. Ferrara: Phys. Rev. D**13**, 3214 (1976)
130. D.Z. Freedman, P. van Nieuwenhuizen: Phys. Rev. D**14**, 912 (1976)
131. S. Deser, B. Zumino: Phys. Lett. B**62**, 335 (1976)
132. A. Salam, J. Strathdee: Nucl. Phys. B**76**, 477 (1974)
133. A. Salam, J. Strathdee: Phys. Rev. D**11**, 1521 (1975)
134. A. Salam, J. Strathdee: Nucl. Phys. B**86**, 142 (1975)
135. A. Salam, J. Strathdee: Fortschr. Phys. **26**, 57 (1978)
136. M.T. Grisaru, W. Siegel, M. Roček: Nucl. Phys. B**159**, 429 (1979)
137. S.V. Ketov: Fortschr. Phys. **36**, 361 (1988)
138. P. Fayet, S. Ferrara: Phys. Rep. **32**, 249 (1977)
139. P. van Nieuwenhuizen: Phys. Rep. **68**, 189 (1981)
140. S.J. Gates Jr., M.T. Grisaru, M. Roček, W. Siegel: *Superspace or One Thousand and One Lessons in Supersymmetry* (Benjamin/Cummings Publ., 1983)
141. J. Wess, J. Bagger: *Supersymmetry and Supergravity* (Priceton Univ. Press, 1983)
142. M.F. Sohnius: Phys. Rep. **128**, 39 (1985)
143. P.C. West: *Introduction to Supersymmetry and Supergravity* (World Scientific, 1986)
144. P. Freund: *Introduction to Supersymmetry* (Cambridge Univ. Press, 1986)
145. *Supersymmetry*, ed. by S. Ferrara et al. (World Scientific, 1987)
146. S.V. Ketov: 'Introduction to Supersymmetry'. In: *General Relativity and Gravitation*, Kazan' Univ. Press **26**, 3 (1988)
147. S.V. Ketov: 'Introduction to Supergravity'. In: *General Relativity and Gravitation*, Kazan' Univ. Press, **27**, 3 (1989)
148. I.L. Buchbinder, S.M. Kuzenko: *Ideas and Methods of Supersymmetry and Supergravity* (IOP Publ., 1995)
149. L. Alvarez-Gaumé, D.Z. Freedman: Commun. Math. Phys. **80**, 443 (1981)
150. C.M. Hull: Actions for $(2,1)$ Sigma Models and Strings. London preprint QMW–97–2 (1997); hep-th/9702067
151. M. Abou Zeid, C.M. Hull: Phys. Lett. B**513**, 490 (1998)
152. E. Kähler: Abh. Math. Sem. Univ. Hamburg, **9**, 173 (1933)
153. B. Zumino: Phys. Lett. B**87**, 203 (1979)
154. T. Kaluza: Sitzber. Preuss. Akad. Wiss. Math.-Phys. Berlin, **1**, 966 (1921)
155. O. Klein: Z. Phys. **37**, 895 (1926)
156. L. Brink, J. Schwarz, J. Scherk: Nucl. Phys. B**121**, 77 (1977)
157. M. Duff, B.E.W. Nilsson, C.N. Pope: Phys. Rep. **130**, 1 (1986)
158. F.A. Berezin: *Method of Second Quantization* (Nauka Publ., 1965)
159. F.A. Berezin: Yadern. Fiz. **29**, 1670 (1979)
160. F.A. Berezin: *An Introduction into Algebra and Analysis with Anticommuting Variables* (Moscow State Univ. Press, 1983)
161. A.A. Deriglazov, S.V. Ketov: Nucl. Phys. B**359**, 498 (1990)
162. R.W. Allen, D.R.T. Jones: Nucl. Phys. B**303**, 291 (1988)
163. V.K. Krivoshchekov: Teor. Mat. Fiz. **303**, 291 (1978)
164. W. Siegel: Phys. Lett. B**84**, 193 (1979)
165. L.V. Avdeev, A.Yu. Kamenshchik: Phys. Lett. B**122**, 247 (1983)
166. S.V. Ketov, Ya.S. Prager: Pramana. Indian J. Phys. **30**, 173 (1988)
167. P. Di Vecchia, V.G. Knizhnik, J.L. Petersen, P. Rossi: Nucl. Phys. B**253**, 701 (1985)

168. A.P. Foakes, N. Mohammedi, D.A. Ross: Nucl. Phys. **B310**, 335 (1988)
169. L. Alvarez-Gaumé: Nucl. Phys. **B184**, 180 (1981)
170. M.B. Green, J.H. Schwarz, E. Witten: *Superstring Theory* (Cambridge Univ. Press, 1987)
171. M. Kaku: *Introduction to Superstrings* (Springer, 1988)
172. S.V. Ketov: *An Introduction into Quantum Theory of Strings and Superstrings* (Nauka Publ., 1990)
173. M.D. Freeman, C.N. Pope, M.F. Sohnius, K.S. Stelle: Phys. Lett. **B178**, 199 (1986)
174. M.D. Freeman, K.S. Stelle: Phys. Lett. **B174**, 48 (1986)
175. M.T. Grisaru, D. Zanon: Phys. Lett. **B177**, 347 (1986)
176. J. Grundberg, A. Karlhede, U. Lindström, S. Theodoridis: Class. Quantum Grav. **3**, L129 (1986)
177. A.A. Deriglazov, S.V. Ketov, Ya.P. Pugai: Pis'ma ZhETF **50**, 309 (1989)
178. Y. Kikuchi, C. Marzban, Y.J. Ng: Phys. Lett. **B176**, 57 (1986)
179. Y. Kikuchi, C. Marzban: Phys. Rev. **D35**, 1400 (1987)
180. A.A. Tseytlin: Phys. Lett. **B176**, 92 (1986)
181. S. Deser, A.N. Redlich: Phys. Lett. **B176**, 350 (1986)
182. M.C. Bento, N.E. Mavromatos: Phys. Lett. **B190**, 105 (1987)
183. D.R.T. Jones, A.M. Lawrence: Z. Phys. **C42**, 153 (1989)
184. W. Siegel: Phys. Lett. **B94**, 37 (1980)
185. L.V. Avdeev, G.A. Chochia, A.A. Vladimirov: Phys. Lett. **B105**, 272 (1981)
186. L.V. Avdeev: Phys. Lett. **B117**, 317 (1982)
187. J. Goldstone, A. Salam, S. Weinberg: Phys. Rev. **127**, 965 (1962)
188. J. Goldstone: Nuovo Cim. **19**, 154 (1961)
189. P. Candelas, G.T. Horowitz, A. Strominger, E. Witten: Nucl. Phys. **B258**, 46 (1985)
190. L. Castellani, D. Lüst: Nucl. Phys. **B296**, 143 (1988)
191. S.V. Ketov: Sov. J. Nucl. Phys. **47**, 341 (1988)
192. D. Lüst: Nucl. Phys. **B276**, 220 (1986)
193. S. Helgason: *Differential Geometry, Lie Groups and Symmetric Spaces* (Academic Press, 1978)
194. W. Ogura, A. Hosoya: Phys. Lett. **B164**, 329 (1986)
195. J. Bagger: Nucl. Phys. **B211** 302 (1983)
196. S. Kobayashi, K. Nomizu: *Foundations of Differential Geometry* (Interscience, 1969)
197. K. Yano: *Differential Geometry on Complex and Almost Complex Spaces* (Pergamon Press, 1965)
198. B.A. Dubrovin, A.T. Fomenko, S.P. Novikov: *Modern Geometry – Methods and Applications* (Springer, 1992)
199. S. Samuel: Nucl. Phys. **B245**, 127 (1984)
200. Ph. Spindel, A. Sevrin, W. Troost W., A. van Proeyen: Nucl. Phys. **B308**, 662 (1988)
201. H.J. Hitchin, A. Karlhede, U. Lindström, M. Roček: Commun. Math. Phys. **108**, 535 (1987)
202. M. Günaydin, S.V. Ketov: Nucl. Phys. **B467**, 215 (1996)
203. S.J. Gates Jr., C.M. Hull, M. Roček: Nucl. Phys. **B248**, 157 (1984)
204. M.F. Atiyah, N.J. Hitchin: *The Geometry and Dynamics of Magnetic Monopoles* (Princeton Univ. Press, 1988)
205. A. Besse: *Einstein Manifolds* (Springer, 1987)
206. R. Penrose: Gen. Rel. Grav. **7**, 31 (1976)
207. T. Eguchi, A. Hanson: Phys. Lett. **B74**, 249 (1978)
208. G.W. Gibbons, S.W. Hawking: Commun. Math. Phys. **66**, 291 (1979)

209. E. Sokatchev: Nucl. Phys. B**99**, 96 (1975)
210. W. Siegel, S.J. Gates Jr.: Nucl. Phys. B**189**, 295 (1981)
211. V. Rittenberg, E. Sokatchev: Nucl. Phys. B**193**, 477 (1981)
212. G.R.J. Bufton, J.G. Taylor: J. Math. Phys. A: Gen. Math. **16**, 321 (1983)
213. J. Kim: J. Math. Phys. **25**, 2037 (1984)
214. S.V. Ketov, B.B. Lokhvitsky: J. Math. Phys. **29**, 1244 (1987)
215. R.J. Firth, J.D. Jenkins: Nucl. Phys. B**85**, 525 (1975)
216. J. Wess: Acta Phys. Austr. **41**, 409 (1975)
217. S.V. Ketov, I.V. Tyutin: JETP Lett. **39**, 703 (1984)
218. S.V. Ketov, I.V. Tyutin: Sov. J. Nucl. Phys. **41**, 860 (1985)
219. C.M. Hull: Nucl. Phys. B**260**, 182 (1985)
220. A.A. Galperin, E.A. Ivanov, S. Kalitzin, V.I. Ogievetsky, E. Sokatchev: Class. Quantum Grav. **1**, 469 (1984)
221. A.A. Galperin, E.A. Ivanov, V.I. Ogievetsky, E. Sokatchev: Class. Quantum Grav. **2**, 601; 617 (1985)
222. E. Sokatchev, K.S. Stelle: Class. Quantum Grav. **4**, 501 (1987)
223. J. Grundberg, A. Karlhede, U. Lindström, G. Theodoridis: Nucl. Phys. B**282**, 142 (1988)
224. K. Muck: Phys. Lett. B**221**, 314 (1989)
225. S.V. Ketov: Sov. Phys. J. **28**, 690 (1985)
226. E.A. Ivanov: Nucl. Phys. B (Proc. Suppl.) **49**, 350 (1996)
227. G.G. Hartwell, P.S. Howe: Int. J. Mod. Phys. A**10**, 3901 (1995); Class. Quantum Grav. **12**, 1823 (1995)
228. P.S. Howe: 'On Harmonic Superspace'. Contributed to the International Seminar on Supersymmetries and Quantum Symmetries in Dubna, Russia, July 1998; hep-th/9812133
229. A. Karlhede, U. Lindström, M. Roček: Phys. Lett. B**147**, 297 (1984)
230. S.V. Ketov: 'Selfinteraction of $N = 2$ Multiplets in 4d, and the Ultra-Violet Finiteness of Two-Dimensional $N = 4$ σ-Models'. In: *Group Theory Methods in Physics* (Nauka Publ., 1985) p. 87
231. S.V. Ketov, K.E. Osetrin, I.V. Tyutin, B.B. Lokhvitsky: N=2 Matter in N=2 Superspace. Tomsk preprint TF/SB/AS-31 (1985), unpublished
232. S.V. Ketov, B.B. Lokhvitsky, I.V. Tyutin: Theor. Math. Phys. **71**, 496 (1987)
233. S.V. Ketov: Int. J. Mod. Phys. A**3**, 703 (1988)
234. P. Fayet: Nucl. Phys. B**113**, 135 (1976)
235. M.F. Sohnius: Nucl. Phys. B**138**, 109 (1978)
236. A.A. Galperin, E.A. Ivanov, V.I. Ogievetsky: Jadern. Fiz. **35**, 790 (1982)
237. P.S. Howe, K.S. Stelle, P.K. Townsend: Nucl. Phys. B**214**, 519 (1983)
238. P.S. Howe, K.S. Stelle, P.K. Townsend: Nucl. Phys. B**236**, 125 (1984)
239. A.A. Galperin, E.A. Ivanov, V.I. Ogievetsky: Nucl. Phys. B**282**, 74 (1987)
240. S.V. Ketov: Sov. Phys. J. **31**, 824 (1988)
241. S.M. Kuzenko: Int. J. Mod. Phys. A**14**, 1737 (1999)
242. N. Ohta, H. Sugata, H. Yamaguchi: Ann. Phys. (N.Y.), **172**, 26 (1986)
243. E.H. Saidi: Phys. Lett. B**206**, 639 (1988)
244. U. Lindström, M. Roček: Commun. Math. Phys. **115**, 21 (1988)
245. B. de Wit, R. Philippe, A. van Proeyen: Nucl. Phys. B**219**, 143 (1983)
246. U. Lindström, M. Roček: Nucl. Phys. B**222**, 285 (1983)
247. S.J. Gates Jr., T. Hübsch, S.K. Kuzenko: Nucl. Phys. B**557**, 443 (1999)
248. S. Cecotti, S. Ferrara, L. Girardello: Int. J. Mod. Phys. A**4**, 2475 (1989)
249. S.J. Gates Jr., W. Siegel: Nucl. Phys. B**187**, 389 (1981)
250. A.A. Galperin, E.A. Ivanov, V.I. Ogievetsky: Sov. J. Nucl. Phys. **45**, 157 (1987)

251. S.V. Ketov: Superconformal Hypermultiplets in Superspace. Copenhagen preprint, NBI–HE–00–04 (2000); hep-th/0001109
252. I.T. Ivanov, M. Roček: Commun. Math. Phys. **182**, 291 (1996)
253. 'Higher Transcendental Functions'. In: *Bateman Manuscript Project*, ed. by A. Erdelyi (McGraw-Hill, 1953), Vol. II
254. P. Byrd, M. Friedman: *Handbook of Elliptic Integrals for Engineers and Physicists* (Springer, 1954)
255. P. Kronheimer: J. Diff. Geom. **29**, 665 (1989)
256. G. Chalmers, M. Roček, S. Wiles: JHEP **9901**, 009 (1999)
257. S. Cherkis, A. Kapustin: Adv. Theor. Math. Phys. **2**, 1287 (1999)
258. S. Cherkis, A. Kapustin: Commun. Math. Phys. **203**, 713 (1999)
259. F. Gonzalez-Rey, M. Roček, S. Wiles, U. Lindström, R. von Unge: Nucl. Phys. **B516**, 426 (1998)
260. F. Gonzalez-Rey, R. von Unge: Nucl. Phys. **B516**, 449 (1998)
261. W. Siegel: Phys. Lett. **B153**, 51 (1985)
262. S.V. Ketov, B.B. Lokhvitsky: Czech. J. Phys. **39**, 1136 (1989)
263. L. Mezincescu, York-Peng Yao: Nucl. Phys. **B241**, 605 (1984)
264. B. Spence: Nucl. Phys. **B260**, 531 (1985)
265. A.A. Galperin, E.A. Ivanov, V.I. Ogievetsky, E. Sokatchev: Ann. Phys. (N.Y.) **185**, 1; 22 (1988)
266. J.A. Bagger, A.A. Galperin, E.A. Ivanov, V.I. Ogievetsky: Nucl. Phys. **B303**, 522 (1988)
267. G.W. Gibbons, C.N. Pope: Commun. Math. Phys. **66**, 267 (1979)
268. T. Eguchi, P. Gilkey, A. Hanson: Phys. Rep. **66**, 213 (1980)
269. M. Ko, M. Ludvigsen, E.T. Newman, K. Tod: Phys. Rep. **71**, 51 (1981)
270. S.T. Yau: Proc. Nat. Acad. Sci. (USA) **74**, 1748 (1977)
271. C. Boyer, J. Finley: J. Math. Phys. **23**, 1126 (1982)
272. I. Bakas, K. Sfetsos: Int. J. Mod. Phys. **A12**, 2585 (1997)
273. G.W. Gibbons, P.J. Ruback: Commun. Math. Phys. **115**, 267 (1988)
274. M. Saveliev: Commun. Math. Phys. **121**, 283 (1989)
275. D. Page: Phys. Lett. **B100**, 313 (1981)
276. A. Dancer: A Family of Gravitational Instantons. Cambridge preprint DAMTP 92–13 (1992), unpublished
277. D. Olivier: Gen. Rel. Grav. **23**, 1349 (1991)
278. M. Manton: Phys. Lett. **B110**, 54 (1982)
279. R. Grimm, M.F. Sohnius, J. Wess: Nucl. Phys. **B133**, 275 (1978)
280. J. Koller: Phys. Lett. **B124**, 324 (1983)
281. B. Zupnik: Phys. Lett. **B183**, 175 (1987)
282. E.A. Ivanov, S.V. Ketov, B.M. Zupnik: Nucl. Phys. **B509**, 503 (1998)
283. E.I. Buchbinder, I.L. Buchbinder, E.A. Ivanov, S.M. Kuzenko, B.A. Ovrut: Phys. Lett. **B412**, 309 (1997)
284. S.V. Ketov, Ch. Unkmeir: Phys. Lett. **B422**, 179 (1998)
285. A.A. Galperin, E.A. Ivanov, V.I. Ogievetsky, E. Sokatchev: Commun. Math. Phys. **103**, 515 (1986)
286. A.A. Galperin, E.A. Ivanov, V.I. Ogievetsky, P.K. Townsend: Class. Quantum Grav. **3**, 625 (1986)
287. G.W. Gibbons, D. Olivier, P.J. Ruback, G. Valent: Nucl. Phys. **B296**, 679 (1988)
288. T.L. Curtright, D.Z. Freedman: Phys. Lett. **B90**, 71 (1980)
289. L. Alvarez-Gaumé, D.Z. Freedman: Phys. Lett. **B94**, 171 (1980)
290. S.V. Ketov: Phys. Lett. **B469**, 136 (1999)
291. G. Sierra, P.K. Townsend: Nucl. Phys. **B233**, 289 (1984)
292. J.A. Bagger, E. Witten: Nucl. Phys. **B222**, 1 (1983)

293. A. Barut, R. Raczka: *Theory of Group Representations and Applications* (World Scientific, 1986)
294. P. West: An Introduction to String Theory. CERN preprint TH.-5165 (1988), unpublished
295. D. Gepner, E. Witten: Nucl. Phys. **B278**, 493 (1986)
296. H. Sugawara: Phys. Rev. **170**, 1659 (1968)
297. C. Sommerfeld: Phys. Rev. **176**, 2019 (1968)
298. E. Martinec: Nucl.Phys. **B281**, 157 (1987)
299. M. Baranov, I. Frolov, A. Schwarz: Theor. Math. Phys. **79**, 509 (1989)
300. D. Friedan, Z. Qiu, S. Shenker: Phys. Lett. **B151**, 37 (1985)
301. M. Bershadsky, V. Knizhnik, M. Teitelman: Phys. Lett. **B151**, 31 (1985)
302. S. Shenker: 'Introduction to Two-Dimensional Conformal and Superconformal Field Theory'. In: *Unified String Theories* (World Scientific, 1986) p. 141
303. C. Montonen: Nuovo Cim. **19**, 69 (1974)
304. E. Martinec: Phys. Rev. **D28**, 2604 (1983)
305. J.H. Schwarz: Int. J. Mod. Phys. **A4**, 2653 (1989)
306. E. Kiritsis, G. Siopsis: Phys. Lett. **B184**, 353 (1987)
307. S. Nahm: Phys. Lett. **B187**, 340 (1987)
308. K. Schoutens: Nucl. Phys. **B295**, 634 (1988)
309. *Conformal Field Theories and Related Topics*, ed. by P. Binetruy, P. Sorba, R. Stora (Elsevier Publ., 1990)
310. F. Gliozzi, J. Scherk, D. Olive: Nucl. Phys. **B122**, 253 (1977)
311. M. Ademollo, L. Brink, A. D'Adda, R. D'Auria, E. Napolitano, S. Sciuto, E. Del Giudice, P. Di Vecchia, S. Ferrara, F. Gliozzi, R. Musto, R. Pettorino: Phys. Lett. **B62**, 105 (1976)
312. M. Ademollo, L. Brink, A. D'Adda, R. D'Auria, E. Napolitano, S. Sciuto, E. Del Guidice, P. Di Vecchia, S. Ferrara, F. Gliozzi, R. Musto, R. Pettorino, J. Schwarz: Nucl. Phys. **B111**, 77 (1976)
313. P. Di Vecchia, J. Petersen, M. Yu, H. Zheng: Phys. Lett. **B162**, 327 (1985)
314. P. Di Vecchia: Phys. Lett. **144B**, 245 (1984)
315. P. Goddard, W. Nahm, D. Olive: Phys. Lett. **B160**, 111 (1985)
316. P. Windey: Commun. Math. Phys. **105**, 511 (1986)
317. P. Goddard, A. Kent, D. Olive: Commun. Math. Phys. **103**, 105 (1986)
318. I. Antoniadis, C. Bachas, C. Kounnas, P. Windey: Phys. Lett. **B171**, 51 (1985)
319. V. Kač, I. Todorov: Commun. Math. Phys. **102**, 337 (1985)
320. M. Sakamoto: Phys. Lett. **B151**, 115 (1985)
321. C. Hull, E. Witten: Phys. Lett. **B160**, 398 (1985)
322. R. Brooks, F. Muhammad, S.J. Gates Jr.: Nucl. Phys. **B268**, 599 (1986)
323. S.J. Gates Jr., S.V. Ketov, S.M. Kuzenko, O.A. Soloviev: Nucl. Phys. **B362**, 199 (1991)
324. A. Kirillov: *Elements of the Theory of Representations* (Springer, 1975)
325. G.W. Delius, P. van Nieuwenhuizen, V. Rodgers: Int. J. Mod. Phys. **A5**, 3943 (1990)
326. C. Hull, B. Spence: Nucl. Phys. **B345**, 493 (1990)
327. S.J. Gates Jr., S.V. Ketov: Phys. Lett. **B271**, 355 (1991)
328. G.W. Delius: Int. J. Mod. Phys. **A5**, 4753 (1990)
329. P.S. Howe, G. Papadopoulos: Nucl. Phys. **B289**, 264 (1986)
330. P. Goddard, D. Olive: Int. J. Mod. Phys. **A1**, 303 (1986)
331. A. Zamolodchikov, V. Fateev: JETP **63**, 913 (1986)
332. P. Di Vecchia, J. Petersen, M. Yu: Phys. Lett. **B172**, 211 (1986)
333. W. Boucher, D. Friedan, A. Kent: Phys. Lett. **B172**, 316 (1986)
334. T. Eguchi, A. Taormina: Phys. Lett. **B210**, 125 (1988)
335. K. Bardakci, E. Rabinovici, B. Säring: Nucl. Phys. **B299**, 151 (1988)

336. K. Gawedzki, A. Kupiainen: Nucl. Phys. B**320**, 625 (1989)
337. Q.-H. Park: Phys. Lett. B**223**, 422 (1989)
338. D. Karabali, Q.-H. Park, H.J. Schnitzer, Z. Yang: Phys. Lett. B**216**, 307 (1989)
339. H.J. Schnitzer: Nucl. Phys. B**324**, 412 (1989)
340. D. Karabali, H.J. Schnitzer: Nucl. Phys. B**329**, 649 (1990)
341. Y. Kazama, H. Suzuki: Nucl. Phys. B**321**, 232 (1989)
342. A. van Proeyen: Class. Quantum Grav. **6**, 1501 (1989)
343. A. Sevrin, G. Theodoridis: Nucl. Phys. B**332**, 380 (1990)
344. M. Günaydin, S. Hyun: Nucl. Phys. B**373**, 688 (1992)
345. M. Günaydin: Phys. Rev. D**47**, 3600 (1993)
346. S.J. Gates Jr., S.V. Ketov: Phys. Rev. D**52**, 2278 (1995)
347. D. Lüst, S. Theisen: *Lectures in String Theory* (Springer, 1989)
348. M. Kaku: *Strings, Conformal Fields, and Topology. An Introduction* (Springer, 1991)
349. J. Polchinski: *String Theory* (Cambridge Univ. Press, 1998)
350. A.M. Polyakov: Phys. Lett. B**103**, 207; 211 (1981)
351. C.B. Thorn: Phys. Rep. **175**, 1 (1989)
352. *Superstring Construction*, ed. by A. Schellekens (Elsevier Publ., 1991)
353. O. Alvarez: Nucl. Phys. B**216**, 125 (1983)
354. C. Lovelace: Phys. Lett. B**135**, 75 (1984)
355. E.S. Fradkin, A.A. Tseytlin: Nucl. Phys. B**261**, 1 (1985)
356. A.A. Tseytlin: Phys. Lett. B**178**, 34 (1986)
357. G.M. Shore: Nucl. Phys. B**286**, 349 (1987)
358. A.A. Tseytlin: Nucl. Phys. B**286**, 383 (1987)
359. G. Curci, G. Paffuti: Nucl. Phys. B**286**, 399 (1987)
360. G.F. Chapline, N.S. Manton: Phys. Lett. B**120**, 105 (1983)
361. N.E. Mavromatos, J.L. Miramontes: Phys. Lett. B**212**, 33 (1988)
362. J.L. Cardy: Phys. Lett. B**215**, 749 (1988)
363. A.A. Tseytlin: Nucl. Phys. B**276**, 391 (1986)
364. I. Jack, D.R.T. Jones: Nucl. Phys. B**303**, 260 (1988)
365. I. Jack, D.R.T. Jones, D.A. Ross: Nucl. Phys. B**307**, 531 (1988)
366. E. Witten: Phys. Rev. D**44**, 314 (1991)
367. I. Jack, D.R.T. Jones, J. Panvel: Nucl. Phys. B**393**, 95 (1993)
368. R. Dijkgraaf, H. Verlinde, E. Verlinde: Nucl. Phys. B**371**, 269 (1992)
369. L.J. Dixon, M.E. Peskin, J. Lykken: Nucl. Phys. B**325**, 329 (1989)
370. E. Witten: Two-Dimensional String Theory and Black Holes. Princeton preprint IASSNS–HEP–92/24 (1992); hep-th/9206069
371. A.A. Tseytlin: Nucl. Phys. B**399**, 601 (1993)
372. A.A. Tseytlin: Two-Dimensional Conformal Sigma-Models and Exact String Solutions. CERN preprint TH–6820–93 (1993); hep-th/9303054
373. A. Giveon, M. Porrati, E. Rabinovici: Phys. Rep. **244**, 77 (1994)
374. S.V. Ketov: Class. and Quantum Grav. **4**, 1163 (1987)
375. P. Ramond: Phys. Rev. D**3**, 2415 (1971)
376. A. Neveu, J.H. Schwarz: Nucl. Phys. B**31**, 86 (1971)
377. S. Deser, B. Zumino: Phys. Lett. B**65**, 369 (1976)
378. L. Brink, P. Di Vecchia, P.S. Howe: Phys. Lett. B**65**, 471 (1976)
379. L. Brink, J.H. Schwarz: Nucl. Phys. B**121**, 285 (1977)
380. K. Yamagishi: Ann. Phys. (N.Y.) **150**, 439 (1983)
381. B. de Wit, P. van Nieuwenhuizen: Nucl. Phys. B**312**, 58 (1989)
382. P.S. Howe: Phys. Lett. B**70**, 453 (1977)
383. E. Witten, J. Bagger: Phys. Lett. B**115**, 202 (1982)
384. E.S. Fradkin, A.A. Tseytlin: Ann. Phys. (N.Y.) **143**, 413 (1982)
385. D. Nemeschansky, S. Yankielowicz: Phys. Rev. Lett. **54**, 620 (1985)

386. H. Ooguri, C. Vafa: Nucl. Phys. B**361**, 469 (1991)
387. N. Marcus: Nucl. Phys. B**387**, 263 (1992)
388. S.V. Ketov: Phys. Lett. B**395**, 48 (1997)
389. M. Born, L. Infeld: Proc. Roy. Soc. Lond. A**144**, 425 (1934)
390. H. Ooguri, C. Vafa: Nucl. Phys. B**367**, 83 (1991)
391. D. Kutasov, E. Martinec: Nucl. Phys. B**477** 652 (1996)
392. D. Kutasov, E. Martinec, M. O'Loughlin: Nucl. Phys. B**477**, 675 (1996)
393. J. Bischoff, O. Lechtenfeld, S.V. Ketov: Nucl. Phys. B**438**, 373 (1995)
394. N. Berkovits, C. Vafa: Nucl. Phys. B**433**, 123 (1995)
395. S. Donaldson: Proc. London Math. Soc. **50** 1 (1985)
396. V.P. Nair, J. Schiff: Nucl. Phys. B**371**, 329 (1992)
397. C.N. Yang: Phys. Rev. Lett. **38**, 1377 (1977)
398. A.N. Leznov: Theor. Math. Phys. **73**, 1233 (1988)
399. A. Parkes: Phys. Lett. B**286**, 265 (1992)
400. O. Lechtenfeld, W. Siegel: Phys. Lett. B**405**, 49 (1997)
401. Y. Hashimoto, Y. Yasui, S. Miyagi, T. Otsuka: J. Math. Phys. **38**, 5833 (1997)
402. E.J. Flaherty: Gen. Rel. Grav. **9**, 961 (1978)
403. C.P. Boyer, J.F. Plebański: Phys. Lett. A**106**, 125 (1984)
404. C.P. Boyer, J.F. Plebański: J. Math. Phys. **26**, 229 (1985)
405. I. Bakas: Phys. Lett. B**228** 57 (1989)
406. Q.-H. Park: Phys. Lett. B**238**, 287 (1990)
407. D.J. Gross, J.A. Harvey, E. Martinec, R. Rohm: Nucl. Phys. B**256**, 253 (1985); ibid. **267**, 75 (1985)
408. D.J. Gross: 'The Heterotic String'. In: *Unified Field Theories* (World Scientific, 1986) p. 357
409. A. Sen: Phys. Rev. D**32**, 2102
410. C.M. Hull, P.K. Townsend: Phys. Lett. B**178**, 187 (1986)
411. G. Moore, P. Nelson: Nucl. Phys. B**274**, 509 (1986)
412. M. Evans, J. Louis, B.A. Ovrut: Phys. Rev. D**35**, 3045 (1987)
413. C.M. Hull, P.K. Townsend: Nucl. Phys. B**301**, 197 (1988)
414. S.V. Ketov: 'Anomalies of the Heterotic Sigma-Model in a Curved (1,0) Superspace'. In: *Proceedings of the 1989 Spring Workshop on Superstrings* (World Scientific, 1989) p. 376
415. S.V. Ketov, O.A. Soloviev: Phys. Lett. B**232**, 75 (1989)
416. S.V. Ketov, S.M. Kuzenko, O.A. Soloviev: Class. Quantum Grav. **7**, 1403 (1990)
417. S.V. Ketov, O.A. Soloviev: Int. J. Mod. Phys. A**A6**, 2971 (1991)
418. S.J. Gates Jr., M.T.Grisaru, L. Mezincescu, P.K. Townsend: Nucl. Phys. B**286**, 1 (1987)
419. S.J. Gates Jr.: 'Strings, Superstrings and Two-Dimensional Lagrangean Field Theory'. In: *Functional Integration, Geometry and Strings.* (Birkhäuser, 1989) p. 140
420. S.J. Gates Jr.: 'Off-Shell BRST-Invariant Superjacobians'. Invited talk given at the NATO Workshop on Superstrings (Boulder, USA, July 27 – August 1, 1987)
421. S. Adler: Phys. Rev. **177**, 2426 (1969)
422. J.S. Bell, R. Jackiw: Nuovo. Cim. **60**, 47 (1969)
423. B. Zumino, Y.S. Wu, A. Zee: Nucl. Phys. B**239**, 477 (1984)
424. A.Yu. Morozov: Usp. Fiz. Nauk **150**, 337 (1986)
425. R.A. Bertlmann: *Anomalies in Quantum Field Theory* (Clarendon Press. 1996)
426. M.J. Duff: Nucl. Phys. B**215**, 334 (1977)
427. S.M. Christensen, M.J. Duff: Phys. Lett. B**76**, 571 (1978)

428. L. Alwarez-Gaumé, E. Witten: Nucl. Phys. B**234**, 269 (1984)
429. S.V. Ketov: Class. and Quantum Grav. **7**, 1377 (1990)
430. R. Brustein, D. Nemeshansky, S. Yankielowicz: Nucl. Phys. B**301**, 224 (1988)
431. S. Bellucci: Progr. Theor. Phys. **79**, 1288 (1988)
432. S.V. Ketov, O.A. Soloviev: Sov. J. Nucl. Phys. **51**, 567 (1990)
433. U. Ellwanger, J. Fuchs, M.G. Schmidt: Nucl. Phys. B**314**, 175 (1989)
434. M.T. Grisaru, B. Milewski, D. Zanon: 'Superconformal Anomalies and the Adler-Bardeen Theorem'. In: *Supersymmetry and its Applications: Superstrings, Anomalies, and Supergravity* (Cambridge Univ. Press, 1986) p. 61
435. S.L. Adler, W.A. Bardeen: Phys. Rev. **182**, 1517 (1969)
436. M. de Groot, P. Mansfield: Phys. Lett. B**202**, 519 (1988)
437. Q.-H. Park, D. Zanon: Phys. Rev. D**35**, 4038 (1987)
438. S. Bellucci: Z. Phys. C**36**, 229 (1987)
439. D. Gross, J. Sloan: Nucl. Phys. B**291**, 41 (1987)
440. A.P. Foakes, N. Mohammedi, D.A. Ross: Phys. Lett. B**206**, 57 (1988)
441. M.T. Grisaru, D. Zanon: Phys. Lett. B**184**, 209 (1987)
442. D. Zanon: Phys. Lett. B**186**, 309 (1987)
443. I.L. Buchbinder, S.M. Kuzenko, O.A. Soloviev: Phys. Lett. B**228**, 341 (1989)
444. K.S. Narain: Phys. Lett. B**169**, 41 (1986)
445. H. Kawai, D.C. Lewellen, S.-H. Tye: Nucl. Phys. B**288**, 1 (1987)
446. I. Antoniadis, C.P. Bachas, C. Kounnas: Nucl. Phys. B**289**, 87 (1987)
447. W. Siegel: Nucl. Phys. B**238**, 307 (1984)
448. M.T. Grisaru, L. Mezincescu, P.K. Townsend: Phys. Lett. B**179**, 247 (1986)
449. C.M. Hull: Phys. Lett. B**206**, 234 (1988)
450. S.J. Gates Jr., R. Brooks, F. Muhammad: Phys. Lett. B**194**, 37 (1987)
451. S.M. Kuzenko, O.A. Soloviev: Int. J. Mod. Phys. A**5**, 1341 (1990)
452. S. Bellucci, R. Brooks, J. Sonnenschein: Nucl. Phys. B**304**, 173 (1988)
453. S.J. Gates Jr., W. Siegel: Phys. Lett. B**206**, 631 (1988)
454. S. Bellucci, R. Brooks, J. Sonnenschein: Mod. Phys. Lett. A**3**, 1537 (1988)
455. D.A. Depireux, S.J. Gates Jr., B. Radak: Phys. Lett. B**236**, 411 (1990)
456. S. Bellucci, D.A. Depireux, S.J. Gates Jr.: Phys. Lett. B**232**, 67 (1989)
457. E. Bergshoeff, S. Randjbar-Daemi, A. Salam, H. Sarmadi, E. Sezgin: Nucl. Phys. B**269**, 77 (1986)
458. D. Gepner: Nucl. Phys. B**296**, 757 (1988)
459. M. Lynker, R. Schimmrigk: Phys. Lett. B**208** 216 (1988)
460. M. Lynker, R. Schimmrigk: Phys. Lett. B**215** 681 (1988)
461. S.J. Gates Jr., T. Hübsch: Nucl. Phys. B**343** 741 (1990)
462. J. Schwarz: 'Superconformal Symmetry in String Theory'. In: *Proceedings of the Banff Summer Institute on Particles and Fields* (Banff Summer Inst., 1988) p. 55
463. N. Seiberg, E. Witten: 'Gauge Dynamics and Compactification to Three Dimensions'. In: *Saclay'96. The Mathematical Beauty of Physics*; hep-th/9607163
464. E. Witten: Nucl. Phys. B**443**, 85 (1995)
465. J.H. Schwarz, N. Seiberg: String Theory, Supersymmetry, Unification and All That. Princeton and Caltech preprint, IASSNS-HEP-98/27 and CALT-68-2168; hep-th/9803179;
466. G.W. Gibbons: 'Quantum Gravity/String/M-Theory as We Approach the 3rd Millennium'. In: *Proceedings of the GR-15 Conference* (1997); gr-qc/9803065
467. A. Bilal: Duality in N=2 SUSY SU(2) Yang-Mills Theory: a Pedagogical Introduction to the Work of Seiberg and Witten. Paris preprint LPTENS-95/53 (1996); hep-th/9601007
468. A. Alvarez-Gaumé, S.F. Hassan: Fortschr. Phys. **45**, 159 (1997)

469. W. Lerche: Nucl. Phys. B (Proc. Suppl.) **55**, 83 (1997)
470. P. Di Vecchia: Duality in N=2,4 Supersymmetric Gauge Theories. Nordita preprint 98/11-HE (1998); hep-th/9803026
471. S.V. Ketov: Fortschr. Phys. **45**, 237 (1997)
472. S. Kachru, C. Vafa: Nucl. Phys. B**450**, 69 (1995)
473. S. Kachru, A. Klemm, W. Lerche, P. Mayr, C. Vafa: Nucl. Phys. B**459**, 537 (1995)
474. A. Klemm, W. Lerche, P. Mayr, C. Vafa, N. Warner: Nucl. Phys. B**477**, 746 (1996)
475. A. Hanany, E. Witten: Nucl. Phys. B**492**, 152 (1997)
476. E. Witten: Nucl. Phys. B**500**, 3 (1997)
477. A. Giveon, D. Kutasov: Brane Dynamics and Gauge Theory. Chicago preprint EFI-98-06 (1998); hep-th/9802067
478. S.V. Ketov: Fortschr. Phys. **47**, 643 (1999)
479. L.J. Mason, N.M.J. Woodhouse: *Integrability, Self-Duality and Twistor Theory* (Clarendon Press, 1996)
480. S.V. Ketov, S.J. Gates Jr., H. Nishino: Nucl. Phys. B**393**, 149 (1993)
481. E. Witten: Nucl. Phys. B**202**, 253 (1982)
482. I.L. Buchbinder, S.M. Kuzenko, B.A. Ovrut: Phys. Lett. B**433**, 335 (1998)
483. A. Klemm, W. Lerche, S. Yankielowicz, S. Theisen: Phys. Lett. B**344**, 95 (1995)
484. A. Klemm, W. Lerche, S. Theisen: Int. J. Mod. Phys. A**11**, 1929 (1996)
485. N. Dorey, V.V. Khoze, M.P. Mattis: Phys. Rev. D**54**, 2921; 7832 (1996)
486. A. D'Adda, P. Di Vecchia: Phys. Lett. B**73**, 162 (1978)
487. M.F. Atiyah, V.G. Drinfeld, N.J. Hitchin, I.Yu. Manin: Phys. Lett. A**65**, 185 (1978)
488. W. Nahm: Phys. Lett. B**90**, 413 (1980)
489. G. t'Hooft: Phys. Rev. D**14**, 3432 (1976)
490. H. Osborn: Ann. Phys. (N.Y.) **135**, 373 (1981)
491. N. Dorey, V.V. Khoze, M.P. Mattis: Nucl. Phys. B**513**, 681 (1998)
492. D. Finnell, P. Pouliot: Nucl. Phys. B**453**, 225 (1995)
493. N. Dorey, V.V. Khoze, M.P. Mattis: Phys. Lett. B**388**, 324 (1996)
494. N. Dorey, V.V. Khoze, M.P. Mattis: Phys. Lett. B**390**, 205 (1997)
495. F. Fucito, G. Travaglini: Phys. Rev. D**55**, 1099 (1997)
496. H. Farkas, I. Kra: *Riemann Surfaces* (Springer, 1980)
497. M. Douglas, S. Shenker: Nucl. Phys. B**447**, 271 (1995)
498. P. Argyres, A. Faraggi: Phys. Rev. Lett. **74**, 3931 (1995)
499. U. Danielsson, B. Sundborg: Phys. Lett. B**358**, 273 (1995)
500. A. Hanany, Y. Oz: Nucl. Phys. B**452**, 283 (1995)
501. P.C. Argyres, M.R. Plesser, A. Shapere: Phys. Rev. Lett. **75**, 1699 (1995)
502. V. Arnold, A. Gusein-Zade, A. Varchenko: *Singularities of Differentiable Maps* (Birkhäuser, 1985)
503. A. Gorsky, I. Krichever, A. Marshakov, A. Mironov, A. Morozov: Phys. Lett. B**355**, 466 (1995)
504. R. Donagi, E. Witten: Nucl. Phys. B**460**, 299 (1996)
505. A. Marshakov, A. Mironov, A. Morozov: Phys. Lett. B**389**, 43 (1996)
506. R. Dijkgraaf, E. Verlinde, H. Verlinde: Nucl. Phys. B**352**, 59 (1991)
507. E. Witten: Surv. Diff. Geom. **1**, 243 (1991)
508. M. Matone: Phys. Lett. B**357**, 342 (1995)
509. P. Howe, P. West: Nucl. Phys. B**486**, 425 (1997)
510. B. de Wit, M. Grisaru, M. Roček: Phys. Lett. B**374**, 297 (1996)
511. M. Henningson: Nucl. Phys. B**458**, 445 (1996)

512. I.L. Buchbinder, E.L. Buchbinder, S.M. Kuzenko, B,A. Ovrut: Phys. Lett. B417, 61 (1998)
513. S.V. Ketov: Phys. Rev. D57, 1277 (1998)
514. M. Dine, N. Seiberg: Phys. Lett. B409, 239 (1997)
515. N. Seiberg: Phys. Lett. B206, 75 (1988)
516. F. Gonzalez-Rey, M. Roček: Phys. Lett. B434, 303 (1998)
517. I.L. Buchbinder, S.M. Kuzenko: Mod. Phys. Lett. A13, 1623 (1998)
518. A. De Giovanni, M.T. Grisaru, M. Roček, R. von Unge, D. Zanon: Phys. Lett. B409, 251 (1997)
519. N. Dorey, V.V. Khoze, M. Mattis, M. Slater, W. Weir: Phys. Lett. B408, 213 (1997)
520. D. Bellisai, F. Fucito, M. Matone, G. Travaglini: Phys. Rev. D56, 5218 (1997)
521. M. Matone: Phys. Rev. Lett. 78, 1412 (1997)
522. A. Yung: Nucl. Phys. B485, 38 (1997)
523. A. Yung: Nucl. Phys. B512, 79 (1998)
524. J. de Boer, K. Hori, H. Ooguri, Y. Oz: Nucl. Phys. B518, 173 (1998)
525. S.V. Ketov: Phys. Lett. B399, 83 (1997)
526. M.T. Grisaru, M. Roček, R. von Unge: Phys. Lett. B383, 415 (1996)
527. B.M. Zupnik: Theor. Math. Phys. 69, 1101 (1986)
528. R. Sorkin: Phys. Rev. Lett. 51, 87 (1983)
529. D.J. Gross, M.J. Perry: Nucl. Phys. B226, 29 (1983)
530. S.K. Donaldson: Commun. Math. Phys. 96, 387 (1984)
531. J. Hurtubise: Commun. Math. Phys. 92, 195 (1983); ibid. 100, 191 (1985)
532. I. Bakas: Fortschr. Phys. 48, 9 (2000)
533. P.C. Argyres, M.R. Plesser, N. Seiberg: Nucl. Phys. B471, 159 (1996)
534. L. Alvarez-Gaumé, M. Marino. F. Zamore: Int. J. Mod. Phys. A13, 403; 1847 (1998)
535. G. t'Hooft: Commun. Math. Phys. 81, 267 (1981)
536. Z. Guralnik, S. Ramgoolam: Nucl. Phys. B499, 241 (1997); ibid. B521, 129 (1998)
537. A. Zhitnitsky: Nucl. Phys. B340, 56 (1990)
538. V. Novikov, M. Shifman, A. Vainstein, V. Zakharov: Nucl. Phys. B229, 381; 394; 407 (1983)
539. A. Mikhailov: Nucl. Phys. B533, 243 (1998)
540. P.S. Howe, N.D. Lambert, P.C. West: Phys. Lett. B418, 85 (1998)
541. P.K. Townsend: 'M-Theory from its Superalgebra'. In: Proceedings of the Cargèse Summer School (1997); hep-th/9712004
542. K.S. Stelle: 'BPS Branes in Supergravity'. In: Proceedings of the ICTP Summer Schools (1996 and 1997); hep-th/9803116
543. J.P. Gauntlett: 'Intersecting Branes'. In: Proceedings of the APCTP Winter School (1997); hep-th/9705011
544. W. Nahm: Nucl. Phys. B135, 149 (1978)
545. E. Cremmer, B. Julia, J. Scherk: Phys. Lett. B76, 409 (1978)
546. S. Deser: Nonrenormalizability of (Last Hope) D=11 Supergravity, with a Terse Survey of Divergences in Quantum Gravities. Brandeis preprint BRX–TH–457 (1999); hep-th/9905017
547. C.M. Hull: Phys. Lett. B139, 39 (1984)
548. J.W. van Holten, A. van Proeyen: J. Phys. A15, 3763 (1982)
549. J.A. Azćarraga, J.P. Gauntlett, J.M. Izquierdo, P.K. Townsend: Phys. Rev. Lett. 63, 2443 (1989)
550. E. Bergshoeff, E. Sezgin, P.K. Townsend: Phys. Lett. B189, 75 (1987)
551. R. Güven: Phys. Lett. B276, 49 (1991)
552. G.W. Gibbons, P.K. Townsend: Phys. Rev. Lett. 71, 3754 (1993)

553. M.J. Duff, G.W. Gibbons, P.K. Townsend: Phys. Lett. B**332**, 321 (1994)
554. E. Witten: Nucl. Phys. B**460**, 335 (1996)
555. G.T. Horowitz, A. Strominger: Nucl. Phys. B**360**, 197 (1991)
556. P.K. Townsend: Phys. Lett. B**350**, 184 (1995)
557. A.A. Tseytlin: Nucl. Phys. B**475**, 149 (1996); ibid. B**487**, 141 (1997)
558. A. Abouelsaood, C.G. Callan, C.R. Nappi, S.A. Yost: Nucl. Phys. B**280**, 599 (1987)
559. R.G. Leigh: Mod. Phys. Lett. A**4**, 2767 (1989)
560. D. Kaplan, J. Michelson: Phys. Rev. D**53**, 3474 (1996)
561. I. Bandos, K. Lechner, A. Nurmagambetov, P. Pasti, D. Sorokin, M. Tonin: Phys. Rev. Lett. **78**, 4332 (1997); Phys. Lett. B**408**, 135 (1977)
562. M. Aganagic, J. Park, C. Popescu, J.H. Schwarz: Nucl. Phys. B**496**, 191 (1997)
563. S. Mandelstam: Phys. Rep. **23**, 245 (1976)
564. G. t'Hooft: 'Gauge Fields with Unified Weak, Electromagnetic, and Strong Interactions'. In: *High Energy Physics*, Vol. 1, ed. by A. Zichichi (Bologna, 1976)
565. G. t'Hooft: Phys. Scripta **25**, 133 (1982)
566. M. Peskin: Duality in Supersymmetric Gauge Theories. SLAC preprint PUB-7393 (1997); hep-th/970294
567. E. D'Hoker, Y. Mimura, N. Sakai: Phys. Rev. D**54**, 7724 (1996)
568. A.M. Polyakov: Int. J. Mod. Phys. A**14**, 645 (1999)
569. A.A. Migdal: Phys. Rep. **102**, 199 (1983)
570. Y. Makeenko, A.A. Migdal: Nucl. Phys. B**188**, 269 (1981)
571. J. Maldacena: Adv. Theor. Math. Phys. **2**, 231 (1998)
572. L. Susskind: J. Math. Phys. **36**, 6377 (1995)
573. S.S. Gubser, I.R. Klebanov, A.M. Polyakov: Phys. Lett. B**428**, 105 (1998)
574. J.-L. Petersen: An Introduction to the Maldacena Conjecture on AdS/CFT. Copenhagen preprint NBI–HE–99–05 (1999); hep-th/9902131
575. O. Aharony, S.S. Gubser, J. Maldacena, H. Ooguri, Y. Oz: Large N Field Theories, String Theory and Gravity. CERN preprint TH/99–122 (1999); hep-th/9905111
576. E. Witten: Adv. Theor. Math. Phys. **2**, 505 (1998)
577. S.W. Hawking, D. Page: Commun. Math. Phys. **87**, 577 (1983)
578. C. Csaki, H. Ooguri, Y. Oz, J. Terning: JHEP **9901**, 017 (1999)
579. S.V. Ketov: Nucl. Phys. B**544**, 181 (1999)
580. S.V. Ketov: 'Making Manifest the Symmetry Enhancement for Coincident BPS Branes'. In: *Proceedings of the STRINGS'98 Conference* (1998); http://www.itp.ucsb.edu/online/strings98/; hep-th/9806130
581. S.V. Ketov: Nucl. Phys. B**553**, 250 (1999)
582. I.L. Buchbinder, S.V. Ketov: Fortschr. Phys. **39**, 1 (1991)
583. C.M. Hull, P.K. Townsend: Nucl. Phys. B**451**, 525 (1995)
584. L.D. Landau, I. Pomeranchuk: Dokl. Akad. Nauk USSR. Ser. Fiz. **102**, 489 (1955)
585. E. Cremmer, B. Julia: Nucl. Phys. B**159**, 141 (1979)
586. B. de Wit, H. Nicolai: Nucl. Phys. B**208**, 323 (1982)
587. A. d'Adda, P. Di Vecchia, M. Lüscher: Nucl. Phys. B**146** 63 (1978); ibid. B**152**, 125 (1979)
588. E. Witten: Nucl. Phys. B**149**, 285 (1979)
589. J.W. van Holten: Phys. Lett. B**135**, 427 (1984); Nucl. Phys. B**242**, (1984)
590. G. Moore, N. Nekrasov, S. Shatashvili: Commun. Math. Phys. **209**, 77 (2000)
591. J. Scherk, J.H. Schwarz: Nucl. Phys. B**153**, 61 (1979)
592. P. Di Vecchia, R. Musto, F. Nicodemi, R. Pettorino: Nucl. Phys. B**252**, 635 (1985)

593. L.S. Brown: *Quantum Field Theory* (Cambridge Univ. Press, 1982), sect. 3.5

594. E.A. Ivanov, B.M. Zupnik: Modified N=2 Supersymmetry and Fayet-Iliopoulos Terms. Dubna preprint E2–97–322 (1997); hep-th/9710236

595. P. Ruback: Commun. Math. Phys. **107**, 93 (1986)

596. M. Henningson, P. Yi: Phys. Rev. D**57**, 1291 (1998)

597. A. Fayyazuddin, M. Spalinski: Nucl. Phys. B**508**, 219 (1997)

598. H. Ooguri, C. Vafa: Nucl. Phys. B**463**, 55 (1996)

599. A. Sen: Adv. Theor. Math. Phys. **1** 115 (1997); JHEP **9709**, 001 (1997)

600. M. Aganagic, C. Popescu, J.H. Schwarz: Nucl. Phys. B**495**, 99 (1997)

601. M. Cederwall, A. von Gussich, B.E.W. Nilsson, A. Westerberg: Nucl. Phys. B**490**, 163 (1997);
    M. Cederwall, A. von Gussich, B.E.W. Nilsson, P. Sundell, A. Westerberg: Nucl. Phys. B**490**, 179 (1997)

602. E. Bergshoeff, P.K. Townsend: Nucl. Phys. B**490**, 145 (1997)

603. I.A. Bandos, D.P. Sorokin, M. Tonin: Nucl. Phys. B**497**, 275 (1997)

604. R. Kallosh: 'Volkov-Akulov Theory and D-Branes'. In: *Supersymmetry and Quantum Field Theory*, 49 (1997); hep-th/9705118

605. P.K. Townsend: 'Four Lectures about M Theory'. In: *High Energy Physics and Cosmology*, 385 (1996); hep-th/9612121

606. D.V. Volkov, V.P. Akulov: Phys. Lett. B**46**, 109 (1973)

607. J. Bagger, A. Galperin: 'Linear and Non-Linear Supersymmetries'. In: *Proceedings of the Dubna Seminar on Supersymmetries and Quantum Symmetries* (1997); hep-th/9810109

608. S. Bellucci, E. Ivanov, S. Krivonos: 'Partial Breaking N=4 to N=2: Hypermultiplet as a Goldstone Superfield'. In: *Proceedings of the 11th Dubna Conference on Problems of Quantum Field Theory* (1998); hep-th/9809190;

609. S.S. Gubser, I.R. Klebanov, A.A. Tseytlin: Nucl. Phys. B**499**, 217 (1997)

610. A. Karlhede, U. Lindström, M. Roček, G. Theodoridis: Nucl. Phys. B**294**, 498 (1987)

611. S.J. Gates Jr.: Phys. Lett. B**365**, 132 (1996); Nucl. Phys. B**485**, 145 (1996)

612. M.K. Gaillard, B. Zumino: 'Self-Duality in Non-Linear Electromagnetism'. In: *Supersymmetry and Quantum Field Theory*, 121 (1997); hep-th/9705226;
    'Non-Linear Electromagnetic Self-Duality and Legendre Transformations'. In: *Proceedings of the Newton Institute Euroconference on Duality and Supersymmetric Theories* (1997); hep-th/9712103

613. S.M. Kuzenko, S. Theisen: Supersymmetric Duality Rotations. München preprint LMU–TPW–00–03 (2000); hep-th/0001068

614. J. Plebański: Nordita Lectures on Non-Linear Electrodynamics. Copenhagen preprint (1968)

615. G.W. Gibbons, D.A. Rasheed: Nucl. Phys. B**454**, 185 (1995)

616. J. Hughes, J. Liu, J. Polchinski: Phys. Lett. B**180**, 370 (1986)

617. J. Hughes, J. Polchinski: Nucl. Phys. B**278**, 147 (1986)

618. E. Bergshoeff, E. Sezgin, C.N. Pope, P.K. Townsend: Phys. Lett. B**188**, 70 (1987)

619. R.R. Metsaev, M.A. Rachmanov, A.A. Tseytlin: Phys. Lett. B**193**, 207 (1987)

620. D.V. Volkov: JETP Lett. **18**, 312 (1973)

621. V.I. Ogievetsky: In: *Proceedings of X-th Winter School of Theoretical Physics in Karpacz*, 227 (1974)

622. J. Bagger, A. Galperin: Phys. Rev. D**55**, 1091 (1997)

623. M. Roček, A.A. Tseytlin: Phys. Rev. D**59**, 106001 (1999)

624. S.V. Ketov: Mod. Phys. Lett. A**14**, 501 (1999)

625. A.A. Tseytlin: Nucl. Phys. B**469**, 51 (1996)

626. S. Cecotti, S. Ferrara: Phys. Lett. B**187**, 335 (1986)

414   References

627. M.J. Duff, C.J. Isham: Phys. Lett. B**86**, 157 (1979)
628. I. Antoniadis, H. Partouche, T.R. Taylor: Phys. Lett. B**372**, 83 (1996)
629. J. Bagger, A. Galperin: Phys. Lett. B**336**, 25 (1994); ibid. B**412**, 296 (1997)
630. F. Gonzalez-Rey, I.Y. Park, M. Roček: On Dual 3-Brane Actions with Partially Broken N=2 Supersymmetry, Stony Brook preprint (1998); hep-th/9811130
631. A. A. Tseytlin: Born-Infeld Action, Supersymmetry and String Theory, London preprint Imperial/TP/98–99/67 (1999); hep-th/9908105
632. L. Mezincescu: On the Superfield Formulation of O(2) Supersymmetry. Dubna preprint P2–12572 (1979), unpublished
633. A.A. Deriglazov, S.V. Ketov: Theor. Math. Phys. **77**, 1160 (1988)
634. A.A. Deriglazov, S.V. Ketov: Izv. VUZov USSR, Fizika **32**, 112 (1989)
635. K.S. Stelle: Phys. Rev. D**16**, 953 (1977)
636. A. Salam, J. Strathdee: Phys. Rev. D**18**, 4480 (1978)
637. B.L. Voronov, I.V. Tyutin: Yadern. Fiz. **39**, 998 (1984)
638. E.S. Fradkin, A.A. Tseytlin: Phys. Rep. **119**, 233 (1985)
639. E.S. Fradkin, A.A. Tseytlin: Phys. Lett. B**104**, 377 (1981)
640. I.G. Avramidi, A.O. Barvinsky: Phys. Lett. B**159**, 269 (1985)
641. E.T. Tomboulis: Nucl. Phys. B**97**, 77 (1980)
642. D.G. Boulware: 'Quantization of Higher Derivative Theories of Gravity'. In: *Quantum Theory of Gravity*, 267 (Adam Hilger Publ., 1984)
643. I. Antoniadis, E.T. Tomboulis: Phys. Rev. D**33**, 2756 (1986)
644. D.A. Johnston: Nucl. Phys. B**297**, 721 (1988)
645. A.A. Deriglazov, S.V. Ketov: Sov. Phys. J. **33**, 18 (1990)
646. I.L. Buchbinder, S.V. Ketov: Renormalization of the 4d NLSM with Higher Derivatives. Tomsk preprint TF/SB/AS-42 (1986)
647. A.O. Barvinsky, G.A. Vilkovisky: Phys. Rep. **119**, 1 (1985)
648. P. Hasenfratz: Nucl. Phys. B**321**, 139 (1989)
649. S. Theisen: Nucl. Phys. B**263**, 687 (1986)
650. E.A. Bergshoeff, R.I. Nepomechie, H.J. Schnitzer: Nucl. Phys. B**249**, 93 (1985)
651. A.A. Deriglazov, S.V. Ketov: Supersymmetric NLSM with Higher Derivatives in 4d. Tomsk preprint TF/SB/AS-5 (1987)
652. I.L. Buchbinder, S.M. Kuzenko: Nucl. Phys. B**274**, 653 (1986)
653. J. Schwinger: Phys. Rev. **82**, 664 (1951)
654. B.S. De Witt: *Dynamical Theory of Groups and Fields* (Gordon and Breach, 1964)
655. S.V. Ketov: Sov. Phys. J. **33**, 355 (1990)
656. S. Ferrara, E. Cremmer, J. Scherk: Phys. Lett. B**74**, 61 (1978)
657. A. Font, L.E. Ibanez, D. Lüst, F. Quevedo: Phys. Lett. B**245**, 401 (1990)
658. C.M. Hull: 'Duality, Enhanced Symmetry, and Massless Black Holes'. In: *Proceedings of the Strings'95 Conference*, 230 (World Scientific, 1996)
659. H. Nicolai, H. Samtleben: Nucl. Phys. B**533**, 210 (1998)
660. B. Julia: 'Infinite Lie Algebras in Physics'. In: *Proceedings of the Johns Hopkins Workshop on Current Problems in Particle Theory*, 23 (Baltimore, 1981)
661. O. Barwald, R.W. Gebert, M. Günaydin, H. Nicolai: Commun. Math. Phys. **195**, 29 (1998).

Text Abbreviations

ADHMN	Atiyah-Drinfeld-Hitchin-Manin-Nahm
AdS	Anti-de-Sitter
AH	Atiyah-Hitchin
AKM	Affine Kač-Moody
ALE	Asymptotically Locally Euclidean
ALF	Asymptotically Locally Flat
APT	Antoniadis-Partouche-Taylor
BCS	Bardeen-Cooper-Shriffer
BI	Born-Infeld
BING	Born-Infeld-Nambu-Goto
BPS	Bogomolnyi-Prasad-Sommerfeld
BRST	Becchi-Rouet-Stora-Tyutin
CFT	Conformal Field Theory
D	Dirichlet
DNS	Donaldson-Nair-Schiff
DVVW	Dijkgraaf-Verlinde-Verlinde-Witten
EH	Eguchi-Hanson
e.-m.	electric-magnetic
EuH	Euler-Heisenberg
FI	Fayet-Iliopoulos
FP	Faddeev-Popov
FS	Fayet-Sohnius
GKO	Goddard-Kent-Olive
GSO	Gliozzi-Scherk-Olive
h.c.	Hermitian conjugation
HP	t'Hooft-Polyakov
HSS	Harmonic Superpace
HST	Howe-Stelle-Townsend
HVB	Hooft-Veltman-Bos
IR	Infrared
KK	Kaluza-Klein
KS	Kazama-Suzuki
LEEA	Low-Energy Effective Action
LP	Leznov-Parkes

MA	Monge-Ampère
MM	Makeenko-Migdal
MW	Majorana-Weyl
NG	Nambu-Goto
NLSM	Non-Linear Sigma-Model
NS	Neveu-Schwarz
NSR	Neveu-Schwarz-Ramond
1PI	1-Particle-Irreducible
OPE	Operator Product Expansion
PG	Pauli-Gürsey
PSS	Projective Superspace
QCD	Quantum Chromodynamics
QED	Quantum Electrodynamics
QFT	Quantum Field Theory
R	Ramond
RG	Renormalization Group
RR	Ramond-Ramond
SBI	Super-Born-Infeld
SCA	Superconformal Algebra
SCFT	Superconformal Field Theory
SDG	Self-Dual Gravity
SDR	Supersymmetric Dimensional Regularization
SDYM	Self-Dual Yang-Mills
SM	Standard Model
SS	Sugawara-Sommerfeld
susy	Supersymmetry
SW	Seiberg-Witten
SYM	Super-Yang-Mills
UV	Ultraviolet
VEV	Vacuum Expectation Value
WZ	Wess-Zumino
WZNW	Wess-Zumino-Novikov-Witten

Index

Printing: Saladruck, Berlin
Binding: Buchbinderei Lüderitz & Bauer, Berlin